Fast Pyrolysis of Biomass:

A Handbook

Volume 2

Contributors:

J Arauzo	L Garcia	H Pakdel
R Bilbao	P Girard	GVC Peacocke
O Boutin	M Lauer	J Piskorz
AV Bridgwater	J Lédé	D Radlein
S Czernik	R Maggi	C Roy
C Di Blasi	P Majerski	ML Salvador
JP Diebold	D Meier	DS Scott
C Diez	JN Murwanashyaka	P Thornley
G Dobele	A Oasmaa	E Wright

Editor:

AV Bridgwater

CPL Press, Liberty House, The Enterprise Centre
New Greenham Park, Newbury RG19 6HW, UK
Online Bookshop: www.cplbookshop.com

©   Aston University, Bio-Energy Research Group, UK
       May 2002

Published in the United Kingdom by:

CPL Press, Liberty House, The Enterprise Centre
New Greenham Park, Newbury, Berks RG19 6HW, UK
Tel: +44 1635 817408   Fax: +44 1635 817409
Email: press@cplbookshop.com   Website: www.cplpress.com

for:

Aston University, Bio-Energy Research Group
Aston Triangle, Birmingham B4 7ET, UK
Email: a.v.bridgwater@aston.ac.uk   Fax: +44 121 359 6814

Sponsored by IEA Bioenergy, the European Commission FAIR Programme and PyNe

CPL Press Online Bookshop

Numerous publications covering pyrolysis, gasification and many other aspects of renewable energy, as well as the life and chemical sciences, are available at: **www.cplbookshop.com**

Printed in the UK by Antony Rowe Ltd, Chippenham, UK

ISBN 1 872691 47 1

EUR 20341

PREFACE

This handbook is an edited version of the final report of the European Commission and IEA Bioenergy sponsored Pyrolysis Network that officially finished in 2001. It provides a companion volume to the first handbook published in 1999 and it is again intended that this will provide a useful guide both to newcomers to the subject area as well as those already involved in research, development and implementation. A significant feature of this second volume is the greater attention paid to addressing commercial issues such as marketability, transportation and safety.

The benefits from the PyNe Network have been in drawing together the hard core of scientists and engineers involved in developing fast pyrolysis from Europe and North America on a regular basis to share information and help develop the technology. This small group has, however, not been the only beneficiaries as all the meetings have been open to guests and visitors and some meetings have attracted up to 80 participants, all of whom have both contributed to and/or benefited from the activities of the Network.

PyNe is continuing for a third term, again sponsored by the European Commission and IEA Bioenergy and again as a firmly integrated activity between the two organisations. It is now associated with an analogous network on gasification, GasNet, and the two networks will work closely together as ThermoNet to maximise the benefits from integration.

PyNe has been a genuine team effort with contributions by all the members and guests throughout the life of the Network, but the Subject Group leaders in particular have taken the burden of preparing the reports and contributions that make up a significant part of this second volume (Chapters 1 to 8). Other organisations have also contributed to the output of PyNe notably the National Renewable Energy Laboratory in the USA who co-sponsored the report by Jim Diebold (Chapter 11), Natural Resources Canada who funded the report on levoglucosan by Desmond Radlein (Chapter 10), and the Institute for Wood Chemistry in Hamburg who redrew many of the figures in the report by Galina Dobele (Chapter 9). As part of the development of the Subject Groups, which formed the main technical contributions of this Network, most Subject Group Leaders commissioned special reports from experts in different areas as a significant contribution to furthering the science and technology (Chapters 11, 12 and 13), while the Science Subject Group organised a number of very successful workshops resulting in the contributions of Chapters 14 to 18.

Particular thanks should also go to the Nina Ahrendt and Claire Humphreys at Aston University for organising the meetings and publications so efficiently.

Tony Bridgwater
January 2002

CONTENTS

Chapter 1.   The Status of Biomass Fast Pyrolysis
                                    A. V. Bridgwater, S. Czernik, J. Piskorz   1

Chapter 2.   Analysis, Characterisation and Test Methods of Fast Pyrolysis
              Liquids   *A. Oasmaa, D. Meier*   23

Chapter 3.   Pyrolysis Liquids Analyses – The Results of IEA-EU Round
              Robin   *A. Oasmaa, D. Meier*   41

Chapter 4.   Summary of the Analytical Methods Available for Chemical
              Analysis of Pyrolysis Liquids   *D. Meier*   59

Chapter 5.   Environment, Health and Safety Aspects Related to Fast
              Pyrolysis Oils – A Guideline to Notify a New Substance
                                                P. Girard, C. Diez   69

Chapter 6.   Implementation Subject Group Report   *M. Lauer*   87

Chapter 7.   Fundamentals, Mechanisms and Science of Pyrolysis   *J. Piskorz*   103

Chapter 8.   Review of Methods for Upgrading Biomass-derived Fast
              Pyrolysis Oils   *S. Czernik, R. Maggi, G.V.C. Peacocke*   141

Chapter 9.   Production, Properties and Use of Wood Pyrolysis Oil - A
              brief review of the work carried out at research and production
              centres of the former USSR from 1960 to 1990   *G. Dobele*   147

Chapter 10.   Study of Levoglucosan Production – A Review   *D. Radlein*   205

Chapter 11.   A Review of the Chemical and Physical Mechanisms of the
              Storage Stability of Fast Pyrolysis Bio-oils   *J.P. Diebold*   243

Chapter 12.   Transport, Handling and Storage of Fast Pyrolysis Liquids
                                                G.V.C. Peacocke   293

Chapter 13.   Evaluation of Bio-Energy Projects   *P. Thornley, E. Wright*   339

Chapter 14.   Radiant Flash Pyrolysis of Cellulose – Evidence for the
              Formation of Short Life Time Liquid Species and
              Experimental Determination of Mass Balances   *O. Boutin, J. Lédé*   363

Chapter 15.   Formulation and Application of Biomass Pyrolysis Models for
              Process Design and Development   *C. Di Blasi*   371

Chapter 16.   Pyrolysis of Cellulose – From Oligosaccharides to Synthesis
              Gas   *J. Piskorz, D. Radlein, P. Majerski, D.S. Scott*   381

Chapter 17.   Gas Production from Catalytic Pyrolysis of Biomass
                          L. Garcia, M.L. Salvador, R. Bilbao, J. Arauzo   393

Chapter 18.   Fractional Vacuum Pyrolysis of Biomass and Separation of
              Phenolic Compounds by Steam Distillation
                          J Népo Murwanashyaka, H Pakdel and C Roy   407

Author index   419

Subject index   421

THE STATUS OF BIOMASS FAST PYROLYSIS

AV Bridgwater
Bio-Energy Research Group, Aston University, Birmingham B4 7ET, UK
S Czernik
NREL, 1617 Cole Boulevard, Golden, Colorado 80401, USA
J Piskorz
RTI Ltd, 110 Baffin Place, Unit 5, Waterloo, Ontario, N2V 1Z7, Canada

ABSTRACT

The process of fast pyrolysis is one of the most recent renewable energy processes to have been introduced and offers the advantages of a liquid product, bio-oil, that can be readily stored and transported and that can also be used for production of chemicals as well as being a fuel. Thermal biomass conversion has been investigated for many years as a source of renewable solid, gaseous and liquid fuels. Compared to combustion, which is widely practised commercially and gasification, which is being extensively demonstrated around the world, fast pyrolysis is at a relatively early stage of development. The technology has now achieved some commercial success for production of chemicals and is being actively developed for producing liquid fuels. Bio-oils have been successfully tested in engines, turbines and boilers, and have been upgraded to high quality hydrocarbon fuels although at an unacceptable energetic and financial cost. The paper critically reviews scientific and technical developments and applications to date paying particular attention to the research and developments reported in this book. It concludes with some suggestions for strategic developments.

INTRODUCTION

Renewable energy is of growing importance in satisfying environmental concerns over fossil fuel usage. Wood and other forms of biomass are some of the main renewable energy resources available and provide the only source of renewable liquid, gaseous and solid fuels. Wood and biomass can be used in a variety of ways to provide energy:

- by direct combustion to provide heat. This technology is commercially available and presents minimum risk to investors. The product is heat, which must be used immediately for heat and/or power generation. Overall efficiencies to power tend to be rather low, although in many California power plants high efficiencies have been reported but there have been problems with ash build-up in many installations. The status is reviewed in these proceedings (1).
- by gasification to provide a fuel gas for combustion for heat, or in an engine or turbine for electricity generation. The fuel gas quality requirements, for turbines in particular, are very high, although there is now extensive experience available from the Varnamo plant (2). The gas is very costly to store or transport so it has to be used immediately. Hot gas efficiencies (total energy in raw product gas divided by energy in feed) can be as high as 95-97% for close coupled turbine and boiler applications, and up to 85% for cold gas efficiencies. The status is reviewed in (3). There is renewed interest in synthesis of liquid fuels from the product gas, but costs are still very high.
- by fast pyrolysis to provide a liquid fuel that can substitute for fuel oil in any static heating or electricity generation application. The liquid can also be used to produce a range of speciality and commodity chemicals. The key advantage is that a liquid can be readily stored and/or transported.

This review explains how the technology is developing and how it depends on an improved understanding of the underlying science as reviewed four years ago (4).

FAST PYROLYSIS

Pyrolysis is by definition thermal decomposition occurring in the absence of oxygen. It is always also the first step in combustion and gasification processes where it is followed by total or partial oxidation of the primary products.

Fast pyrolysis occurs in time of few seconds or less. Therefore, not only chemical reaction kinetics but also heat and mass transfer processes, as well as phase transition phenomena, play important roles. The critical issue is to bring the reacting biomass particle to the optimum process temperature and minimise its exposure to the intermediate (lower) temperatures that favour formation of charcoal. This objective can be achieved by using small particles, thus reducing the time necessary for heat up. This option is used in fluidised bed processes that are described later. Another possibility is to transfer heat very fast only to the particle surface that contacts the heat source. Because of the low thermal conductivity the deeper parts of the particles will be maintained at temperatures lower than necessary for char production. The products that form on the surface are immediately removed exposing that way consecutive biomass layers to the contact with the heat source. This second method is applied in ablative processes that are described later.

Fast pyrolysis is not an equilibrium process. During fast pyrolysis dramatic changes occur in specific volume between the reactants (biopolymers) and the products (by a factor of x 500) causing the volatile products to leave the pyrolysis zone at considerable velocities. This results in the entrainment of solid particles and aerosols, which normally would not volatilise at the process temperature. All these phenomena have important implications on pyrolysis technologies and are discussed in this paper.

Mechanisms

Biomass is a complex polymeric material and its thermal decomposition is a multistage complicated process. Many pathways and mechanisms have been proposed to illustrate/explain the fundamental steps in pyrolysis (e.g. 5, 6, 7, 8, 9, 10, 11, 12, 13). Broido-Shafizadeh type kinetic models are perhaps the most widely used for cellulose pyrolysis but they can be also applied, at least qualitatively, to the whole biomass (see Figure 1).

Correct estimation of the kinetic constants is essential in optimising process parameters for maximizing liquids production. Nevertheless, kinetic data does not contribute any predictive powers aiming to optimise yields of specific chemicals or their classes. As fast pyrolysis is a non-steady state process, the derivation of isothermal kinetic data for modelling is extremely difficult, except in very small samples, as mass transfer processes, phase transition and heat transfer phenomena play important roles. The so-called "black box" engineering approach dominates our understanding in this area at present.

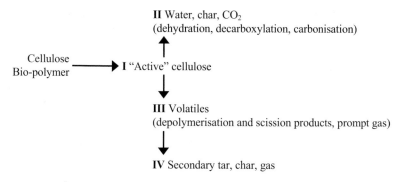

Figure 1 Typical cellulose decomposition model

As shown in the model, pyrolysis of cellulose results in solid, liquid and gaseous products. However, the proportions of the product yields can change depending on the process conditions. The knowledge of thermodynamics and kinetics of the reaction pathways allows us to adjust the conditions to maximize the yield of the desired products.

Dehydration of cellulose is exothermic while depolymerisation and secondary vapour cracking are endothermic and have higher activation energy than dehydration. Therefore, lower process temperature and longer vapour residence times will favour the production of charcoal. High temperature and longer residence time will increase the biomass conversion to gas and moderate temperature and short vapour residence time, necessary to minimize secondary cracking, are optimum for producing liquids. Table 1 provides data on the product distribution obtained from different modes of pyrolysis process.

Table 1 Typical product yields obtained by different modes of pyrolysis of wood

	Liquid	Char	Gas
FAST PYROLYSIS	75%	12%	13%
moderate temperature and short residence time, particularly vapour			
CARBONISATION	30%	35%	35%
low temperature and long residence time			
GASIFICATION	5%	10%	85%
high temperature and long residence times			

Principles

Fast pyrolysis is a high temperature process in which biomass is rapidly heated in the absence of oxygen. As a result it decomposes to generate mostly vapours and aerosols and some charcoal. After cooling and condensation, a dark brown mobile liquid is formed which has a heating value about half that of conventional fuel oil. While it is related to the traditional pyrolysis processes for making charcoal, fast pyrolysis is an advanced process, with carefully controlled parameters to give high yields of liquid.

The essential features of a fast pyrolysis process for producing liquids are:

- very high heating and heat transfer rates at the reaction interface, which usually requires a finely ground biomass feed,
- carefully controlled pyrolysis reaction temperature of around 500°C and vapour phase temperature of 400-450°C,
- short vapour residence times of typically less than 2 seconds,
- rapid cooling of the pyrolysis vapours to give the bio-oil product.

3

The main product, bio-oil, is obtained in yields of up to 75% wt on dry feed basis, together with by-product char and gas which are used within the process so there are no waste streams other than flue gas and ash. The variation of spectrum of products with temperature is shown in Figure 2.

Yield on dry feed, wt%

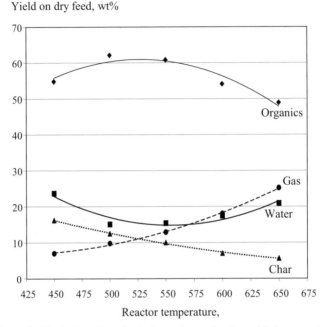

Reactor temperature,

Figure 2   Variation of products from Aspen Poplar with temperature (14)

Figure 3 shows a conceptual schematic of fast pyrolysis process that includes the necessary steps of drying the feed to typically less than 10% water to minimise the water in the product liquid oil (although up to 15% can be acceptable), grinding the feed (to around 2 mm in the case of fluid bed reactors) to give sufficiently small particles to ensure rapid reaction, pyrolysis reaction, separation of solids (char), and collection of liquid product (bio-oil).

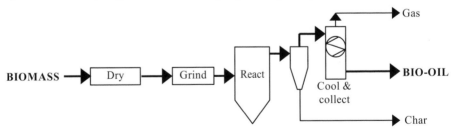

Figure 3   Conceptual Fast Pyrolysis Process

Reactors

The heart of a fast pyrolysis process is the reactor. Although it probably represents at most only about 20% of the total capital cost, almost all research and development has focussed on the reactor. The rest of the process consists of biomass reception, storage and handling, biomass drying and comminution, and product collection, storage and, when relevant, upgrading. The key aspects of these peripheral steps are described later.

The critical features of successful pyrolysis reactors have been defined above as very high heating rates, carefully controlled temperatures and rapid cooling or quenching of the gaseous product which includes vapours, aerosols and gases.

A comprehensive survey of fast pyrolysis processes has recently been published that lists and describes all the pyrolysis processes for liquids production that have been built and tested in the last 10-15 years (15). Table 2 summarises the current significant operational fast pyrolysis plants for production of liquids.

**Table 2 Operational pyrolysis units for liquids (see (15) for further details).
Throughputs are on a dry wood basis.**

Fluid bed	250 kg/h at Wellman
	400 kg/h at Dynamotive with 2000 kg/h under design
	20 kg/h at RTI
	Many research units (15)
Transported bed	about 2 x 1500 kg/h at Red Arrow (Ensyn)
	650 kg/h at ENEL (Ensyn)
	20 kg/h at VTT (Ensyn)
CFB	10 kg/h at CRES (23)
Rotating cone	150 kg/h at BTG (29)
Ablative	20 kg/h at NREL
	20 kg/h at Aston
Vacuum	3500 kg/h at Pyrovac (30)
Unspecified	350 kg/h at Fortum (16)

Bubbling fluid beds

Bubbling fluid beds have been selected for further development by several companies including Union Fenosa (17) who have built and operated a 200 kg/h pilot unit in Spain, Dynamotive who have a 75 kg/h unit in Canada based on a RTI design (18) with a 10 t/d plant commissioned recently (19) and Wellman who have built a 250 kg/h unit in the UK. Many research units have been built as they are relatively easy to construct and operate and give good results (15, and in these proceedings).

Bubbling fluid beds have many attractive features, particularly for research and development, which include:

- Simple construction and operation,
- Good temperature control,
- Very efficient heat transfer to biomass particles due to high solids density,
- Easy scaling,
- Well understood technology,
- Good and consistent performance with high liquid yields: of typically 70-75%wt. from wood on a dry feed basis,

Particular features that require consideration in design and operation include:

- Heating can be achieved in a variety of ways as shown in Figure 3,

- Residence time of solids and vapours is controlled by the fluidising gas flow rate and is higher for char than for vapours,
- Char acts as an effective vapour cracking catalyst at fast pyrolysis reaction temperatures so rapid and effective char separation/elutriation is important,
- Small biomass particle sizes are needed to achieve high biomass heating rates, and particle sizes generally need to be less than 2 to 3 mm,
- Good char separation is important and this is usually achieved by ejection and entrainment followed by separation in one or more cyclones,
- Heat transfer to bed at large scale has to be considered carefully due to scale-up limitations (see also Figure 4).

A typical plant configuration is shown in Figure 5, in this case based on a bubbling fluid bed, but similar principles apply to most reactors.

Circulating fluid beds and transported bed

Circulating fluid beds and transported bed reactors have been developed to commercial status by Ensyn and their process is used commercially by Red Arrow in the USA for food flavourings in several plants of 1 to 1.5 t/h (20). Ensyn have also supplied a 650 kg/h unit to ENEL in Italy (21) and a 20 kg/h system to VTT in Finland (22). In some of these plants it is believed that only sand is recycled while char is separated and collected, in others, operating in a dual reactor configuration, char is also recycled and burned in the second reactor to provide process energy by reheating and recycling the sand.

CRES is operating a 10 kg/h circulating fluid bed unit with the char combustor integrated into the base of the CFB riser as a bubbling fluid bed (23) as indicated in Figure 6. Liquid yields of 60-70% wt on a dry feed basis have been achieved.

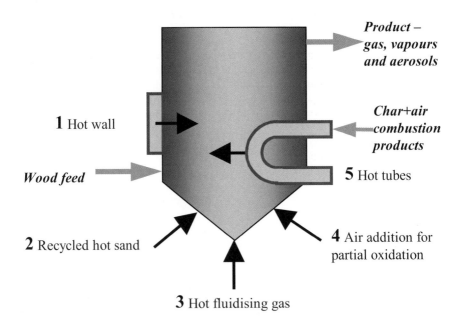

Figure 4   Means of providing heat to a fast pyrolysis reactor

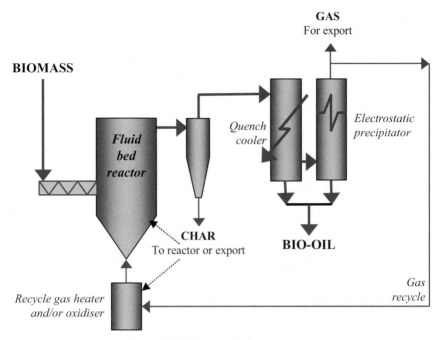

BIOMASS

GAS
For export

Fluid bed reactor

Quench cooler

Electrostatic precipitator

CHAR
To reactor or export

BIO-OIL

Recycle gas heater and/or oxidiser

Gas recycle

Figure 5   Bubbling fluid bed reactor

The principles of a circulating fluid bed system are shown in Figure 6. As discussed above, a second rector can be required for char combustion and sand reheating (Ensyn) or the char combustion can occur in the lower part of the pyrolyser (CRES) thus eliminating the need for the second vessel.

Particular features of circulating fluid bed and transported bed reactors include:

*   Good temperature control can be achieved in reactor,
*   Residence time for the char is almost the same as for vapours and gas,
*   CFBs are suitable for very large throughputs,
*   Well understood technology,
*   Hydrodynamics more complex,
*   Char is more attrited due to higher gas velocities; char separation is by cyclone,
*   Closely integrated char combustion in a second reactor requires careful control,
*   Heat transfer at large scale has to be proven.

Ablative pyrolysis

Ablative pyrolysis is substantially different in concept compared to the other methods of fast pyrolysis. In all these other methods, the rate of reaction is limited by the rate of heat transfer through a biomass particle, which is why small particles are required. The mode of reaction in ablative pyrolysis is analogous to melting butter in a frying pan, when the rate of melting can be significantly enhanced by pressing down and moving the butter over the heated pan surface. In ablative pyrolysis heat is transferred from the hot reactor wall to "melt" wood that is in contact with it under pressure. The pyrolysis front thus moves unidirectionally through the biomass particle.   As the wood is mechanically moved away, the residual oil film both provides lubrication for successive biomass particles and also rapidly evaporates to give pyrolysis vapours for collection in the same way as other processes. The rate of reaction is strongly influenced by pressure, the relative velocity of wood on the heat exchange surface and the reactor surface temperature.

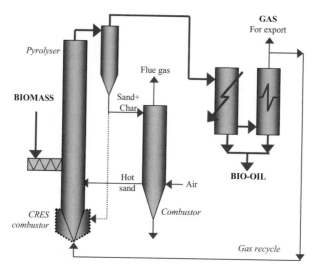

Figure 6   Circulating fluid bed reactor

The key features of ablative pyrolysis are therefore as follows:

- High pressure of particle on hot reactor wall, achieved due to centrifugal force (NREL) or mechanically (Aston)
- High relative motion between particle and reactor wall,
- Reactor wall temperature less than 600°C.

Particular features of ablative reactor systems include:

- Use of large feed sizes,
- Inert gas is not required, so the processing equipment is smaller, (in case of mechanically applied pressure)
- The reaction system is more intensive,
- The process is limited by the rate of heat supply to the reactor rather than the rate of heat absorption by the pyrolysing biomass as in other reactors.
- Reaction rates are limited by heat transfer to the reactor, not to the biomass,
- The process is surface area controlled so scaling is more costly,
- The process is mechanically driven so the reactor is more complex.

Ablative pyrolysis is interesting as much larger particle sizes can be employed than in other systems and there is no requirement for inert gas. Both lead to a potentially lower cost system.

Much of the pioneering work on ablative pyrolysis reactors was performed by CNRS at Nancy where extensive basic research has been carried out onto the relationships between pressure, motion and temperature (24). NREL developed an ablative vortex reactor (see Figure 7 and (15)), in which the biomass is accelerated to supersonic velocities to derive high tangential pressures inside a heated cylinder. Unreacted particles are recycled and the vapours and char

8

fines leave the reactor axially for collection. Liquid yields of 60-65%wt. on dry feed basis are typically obtained.

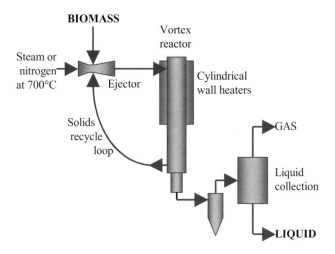

Figure 7 NREL Vortex ablative reactor

More recent developments have been carried out at Aston University with a prototype rotating blade reactor in which pressure and motion is derived mechanically thus obviating the need for a carrier gas (see Figure 8 and (15)). Liquid yields of 70-75%wt. on dry feed are typically obtained. A second-generation reactor has recently been built and commissioned. Other configurations include the coiled tube at Castle Capital (15) (now owned by Enervision) and cyclonic reactors (25).

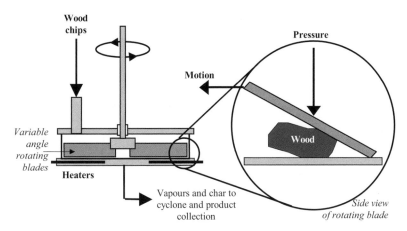

Figure 8   Aston University rotating blade ablative reactor

Entrained flow

Entrained flow fast pyrolysis was developed at Georgia Tech Research Institute (26) and scaled up by Egemin (27). Further details are available in (15). Neither the GTRI nor Egemin process is now operational and there are no known plans for further development, probably because of the difficulties that have been encountered in achieving good heat transfer from a gaseous heat carrier to solid biomass. Liquid yields of 50-60%wt. on dry feed have been obtained.

Rotating cone

The rotating cone reactor, invented at the University of Twente (28) and being developed by BTG, is a recent development and effectively operates as a transported bed reactor, but with transport effected by centrifugal forces rather than gas (29). A 250 kg/h unit is now operational, and plans for scale-up to 10 t/d have recently been announced. The basic principle is shown in Figure 9 and a flowsheet is given in Figure 10. Further details are available in (15) and (29).

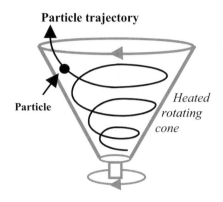

Figure 9   Principle of rotating cone pyrolysis reactor

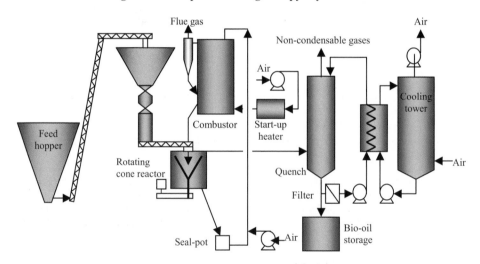

Figure 10   BTG Rotating Cone pilot plant

The key features of this technology are:

- Centrifugation at 600 rpm drives hot sand and biomass up a rotating heated cone,
- Vapours are collected and processed conventionally.
- Char is burned in a secondary bubbling fluid bed combustor. The hot sand is recirculated to the pyrolyser.
- Carrier gas requirements in the pyrolysis reactor are much less than for fluid bed and transported bed systems (however, more gases are needed for char burn off and for sand transport).

- Complex integrated operation of three subsystems: rotating cone pyrolyser, bubbling bed char combustor, and riser for sand recycling.
- Liquid yields of 60-70% on dry feed are typically obtained.

Vacuum pyrolysis

Vacuum pyrolysis is arguably not a true fast pyrolysis as the heat transfer rate to and through the solid biomass is much slower than in the previously described reactors, but the vapour residence time is comparable.

The basic technology was developed at the University of Laval using a multiple hearth furnace but is now based on a purpose-designed horizontal moving bed. The flowsheet is shown in Figure 11. A 50 kg/h unit is available for research and the technology has been scaled up by Pyrovac to a 3.5 t/h unit, which is currently operating at Jonquiere in Canada (30). The technology was selected for some of the Non Fossil Fuel Obligation contracts in the UK where six plants totalling about 70MWe are at various stages of permitting and design (31).

Key features of the process include:

- It can process larger particles than most fast pyrolysis reactors.
- There is less char in the liquid product due to lower gas velocities.
- There is no requirement for a carrier gas.
- Liquid yields of 35-50% on dry feed are typically obtained with higher char yields than fast pyrolysis systems. Conversely, the liquid yields are higher than in slow pyrolysis technologies because of fast removal of vapours from the reaction zone.
- The process is relatively complicated mechanically.

Char removal

Char acts as a vapour cracking catalyst so rapid and effective separation from the pyrolysis product vapours is essential. Cyclones are the usual method of char removal and two are usually provided – the first to remove the bulk of the material and the second to remove as much of the residual fines as possible. However, some fines always pass through the cyclones and collect in the liquid product where they accelerate aging and exacerbate the instability problem, which is described below.

There are several solutions to the problem of char entrainment. NREL has shown that hot vapour filtration, analogous to hot gas filtration in gasification processes, gives a high quality char free product (32), however the liquid yield is reduced by about 10-20% due to the char accumulating on the filter surface that cracks the vapours. VTT has also been developing hot vapour filtration (33).

Pressure filtration of the liquid is very difficult due to the complex interaction of the char and pyrolytic lignin, which appears to form a gel-like phase that rapidly blocks the filter. Modification of the liquid micro-structure by addition of solvents such as methanol or ethanol that solubilise the less soluble constituents will improve this problem and also contribute to improvements in liquid stability as described below.

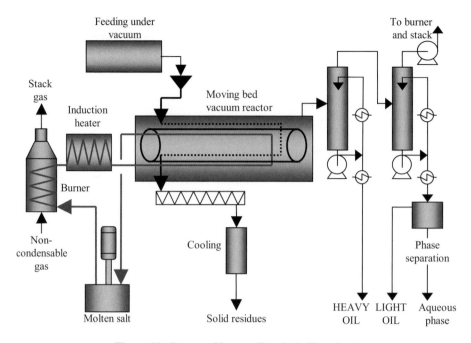

Figure 11 Pyrovac Vacuum Pyrolysis Flowsheet

Liquid collection

The product of fast pyrolysis is vapours, aerosols and gases from decomposition of holocellulose and lignin with any carrier gases from fluidisation or transport. Aerosols consist of sub-micron liquid droplets and they present a severe problem in the successful recovery of the pyrolysis oils. These aerosols appear visually as smoke. The aerosols are probably formed directly from pyrolysing biomass, especially from submicron biomass particles that are rapidly depolymerised. The liquid product can then be entrained out of the reactor before it is vaporised. Another mechanism proposed for the formation of aerosols in the pyrolysis reactor involves the ejection of liquid droplets from internally pressurised cell capillaries of a pyrolysing particle (34).

The vapours require rapid cooling to minimise secondary reactions and to condense the true vapours, while the aerosols require coalescence or agglomeration. Simple heat exchange can cause preferential deposition of lignin derived components leading to liquid fractionation and eventually blockage.

Quenching in product oil or in an immiscible hydrocarbon solvent is widely practiced. Orthodox aerosol capture devices such as demisters and other commonly used impingement devices are not very effective and electrostatic precipitation is currently the preferred method.

The vapour product from fluid bed and transported bed reactors has a low partial pressure of collectible products due to the large volumes of fluidising gas, and this is an important design consideration in liquid collection.

The exit gas temperature from the liquid condensation train has a definite effect on the properties of the recovered oil product. As the exit gas temperature is lowered, progressively more of the volatile organic vapours and water will be condensed to become the pyrolysis oil product. These volatiles can be considered solvents as they have very low viscosities and thus have a very beneficial impact on lowering the viscosity of the oil, even at low concentrations. The recovery of these volatile, oxygenated organic solvents is thought to decrease the tendency of the oil to phase separate. Use of higher collection temperatures to minimise water condensation will result in lower yields and in higher viscosity of less stable bio-oil.

PYROLYSIS LIQUID - BIO-OIL

Pyrolysis liquid is referred to by many names including pyrolysis oil, bio-oil, bio-crude-oil, bio-fuel-oil, wood liquids, wood oil, liquid smoke, wood distillates, pyroligneous tar, pyroligneous acid, and liquid wood. The crude pyrolysis liquid is dark brown and approximates to biomass in elemental composition. It is composed of a very complex mixture of oxygenated hydrocarbons with an appreciable proportion of water from both the original moisture and reaction product. Solid char and dissolved alkali metals from ash (35) may also be present.

The complexity arises from the degradation of lignin, cellulose, hemicellulose and any other organics in the feed material, giving a broad spectrum of phenolic and many other classes of compounds that result from uncontrolled degradation as described below. The liquid from fast or flash pyrolysis has significantly different physical and chemical properties compared to the liquid from slow pyrolysis processes, which is more like a tar.

Liquid Product Characteristics

Composition

The liquid is formed by rapidly quenching and thus "freezing" the intermediate products of flash degradation of hemicellulose, cellulose and lignin as shown in Table 3 and Table 4. The liquid thus contains many reactive species, which contribute to its unusual attributes. Bio-oil can be considered a micro-emulsion in which the continuous phase is an aqueous solution of holocellulose decomposition products, that stabilizes the discontinuous phase of pyrolytic lignin macro-molecules through mechanisms such as hydrogen bonding. Aging or instability is believed to result from a breakdown in this emulsion. In some ways it can be considered analogous to asphaltenes.

Table 3   Degradation products of fast pyrolysis of biomass constituents (12)

Hemicellulose	produces acetic acid, furfural, furan
Cellulose	produces levoglucosan, 5-hydroxymethylfurfural, hydroxyacetaldehyde, acetol, formaldehyde
Lignin	produces small amount of monomeric phenols (including phenols, cresols, guaiacols, syringols) but mostly oligomeric product of molecular mass ranging from few hundred to several thousand Da.
Extractives	produces molecules of waxy components such as fatty acids and rosin acids that are difficult to identify and are mostly immiscible with bio-oil. They are relatively thermally stable and volatile. The content depends on the biomass feed and can reach 5-15wt% levels
Compounds classes in bio-oil	
C1	formic acid, methanol, formaldehyde
C2-C4	linear hydroxyl and oxo substituted aldehydes and ketones
C5-C6	hydroxyl, hydroxymethyl and/or oxo substituted furans, furanones and pyranones
Anhydrosugars	e.g. levoglucosan, anhydro-oligosaccharides
Substituted phenols	Monomeric and dimeric methoxyl substituted phenols
Pyrolytic lignin	
Waxes, resins, fatty acids, terpenoid derivatives	

Table 4 Representative chemical composition of fast pyrolysis liquid

Major Components	Mass %
Water	20-30
Lignin fragments: insoluble pyrolytic lignin	15-30
Aldehydes: formaldehyde, acetaldehyde, hydroxyacetaldehyde, glyoxal, methylglyoxal	10-20
Carboxylic acids: formic, acetic, propionic, butyric, pentanoic, hexanoic, glycolic, (hydroxy acetic)	10-15
Carbohydrates: cellobiosan, α-D-levoglucosan, oligosaccharides, 1.6 anhydroglucofuranose	5-10
Phenols: phenol, cresols, guaiacols, syringols	2-5
Furfurals	1-4
Alcohols: methanol, ethanol	2-5
Ketones: acetol (1-hydroxy-2-propanone), cyclo pentanone	1-5

Fast pyrolysis liquid has a higher heating value of about 17 MJ/kg as produced with about 25% wt. water that cannot readily be separated. The liquid is often referred to as "oil" or "bio-oil" or "bio-crude" although it will not mix with any hydrocarbon liquids. It is composed of a complex mixture of oxygenated compounds that provide both the potential and challenge for utilisation. There are some important characteristics of this liquid that are summarised in Table 5 and discussed briefly below.

Table 5 Typical properties and characteristics of wood derived crude bio-oil

Physical property		*Typical value*
Moisture content		15-30%
pH		2.5
Specific gravity		1.20
Elemental analysis	C	55-58%
	H	5.5-7.0%
	O	35-40%
	N	0-0.2%
	Ash	0-0.2%
HHV as produced (depends on moisture)		16-19 MJ/kg
Viscosity (at 40°C and 25% water)		40-100 cp
Solids (char)		1%
Vacuum distillation residue		up to 50%

Characteristics
- Liquid fuel,
- Ready substitution for conventional fuels in many static applications such as boilers, engines, turbines,
- Heating value of 17 MJ/kg at 25% wt. water, is about 40% that of fuel oil / diesel
- Does not mix with hydrocarbon fuels,
- Not as stable as fossil fuels,
- Quality needs definition for each application,

Yield of products

The variation of organic liquid yield with temperature for different feedstocks is shown in Figure 12.

Organics yield, wt% on dry feed

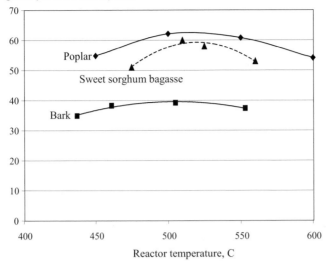

Reactor temperature, C

Figure 12 Variation of organic liquid yield with temperature for a variety of feedstocks
(Poplar (14) , Sweet sorghum bagasse (36), Bark (37))

Appearance

Pyrolysis oil typically is a dark brown free flowing liquid. Depending upon the initial feedstock and the mode of fast pyrolysis, the colour can be almost black through dark red-brown to dark green, being influenced by the presence of micro-carbon in the liquid and by the chemical composition. Hot vapour filtration gives a more translucent red-brown appearance due to the absence of char. High nitrogen contents in the liquid can give it a dark green tinge.

Odour

The liquid has a distinctive odour - an acrid smoky smell, which can irritate the eyes if exposed for a prolonged period to the liquids. The cause of this smell is due to the low2 molecular weight aldehydes and acids. The liquid contains several hundred different chemicals in widely varying proportions, ranging from formaldehyde and acetic acid to complex high molecular weight phenols, anhydrosugars and other oligosaccharides.

Miscibility

The liquid contains varying quantities of water which forms a stable single phase mixture, ranging from about 15 wt% to an upper limit of about 30-50wt% water, depending on how it was produced and subsequently collected. Pyrolysis liquids can tolerate the addition of some water, but there is a limit to the amount of water, which can be added to the liquid before phase separation occurs, in other words the liquid cannot be dissolved in water. It is miscible with polar solvents such as methanol, acetone, etc. but totally immiscible with petroleum-derived fuels.

Density

The density of the liquid is very high at around 1.2 kg/litre compared to light fuel oil at around 0.85 kg/litre. This means that the liquid has about 42% of the energy content of fuel oil on a weight basis, but 61% on a volumetric basis. This has implications on the design and specification of equipment such as pumps.

Viscosity

The viscosity of the bio-oil as produced can vary from as low as 25 cSt to as high as 1000 cSt (measured at 40 °C) or more depending on the feedstock, the water content of the oil, the amount of light ends that have been collected and the extent to which the oil has aged. Viscosity is important in many fuel applications (38).

Distillation

Pyrolysis liquids cannot be completely vaporised once they have been recovered from the vapour phase. If the liquid is heated to 100°C or more to try to remove water or distil off lighter fractions, it rapidly reacts and eventually produces a solid residue of around 50wt% of the original liquid and some distillate containing volatile organic compounds and water. The liquid is, therefore, chemically unstable, and the instability increases with heating, so it is preferable to store the liquid at room temperature. These changes do also occur at room temperature, but much more slowly and can be accommodated in a commercial application.

Unusual Properties

The complexity and nature of bio-oil causes some unusual behavior, specifically that the following properties tend to change with time:

- Viscosity increases,
- Volatility decreases,
- Phase separation and deposition of gums can occur.

This is due to a complex interaction of physical and chemical processes such as:

- Polymerization/condensation
- Esterification and etherification
- Agglomeration of oligomeric molecules

Upgrading of pyrolysis liquid

The properties that negatively affect bio-oil fuel quality are foremost low heating value, incompatibility with conventional fuels, solids content, high viscosity, and chemical instability. The heating value can be significantly increased, but it requires extensive changes to the chemical structure of bio-oils, which is technically feasible but not economic. The other undesired characteristics can be improved using simpler, physical methods. Both options are reviewed below.

Physical Methods

Hot-gas filtration can reduce the ash content of the oil to less than 0.01% and the alkali content to less than 10 ppm - much lower than reported for biomass oils produced in systems using only cyclones. Diesel engine tests performed on crude and on hot-filtered oil showed a substantial increase in burning rate and a lower ignition delay for the latter, due to the lower average molecular weight for the filtered oil (39). Hot gas filtration has not yet been demonstrated over a long-term process operation. Pyrolysis oils are not miscible with hydrocarbon fuels but with the aid of surfactants they can be emulsified with diesel oil.

A process for producing stable micro-emulsions with 5-30% of bio-oil in diesel has been developed at CANMET (40). The resultant emulsions showed promising ignition characteristics. A drawback of this approach is the cost of surfactants and the high energy required for emulsification. Other work on bio-emulsions is being carried out by the University of Florence, Italy, in an EC sponsored project, using proportions of bio-oil ranging from 5% to 75% (41).

16

Polar solvents have been used for many years to homogenize and to reduce viscosity of biomass oils. The addition of solvents, especially methanol, also showed a significant effect on the oil stability. It was observed (42) that the rate of viscosity increase ("aging") for the oil with 10 wt. % of methanol was almost 20 times less than for the oil without additives.

Chemical Methods

The chemical reactions that can occur between the bio-oil and methanol or ethanol are esterification and acetalization. Although not favoured thermodynamically, they can proceed to a significant extent if appropriate conditions are applied. For example, in the presence of an acid catalyst and molecular sieves (to adsorb water and to shift the reaction equilibria), bio-oil reacted with ethanol forming ethyl acetate, ethyl formate, and diethoxyacetal of hydroxyacetaldehyde at the expense of formic acid, acetic acid, and hydroxyacetaldehyde (43). Eventually, in addition to the decrease in viscosity and in the aging rate, other desirable changes such as reduced acidity, improved volatility, heating value, and miscibility with diesel fuels were also achieved. This field has been thoroughly reviewed (44).

The chemical/catalytic upgrading processes aim at the removal of oxygen, which is the main cause of instability and other unwanted characteristics of bio-oils. They are more complex and expensive than physical methods, but offer significant improvements ranging from simple stabilization to high-quality fuel products (45). Full deoxygenation to high-grade products such as transportation can be accomplished by two main routes: hydrotreating and catalytic vapour cracking. Chemical methods for upgrading bio-oil by hydro-treating and zeolite cracking have been reviewed (46, 47).

Hydrotreating of bio-oil carried out at high temperature, high hydrogen pressure, and in the presence of catalysts results in elimination of oxygen as water and in hydrogenation-hydrocracking of large molecules. The catalysts (sulphided CoMo or NiMo supported on alumina) and the process conditions are similar to those used in the refining of petroleum cuts (48).

Catalytic vapour cracking makes deoxygenation possible through simultaneous dehydration-decarboxylation over acidic zeolite catalysts. At 450°C and the atmospheric pressure oxygen is rejected as H_2O, CO_2, and CO producing mostly aromatics (49). The low H/C ratio in the bio-oils imposes a relatively low limit on the hydrocarbon yield and, in addition, the technical feasibility is not yet completely proven. The catalyst deactivation still raises many concerns for both routes. The processing costs are high and the products are not competitive with fossil fuels (50).

APPLICATIONS FOR BIO-OIL

Bio-oil can substitute for fuel oil or diesel in many static applications including boilers, furnaces, engines and turbines for electricity generation. The possibilities are summarised in Figure 13. There is also a range of chemicals that can be extracted or derived including food flavourings, specialities, resins, agri-chemicals, fertilisers, and emissions control agents. Upgrading bio-oil to transportation fuels is feasible but currently not economic.

Electricity production

At least 500 hours operation has been achieved in the last few years on various engines from laboratory test units to 1.4 MWe modified dual fuel diesel engines. One such engine is a 250 kWe dual fuel engine on which nearly 400 hours have been logged in total, including several runs of over 9 hours, and with electricity being generated for 320 hours (51). A 2.5 MWe gas turbine has been modified and successfully run on bio-oil (52), and work is being carried out with bio-oil in gas turbine combustors (53).

**Table 6 Demonstrated applications
(with references to papers in these proceedings. See also Table 7)**

Electricity	diesel, turbine, Stirling (54)
Heat	CHP and boiler (55)
Transport fuels	upgrading, emulsions (41)
Bulk chemicals	e.g. resins (56, 57), fertilizers, hydrogen (58)
Fine chemicals	e.g. levoglucosenone (59)
Emissions control	Calcium enriched bio-oil (60)

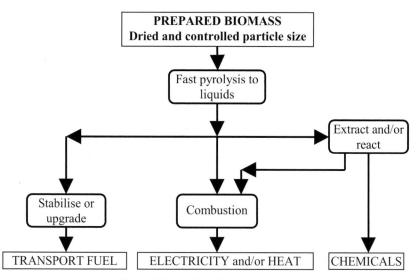

Figure 13 Applications for Bio-oil.

Chemicals

A range of chemicals can also be produced from specialities such as levoglucosan to commodities such as resins and fertilisers as summarised in Table 7. Food flavourings are commercially produced from wood pyrolysis products in many countries. All chemicals are attractive possibilities due to their much higher added value compared to fuels and energy products, and lead to the possibility of a bio-refinery concept in which the optimum combinations of fuels and chemicals are produced.

Table 7 Chemicals from fast pyrolysis (with examples of references)

Acetic acid (61)	Adhesives (56, 57)	Calcium enriched bio-oil (60)
Food flavourings (20)	Hydrogen (58)	Hydroxyaceladehyde (62)
Levoglucosan (63)	Levoglucosenone (59)	Preservatives (64)
Resins (56, 57)	Slow release fertilisers (65)	Sugars (7)

SUMMARY AND CONCLUSIONS

The liquid bio-oil product from fast pyrolysis has the considerable advantage of being storable and transportable as well as the potential to supply a number of valuable chemicals, but there are many challenges facing fast pyrolysis that relate to technology, product and applications. The problems facing the sector include the following:

- Cost of bio-oil, which is 10% to 100% more than fossil fuel.
- Availability: there are limited supplies for testing.
- There is a lack of standards for use and distribution of bio-oil and inconsistent quality inhibits wider usage. Considerable work is required to characterise and standardise these liquids and develop a wider range of energy applications.
- Bio-oil is incompatible with conventional fuels.
- Users are unfamiliar with this material.
- Dedicated fuel handling systems are needed.
- Pyrolysis as a technology does not enjoy a good image.

Therefore, more research is needed in the field of fast pyrolysis. The most important issues that need to be addressed seem to be:

- Scale-up.
- Cost reduction.
- Improving product quality including setting norms and standards for producers and users.
- Environment health and safety issues in handling, transport and usage.
- Encouragement for developers to implement processes; and users to implement applications.
- Information dissemination.

There is much potential for further development and optimisation. Chemicals offer more interesting commercial opportunities and are likely to be a major focus of continuing research and development effort.

REFERENCES

1   Baxter, L, "Biomass combustion: Status quo and challenges", in Progress in Thermochemical Biomass Conversion, Ed. Bridgwater AV, (Blackwell, Oxford, UK, 2001)
2   Stahl K, Neergaard M, Nieminen J, "Final report: Varnamo demonstration prógramme", in Progress in Thermochemical Biomass Conversion, Ed. Bridgwater AV, (Blackwell, Oxford, UK, 2001)
3   Maniatis K, "Progress in biomass gasification: An overview", in Progress in Thermochemical Biomass Conversion, Ed. Bridgwater AV, (Blackwell, Oxford, UK, 2001)
4   Diebold JP and Bridgwater AV, "Overview of Fast Pyrolysis of Biomass for the production of Liquid Fuels", pp 5-26, Developments in Thermochemical Biomass Conversion, Bridgwater, AV and Boocock, DGB (Eds.) (Blackie Academic & Professional, London 1997)
5   Bradbury AGW, Sakai Y, Shafizadeh F, "Kinetic Model for Pyrolysis of Cellulose", J. Appl. Polym. Sci., Vol. 23, pp.3271, 1979.
6   Boroson ML, Howard JB, Longwell JP, Peters WA, "Heterogeneous Cracking of Wood Pyrolysis Tars over Fresh Wood Char Surfaces", Energy & Fuels, 3, 735-740, 1989.
7   Radlein D, Piskorz J, Scott DS, " Fast Pyrolysis of Natural Polysaccharides as a Potential Industrial Process", J. Anal. Appl. Pyrol., 19, 41-63, 1991.

8   Vladars-Usas A, "Thermal Decomposition of Cellulose", M.A.Sc. Thesis, University of Waterloo, Ont. Canada, 1993.

9   Diebold J, "A Unified Global Model for the Pyrolysis of Cellulose", Biomass and Bioenergy Vol. 7, Nos. 1-6, pp.75-85, 1994.

10 Piskorz J, Scott D, Radlein D, "Mechanisms of the Fast Pyrolysis of Biomass: Comments on Some Sources of Confusion", presented at Frontiers of Pyrolysis: Biomass Conversion and Polymer Recycling Conference, Breckenridge, 1995, included in written form in Minutes of the 2nd PYRA Meeting, La Coruna, 1995.

11 Narayan R, Antal MJ Jr, "Thermal Lag, Fusion and the Compensation Effect during Biomass Pyrolysis", Ind. Eng. Chem. Res. 35, pp.1711-1721. 1996.

12 Radlein D, "The Production of Chemicals from Fast Pyrolysis Bio-oils", Fast Pyrolysis of Biomass: A Handbook, A. Bridgwater et al. pp164, CPL Press 1999.

13 Boutin O, Lede J, "Use of a Concentrated Radiation for the Determination of the Elementary Mechanisms of Cellulose Thermal Decomposition", in Progress in Thermochemical Biomass Conversion, Ed. Bridgwater AV, (Blackwell, Oxford, UK, 2001)

14 Scott DS and Piskorz J, "The flash pyrolysis of Aspen-Poplar wood", The Canadian journal of chemical engineering, Vol. 60, October (1982) 666-674

15 Bridgwater AV and Peacocke GVC, "Fast pyrolysis processes for biomass", Sustainable and Renewable Energy Reviews, 4 (1) 1-73 (Elsevier, 1999)

16 See Fortum Website: http://www.fortum.com

17 Cuevas A, Reinoso C and Scott DS, "Pyrolysis oil production and its perspectives", in Proc. Power production from biomass II, Espoo, March 1995 (VTT)

18 PyNe Newsletter 4, September 1997, Aston University, UK

19 Dynamotive Press release, 3 October 2000.

20 Underwood G, "Commercialisation of fast pyrolysis products", in 'Biomass thermal processing', Eds. Hogan E, Robert, J, Grassi G and Bridgwater AV, pp 226-228, (CPL Press, 1992)

21 Rossi C and Graham RG. "Fast pyrolysis at ENEL". In: Kaltschmitt MK and Bridgwater AV, editors, Biomass gasification and pyrolysis: State of the art and future prospects, CPL Press, 1997, p 300-306.

22 PyNE Newsletter No. 4, September 1997, Aston University, UK

23 Boukis I, Gyftopoulou ME, Papamichael I, "Biomass fast pyrolysis in an air-blown circulating fluidized bed reactor", in Progress in Thermochemical Biomass Conversion, Ed. Bridgwater AV, (Blackwell, Oxford, UK, 2001)

24 Lédé J, Panagopoulos J, Li HZ and Villermaux J, "Fast Pyrolysis of Wood: direct measurement and study of ablation rate", in Fuel, 1985, vol. 64, pp. 1514-1520.

25 Lédé J, "The Cyclone: A Multifunctional Reactor for the Fast Pyrolysis of Biomass", Ind. Eng. Chem. Res., 2000, 39, 893-903.

26 Kovac R.J. and O'Neil DJ, "The Georgia Tech Entrained Flow Pyrolysis Process," Pyrolysis and Gasification, G.L. Ferrero, K. Maniatis, A. Buekens, and A.V. Bridgwater, eds., Elsevier Applied Science 1989, pp. 169-179.

27 Maniatis K, Baeyens J, Peeters H and Roggeman G, "The Egemin flash pyrolysis process: commissioning and results", pp 1257-1264 in Advances in thermochemical biomass conversion, Ed. AV Bridgwater (Blackie 1993)

28 Prins W and Wagenaar BM, In Biomass gasification and pyrolysis, Eds. Kaltschmitt MK and Bridgwater AV, pp 316-326 (CPL Press 1997)

29 Wagenaar BM, Venderbosch RH, Carrasco J, Strenziok R, van der Aa BJ, "Rotating cone bio-oil production and applications", in Progress in Thermochemical Biomass Conversion, Ed. Bridgwater AV, (Blackwell, Oxford, UK, 2001)

30 Yang J, Blanchette D, de Caumia B, Roy C "Modelling, scale-up and demonstration of a vacuum pyrolysis reactor", in Progress in Thermochemical Biomass Conversion, Ed. Bridgwater AV, (Blackwell, Oxford, UK, 2001)

31 OFFER announcement, February 1997, (Office of Electricity Regulation, UK)

32   Diebold JP, Czernik S, Scahill JW, Philips SD and Feik CJ. "Hot-gas filtration to remove char from pyrolysis vapours produced in the vortex reactor at NREL", in: Milne, TA, editor, Biomass Pyrolysis Oil Properties and Combustion Meeting, NREL, 1994, p. 90-108.

33   PyNe Newsletter 3, March 1997, Aston University, UK pp21

34   Lédé J, Diebold JP, Peacocke GVC, Piskorz J, (1996) "The Nature and Properties of Intermediate and Unvaporized Biomass Pyrolysis Materials," in Developments In Thermochemical Biomass Conversion, Bridgwater AV and Boocock DGB, (Eds.) (Blackie Academic & Professional, London 1997).

35   Huffman, D.R., Vogiatzis, A.J. and Bridgwater, A.V., "The characterisation of RTP bio-oils", Advances in Thermochemical Biomass Conversion, Ed. A V Bridgwater, (Elsevier, 1993)

36   Piskorz J et al. "Fast pyrolysis of sweet sorghum and sweet sorghum bagasse", JAAP, 46 (1998) 15-29

37   Bridgwater AV, Peacocke GVC, Dick CM, Cooke LA, Tiplady IR, Hague RA, Grimwood M, Overton N, "Preliminary evaluation of the fast pyrolysis of pine bark", Final Report, New Zealand Forest Research Institute, Private Bag 3020, Rotorua, New Zealand

38   Diebold JP, Milne TA, Czernik S, Oasmaa A, Bridgwater AV, Cuevas A, Gust S, Huffman D, Piskorz J, "Proposed Specifications for Various Grades of Pyrolysis Oils", In Developments in Thermochemical Biomass Conversion; Bridgwater AV and Boocock DGB, Eds. (Blackie Academic & Professional, London 1997), pp. 433-447.

39   Shihadeh AL, Rural Electrification from Local Resources: Biomass Pyrolysis Oil Combustion in a Direct Injection Diesel Engine, Ph.D. Thesis, (Massachussetts Institute of Technology, 1998)

40   Ikura M, Slamak M, Sawatzky H, Pyrolysis Liquid-in-Diesel Oil Microemulsions, US Patent 5,820,640, (1998)

41   Baglioni P, Chiaramonti D, Bonini M, Soldaini I, Tondi G, "BCO/Diesel oil emulsification: Main achievements of the emulsification process and preliminary results of tests on diesel engine", in Progress in Thermochemical Biomass Conversion, Ed. Bridgwater AV, (Blackwell, Oxford, UK, 2001)

42   Diebold, J.P. and Czernik, S., Additives to Lower and Stabilize the Viscosity of Pyrolysis Oils during Storage, Energy & Fuels, (1997), 11, 1081-1091.

43   Radlein D, Piskorz J, Majerski P, Method of upgrading biomass pyrolysis liquids for use as fuels and as a source of chemicals by reaction with alcohols, European Patent # 0718392, (1999.

44   Diebold JP, "A review of the chemical and physical mechanisms of the storage stability of fast pyrolysis bio-oils", Report for PyNe, 1999. (To be published in 2001)

45   Maggi R and Elliott D, In Developments in Thermochemical Biomass Conversion, Bridgwater AV and Boocock DGB Eds. (Blackie Academic & Professional, London 1997), pp. 575-588.

46   Bridgwater AV, "Production of high grade fuels and chemicals from catalytic pyrolysis of biomass", Catalysis Today, 29, 285-295 (1996)

47   Bridgwater AV, "Catalysis in thermal biomass conversion", Applied Catalysis A, 116, (1-2), pp5-47, (1994)

48   Elliott DC and Baker E, In Energy from Biomass and Wastes X, Klass D, Ed., (IGT, 1983), pp. 765-782.

49   Chang C and Silvestri AJ, Catalysis, (1977), 47, p. 249.

50   Bridgwater AV and Cottam ML, "Costs and Opportunities for Biomass Pyrolysis Liquids Production and Upgrading", Proc 6th conference on Biomass for Energy, Industry and the Environment, Athens, (April 1991).

51   Leech J, "Running a dual fuel engine on pyrolysis oil", pp 495-497, In: Kaltschmitt MK and Bridgwater AV, editors, Biomass gasification and pyrolysis: State of the art and future prospects, (CPL Press, 1997).

52   Andrews R., Patnaik PC, Liu Q. and Thamburaj, "Firing Fast Pyrolysis Oils in Turbines", Proceedings of the Biomass Pyrolysis Oil Properties and Combustion Meeting, Eds. Milne, T., (NREL/CP-430-7215, 1994), pp. 383-391.

53   Strenziok R, Hansen U, Kunstner H, "Combustion of bio-oil in a gas turbine", in Progress in Thermochemical Biomass Conversion, Ed. Bridgwater AV, (Blackwell, Oxford, UK, 2001)

54   Bandi A, Baumgart F, "Stirling engine with flox burner fuelled with fast pyrolysis with fast pyrolysis liquid", in Progress in Thermochemical Biomass Conversion, Ed. Bridgwater AV, (Blackwell, Oxford, UK, 2001)

55   Oasmaa A, Kyto M, Sipila K, "Pyrolysis oil combustion tests in an industrial boiler", in Progress in Thermochemical Biomass Conversion, Ed. Bridgwater AV, (Blackwell, Oxford, UK, 2001)

56   Pakdel H, Murwanashyaka J N, Roy C, "Fractional vacuum pyrolysis of biomass for high yields of phenolic compounds", in Progress in Thermochemical Biomass Conversion, Ed. Bridgwater AV, (Blackwell, Oxford, UK, 2001)

57   Himmelblau A "Combined chemicals and energy production from biomass pyrolysis", in Progress in Thermochemical Biomass Conversion, Ed. Bridgwater AV, (Blackwell, Oxford, UK, 2001)

58   Czernik S, French R, Feik C, Chornet E, "Production of hydrogen from biomass-derived liquids", in Progress in Thermochemical Biomass Conversion, Ed. Bridgwater AV, (Blackwell, Oxford, UK, 2001)

59   Dobele G, Rossinskaja G, Telysheva G, Meier D, Radtke S, Faix O, "Levoglucosenone - A product of catalytic fast pyrolysis of cellulose", in Progress in Thermochemical Biomass Conversion, Ed. Bridgwater AV, (Blackwell, Oxford, UK, 2001)

60   Venderbosch RH, Wagenaar BM, Gansekoele E, Sotirchos S, Moss HDT, "Co-firing of bio-oil with simultaneous SOx and NOx reduction", in Progress in Thermochemical Biomass Conversion, Ed. Bridgwater AV, (Blackwell, Oxford, UK, 2001)

61   PyNe Newsletter 4, pp 21, (September 1997), Aston University, UK

62   Radlein, D. and Piskorz, J., "Production of chemicals from bio-oil". In: Kaltschmitt M, Bridgwater AV, editors, Biomass Gasification and Pyrolysis, (CPL Press, 1997), p 471-481.

63   Pernikis P, Zandersons J, and Lazdina B, "Obtaining Levoglucosan by Fast Thermolysis of Cellolignin-Pathways of Levoglucosan Use," in Developments In Thermochemical Biomass Conversion, Bridgwater AV and Boocock DGB, (Eds.) (Blackie Academic & Professional, London 1997)

64   Meier D, Andersons B, Irbe I, Tshirkova J, Faix O, "Preliminary study of fungicide and sorption effects of fast pyrolysis liquids used as wood preservative", These proceedings

65   Radlein, D, Piskorz J and Majerski, P, "Method of Producing Slow-Release Nitrogenous Organic Fertilizer from Biomass" US Patent 5,676,727, 1997 and European Patent Application 0716056.

ANALYSIS, CHARACTERISATION AND TEST METHODS OF FAST PYROLYSIS LIQUIDS

A. Oasmaa
VTT Energy, P.O.Box 1601, FIN-02044 VTT, Finland
D. Meier
Federal Research Centre for Forestry and Forest Products, Institute for Wood Chemistry
and Chemical Technology of Wood, D-21027 Hamburg, Germany

ABSTRACT

The PyNe (Pyrolysis Network) subject group on analysis, characterisation and test methods has continued the work carried out in IEA PYRA (International Energy Agency, Pyrolysis Activity) and EC PyNE (European Commission, Pyrolysis Network in Europe) Activities in 1995-97. Both EC and IEA have funded the project. In total 8 objectives were established comprising data collection and distribution in feedback analyses, evaluation of standard and new analysis methods, organization of a round robin and of workshops. The main results of the project are: organising two workshops, collation and assessment of information and publications on analysing biomass feedstocks and pyrolysis liquids, assessment a suggestion for fuel oil specifications for pyrolysis liquids with the liquid end-users, reporting methods development, and organizing a round robin analysis for various pyrolysis liquids.

1   INTRODUCTION

The PyNe (Pyrolysis Network) subject group in analysis, characterisation and test methods has continued the work carried out in IEA PYRA (International Energy Agency, Pyrolysis Activity) and EC PyNE (European Commission, Pyrolysis Network in Europe) in 1995-97. In IEA PYRA the main results on this area were a collation on conventional fuel standards and recommendations for bio-oil standards, a state of the art on the suitability of standard test methods for pyrolysis liquids, and summary of new methods for measuring important properties of pyrolysis liquids like stability and water insolubles. This work has been published in the IEA Bioenergy Handbook on Biomass Pyrolysis (Bridgwater et al. 1999). In EC PyNE the characterisation group focused on reviewing various analytical methods and their usability for pyrolysis liquids. The main task for PyNe is to discuss and assess the key issues on characterisation of pyrolysis liquids. The methodology of the Characterisation subject group is shown in Figure 1. Feedback from application people is one of the key issues.

The objectives of the PyNe Subject group on analysis and characterisation are:

- Provide data on feedstock analyses and methods of analysis. Include a few different examples (untreated hardwood and softwood, one agricultural feedstock e.g. straw) including analyses of feedstock and pyrolysis products, and including process type, conditions (pyrolysis temperature, residence time) and char removal method,
- Collect data on the effect of pretreatment (e.g. washings) of the feedstock on pyrolysis products,
- Collect feedback from pyrolysis liquid end-users (such as Ormrod, Neste, and Orenda) for identifying important properties and evaluating and/or developing suitable test methods,
- Evaluate alternative standard methods and new methods for measuring physical properties,
- Evaluate new methods for chemical characterisation (HPLC, SPE, i.e., solid phase extraction, 13C-NMR, FTIR, HS i.e. head space analysis, GPC),

- Conduct a new Round Robin on physical (water, solids, particles, viscosity etc.) and chemical characterisation (pyrolytic lignin, water solubles). Suggested samples: fast, slow, and vacuum pyrolysis liquids produced under well-defined and documented conditions,
- Study and evaluate methods for measuring pyrolysis liquid quality,
- Organise a workshop to review progress and present developments in these areas.

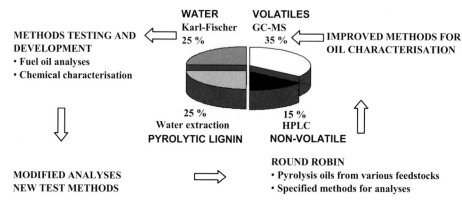

Figure 1   Methodology of the characterisation subject group

2   PROPERTIES OF PYROLYSIS LIQUIDS

Bio-oils (pyrolysis oils or pyrolysis liquids), are usually dark brown organic liquids.   The physical properties of the oils are shown in Table 1. These properties result from the chemical composition of the oils, which is significantly different from that of petroleum-derived fuels (Oasmaa & Czernik 1999).

In contrast to petroleum fuels, bio-oils contain a large amount of oxygen, usually 45-50 wt. %. This oxygen is present in most of the more than 200 compounds that have been identified (Soltes & Elder 1981) in the oils. The single most abundant bio-oil component is water.   The other major groups of compounds identified are hydroxyaldehydes, hydroxyketones, sugars, carboxylic acids, and phenolics (Piskorz et al. 1988).   The presence of oxygen is the primary reason for differences in the properties and behavior between hydrocarbon fuels and biomass pyrolysis oils (Oasmaa & Czernik 1999).

The lower heating value (LHV) of bio-oils is only 40-45% of that for hydrocarbon fuels. Because of the higher density bio-oil heating value on the volumetric basis is about 60% of that for diesel oil. Water content varies in a wide range (15-30%) depending on the feedstock and process conditions. Biomass pyrolysis oils contain substantial amount of non-volatile materials some of which under heating polymerize. Thus, bio-oils can not be used for applications requiring complete evaporation before combustion. Biomass oils contain 0.01-3 wt.% solids as char and/or fine sand or other heat transfer medium entrained from the reactor. In most cases solids are highly undesirable because they tend to settle at the bottom of the vessel in a form of sludge, can cause erosion and block injection nozzles. Presence of char also seems to catalyze reactions leading to the increase of viscosity of the oil and, eventually, formation of gummy tars (Oasmaa & Czernik 1999).

Biomass oils contain compounds that, during storage or handling, can react with themselves to form larger molecules (Oasmaa & Czernik 1999, Diebold 1999). These reactions result in undesirable changes in physical properties, such as increase of viscosity and water content with a corresponding decrease of volatility. The most important factor for this "aging" is temperature (Czernik 1994). Phase separation can take place at higher water and/or lignin-derived material concentrations. It can be a result of a long-term storage of the oil. Typically, two layers of different properties are formed (Oasmaa et al. 1997). The oil may be homogenized for limited time by addition of a common polar organic solvent such as alcohol (Oasmaa & Kytö 2000). Biomass oils, due to differences in polarity do not mix with light hydrocarbon fuels. In Table 2 variation of properties for wood (hardwood, softwood) pyrolysis liquids are presented.

Table 1   Pyrolysis liquid properties

	PYROLYSIS LIQUIDS			MINERAL OILS	
Feedstock Char removal	Bircha) Char cyclone	Pinea) Char cyclone	Poplarb) Hot-vapor filtered	POK 15	POR2000
Solids, wt-%	0.06	0.03	0.045	-	-
pH	2.5	2.4	2.8	-	-
Water, wt-%	18.9	17.0	18.9	0.025	Max. 7
Viscosity (50°C), cSt	28	28	13.5	6	140-380
Density (15°C), kg/dm3	1.245	1.243	1.200	0.890	Max. 0.900-1.020
LHV, MJ/kg	16.5	17.2	17.4	40.3	-
Ash, wt-%	0.004	0.03	0.01	0.01	0.1
CCR, wt-%	20	16		0.2	-
C, wt-%	44.0	45.7	46.5	-	-
H, wt-%	6.9	7.0	7.2	-	-
N, wt-%	<0.1	<0.1	0.15	-	-
S, wt-%	-	0.020	0.02		1.0
O, wt-%	49	47	46.1	0	-
Na+K, ppm	29	22	6	-	-
Ca, ppm	50	23	4	-	-
Mg, ppm	12	5	3	-	-
Flash point, °Cd)	62	95	64	60	Min. 65
Pour point, °C	-24	-19	NA	-15	Min. 15

a)   Produced at
b)   Produced at NREL,
c)   About 50
d)   Flash point does not correlate directly with ignition and   pyrolysisliquids
    POK 15: Light-Medium Fuel Oil in Finland. Comparable to #4FO in U.S. and
    POR 2000: Heavy Fuel Oil In Finland. Comparable to #6FO and #4-GT in U.S. and Canada, HFO S6

Table 2   Physical properties for wood (hardwood, softwood) pyrolysis liquids

Property (wet basis)	Range
Density (15°C), kg/dm3	1.11 - 1.30
Lower heating value, MJ/kg	13 – 18
Viscosity @50 °C, cSt	10 – 80
Pour point, °C	‾9 - ‾36
Coke residue, wt-%	14 – 23
Flash point, °C	50 – 110
Water, wt-%	20 – 35
pH	2.0 - 3.7
Char/solids, wt-%	0.01 – 1
C, wt-% (dry basis in parentheses)	32 - 49 (48 - 60)
H, wt-%	6.9 - 8.6 (5.9 - 7.2)
N, wt-%	0 – 0.3
O, wt-%	44 - 60 (34 - 45)
S, ppm	60 – 500
Cl, ppm	3 – 100
Ash, wt-%	0.01 - 0.20
K+Na, ppm	10 – 330

3   RESULTS

3.1   Feedstock Analyses

Analyses of different Scandinavian and European biomass feedstocks have been summarised in the report by Wilen et al. (1996). The fuels for the additional analyses were chosen to represent typically three significant biomass sources (Table 3). The analyses included proximate and ultimate analyses, trace compounds, ash composition and ash fusion behaviour in oxidizing and reducing atmosphere.

Table 3   Feedstocks chosen for additional analyses.

Scandinavian biomass feedstocks		European biomasses
Woody biomasses	Agricultural biomasses	
Wood chips	Barley straw	Sweet Sorghum
Forest residue chips	Wheat straw	Miscanthus
Pine bark	Rapeseed	Kenaf
Spruce bark	Flax (Linum)	Cane
Willow (Salix)	Reed canary grass	
Sawdust (pine)		

Whole biomass feedstock analyses can be performed by combined analytical pyrolysis-gas chromatography connected either to a flame ionization detector or to a mass spectrometer. The resulting peaks should be well resolved. After careful assignment of the peaks to carbohydrate and lignin derived products and the calculation of normalized area percent the lignin and carbohydrate composition can be determined. Furthermore, the type of lignin, e.g. annual plants: more H-units, softwood: almost only G-units, hardwood: more S-units, can also be determined.

Other wood-analytical methods always comprise the isolation and purification of components leading often to alteration of the product. In most cases, the non-analyzed component is chemically degraded in such a way, that the component to be analyzed is left. This is true for the determination of holocellulose and Klason lignin.

Many analyses methods for the carbohydrates and lignins are described by TAPPI, the Technical Association of the Pulp and Paper Industry.

Further analysis of the separated compounds is mostly done to determine more details such as molecular weight and functional groups. A comprehensive overview is given by Fengel und Wegener (1984). A review of methods of analysis for various biomass feedstocks has been given by Milne et al. (1990).

Hague has recently carried out work on pretreatment of biomass (1998). The results were analogous to those previously presented by, i.a., Jan Piskorz et al. (Piskorz et al. 1988, Radlein et al. 1987).

3.2   Feedback From Pyrolysis Liquid End-Users

3.2.1   Fuel oil use

Biomass pyrolysis oils have potential to be used as a fuel oil substitute (Oasmaa & Czernik 1999). Combustion tests in a 8 MW test furnace of Oilon Oy´s R&D Centre have shown that the oils can be burnt efficiently in standard or slightly modified equipment (Kytö 1999, Oasmaa et al. 2000). However, these tests also identified several challenges in the bio-oils applications resulting from their properties. The oils have heating values of only 40-50% of that for hydrocarbon fuels. High water content is detrimental for ignition rates. Organic acids cause corrosiveness of the oils to some common construction materials. Solids (char) can block injectors or erode turbine blades. Reactivity of some components leads to formation of larger molecules that results in high viscosity and in slower combustion chemistry. Additional results related to fuel quality of the Oilon combustion tests were:

- Handling and pumping of pyrolysis oil should be performed according to exact recommendations.
- Quality specifications should be defined for pyrolysis oil, including especially water and solids contents.
- Viscosity range is significant for good atomisation.
- The quality of pyrolysis oil has a strong impact on emissions. High solids content in pyrolysis oil yields high particulate emissions. Hence, solids removal from pyrolysis vapours or oil is highly recommended. High (above 30 wt-%) water content also yields high particulate emissions. These emissions can be decreased to a certain extent by using a support fuel and optimizing the atomization viscosity. Methanol addition (max 10 wt%) homogenizes the inferior-grade oil and decreases particulate emissions. The costs for methanol addition and oil combustion in a commercial boiler are most probably lower than those of incinerating poor-quality oil in a special incineration plant for hazardous wastes.
- Further research is required on combustion properties of various commercial pyrolysis oils in order to identify the reasons for emission behaviour, nozzle blockages and related phenomena.

The end-users for pyrolysis liquids need to know certain properties of the liquids and their behaviour in boilers, engines, and turbines. Feedstock has probably the greatest effect on pyrolysis liquid quality. However, feedstock pretreatment, process type, process conditions, liquid recovery method, upgrading, and handling and storage of the liquid may also significantly influence pyrolysis liquid fuel properties and thus indirectly e.g. emissions.

A summary of the prioritised properties assessed by the industry is shown in Table 4. Stability, homogeneity, solids, viscosity and ignition are perhaps the most notable properties. Reliable ways of measuring water and solids content are needed. The solids content in the oil affect directly particulate emissions. Properties like heating value, vapour pressure, flash point, acidity, and miscibility can be influenced only in a limited scale without using any expensive upgrading. Companies designing pyrolysis units and recovery systems need to know certain design values, like vapour pressure, viscosity versus temperature, ratio of aerosols and gas, thermal conductivity, and specific heat capacity. Standard test methods are a tool for estimating the quality of pyrolysis liquids from different feedstocks and processes. Methods should be developed taken into account the special chemical composition of pyrolysis liquids. A simple test method (Oasmaa et al. 1997) for storage/thermal stability is based on an increase in viscosity (24 hours @ 80°C). In the method standard ASTM D445-88 is used. For a pine pyrolysis liquid this test corresponds to a one year storage at room temperature.

Table 4. Important properties of pyrolysis liquids in prioritised order defined by industry August 1999

BOILERS	DIESEL ENGINES	GAS TURBINES
Fortum, Oilon, SEAb	*Ormrod*	*Orenda*
Viscosity	Homogeneity	Homogeneity
Stability, homogeneity	Water	Trace elements
Ignition, flash point	Solids	Viscosity
Solids	Viscosity	Particle size distribution
HHV & water	HHV	Ash, solids/char
EHS: flash point & toxicity	Ignition, flash point	HHV, LHV
Acidity	Acidity	Acidity
CL	Lubricity	Pour point
Smell	Heavy metals, Cl	CONH
Heavy metals	Silicon	Water

Additional properties that may influence applications listed in the Montpellier workshop are:
- Molecular weight distribution
- Chemical composition (ratios of water, pyrolytic lignin, GC-eluted fraction, HPLC fraction)
- Oxygen-containing functionalities
- Surface tension
- Odour

Chemical composition, oxygen-containing functionalities, and molecular weight distribution play a major role in the instability of pyrolysis liquids. However, a simple test method should be developed for predicting the behaviour of pyrolysis liquids during its applications.

Surface tension of pyrolysis liquids may effect to the behaviour of pyrolysis liquids around nozzles, especially in boiler applications. Odour may prevent the commercialisation of pyrolysis liquids by the consumers.

The specifications and norms for the quality of pyrolysis liquids have to be developed before the commercialisation of the applications can take place. However, this time frame may be long. The most essential issue is close co-operation between the research community and the industry.

To improve combustion, the following items may be of significance:
- Homogeneity of pyrolysis liquid on a microscopic level. Slight phase-separation may yield in poor combustion.

28

- Heating value of the light fraction of pyrolysis oil may correlate with combustion properties.
- Flash point seems to have no correlation with combustion properties. Hence, the flash point may be excluded from the safety regulations after the tests needed.
- High solids content in pyrolysis oil yields high particulate emissions.
- High (above 30 wt-%) water content also yields high particulate emissions. These emissions can be decreased to a certain extent by using a support fuel and optimizing the atomization viscosity.

The most important issue in commercialisation will be the need for pyrolysis liquid specifications. Elliott (1983) has assessed the specification standards for various pyrolysis liquids in IEA BLTF (Biomass Liquefaction Test Facility) project. The classification was based on ASTM standards D-396 for fuel oils, D-975 for diesel fuels, and D-2880 for gas turbine fuels. The IEA PYRA group suggested (Diebold et al. 1997) specifications shown in Table 5.

Based on the feedback from end-users following changes to the specifications should be discussed:
- Specs should be tighter.
- Water should be lower, because of poor ignition properties and bad emissions of high-water content pyrolysis oils.
- Solids should be specified as methanol-methylene chloride insolubles.
- Viscosity should be determined at two temperatures which are usable for end-users, as 20 and 40 °C.
- The meaning of flash point should be clarified. It does not correlate with the ignition properties of pyrolysis oils.

3.2.2 Chemicals production

Properties important for chemical production are listed in Table 6 with possible analytical/test methods.

Flavour and odour are qualitatively tested by nose. Specific compounds can be identified by GC and groups of compounds by HPLC. A quantitative analysis of compounds is also possible by using SPME (Solid Phase Micro Extraction). Henriksen (AP Broste, Denmark) has experience in analysis of smoke flavours. The coloring properties are measured by determination of carbonylic compounds and the Staining Index (similar to Browning Index). Sensory properties are determined based on phenolic compounds, acid content and sensory evaluations. A uniform analysis with a practical standard for Staining/Browning Index used for pyrolysis liquids could be useful.

A quick method for determining the chemical groups in pyrolysis liquids is TLC (Thin Layer Chromatography). A more specific group analysis can be carried out by HPLC. Individual compounds are typically analysed by GC/MS. In quantitative analyses the use of response factors is a must. Functional groups of pyrolysis liquids can be measured by ^{13}C-NMR or by analytical methods in wet chemistry like carbonyls, methoxyls etc. The molecular weight distribution is analysed by GPC.

Reactivity and stability are also critical considering the chemicals production. The change in the ratio of functional groups can be checked by ^{13}C-NMR or by wet chemistry. Different separation and fractionation methods can also be used.

Table 5. Proposed specifications for biomass pyrolysis liquids (Diebold et al. 1997)

	Light bio-oil (~ASTM #2)	Light-Medium bio-oil (~ASTM #4)	Medium bio-oil (~PORL100)	Heavy bio-oil (~Can. #6)
Viscosity, cSt	1.9-3.4 FO 1.9-4.1 D 1.9-4.1 GT @ 40 °C	5.5-24 @ 40 °C	17-100 @ 50 °C	100-638 @ 50 °C
Ash, wt%	0.05 FO 0.01 D 0.01 GT	0.05 FO 0.10 D	0.10 FO	0.10 FO
Pour point, °C min	report	report	report	report
CCR, wt%	report	report	report	report
Max. 0.1 μm filtered ethanol insolubles	0.01 FO	0.05	0.10	0.25
Accelerated aging rate @ 90 °C cSt/h	report	report	report	report
Water, wt% of wet oil, max.	32	32	32	32
LHV, MJ/L min., wet oil	18	18	18	report
C, wt% dry	report	report	report	report
H, wt% dry	"	"	"	"
O, wt% dry	"	"	"	"
S, wt% dry	0.1 max.	0.1 max.	0.2 max.	0.4 max.
N, wt% dry	0.2 max.	0.2 max.	0.3 max.	0.4 max.
K + Na, ppm	report 0.5 GT	report	report	report
Phase Stability @ 20 °C 8 h @ 90 °C	single phase	single phase	single phase	single phase
Flash point, °C min.	52	55	60	60
Density, kg/m3	report	report	report	report

Table 6. Important properties of pyrolysis liquids considering chemical production

Property	Analysis/test method
Flavour, odour	Nose, GC, HPLC, SPME
Composition	TLC, HPLC, GC/MS, ^{13}C-NMR, GPC
Reactivity, stability	^{13}C-NMR, wet chemistry, fractionation technologies

3.3   Evaluation Of Standard And New Methods For Physical Characterisation

Testing of standard methods has been summarised by Oasmaa et al. (1997) and by Bridgwater et al. (1999). The results are shown in Appendix 2. Detailed information on the usability of the method can be found in the suggested references.

The main purpose of the study was to test the applicability of standard fuel oil methods developed for petroleum-based fuels to pyrolysis liquids. Research on sampling, homogeneity, stability, miscibility and corrosivity has been carried out. The standard methods have been tested for several different pyrolysis liquids. Recommendations on sampling, sample size and small modifications of standard methods are presented. In general, most of the methods can be used as such but the accuracy of the analysis can be improved by minor modifications. Fuel oil analyses not suitable for pyrolysis liquids have been identified. Homogeneity of the liquids is the most critical factor in accurate analysis. New feedstocks like forest residues and bark yield to pyrolysis oils possessing different solubility than white wood oils. Solids content cannot be determined as ethanol insolubles, because extractives do not dissolve in alcohols. A stronger solvent, like a mixture of methanol and methylene chloride can be used. Strenziok in University of Rostock, Germany, has studied surface tension of pyrolysis liquids. The measurement was carried out as a function of temperature with a LAUDA Tensiometer TD1. Results were compared to the values obtained by Adjaye & Bakhshi and Andrews & Patnaik. Flash point and density of analysed samples were atypically low. It was pointed out that the flash point measurement may not be accurate and needs more testing. An updated version of the VTT Publication 306 (Oasmaa et al. 1997) will be published 2001.

3.4   Methods For Chemical Characterisation

A "cake" Figure 2 reflects typical portions of the main fractions. About 40% of the compounds are detectable by GC (Gas Chromatography), 15% can be analysed by HPLC (High Performance Liquid Chromatography), 25% is high-molecular-weight "pyrolytic lignin" measured as water-insolubles, and 20% consists of water. Whole pyrolysis liquid can be analysed, i.e., by GC-MS (volatile compounds), HPLC (non-volatile compounds), FTIR (functional groups), GPC (molecular weight), and NMR (functional groups, bonds).

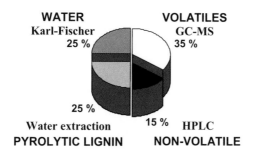

Figure 2   Composition of pyrolysis liquids

Gas chromatographic methods are quite developed and operable. In quantitative analyses, response factors have to be used. A suggestion for the method is shown in Appendix 4. However, it must be reminded that the instability of pyrolysis liquids may cause some erroneous results if temperatures much above room temperature are used at the injection port of GC. However, if this may occur it is compensated by the response factor.

High Performance Liquid Chromatography (HPLC) is one of the best ways forward for polar and higher molecular weight compounds. However, it needs more development, especially the columns. RTI have a wide experience in using HPLC in pyrolysis liquid analyses. By using HPLC more information is obtained from the non-volatile fraction of pyrolysis liquids which has definitely an important role on the behaviour and stability of pyrolysis liquids. Numerous HPLC columns have been tested and the best one (Aminex HPX-87) is presently in use. Co-operation with HP for new column development was suggested. Several fractionation solvent series have been used to separate the pyrolysis liquids into compound classes prior to HPLC analysis and even gas chromatography.

3.5 Evaluation Of New Methods For Chemical Characterisation

3.5.1 Carbonyl group determination by oximation

The change in carbonyl groups may be used as an indication of aging of pyrolysis liquids. Carbonyl groups are known to participate in aging reactions.

Hydroxylamine hydrochloride in the presence of pyridin reacts quantitatively with a variety of aldehydes and ketones. The stoichiometry of the oximation reaction is as follows:

$$
\underset{R-\overset{\overset{\displaystyle O}{\|}}{C}-R' \ + \ NH_2OH \cdot HCl \ + \ C_5H_5N}{} \longrightarrow
$$

$$
\underset{R-\overset{\overset{\displaystyle N-OH}{\|}}{C}-R' \ + \ H_2O \ + \ C_5H_5N \cdot HCl}{}
$$

The function of pyridine in the reaction system is to force oxime formation to completion. The acid liberated in the form of pyridine hydrochloride is determined by titration and is a direct measure of the amount of carbonyl groups originally present in the sample.

Typically, the carbonyl content is in the range of 4 - 6 mol carbonyl/kg oil. The method can be easily validated by model compounds such as furfural or vanillin. Due to the specific oximation reaction other oil compounds do not interfere and a routine analysis is possible.

3.5.2 Head space analysis

The light volatile compounds of pyrolysis oils can be quantitatively analysed by head space analysis. This is a simple tool for safety measurements in storage of pyrolysis oils.

Solid Phase Micro Extraction (SPME) has two important functions: extracting analyses and desorbing them into analytical instruments. A fused silica fiber, coated with an adsorbing material, is exposed to the head space of the sample. After the sampling time is over, the fiber is drawn back into the needle and then the needle is introduced into the hot injector of a gas chromatograph. Here, the fused silica fiber is pushed out of the needle so that adsorbed compounds can be desorbed from the fiber through the heat of the injector. A typical chromatogram is shown in Figure 3.

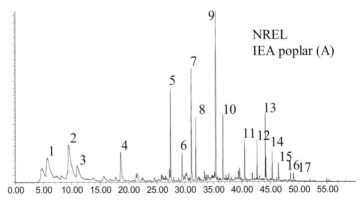

Figure 3   Head space analysis of volatiles of a liquid from fast pyrolysis.

Table 7   Identification of compounds

No.	compound
1	acetic acid methyl ester
2	acetic acid
3	acetol
4	furfural
5	unknown silicon compound from SPME needle
6	2-hydroxy-3 methyl-2-cyclopentene
7	phenol
8	guaiacol
9	guaiacol TMS derivative
10	4-methyl-guaiacol
11	4-ethyl-guaiacol
12	unknown silicon product from SPME needle
13	cis-isoeugenol
14	syringol
15	eugenol
16	trans-isoeugenol
17	4-methyl-syringol

3.6   Methods For Measuring Pyrolysis Liquid Quality

Bio-oils have some undesired properties for fuel applications, such as high water content, high viscosity, poor ignition characteristics, corrosiveness, and instability. Various fuel oil analyses with minor modifications can be used as described earlier. For storage stability an enhanced stability analysis (Oasmaa et al. 1997) can be used. However, the combustion quality of pyrolysis oil and the different chemical fractions in the oil should be studied more closely.

3.6.1   On-line monitoring/quality control

On-line monitoring is highly needed in the quality control of bigger pyrolysis liquids production plants. The key parameters are water and solids. Water may be possible by using FTIR. On-line solids analyses should be developed. Viscosity is also possible to measure on-line. However, it needs more testing.

A "cake" relationship may be further developed. As a rule of thumb mentioned by Jan Piskorz the ratio of water, water-solubles and lignin should be 25:50:25. Other ratios may result in problems with phase stability. Lignin detection is possible by NIR (Near Infra Red). On-line analyses in toxic compounds is also requested.

4   SUMMARY

The main results of the task are:
- Organizing two workshops and collation and assessment of information and publications on analysing biomass feedstocks and pyrolysis liquids
- Discussions with pyrolysis liquid end-users and feedback of the important properties of pyrolysis liquid for them
- Reporting methods development
- Organizing a Round Robin for various pyrolysis liquids

The main topics for the next project include:
- Updating the specifications for pyrolysis liquids based on discussions with end-users and oil producers
- Evaluating simple test methods for measuring the quality of pyrolysis liquids for various applications.
- Testing and developing the stability index
- Assessing methods for on-line monitoring of pyrolysis liquid quality during processing
- Suggesting quality criteria for pyrolysis liquids
- Analytical methods needed for chemical production

5   ACKNOWLEDGMENTS

The authors wish to acknowledge the valuable participation of all PyNe members and observers during the whole three-year period. Thanks are also due to oil producers and end-users for their valuable comments.

6   REFERENCES

Bridgwater, A.V., Czernik, S., Diebold, J., Meier, D., Oasmaa, A., Peacocke, C., Piskorz, J., Radlein, D. Fast pyrolysis of biomass: A handbook. CPL Press, Newbury, UK, 1999. 188 p. ISBN 1-872691-07-2

Czernik, S., Storage of Biomass Pyrolysis Oils. In Proceeding of Specialist Workshop on Biomass Pyrolysis Oil Properties and Combustion, Estes Park, CO, Sept. 26-28 1994, NREL CP-430-7215, pp. 67-76.

Diebold, James; Milne, T.; Czernik, S.; Oasmaa, Anja; Bridgwater, Anthony; Cuevas, A.; Gust, Steven; Huffman, D.; Piskorz, Jan. Proposed specifications for various grades of pyrolysis oils, Developments in Thermochemical Biomass Conversion. Banff, 20 - 24 May 1996. Bridgwater, A. & Boocock, D. (eds.). Vol. 1. Blackie Academic & Professional (1997), s. 433 - 447.

Elliott, D. C. 1983. Analysis and upgrading of biomass liquefaction products. Final report. Vol. 4, IEA Co-operative project D1 Biomass Liquefaction Test Facility Project. Richland, Washington: Pacific Northwest Laboratory. 87 p. + app.

Fengel, D. and Wegener, G. Wood Chemistry, Ultrastructure, Reactions. Walter de Gruyter, Berlin,1984.

Hague, Bob. The pre-treatment and pyrolysis of biomass for the production of liquids for fuels and speciality chemicals, Thesis, September 1998, Aston University, Birmingham

Kytö, M. Burning of pyrolysis oil will not increase the net quantity of carbon dioxide in the atmosphere, Oilon Flame, The Oilon News Magazine 2, 1999, p. 8.

Milne, T. A., Brennan, A. H. & Glenn, B. H. 1990. Sourcebook of methods of analysis for biomass and biomass conversion processes. London: Elsevier Appl. Sci. 327 p. + app.

Oasmaa, Anja. Fuel testing methods for bio-oil, PyNe (Pyrolysis Network for Europe). (1997) 4/Sept., s. 17.

Oasmaa, Anja; Czernik, Stefan. Fuel oil quality of biomass pyrolysis oils. Biomass. A Growth Opportunity in Green Energy and Value-added Products. Vol. 2. Elsevier Science (1999), pp. 1247 - 1252.

Oasmaa, Anja; Meier, Dietrich. Analysis, characterization, and test methods of fast pyrolysis liquids. Biomass. A Growth Opportunity in Green Energy and Value-added Products. Vol. 2. Elsevier Science (1999), pp. 1229 - 1234.

Oasmaa, A., Leppämäki, E., Koponen, P., Levander, J. and Tapola, E. Physical Caracterisation of Biomass-based Pyrolysis Liquids. Application of Standard fuel oil Analyses. VTT Publication 306. Technical Research Centre of Finland, Espoo. 1997.

Piskorz, J., Radlein, D., Scott, D., Czernik, S. Liquid products from the fast pyrolysis of wood and cellulose, Bridgwater,A.V., Kuester, J.L.,Eds.: Research in thermochemical biomass conversion, 557-571.

Piskorz, J.; Scott, D.S.; Radlein, D., Composition of Oils Obtained by Fast Pyrolysis of Different Woods. In Pyrolysis Oils from Biomass: Producing, Analyzing, and Upgrading; Soltes, E.J, Milne, T.A., Eds.; ACS Symposium Series 376, ACS, Washington, D.C. 1988; pp.167-178.

Radlein, D., Piskorz, J., Grinshpun, A., Scott, D.S. Fast pyrolysis of pretreated wood and cellulose. ACS Preprints, Div. of Fuel Chemistry, 1987, 32, No. 2, 29-36.

Soltes, E.J.; Elder, T.J., Pyrolysis. In Organic Chemicals from Biomass, Goldstein, I.S., Ed., 1981, CRC Press, Boca Raton, FL., pp. 63

Wilen, C., Moilanen, A. and Kurkela, E. Biomass feedstock analyses. VTT Publications 282. Technical Research Centre of Finland, Espoo. 1996.

APPENDIX 1

Instructions for Round Robin   20.12.99

D. Meier and A. Oasmaa

BACKGROUND

Pyrolysis liquids for Round Robin

provider	feedstock	additional information
BTG	softwood mixture (spruce and fir) moisture content 10-12 % particle size 0.8-1.1 mm	production date: 11.01.2000 pyrolysis temp. 500 °C
Dynamotive	85% pine, 15% spruce feedstock moisture 7 %	production date; 07.10.99 pyrolysis temp. 460 °C
ENEL/Ensyn		none provided
Pyrovac	softwood bark (1/3 fir and 2/3 spruce with traces of hard wood bark moisture content 12 %	production date 29.09.99 operation temp. 510 °C

The criterion of oil quality is a reasonable homogeneity. In case of non-homogeneity the oil will be excluded from the Round Robin.

Data on feedstock, production conditions, production dates should be provided.

The sample size for each laboratory was set as one litre.

17 laboratories participated.

INSTRUCTIONS
1.   After receiving the sample please indicate date of arrival.
2.   Assure that the sample is being stored at refrigerator conditions.
3.   Stability tests with analyses should be carried out within a week after receiving the sample and all other analyses within a month. Please indicate date of analyses.
4.   Before sampling let the sample reach room temperature, then shake it to ensure homogeneous sampling.
5.   After usage please store the sample again in a refrigerator.
6.   Please report dates of arrival and analyses, sample size used, all duplicates, methods used, possible difficulties, and suggestions
7.   Send the results to IWC and VTT:

Dr. Dietrich Meier   Anja Oasmaa
BFH-Institute for Wood Chemistry   VTT Energy
Leuschnerstrasse 91   P.O.Box 1601
D-21031 Hamburg, GERMANY   FIN-02044 VTT, FINLAND

ROUND ROBIN ANALYSES

Property	Method	Reporting unit
water content	Karl Fischer Titration	wt.% water based on wet oil
viscosity	capillary or rotary viscosimeter, 2 temp. @ 20 and 40°C	cSt @ 20°C and 40 °C
solids	insolubles in ethanol, filter pore size 3μm or lower	wt% based on wet oil
pH	use pH-meter	pH unit
stability (see Appendix 1,3/3)	store samples for 1) 6 h @ 80 °C, 2) 24 h @ 80 °C, and 3) 7 days @50 °C, viscosity @ 20 and 40 °C and water by K-F titration	cSt wt.% water based on wet oil
elemental analysis	elemental analyzer (complete oxidation)	wt%C, wt%H, wt%N, wt%O, based on wet oil
pyrolytic lignin	add 60 ml oil to 1 L of ice-cooled water under stirring, filter and dry precipitate below 60 C	wt% based on wet oil
GPC	not yet established	
HPLC	not yet established	
GC	column type DB 1701 dimensions: 60m x 0.25 mm film thickness: 0.25 μm injector: 250 °C, split 1:30 FID detector: 280 °C oven programme: 45 °C, 4 min const., 3 °C/min. to 280 °C, hold 20 min. sample conc.: 6 wt%, solvent acetone	

STABILITY TEST METHOD

Ref. Oasmaa, Anja; Leppämäki, Eero; Koponen, Päivi; Levander, Johanna; Tapola, Eija, Physical characterisation of biomass-based pyrolysis liquids Application of standard fuel oil analyses, VTT, Espoo. 1997. - 46 p. + app. 30 p. VTT Energia; VTT Energy, VTT Publications, ISBN: 951-38-5051-X

Description of method

The pyrolysis liquid sample is mixed properly and left to stand until the air bubbles are removed. 90 ml of the sample is poured in 100 ml tight glass bottles (or 45 ml in 50 ml bottles). The bottles are firmly closed and pre-weighed before placing in a heating oven for a certain time. The bottles are re-tightened a few times during the heating-up period. After a certain time the closed sample bottles are cooled rapidly under cold water, weighed, and analyses are performed. The possible difference in the weights before and after the test is an indication of leakage and the test should be repeated if the net weight loss is above 0.1 wt% of original weight.

The samples are mixed and measured for viscosity and water. The viscosity of the liquid at 20 and 40 °C is measured as kinematic viscosity by a standard method (ASTM D 445). The water content is analysed by Karl Fischer titration according to ASTM D 1744.

$$\text{Viscosity Index} = \frac{(v_2 - v_1)}{v_1}$$

$$\text{Water Index} = \frac{(\omega_2 - \omega_1)}{\omega_1}$$

v_1 = viscosity of the original sample, cSt
v_2 = viscosity of the aged sample, cSt
ω_1 = water content of the original sample, wt%
ω_2 = water content of the aged sample, wt%

APPENDIX 2
Applicability of methods to the analysis of pyrolysis liquids

Property	Standard Method	Suitability of the method (notes)	Ref.
Physical/chemical			
Density	ASTM D4052	Can be used (a)	3,12,13
	ASTM D941	Can be used (a)	3,8,12,14
	ASTM D1298-85	Can be used (a)	3,6,12
Heating value	ASTM D2382 DIN 51900	Can be used (b)	3,6,8,12,13
Viscosity-kinematic	ASTM D445-88	Can be used (c)	3,8,12,13
Viscosity-dynamic	No standard	Can be used (d)	6,8,9,12-14
Thermal conductivity	No standard	Rough estimate (e)	12,14
Specific heat capacity	No standard	Rough estimate (e)	12,14
Pour point	ASTM D97-87	Can be used (f)	3,6,8,12,13
Setting point	DIN51583	Can be used (f)	12
Cloud point	IP 219/82	No information (g)	3,12,13
Combustion Technology			
Boiling curve	ASTM D86-82	Cannot be used (h)	1,3,6,8,12
Coke residue	ASTM D189-88	Can be used (å)	3,8,12,13
	ASTM D524-88	Can be used (å)	3,8,12
Miscibility	No standard	-	1,6,8,12,13
Lubricity	IP 300-82	Cannot be used (i)	12
	IP 239/69T	Comparative info (j)	12,13
Corrosion	ASTM D130-88	No information (k)	12,13
	ASTM D665A	Comparative info 9 (l)	12,13
Safety Technology			
Flash point	ASTM D93-90	Can be used (m)	3,6,8,12,13
Ignition limit	DIN51603	Limited testing (n)	12,15
Ignition temperature	DIN 57194	Not tested (n)	12,15
Composition			
Water	ASTMD1744	Can be used (o)	3,6,12,13
	ASTM E203	Can be used (o)	12,13
	ASTM D95	Not acceptable (p)	6,12,13
Char	No standard	Can be used (q)	6,8,12,13
Particle size distribution	No standard	Methods available (r)	13
Vapour pressure	IP69/89	Can be used	12
Surface tension	ASTM D 971-50	Limited testing (z)	1,12
CHN	ASTM D5291-92	Can be used	6,8,12,13
O	As difference	Reasonable	8,12,13
S	ASTM D4208	Can be used (s)	12,13
	IP244/71+EF	Can be used (t)	12,13
Cl	IP244/71+EF	Can be used (t)	12,13
Ash	EN7	Can be used (u)	3,6,8,12,13
pH	No standard	Can be used (v)	8,12,13
Na, K, Ca, Mg	No standard	Can be used (x)	4,6,8,12,13
Other metals	No standard	Can be used (y)	6,12,13

NOTES

a Particles may disturb the analysis

b Use of a fine cotton thread for ignition. Lower heating value (LHV) obtained from calorimetric heating value and hydrogen analysis.

c Cannon-Fenske viscometer tubes at room temperature and for non-transparent liquids, Ubbelohde tubes may also be used for more transparent liquids (hot-filtered liquids). No prefiltration of the sample if visually homogenous. Elimination of air bubbles before sampling. Equilibration time in the viscometer bath 30 minutes at 20 °C, 20 minutes at 80 °C. A high amount of particles and especially larger particles may cause an uneven flow of liquid. In this case the filtration suggested in the standard should be used. The original pyrolysis liquid has to be a homogenous single-phase liquid. Otherwise a phase separation may take place during the equilibration time and lead to erroneous results.

d Closed-cup systems are recommended because of the evaporation of volatiles. Above 50 °C error caused by evaporation.

e Preliminary testing carried out.

f No preheating of the sample.

g No visual observations cannot be made because of the dark colour of liquids.

h Pyrolysis liquids cannot be distilled because of their thermal instability.

i Open systems cannot be used because of the volatility of pyrolysis liquids.

j Room temperature measurements can be made. References should have a viscosity similar to that of the sample tested.

k Standard copper rods are resistant to different types of pyrolysis liquid in the standard test conditions. No information is obtained.

l A standard steel corrosion test may be used for comparative measurements.

m Elimination of air bubbles before sampling. Water vapours disturb analysis at 70-100 °C.

n The test method gives values which tend to be lower than those achieved in practice.

o Karl-Fischer titration. Chloroform-methanol (1:3) or methanol as a solvent. HYDRANAL-K reagents (Composite 5K and Working Medium K) in case of a fading titration end-point. Other solvents may also be used but their suitability should be tested. Confirmation of the results may be performed by successive water additions to the pyrolysis liquids and remeasurement of the water content.

p The method cannot be used due to the presence of organic material soluble in water.

q Solvent and filter pore size used are important. Possible solvents for wood pyrolysis liquids: methanol, ethanol, acetone, THF. For straw pyrolysis liquids, acetone cannot be used because of poor solubility/precipitation. Filter pore size below 3 μm.

r Optical methods available for measuring solids in liquids.

s Detection limit 0.1 wt-%.

t Sample pretreatment by combustion according to ASTM D 4208. EF = Electro phoresis.

u Controlled evaporation of water to avoid foaming.

v Frequent calibration of the pH meter.

x Wet combustion as a pretreatment method. If trace levels of alkali metals are measured the contamination from glass containers or dust in the air can be significant, and care has to be taken during sample handling. The use of Teflon (polytetrafluoroethylene, PTFE) bombs for sample pretreatment should be considered. A method requiring no sample pretreatment may also be advantageous. Neutron activation or atomic emission spectrometer may be used.

y Metals can be analysed by AAS, ICP, neutron activation analyses etc. No standardised method available.

z A modification of the standard ASTM D 971-50. Ring Method (platinum-iridium ring) was used. The instrument calibration has been carried out using ethanol, acetone, and methanol.

å The implication of carbon residue is still unclear. The method can be used to satisfy the end user, as is can depend on the application what degree of residue forms.

PYROLYSIS LIQUIDS ANALYSES
THE RESULTS OF IEA-EU ROUND ROBIN

Anja Oasmaa
VTT Energy, P.O.Box 1601, FIN-02044 VTT, Finland
Dietrich Meier
BFH-IWC, Leuschnerstrasse 91, D-21031 Hamburg, Germany

ABSTRACT

Round robin testing was carried out in 2000 by 12 laboratories using four different pyrolysis liquids (pine, spruce, hardwood mix, bark) produced by different pyrolysis processes (transported bed, bubbling fluidized bed, rotating cone, and vacuum). Both physical (water, pH, viscosity, stability index, CHN, solids) and chemical (pyrolytic lignin, chemical composition) characterization were performed. In general, the accuracies of all the physical analyses, except stability index, were good. Good laboratory practice, including proper calibration of equipment, adherence to instructions, proper homogenisation, homogenity verification, and proper sampling of the liquid, prevents systematic errors.

Karl-Fischer titration is recommended for analysing water in pyrolysis liquids and the water addition method for the calibration. pH measurements provide only rough estimates of pH levels. Kinematic viscosity is an accurate method for pyrolysis liquids. For bark/forest residue liquids, Newtonian behaviour (viscosity - shear rate correlation) should be checked by closed-cup rota-viscotester. An error in viscosity causes an error in the stability index. Stability index determination needs more detailed method description and more testing. For elemental analysis, large sample size and at least triplicates are recommended. Solids content as ethanol insolubles is accurate for white wood liquids but not for extractive-rich liquids. For bark and forest residue liquids, a mixture of a polar (methanol, ethanol) and a neutral (dichloromethane) solvent is recommended.

The method for determining pyrolytic lignin needs more testing and more defined method description. Only four laboratories performed chemical analyses by using GC and HPLC. Derivatization of organic acids and carbonyl compounds seems to be favourable for obtaining a complete list of these compounds. In general, the results of chemical characterization were not very consistent. As a consequence, it might be necessary to distribute solutions with known amounts of standards to crosscheck methods for quantitative analysis.

INTRODUCTION

Biomass pyrolysis liquids differ significantly from petroleum-based fuels in both physical properties and chemical composition. These liquids are typically high in water and solids, acidic, have a heating value of about half of that of mineral oils, and are unstable when heated, especially in air. Pyrolysis liquids contain about 50 wt% oxygen (ca. 40 wt% of dry matter), while mineral oils contain oxygen in ppm levels. Due to these differences, the standard fuel oil methods developed for mineral oils are not always suitable as such for pyrolysis liquids.

Research on analysing physical properties of pyrolysis liquids has been carried out since the 1980s (Elliott 1983, Chum & McKinley 1988, McKinley 1989, Milne et al. 1990, Czernik et al. 1994, McKinley et al. 1994). The first IEA (International Energy Agency) thermochemical round robin was organised in 1988 as part of the IEA Voluntary Standards Activity led by BC Research (McKinley et al. 1994). The main conclusions were: the precision for carbon was excellent, while hydrogen, oxygen by difference and water were more variable, and oxygen by direct determination was poor. It was recommended to use a wider variety of samples in the future studies.

Since then a lot of progress has been made both in the field of pyrolysis liquid production (Bridgwater & Peacocke 2000) and liquid analysis (Oasmaa et al. 1997, Meier & Scholze 1997, Sipilä et al 1998, Oasmaa & Meier 1999, Meier 1999, Bridgwater et al. 1999, Oasmaa & Peacocke 2001).

Therefore, two separate round robins were initiated in 1997: one within EU PyNe (Pyrolysis Network) and the other within IEA PYRA (Pyrolysis Activity). The objective of the EU PyNe round robin was to compare existing analytical methods without any restrictions. Two pine pyrolysis liquids were analysed by eight laboratories for viscosity, water, heating value, elemental analysis, pH, solids, and density. The accuracy for hydrogen, water by Karl-Fischer, and density were good. The xylene-distillation method was stated to yield erroneous results. High variations were obtained for nitrogen, viscosity, pH, and solids. Ethanol was concluded to be more suitable for solids determination than acetone (Meier 1998).

The main objective of the IEA PYRA round robin was to determine the inter-laboratory precision and methods applied for elemental composition, water, pyrolytic lignin and main compounds. Two poplar liquids were analysed by the IEA PYRA participants. It was concluded that the precision of carbon and hydrogen was very good, sample handling plays a very important role in the C, H analysis, water by Karl-Fischer titration was acceptable, but should be checked carefully, and the method for the determination of pyrolytic lignin should be improved (Bridgwater et al. 1999).

The latest round robin within PyNe-IEA was carried out by 12 laboratories during January - March 2000 aimed at comparing the analyses, not the pyrolysis liquids. Four different types of pyrolysis liquids were provided from various feedstocks: BTG (Netherlands) softwood mixture (fir, spruce), Dynamotive (Canada) pine (85% pine, 15% spruce), Ensyn (Canada) hardwood mixture, and Pyrovac (Canada) softwood bark (1/3 fir, 2/3 spruce, traces of hardwood bark). Based on the feedback from previous round robins it was decided to add instructions for handling and analysis.

MATERIALS AND METHODS

Pyrolysis liquids for Round Robin are shown in Table 1.

Table 1 Pyrolysis liquids for RR

Producer	Feedstock	Additional information
BTG Netherlands	softwood mixture (spruce and fir) moisture content 10-12 % particle size 0.8-1.1 mm	production date: 11.01.2000 pyrolysis temp. 500 °C rotating cone
Dynamotive Canada	85% pine, 15% spruce feedstock moisture 7 %	production date; 07.10.1999 pyrolysis temp. 460 °C fluid bed
Ensyn Canada	hardwood mix	transported bed
Pyrovac Canada	softwood bark (1/3 fir and 2/3 spruce with traces of hard wood bark moisture content 12 %	production date 29.09.1999 pyrolysis temp. 510 °C vacuum pyrolysis

The criteria of oil quality was a reasonable homogeneity. In case of inhomogeneity the liquid would be excluded from the round robin. The sample size for each laboratory was set as one litre.

Instructions were as follows:
1. After receiving the sample please indicate date of arrival.
2. Assure that the sample is being stored at refrigerator conditions.
3. Stability tests with analyses should be carried out within a week after receiving the sample and all other analyses within a month. Please indicate date of analyses.
4. Before sampling let the sample reach room temperature, then shake it to ensure homogeneous sampling.
5. After usage please store the sample again in a refrigerator.
6. Please report dates of arrival and analyses, sample size used, all duplicates, methods used, possible difficulties, and suggestions
7. Send the results to IWC (Dr. Dietrich Meier, BFH-Institute for Wood Chemistry, Leuschnerstrasse 91, D-21031 Hamburg, GERMANY) and VTT (Anja Oasmaa, VTT Energy, P.O. Box 1601, FIN-02044 VTT, FINLAND)

Analytical methods for Round Robin are shown in Table 2.

Table 2 Analytical methods for Round Robin

Property	Method	Reporting unit
Water content	Karl Fischer Titration	wt.% water based on wet oil
Viscosity	capillary or rotary viscometer, 2 temp. @ 20 and 40°C	cSt @ 20°C and 40 °C
Solids	insolubles in ethanol, filter pore size 3μm or lower	wt% based on wet oil
pH	use pH-meter	pH unit
Stability[1]	store samples for 1) 6 h @ 80 °C, 2) 24 h @ 80 °C, and 3) 7 days @50 °C, viscosity @ 20 and 40 °C and water by K-F titration	cSt wt.% water based on wet oil
Elemental analysis	elemental analyser (complete oxidation)	wt%C, wt%H, wt%N, wt%O, based on wet oil
Pyrolytic lignin	add 60 ml oil to 1 L of ice-cooled water under stirring, filter and dry precipitate below 60 °C	wt% based on wet oil
GPC	not yet established	
HPLC	not yet established	
GC	column type DB 1701 dimensions: 60m x 0.25 mm film thickness: 0.25 μm injector: 250 °C, split 1:30 FID detector: 280 °C oven programme: 45 °C, 4 min const., 3 °C/min. to 280 °C, hold 20 min. sample conc.: 6 wt%, solvent acetone	

Note [1] Viscosity Index and Water Index is described in Appendix 1 of Chapter 2

RESULTS AND DISCUSSION

Sample Preparation

The homogeneity of pyrolysis liquids as received in their individual containers was verified by analysing the water content at three levels (Table 3). After that the liquids were mixed thoroughly by intensive shaking and divided into 1 litre sample bottles for shipping to the laboratories.

The variation in water content inside the barrel for Pyrovac, Dynamotive, and Ensyn liquids was below the accepted 10 wt% for homogenous liquids. The microscopic images (Figure 1) showed that they were homogenous liquids. BTG pyrolysis liquid was inhomogeneous (Figure 2), due to a high water content causing phase separation. The liquid producer pointed out some problems during production, which now have been at least partly solved. However, this liquid was included to the Round Robin testing as a difficult liquid.

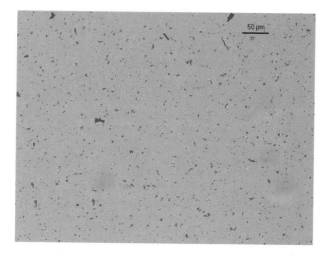

Figure 1 Microscopic image of Dynamotive pine liquid

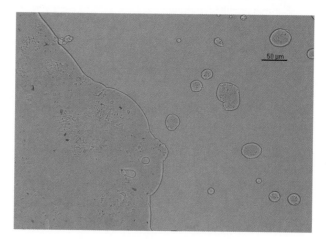

Figure 2 Microscopic image of BTG spruce-fir liquid

**Table 3   Water determination of the RR oils
at different levels in the shipping containers**

		Water, wt%	
		before mixing	after mixing containers
Pyrovac I	top	13.5	15.7
	middle	13.9	
	bottom	13.7	
Pyrovac II	top	15.2	
	middle	14.9	
	bottom	15.9	
Dynamotive I	top	20.9	21.1
	middle	21.1	
	bottom	20.7	
Dynamotive II	top	19.1	
	middle	20.2	
	bottom	21.6	
BTG	top	32.1	28.3
	middle	32.4	
	bottom	20.3	
ENSYN	top	20.8	20.4
	middle	20.5	
	bottom	19.0	

The content of containers I and II from Pyrovac and those
from Dynamotive were combined before shipping.

Physical Properties

Water

All laboratories used Karl-Fischer titration for water content determination. The results are
shown in Table 4. Some laboratories got systematically high or low results. Possible reasons may
be: inadequate sample homogenisation or sampling, burette reading error, dirty electrode,
variation in water equivalent, or fading titration end-point. Possible solutions include: repeating
homogenisation and sampling, calibration of burette, cleaning of electrode, checking the water
equivalent (Oasmaa & Peacocke 2001). Water addition method (Oasmaa et al. 1997, Oasmaa &
Peacocke 2001) is suggested for calibrating the Karl-Fischer titration method for pyrolysis
liquids.

Table 4   Water content determined by Karl-Fischer titration

Water, wt%	Average all results	Min	Max	Count	Stdev
Dynamotive	21.1	20.20	21.80	11	0.47
BTG	30.4	29.40	31.10	10	0.61
Pyrovac	15.7	14.50	16.50	11	0.56
Ensyn	20.3	19.40	21.10	10	0.48

pH

The pH of RR pyrolysis liquids is shown in Table 5. pH results varied a lot and hence it is
recommended to be used for rough checking of pH levels. pH is recommended to be published to
one decimal.

Table 5   pH of RR pyrolysis liquids

pH	Average all results	Min	Max	Count	Stdev
Dynamotive	2.3	1.86	2.40	9	0.17
BTG	2.5	2.30	2.70	8	0.12
Pyrovac	2.8	2.44	3.00	9	0.17
Ensyn	2.5	2.40	2.70	7	0.11

Solids

Solids content was analysed as ethanol insolubles (Table 6). Microscopic images of Ensyn and Pyrovac liquids (Figures 3 and 4) show that there are much more solids (about 50 times) in Ensyn than in Pyrovac liquid even though the solids content for Pyrovac liquid is over twice as big as for Ensyn liquid. This is due to the feedstock. Ethanol is not powerful enough for liquids from bark (Pyrovac liquid) or forest residues, because extractives as neutral substances do not dissolve well in polar solvents like alcohols. Solids content for Pyrovac liquid determined using a mixture of methanol and dichloromethane (1:1) was 0.02 wt%, which is reasonable.

Table 6   Solids content of pyrolysis liquids measured as ethanol insolubles

Solids, wt%	Average All results	Min	Max	Count	Stdev
Dynamotive	0.10	0.05	0.27	7	0.08
BTG	0.08	0.03	0.26	6	0.09
Pyrovac	1.02	0.29	1.52	7	0.43
Ensyn	0.43	0.39	0.47	6	0.04

Solids of Pyrovac liquid measured as insolubles in methanol-dichloromethane 0.02 wt%.

Even though the sub-micron particles of char are not included into solids, solids content analysis using ethanol for white wood liquids can be accepted being accurate enough for its present purpose. Microscopic analysis of the liquids showed a high amount of particles about 1μm. Hence, the pore size of filter paper is recommended to be reduced down to 1 μm. A mixture of a polar and a neutral solvent, like methanol (or ethanol) and dichloromethane, is recommended (Oasmaa & Peacocke 2001). Methanol is typically little more effective solvent than ethanol, but because of health and safety reasons the use of ethanol is recommended when possible.

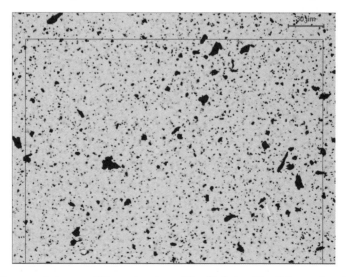

Figure 3 Microscopic image of Ensyn hardwood liquid. 0.4 wt% solids as ethanol insolubles

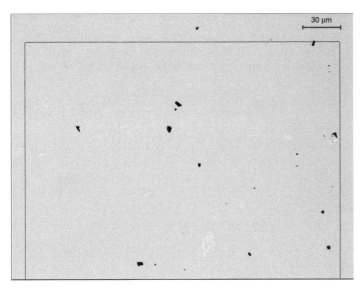

**Figure 4   Microscopic image of Pyrovac bark liquid. 1 wt% solids as
ethanol insolubles, 0.02 wt% solids as methanol-dichloromethane insolubles**

Carbon, hydrogen, nitrogen

The results of elemental analysis (Table 7) were good. Variation in nitrogen is due to low
detection limits for nitrogen (Oasmaa et al. 1997). The sample size for CHN analysis is
suggested to be as large as possible and at least triplicates should be carried out.

Table 7   Elemental analysis of RR liquids

Elemental analysis, wt%	Average All results	Min	Max	Count	Stdev
Carbon					
Dynamotive	44.7	43.9	46.3	7	0.82
BTG	37.1	36.3	37.5	6	0.45
Pyrovac	51.4	50.0	54.0	7	1.63
Ensyn	47.2	46.8	47.9	6	0.41
Hydrogen					
Dynamotive	7.2	6.6	7.6	7	0.34
BTG	7.6	7.2	8.0	6	0.28
Pyrovac	7.0	6.6	7.3	7	0.23
Ensyn	6.9	6.3	7.2	6	0.35
Nitrogen					
Dynamotive	0.1	0.0	0.2	5	0.09
BTG	0.1	0.1	0.3	4	0.13
Pyrovac	0.3	0.2	0.5	5	0.15
Ensyn	0.1	0.1	0.3	4	0.13
Oxygen (diff.)					
Dynamotive	48.1	47.0	48.7	5	0.73
BTG	55.1	54.7	55.3	4	
Pyrovac	41.6	40.4	42.6	5	1.17
Ensyn	45.6	45.2	46.4	4	

Viscosity

The viscosity results (see Table 8) at 20 and 40 °C were very consistent. The small standard deviation at 40 °C is possibly due to the high temperature dependency of pyrolysis liquids. One degree error in temperature causes a higher error in viscosity at 20 °C than at 40 °C (Figure 5).

Table 8 Viscosity of RR liquids

	Average all results	Min	Max	Count	Stdev
Viscosity @20°C, cSt					
Dynamotive	97	89	130	7	14
BTG	26	24	29	6	2
Pyrovac	1271	1075	1406	6	120
Ensyn	1713	1379	2245	6	315
Viscosity @40°C, cSt					
Dynamotive	30	27	33	9	2
BTG	11	9	13	8	1
Pyrovac	206	189	229	9	15
Ensyn	226	202	265	8	22

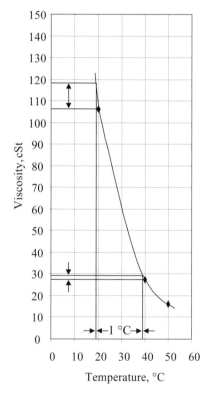

Figure 5 Viscosity of pyrolysis liquid (Dynamotive pine liquid)

White wood pyrolysis liquids are Newtonian liquids (Leroy et al. 1988, Oasmaa et al. 1997) and hence, kinematic viscosity is applicable for them. Though, for bark liquids the Newtonian behaviour should be checked, because extractive-rich fraction (top phase) of forest residue liquid behaves as a non-Newtonian fluid i.e. the viscosity changes by the shear rate (Figure 6).

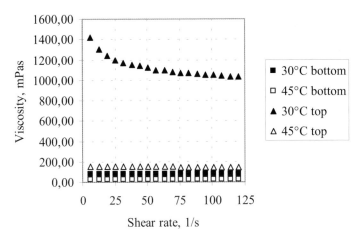

Figure 6 **Dynamic viscosity as a function of shear rate for the two phases (top and bottom) of forest residue liquid (water content: top 6.8 wt%, bottom 24.1 wt%). (Oasmaa & Peacocke 2001)**

Inhomogeneity of the liquid may lead to phase-separation in the capillary tube and hence, to erroneous results. Measuring the viscosity of inhomogeneous liquid in a closed-cup rotavisco-tester is suggested.

Stability index

The stability results varied a lot (see Table 9, Table 10, Table 11). Because the stability is measured as a change in viscosity, errors in viscosity results yields errors in the stability index.

Table 9 Viscosity index. Test conditions: 6 hours at 80 °C

Viscosity index 6h @80°C	Average All results	Min	Max	Count	Stdev
Viscosity @20 °C					
Dynamotive	0.30	0.06	0.78	5	0.28
Pyrovac	0.03	-0.07	0.08	4	
Ensyn	0.14	0.01	0.24	4	
Viscosity @40 °C					
Dynamotive	0.30	0.07	0.65	6	0.26
Pyrovac	0.19	-0.03	0.96	6	0.38
Ensyn	0.15	0.00	0.52	6	0.19

Table 10 Viscosity index. Test conditions: 24 hours at 80 °C

Viscosity index 24h @80°C	Average all results	Min	Max	Count	Stdev
Viscosity @20 °C					
Dynamotive	0.90	0.65	1.63	5	0.41
Pyrovac	0.52	0.33	0.66	3	
Ensyn	1.56	0.72	2.22	4	
Viscosity @40 °C					
Dynamotive	0.54	0.33	0.69	5	0.13
Pyrovac	0.43	0.06	1.01	3	0.31
Ensyn	0.68	0.48	0.86	4	0.16

Table 11   Viscosity index. Test conditions: 7 days at 50 °C

Viscosity index 7 days @50°C	Average All results	Min	Max	Count	Stdev
Viscosity @20 °C					
Dynamotive	0.61	0.17	1.12	4	
Pyrovac	0.35	0.09	0.62	3	
Ensyn	0.45	0.07	0.65	3	
Viscosity @40 °C					
Dynamotive	0.36	0.03	0.49	5	0.19
Pyrovac	0.28	0.04	0.49	4	
Ensyn	0.39	0.05	0.57	5	0.21

Solvents including water stabilise pyrolysis liquid. The stability index should hence be specified more precisely and it cannot be recommended as the only measure of the storage stability of liquid. The authors wish to point out that, in some cases, the instructions had not been delivered to technicians, which has led to erroneous results.

CHEMICAL CHARACTERIZATION

Chemical characterization was performed by four laboratories only. As pyrolysis liquids are very complex mixtures of the thermal degradation products of the biopolymers cellulose, hemicelluloses, and lignin they are difficult to analyse. Basically, the liquids can be divided into a water soluble and a water insoluble fraction. The water insoluble fraction represents higher molecular weight material derived from lignin and is therefore called "pyrolytic lignin". It was recently described by Scholze & Meier 2001, and Scholze et al., 2001)). The water soluble part consists of low-molecular-weight carboxylic acids, alcohols, aldehydes and ketones, hydrocarbons ("sugars"), and phenolic compounds (
Figure 7).

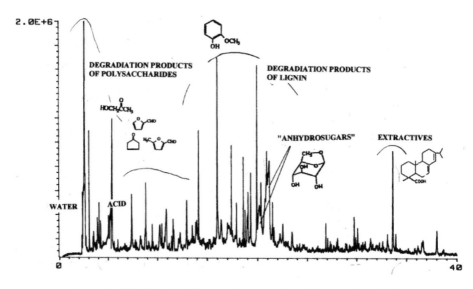

Figure 7 GC-MS of RR Dynamotive pine liquid (analysed at VTT)

50

Pyrolytic lignin

Table 12   Pyrolytic lignin fraction of RR liquids

Pyrolytic lignin, wt%	Average all results	Min	Max	Count	Stdev
Dynamotive	25	13	32	7	6.7
BTG	8	2	18	5	12.7
Pyrovac	44	30	59	7	9.5
Ensyn	54	32	84	7	17.1

The determination of pyrolytic lignin by adding water to the liquid in order to provoc phase separation can be erroneous if bitumen-like sticky material is formed, which privents efficient washing of the sample. This can be overcome by adding the pyrolysis liquid dropwise into chilled water under vigorous mixing (6,000 rpm) with an ultra turrax. The drying of the residue should be carried out below 45 °C in order to prevent the oxydation of lignin (Scholze & Meier 2001).

There may be several reasons for the large variation in pyrolytic lignin determination. The most possible reason for problems is the behaviour of poorly water-soluble material and extractives (Oasmaa & Kuoppala 2001b), which, without vigorous mixing, separate out from the aqueous phase with pyrolytic lignin. These sticky compounds prevent efficient separation and drying of the residue.

Further determination of the main single compounds were all done by the use of either gas chromatography (GC) or high performance liquid chromatography (HPLC).

Organic Acids

Three laboratories used GC and one HPLC for the determination of organic acids. Two laboratories using GC, derivatized the samples to their benzylic esters prior to analysis. This method gave the most comprehensive list of compounds (see labs. 9 and 12). However, the absolute values are still very inconsistent. The quantitative analyses of organic acids are shown in Table a. As expected, formic and acetic acid form the bulk of the acids at 70-80 %.

It is concluded from the results that derivatization of the acids to benzylic esters (Alen et al. 1985) is recommended prior to GC analysis because this technique increases the volatilization of the acids during injection into the gas chromatograph. However, with respect to acetic acid it can be concluded that derivatization is not required and lab. 9 (with derivatization) got similar results as lab. 3 (without derivatization).

The acid number (see Table   13) is a measure of the acidity of the oil and might correlate with corrosion problems. It was determined by one laboratory as follows:

Table   13 Determination of the acid number

BTG	60.3 mg KOH/g oil
Dynamotive	67.9 mg KOH/g oil
Pyrovac	72.1 mg KOH/g oil
Ensyn	80.3 mg KOH/g oil

From these results it is concluded that the acid number does not correlate with the content of organic acids because phenolic compounds are also neutralized with KOH.

Aldehydes, ketones and alcohols

These compounds were analysed both by GC and HPLC (see Table 15a). Laboratory 9 transformed the aldehydes and ketones into hydrazones by derivatization with dinitrophenylhydrazine (DNPH) prior to analysis with GC and HPLC. Obviously, low volatile aldehydes such as formaldehyde and acetaldehyde could be detected. Laboratory 12 used a packed column to detect formaldehyde and acetaldehyde. Again, there is a big inconsistency between the results.

Sugars

Determination of sugars (Table 14) was performed by three laboratories only. Laboratories 12 and 5 used HPLC and laboratory 3 GC for the determination of levoglucosan, which is the most important anhydrosugar in pyrolysis liquids. There is some consistency between laboratory 12 and 3 although they used different methods.

Table 14 Determination of sugars (wt.% based on wet liquid)

	Dynamotive			Pyrovac			Ensyn			BTG		
Laboratory no.	12	3	5	12	3	5	12	3	5	12	3	5
Levoglucosan	3.98	4.83	7.5	4.59	4.74	8.4	3.06	4.14	2.9	4.41	3.31	5.5
Glucose									0		0	
Xylose	0.14								0			
Cellobiosan			2.3			0.7						1.8
TOTAL	4.1	4.8	9.8	4.6	4.7	9.1	3.1	4.1	2.9	4.4	3.3	7.3

Phenols

Phenols were analysed by three laboratories using GC with the internal standard calibration method (Table 15). Number 9 and 12 extracted phenols with ethyl acetate prior to analysis, whereas laboratory number 3 injected the pyrolysis directly. There is fairly good consistency between lab. no. 9 and 3 although they used different methods.

Table 15 Determination of phenols (wt.% based on wet liquid)

	Dynamotive			Pyrovac			Ensyn			BTG		
Laboratory no.	9	12	3	9	12	3	9	12	3	9	12	3
Phenol	0.1	0.07	0.07	0.44	0.16	0.31	0.26	0.1	0.14	0.15	0.04	0.06
Guaiacol	0.54	0.16		0.37	0.17	0.37	0.18	0.05	0.16	0.51	0.11	0.38
o,m,p-Cresols	0.23	0.17	0.11	0.75	0.4	0.49	0.32	0.13	0.16	0.2	0.06	0.04
4-Methylguaiacol	0.83		0.8	0.36		0.61	0.13		0.18	0.59		0.48
4-Ethylguaiacol	0.24		0.24	0.14		0.15	0.05		0.08	0.11		0.12
Vinylguaiacol			0.13			0.07						0.1
Eugenol	0.29	0.06	0.22	0.12	0.02	0.06	0.07	0	0.06	0.21	0.03	0.15
4-Propylguaiacol	0.2		0.07	0.05					0.07	0.11		0.03
1,2-Benzenediol	0.13			0.91			0.1					
Iso-Eugenol	0.79	0.54	0.69	0.46	0.5	0.36	0.07	0.05		0.34		0.36
Syringols		0.13		0		0.58	0.26	0	1.81		0.09	
Vanillin	0.23	0.31	0.34	0.17	0.19	0.17	0.07	0.32	0.17	0.16	0.25	0.33
Coniferylaldehyde	0.36		0.36	0.09		0.07	0.06		0.11	0.27		0.42
TOTAL	3.9	1.4	3.0	3.9	1.4	3.2	1.6	0.7	2.9	2.7	0.6	2.5

Polyaromatic hydrocarbons (PAH)

PAH's were determined only by laboratory 9 (Table 16) using both HPLC and GC. Samples were fractionated on silica with different solvent. The diethyl ether fraction was used for analysis. The data show the big range of PAH content which can be attributed to the pyrolysis process conditions such as temperature and residence time. The knowledge of the PAH content is absolutely necessary in order to use the pyrolysis liquids in the market.

Generally, the results of chemical characterization were not very consistent. It should be necessary to prepare standard solutions with known amounts of compounds for quantitative analyses. A recommendation for a "best method" cannot yet be given. It seems that each laboratory has its own technique and a lot of work and adaptation is necessary to harmonize the methods. The amount of PAH (polyaromatic hydrocarbons) was on high side for Ensyn pyrolysis liquid, and it is pointed out to pay more attention to the analysis of toxic compounds in the liquids. Ensyn commented later that the high PAH may be due to contamination of another fuel and this will be checked.

Table 16 Determination of polyaromatic hydrocarbons (PAH) in ppm

	Dynamotive	Pyrovac	Ensyn	BTG
Laboratory no.	**9**	**9**	**9**	**9**
Acenaphtylene	0.3	1.3	34	0.1
Acenaphtene	0.2	1.3	8.7	0.1
Fluorene	2.2	7.8	30	0.5
Phenanthrene	2	8.4	52	0.5
Anthracene	0.8	2.7	16	0.1
Fluoranthene	0.6	2.8	39	0.4
Pyrene	0.8	0.3	40	0.4
Benzo(a)anthracene & chrysene	0.7	2.5	37	0.4
Benzo(b)- and benzo(k)fluoranthene	0.2	0.2	23	0.2
Benzo(a)pyrene	0.3	0.6	20	0.1
Indeno(1.2.3cd)pyrene	0.2	0.1	16	0.1
Benzo(ghi)perylene	0.1	0.1	11	0.1
Dibenzo(ah)anthracene	0.1	0.1	7.4	<0.1
Total	**8**	**28.2**	**334.1**	**3**

Table 17 Determination of acids (wt.% based on wet liquid)

Laboratory no.	Dynamotive				Pyrovac				Ensyn				BTG			
	9	12	3	5	9	12	3	5	9	12	3	5	9	12	3	5
Formic acid	0.29	9.35			0.48	8.26		3.3	0.52	11.32		2.3	0.91	13.55		
Acetic acid	2.7	7.84	3.31	5.3	2.05	5.26	2.17	2.5	4.6	11.2	5.27	3.9	4.01	8.98	3.23	4.8
Acrylic acid	0.05	0		5.0	0.08				0.05	0.02			0.06	0.22		5
Propionic acid	0.17	0.63			0.19	0.35			0.31	0.26			0.21	0.64		
Isobutyric acid	0.02	0.35			0.03	0.32			0.02	0.1			0.02	0.11		
Methacrylic acid	0.01				0.01				0.01				0.01			
N-Butyric acid	0.07	1.89			0.1	2.75			0.2	1.66			0.08	2.07		
Lactic acid	0.18				0.08				0.09				0.21			
Glycolic acid	0.34	0.62			0.82	1.32			0.36	0.25			0.44	0.69		
Crotonic acid	0.04	0			0.06				0.04	0.05			0.08	0.06		
Valeric acid	0.01	0.66			0.02	0.27			0.01	0.67			0.01	0.09		
Tiglic acid	0.01	0.06			trace	0.26			trace				0.01	0.01		
4-Methylpentanoic acid	0.01				0.02								0.01			
3-Hydroxypropanoic acid	trace				0.04				0.02				0.01			
2-Oxobutanoic acid	0.17				0.15				0.13				0.02			
Levulic acid	0.11				0.23				0.11				0.18			
Benzoic acid	0.02				0.05								0.12			
Hexanoic acid		0.14				0.16				0.16			n.d.	0.05		
TOTAL	4.2	21.5	3.3	10.3	4.4	19.0	2.2	5.8	6.5	25.7	5.3	6.2	6.4	26.5	3.2	9.8

Table 18 Determination of aldehydes, ketones, and alcohols (wt.% based on wet liquid)

Laboratory no.	Dynamotive				Pyrovac				Ensyn				BTG			
	9	12	3	5	9	12	3	5	9	12	3	5	9	12	3	5
Formaldehyde	0.84	8.92		3.3	0.51	2.6		1	0.25	5.23		1.4	1.15	9.37		4.1
Acetaldehyde	0.14	1.88			0.004	1.1			0.01	1.34			0.17	1.67		
Hydroxyacetaldehyde		3.32	6.42	7.7		1.09	3.18	3.2		1.81	3.34	2.9		6.89	8.2	11.1
Glyoxal		0.24		2.4		0.33		1.5		0.67		1		0.91		2.1
Acetol		2.07	7.82	7.1		0.84	3.17	1.8		1.48	3.65	1.8		3.28	7.1	7.3
1-Hydroxy-2-butanone			0.31				0.17				0.17				0.27	
2-Hydroxy-2-cyclopentene 1-one			0.46				0.1				0.06				0.3	
2-Hydroxy-3-methyl-2-cyclopentene-3-one			0.5				0.52				0.32				0.43	
Propionaldehyde	0.05				0.01				0.01				0.03			
Acetone	0.08	0.21			0.01	0.27			0.02	0.27			0.05	0.18		
Furfural	0.49	0.2	0.81		0.39	0.15	0.47		0.36	0.16	0.65		0.31	0.2	0.54	
(5H)-Furan-2-one			0.6				0.53				0.32				0.54	
5-Hydroxymethylfurfural			0.52				0.83				0.23				0.49	
Methanol		1.03				0.07				0.39				0.91		
Ethanol		0.09				0.01				0.06				0		
2-Propanol		0.37				0.06				0				0.25		
Butanol		2.85				0.8				1.29				3.15		
MEK		0.37			0.007	0.46			0.01	0.1			0.02	0.37		
TOTAL	1.6	21.6	17.4	20.5	0.9	7.8	9.0	7.5	0.7	12.8	8.7	7.1	1.7	27.2	17.9	24.6

CONCLUSIONS

In general, the accuracies of the physical analyses were good. Karl-Fischer titration is an accurate method for analysing water in pyrolysis liquids. pH results vary a lot and hence can be used only for rough checking of pH levels.

Kinematic viscosity is applicable to white wood pyrolysis liquids because of its accuracy and the Newtonian behaviour of these liquids. For extractive-rich liquids the Newtonian behaviour should be checked by using a closed-cup rotary viscometer. The error in viscosity also causes an error in the stability index. Stability index needs more specific instructions and its correlation with the water content should be determined. Another simple test method for stability may be needed. In the case of inhomogeneous liquids kinematic viscosity and stability index cannot be determined.

The elemental analysis for carbon and hydrogen is accurate. Nitrogen is not very accurate for white wood liquids due to the low levels and the detection limitations of the equipment. The solids content using methanol as a solvent is accurate for white wood liquids. However, for extractive-rich liquids a mixture of a polar (methanol, ethanol) and a neutral (dichloromethane) solvent should be used.

There was large variation in pyrolytic lignin results. The most likely reason for problems is the behaviour of poorly water-soluble material and extractives, which without vigorous mixing, separate out from the aqueous phase with the pyrolytic lignin. These sticky compounds prevent efficient separation and drying of the residue.

The results of chemical characterization were not very consistent. It may be necessary to prepare standard solutions with known amounts of compounds for quantitative analyses. The complete range of organic acids should be analysed after derivatization of the acids into their benzylic esters. However, for the determination of the main acidic compound, acetic acid, derivatization is not necessary.

RECOMMENDATIONS

Based on the round robin results the following recommendation are made:

- It is recommended to verify homogeneity by water distribution and/or by microscopic determination (Oasmaa & Peacocke 2001).
- Karl-Fischer titration is recommended for analysing water in pyrolysis liquids. For method calibration it is suggested to use the water addition method (Oasmaa et al. 1997)
- pH is recommended to be used only for rough checking of pH levels and to be reported to one decimal place.
- Kinematic viscosity is recommended to be used for viscosity measurement of white wood pyrolysis liquids. The Newtonian behaviour using a closed-cup rotaviscotester should be checked for extractive-rich liquids (Oasmaa & Peacocke 2001).
- Stability test should be carried out each time exactly in the same way, and, in case of weight loss (> 0.1 wt%) during the test, the results should be excluded (Oasmaa & Peacocke 2001). The test is recommended for internal comparison of pyrolysis liquids from one specific process. The best comparison can be made when the differences in the water contents of the samples are small. Viscosity can be measured both at 20 °C and 40 °C, but 40 °C is recommended, because the measuring error is smaller.
- For elemental analysis in cases of inhomogeneity or high solids content, the sample size should be as large as possible and at least triplicates should be carried out.

- Ethanol (or methanol) can be used for solids determination of white wood pyrolysis liquids. For new feedstocks, like bark and forest residue, the solubility of the liquid should be checked, for example, by using solvents of different polarity, for example methanol and mixtures of methanol and dichloromethane (Oasmaa & Peacocke 2001).
- Pyrolytic lignin is recommended to be measured using vigorous mixing (Scholze & Meier 2001).
- For chemical characterization it might be necessary to calibrate the gas and liquid chromatographic systems by preparing standard solutions with known amounts of compounds.

For future round robins it is recommended to include pyrolysis liquids from new feedstocks, like bagasse, olive husk, forest residue, and bark. Instructions for homogenisation, sampling and analyses are highly recommended. Solids and water are most important. Viscosity @40 °C, stability index and water-insoluble determinations are also recommended to be included in the next round robin. However, very precise instructions, including sample amounts, are recommended. Before the round robin, the methods for water insolubles and stability should be compared and specified by two or three laboratories routinely analysing pyrolysis liquids. Reference samples with known amounts of compounds are recommended to be included for chemical characterization.

ACKNOWLEDGEMENT

The authors wish to thank pyrolysis liquid producers Dynamotive, Ensyn, and Pyrovac in Canada, and BTG in The Netherlands for liquid samples for round robin. All laboratories NREL in USA, Orenda, and RTI in Canada, Cirad Foret in France, SINTEF in Norway, Chemviron, Rostock University, and IWC in Germany, INETI in Portugal, Aston University in UK, and Fortum, and VTT in Finland are greatly acknowledged for their participation. The authors wish to acknowledged the help of the whole PyNe group and especially Columba Di Blasi, Stefan Czernik, Jan Piskorz, and Tony Bridgwater for valuable comments.

REFERENCES

Alen, R., Jännäri, P., Sjöström, E., Gas-liquid chromatographic determination of volatile fatty acids C1-C6, and lactic acid as their benzyl esters on a fused-silica capillary column, Finn. Chem. Lett., 1985, 190-192.

Bridgwater, A.V. & Peacocke, G.V.C. Fast pyrolysis processes for biomass, Renewable and Sustainable Energy Reviews, 4 (2000) 1-73.

Bridgwater, A., Czernik, S., Diebold, J.P., Meier, D., Oasmaa, A., Peacocke, G., Piskorz, J. & Radlein, D. Fast pyrolysis of biomass: A handbook . Newbury: CPL Press., 1999. 188 p.ISBN 1-872691-07-2.

Chum, H. L. & Mckinley, J. 1988. Report on characterization of biomass pyrolysis liquid products. In: Bridgwater, A. V. & Kuester, J. L. (eds.). Research in Thermochemical Biomass Conversion, Phoenix, Arizona, April 1988. New York: Elsevier Appl. Sci. Pp. 1177 - 1180.

Czernik, S., Johnson, D. K. & Black, S. 1994. Stability of wood pyrolysis liquid. Biomass and Bioenergy, vol. 7, no. 1 - 6, pp. 187 - 192.

Diebold, J. P., Milne, T. A., Czernik; S., Oasmaa, A., Bridgwater, A. V., Cuevas, A., Gust, S., Huffman, D. & Piskorz, J. Proposed specifications for various grades of pyrolysis liquids. In: Bridgwater, A. V. & Boocock, D. G. B. (eds.). Developments in Thermochemical Biomass Conversion, Banff, 20 - 24 May 1996. Vol. 1. London: Blackie Academic & Professional, 1997. Pp. 433 - 447.

Elliott, D. C. 1983. Analysis and upgrading of biomass liquefaction products. Final report. Vol. 4. IEA Cooperative project D1 Biomass Liquefaction Test Facility Project. Richland, Washington: Pacific Northwest Laboratory. 87 p. + app.

Leroy, J., Choplin, L. & Kallaguime, S. Rheolological characterization of pyrolytic wood derived oils: Existence of a compensation effect. Chem. Eng. Comm., 1988, vol. 71, pp. 157 - 176.

Mckinley, J. W. 1989. Biomass liquefaction: centralized analysis. Final report. Vancouver: B. C. Research. (Project No. 403837. DSS File No. 2321646192.)

Mckinley, J. W., Overend, R. P. & Elliott, D. C. 1994. The ultimate analysis of biomass liquefaction products: The results of the IEA round robin #1. In: Proc. Biomass pyrolysis liquid properties and combustion meeting, 26 - 28 September 1994, Estes Park, CO. Golden, CO: NREL. Pp. 34 - 53.(NREL-CP-430-7215.)

Meier, D., Oasmaa, A. & Peacocke, G. V. C. Properties of fast pyrolysis liquids: status of test methods. Characterization of fast pyrolysis liquids. In: Bridgwater, A. V. & Boocock, D. G. B. (eds.). Developments in Thermochemical Biomass Conversion, Banff, 20 - 24 May 1996. Vol. 1. London: Blackie Academic & Professional, 1997. Pp. 391 - 408.

Meier, D. And B. Scholze 1997. Fast pyrolysis liquid characteristics. In: Biomass Gasification and Pyrolysis - State of the Art and Future Prospects (eds. M. Kaltschmitt and A.V. Bridgwater), CPL Press, Newbury, UK, 431-441.

Meier, D. 1999. New methods for chemical and physical characterization and round robin testing, In: Fast Pyrolysis of Biomass: A Handbook (eds. A. Bridgwater, S. Czernik, J. Diebold, D. Meier, A. Oasmaa, C. Peacocke, J. Piskorz, D. Radlein), CPL Ltd., Newbury, UK, 92-101.

Milne, T. A., Brennan, A. H. & Glenn, B. H. 1990. Sourcebook of methods of analysis for biomass and biomass conversion processes. London: Elsevier Appl. Sci. 327 p. + app.

Oasmaa, A. & Kuoppala, E. Composition and Properties of Forest Residue Liquid. To be submitted to Energy & Fuels.

Oasmaa, A., Leppämäki, E., Koponen, P., Levander, J. & Tapola, E. Physical characterization of biomass-based pyrolysis liquids. Application of standard fuel oil analyses. Espoo: VTT Energy, 1997. 46 p. + app. 30 p. (VTT Publications 306.)

Oasmaa, A & Peacocke, G.V.C. A guide for determining the fuel oil quality of pyrolysis liquids. To be published in VTT Publications series 2001.

Scholze, B. & Meier, D. 2001, Characterization of the water-insoluble fraction from pyrolysis oil (pyrolytic lignin). Part I. PY-GC/MS, FTIR, and functional groups, J. of Anal. and Applied Pyrolysis, 60 (2001) 41-54.

Scholze, B., Hanser, C. & Meier, D. 2001. Characterization of the water-insoluble fraction from fast pyrolysis liquids (pyrolytic lignin). Part II. GPC, carbonyl groups, and ^{13}C-NMR. J. Anal. Appl. Pyrolysis, 58-59, 387-400.

Sipilä, K., Kuoppala, E., Fagernäs, L. & Oasmaa, A. 1998. Characterization of biomass-based flash pyrolysis oils, Biomass and Bioenergy, vol. 14, no. 2, pp. 103 - 113.

SUMMARY OF THE ANALYTICAL METHODS AVAILABLE FOR CHEMICAL ANALYSIS OF PYROLYSIS LIQUIDS

Dietrich Meier

Federal Research Centre for Forestry and Forest Products, Institute for Wood Chemistry and Chemical Technology of Wood, D-21027 Hamburg, Germany

1   GAS CHROMATOGRAPHY

1.1   Introduction

The chromatographic system is composed of the chromatograph and a recorder for plotting chromatograms or a data station for generation and evaluation of chromatograms. The chromatograph consists of the sample injector, gas supplies, oven with temperature control for the chromatographic column and the detector see Figure 1.

Schematic presentation of a GC system

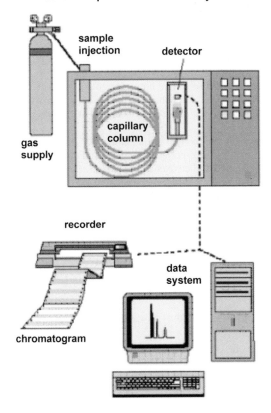

Figure 1   Schematic representation of a GC system

It is not possible to describe all injection techniques in detail. The different basic possibilities of direct and indirect sample injection are just mentioned. With direct injection the sample is introduced into the column without contact with other parts from glass or metal (on column injection). With indirect techniques the sample is injected into an evaporator. The vapour then is transferred into the column either completely or partially (split technique).

With both techniques the injection can take place at low temperatures, at high temperatures or with temperature programming. The gas supply unit of a chromatograph has to provide all gases needed for a separation: the carrier gas and e.g. the burner gases for the flame ionization detector (FID). The detector indicates a substance by generation of a more or less intense electrical signal (response). Analogue (recorder) or digital (computer) processing of these signals produces the chromatogram or a numerical report. Some detectors are specific for certain classes of substances or for certain atoms (P, N, etc.).

1.2   Principle of Operation

The chromatographic separation is achieved by repeated distribution of each sample component between two phases. One of these phases is stationary. The second, the mobile phase moves along this stationary phase. In Gas Chromatography the mobile phase is always a gas (mostly N_2, H_2, He). The stationary phase in GSC (Gas Solid Chromatography or adsorption chromatography) is a porous polymer solid, while in GLC (Gas Liquid Chromatography or partition chromatography) the stationary phase is a mostly viscous liquid based on modified silicon oi. Packed columns are completely filled with a packing, liquid stationary phases being coated onto an inert support. Capillary columns do not require a support, because their inner wall is coated with the stationary phase (WCOT = Wall Coated Open Tubular). Transport of the components is achieved exclusively in the gas phase, separation is accomplished in the stationary phase. The quality of a separation (resolution) depends on how long the components to be separated stay in the stationary phase and on how often they interact with this phase (selectivity). The type of interaction between component and phase (selectivity) is determined by the functional groups. The polarity of the phase is a function of stationary phase substituents.

A chromatogram consists of a base line and a number of peaks. The area of a peak allows quantitative determinations. Starting point of a chromatogram is the time of injection of a dissolved sample. The time interval between a peak and the point of injection is called retention time tR. A component can be identified by its retention time (qualitative determination). The retention time is the sum of the residence time of a solute in the mobile phase (t_0) and in the stationary phase (tR' = net retention time);

$$tR_i = t_0 + tR_i'$$

t_0 is also known as dead time. It is the time required by a component to migrate through the chromatographic system without any interaction with the stationary phase (also called air or gas peak). The net retention time is the difference between total retention time and dead time t_0.

$$tR_i' = tR_i - t_0$$

It indicates how long a substance stays in the stationary phase.

1.3   Derivatization

In gas chromatography it is often advantageous to derivatise polar functional groups (mainly active hydrogen atoms) with suitable reagents. Prerequisite for successful derivatisation is quantitative, rapid and reproducible formation of only one derivative. Aim of this reaction is an improved volatility, better thermal stability or a lower limit of detection due to improved peak symmetry. Elution orders and fragmentation patterns in mass spectroscopy can be influenced by a specific derivatisation.

1.4   Application For Pyrolysis Liquids

GC is the most widely used technique for the analysis of volatile compounds of pyrolysis liquids. They can either be injected as received or after separation into chemical groups such as phenols, aldehydes, acids etc. This separation can be achieved by solvent extraction or solid phase extraction (see chapter below).

About 200 compounds can be separated by GC thanks to the high resolution of modern open tubular quartz columns. The polarity of the stationary phase is of paramount importance for a good separation result. As pyrolysis liquids have a wide distribution of low, medium and high polar compounds the polarity of the column should be middle polar. From the experience of many laboratories involved in GC analysis of pyrolysis liquids using polysiloxane with 14 % Cyanopropylphenyl (1701) as stationary liquid phase gives the best results.

For the detection of acids derivatization is recommended as they are not well separated due to their high polar functional group and hence low volatility. For example derivatization through benzylation using benzyl bromide has proved to work successfully.

2   GAS CHROMATOGRAPHY-MASS SPECTROMETRY (GC/MS)

2.1   Introduction

The appearance of a chromatographic peak at a particular retention time suggests but does not guarantee the presence of a particular compound. The probability of positive identification will depend on factors like the type and complexity of the sample and sample preparation procedures employed. However, the appearance of a GC peak at a known retention time may not be so straightforward in confirming its the presence. Hence, confirmatory evidence is usually sought. A very powerful tool is the combination of gas chromatography with mass spectrometry, a technique known as gas chromatography-mass spectrometry (GC-MS).

Following the first experiences of J. J. Thomson (1912), mass spectrometry has undergone countless improvements. Since 1958, the chromatography-mass spectrometry coupling has revolutionized the analysis of volatile compounds. Combined with the development of the coupling with other separation techniques such as GC and HPLC, the hyphenated instruments allow to obtain significant information from mixtures of natural or synthetic compounds.

2.2   Principles Of Operation

Mass spectrometry is a sophisticated instrumental technique that produces, separates, and detects ions in the gas phase. The basic components of a mass spectrometer are shown in Figure 2. A sample with a moderately high vapour pressure is introduced in an inlet system. operated under vacuum (10^{-4} to 10^{-7} torr) and at high temperature (up to 300 degree C). It vaporizes and is carried to the **ionization source**. Nonvolatile compounds may be vaporized by means of a spark or other source. Analyte molecules are typically neutral and must be ionized. This is accomplished by various means but typically is done by bombarding the sample with high-energy electrons in an **electron-impact source**. The electrons produce a positive ion, for example:

$$M + e^- >> M^- + 2e^-$$

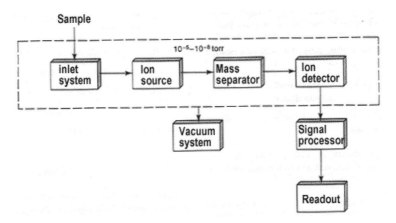

Sample

10^{-5}–10^{-8} torr

inlet system → Ion source → Mass separator → Ion detector

Vacuum system

Signal processor

Readout

Figure 2   Components of a mass spectrometer

M is the analyte molecule and M^+ is called the **molecular ion or parent ion**. The M^+ ions are produced in different energy states and the internal energy (rotational, vibrational, and electronic) is dissipated by fragmentation reactions, producing fragments of lower mass which are themselves ionized or converted to ions by further electron bombardment. The **fragmentation pattern** is fairly consistent for given conditions (electron beam energy). Only a small amount or none of the parent ion may remain.

The ions are separated in the spectrometer by being accelerated through a **mass separator**. Separation is actually accomplished based on the **mass-to-charge (m/e) ratios** of the ions. Various spectrometers are based on magnetic sectors in which ions pass through a magnetic field and are deflected based on their m/e ratio; time-of-flight in which they traverse a long flight tube and arrive at a detector at different times based on their relative kinetic energies after being accelerated through an electrical field (the lighter ones arrive first); or quadrupoles in which the ions pass through an area with four hyperbolic magnetic poles, created by a radio frequency field, and certain ions take a stable path" through the field and others take an "unstable path" and are not detected- the radiofrequency field is scanned rapidly to detect all the ions; or ion traps in which ions are formed, trapped and selectively released from the trap. The **quadrupole mass spectrometer** and ion trap are ideally suited as a gas chromatography detector because it is compact and relatively inexpensive, and a complete scan is achieved in the duration of a GC peak, simply by scanning a voltage. The resolution is more limited than with other mass analyzers, but this is not usually a problem when combined with the gas chromatography information. The separated ions are detected by means of an **electron multiplier**, which is similar in design to photomultiplier tubes. Detection sensitivities at the nanogram level are common.

The effluent from a gas chromatograph may be connected to the sample inlet system of a mass spectrometer, forming a **GC-MS system**. The mass spectrometer then serves as the GC detector with high sensitivity and selectivity. The mass spectrometer may be operated in various modes. In the **total ion current (TIC) monitoring mode**, it sums the currents from all fragment ions as a molecule (or molecules) in a GC peak passes through the detector, to provide a conventional looking gas chromatogram of several GC peaks. In the **selective ion mode (SIM)**, a specific m/e ratio is monitored, and so only molecules that give a molecular or fragment ion at that ratio will be sensed. The **mass spectrum** of each molecule detected is stored in the system's computer, and so the mass spectrum corresponding to a given GC peak can be read out. The mass spectrum is generally characteristic for a given compound (if only one compound is present under the GC peak), giving a certain "fingerprint" of peaks at various (m/e) ratios. Certain peaks will dominate in intensity.

The marriage of capillary gas chromatography with mass spectrometry provides an extremely powerful analytical tool. Capillary GC, with thousands of theoretical plates, can resolve hundreds of molecules into separate peaks, and mass spectrometry can provide identification. Even if a peak contains two or more compounds, identifying peaks can still provide positive identification, especially when combined with retention data.

2.3   Applications For Pyrolysis Liquids

For unambiguous identification of compounds from the complex chromatogram of pyrolysis oils GC/MS is a must. However, the commercial spectral library available today do not contain many of the specific thermal degradation products of biomass. Therefore, the creation of individual libraries is necessary. Some tables with mass spectral data can be found in the literature.

3   HIGH PERFORMANCE LIQUID CHROMATOGRAPHY (HPLC)

3.1   Introduction

High Performance Liquid Chromatography (HPLC) is one mode of chromatography, the most widely used analytical technique. Chromatographic processes can be defined as separation techniques involving mass-transfer between stationary and mobile phases.

3.2   Principle of operation

HPLC utilizes a liquid mobile phase to separate the components of a mixture. These components (or analytes) are first dissolved in a solvent, and then forced to flow through a chromatographic column under a high pressure. In the column, the mixture is resolved into its components. The amount of resolution is important, and is dependent upon the extent of interaction between the solute components and the stationary phase. The stationary phase is defined as the immobile packing material in the column. The interaction of the solute with mobile and stationary phases can be manipulated through different choices of both solvents and stationary phases. As a result, HPLC acquires a high degree of versatility not found in other chromatographic systems and it has the ability to easily separate a wide variety of chemical mixtures.

In the following there are explanations of the separation mechanisms commonly used in HPLC:

In **adsorption chromatography** the stationary phase, properly speaking, is the liquid-solid interface molecules are reversibly bound to this surface by dipole-dipole interactions. Since the strength of interaction with the surface is different for different compounds, residence time at the stationary phase varies for different substances thus achieving separation.

Liquid-solid adsorption chromatography is most often used for polar, non-ionic organic compounds. Partition chromatography is the fundamental distribution mechanism in liquid-liquid chromatography, i. e. when both mobile phase and stationary phase are liquids.

Separation by distribution is based on the relative solubility of the sample in the two phases. In normal phase partition chromatography the stationary phase is more polar than the mobile phase, in **reversed phase** (RP) chromatography the mobile phase is more polar than the stationary phase. Stationary phases may be either coated on to a support, or they may be chemically bonded to the surface.

Normal phase partition chromatography is used for very polar organic compounds, while reversed phase chromatography is commonly used for non-polar or weakly polar substances.

Ionic compounds are often better separated by **ion exchange chromatography** (IEC). In this case, the stationary phase consists of acidic or basic functional groups bonded to the surface of a polymer matrix (resin or silica gel). Charged species in the mobile phase are attracted to appropriate functional groups on the ion exchanger and thereby separated. Ion pairing chromatography is an alternative to ion exchange chromatography. Mixtures of acids, bases and neutral substances are often difficult to separate by ion exchange techniques. In these cases ion pairing chromatography is applied. The stationary phases used are the same reversed phases as developed for reversed phase chromatography. An ionic organic compound, which forms an ion-pair with a sample component of opposite charge, is added to the mobile phase. This ion-pair is, chemically speaking, a salt which behaves chromatographically like a non-ionic organic molecule that can be separated by reversed phase chromatography.

Size exclusion chromatography (SEC) or gel permeation chromatography (GPC) uses as the stationary phase a porous matrix which is permeated by mobile phase molecules. Sample molecules small enough to enter the pore structure are retarded, while larger molecules are excluded and therefore rapidly carried through the column. Thus size exclusion chromatography means separation of molecules by size.

Separation mechanism: The chromatographic column is packed with particles of the same size. For the classification of molecules according to size it is a prerequisite that the particles are totally porous and that the pores have a defined diameter. In Figure 3, the pores are illustrated in a conical shape for demonstration purposes.

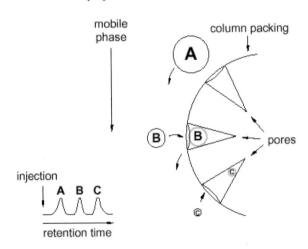

Figure 3   Illustration of the separation mechanism in GPC

Let us assume, that the polymers to be analysed consist of molecules size A, B and C. In the example illustrated, the largest molecules A cannot enter the pores, while molecule B can only half enter and the smallest molecules C can completely enter the pore system. Since adsorptive interactions with the stationary phase can be excluded in GPC by correct choice of the mobile phase, elution of the components is in the order A, B and C, i. e. the largest molecules are eluted first and the smallest last. In order to achieve an optimum separation of oligomers or polymers, packing materials with different pore diameters are available.

3.3 Application for pyrolysis liquids

Generally, the extent of HPLC applications for the analysis of pyrolysis liquid is less compared to GC. There are several reasons for it. Firstly, the resolution of HPLC columns is much smaller than with GC columns leading to overlapping of compounds. The injection of whole oil is not recommended due to insufficient solubility of the sample. High molecular weight compounds will block the solid phase at the beginning of the column. Therefore, fractionation is necessary in order to get less complex mixtures. Some laboratories use the water soluble fraction of pyrolysis liquids for analysis by HPLC. In contrast to GC analysis the number of detected compounds in HPLC is relatively small. The separation of the following compounds has been reported: cellobiosan, glyoxal, hydroxyacetaldehyde, levoglucosan, formaldehyde, formic acid, acetic acid, and acetol (hydroxypropanone).

A special case of HPLC analysis for pyrolysis liquids is SEC. This method has been used to monitor ageing effects by comparing the relative molecular weight distribution. It is assumed, that storage or heating of oils promote recondensation reactions either of low molecular weight products with the polymeric lignin of with lignin itself, leading to an increase in molecular size (and in practice to the formation of a sludge at the end of the oil drum).

4 SOLID PHASE EXTRACTION (SPE)

4.1 Introduction

Solid phase extraction (SPE) is a powerful method for sample preparation and is used by almost half of all chromatographers today. It has capabilities in a broad range of applications such as environmental analyses, pharmaceutical and biochemical analyses, organic chemistry and food analyses. The advantages of SPE compared to classical liquid-liquid extraction are the low solvent consumption, the enormous time saving and the potential for automation. Additionally, a sample preparation task can often be solved more specifically by using SPE, since different interactions of the analyte with the solid phase (adsorbent) are possible, and methods can be optimized by adjusting chromatographic conditions. SPE offers a multitude of adsorbents for polar, hydrophobic and/or ionic interactions, while liquid-liquid extraction is limited to partition equilibriums in the liquid phase. In general, SPE can be used for three important purposes in up-to-date analyses: concentration of the analyte removal of interfering substances changing the matrix of the analyte as needed for subsequent analyses In most cases these three effects occur together. Since analytes can be either adsorbed on the SPE packing material or directly flow through while the interfering substances are retained, two general separation procedures are possible.

4.2 Principle Of Operation

The sample is pressed or drawn with vacuum through the solid phase, which is retained in a cartridge, and the analyte molecules are enriched on the adsorbent. Interfering components and solvent molecules (matrix) are not retained. Then remaining interfering components are washed from the adsorbent with a suitable washing solution. Finally, the analyte is removed from the adsorbent by elution with a suitable solvent. In some cases other interfering components may remain on the adsorbent. Such a strong adsorption of interfering components offers another possibility for the pre-purification of difficult matrices, such as waste oils or sludge. If the analytes show no interaction with the adsorbent and if only the interfering components are retained, the solid phase can be used to simply "filter" the sample. The considerations made above indicate that an optimum SPE presents a poor column-chromatographic separation.

4.3 Application For Pyrolysis Liquids

Only little is known about the separation of whole pyrolysis liquids using SPE. Currently, this method is being developed in order to fractionate liquids prior to GC or HPLC analysis. Generally, SPE is a powerful tool for the enrichment of compounds and often used for the detection of PAHs.

5 SOLID PHASE MICRO EXTRACTION (SPME)

5.1 Introduction

Many headspace studies had the sample preparation step, like the extraction of semi volatile organic compounds from complex samples such as soil, serum, and so forth. These methods require extensive clean-up and evaporative concentration procedures, which may cause loss of the volatile compounds and are time-consuming.

Solid-phase micro-extraction (SPME), is a rapid, inexpensive, solvent less and easily automated technique for the isolation of organic compounds from gaseous and liquid samples. It is based on the enrichment of components on a polymer- or adsorbent-coated fused-silica fiber by exposing the fiber either directly to the sample or to its headspace.

5.2 Principle Of Operation

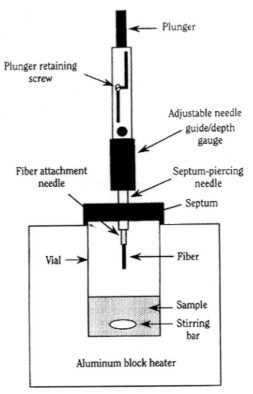

Figure 4   Schematic of the SPME extraction unit

The newly developed solid-phase microextraction (SPME) technique is increasingly being used for the gas chromatographic determination of a wide variety of volatile and semivolatile organic compounds in water or aqueous extracts of different substrates. Basically, it involves extraction of specific organic analyses directly from aqueous samples, or from the headspace of these samples in closed vials, onto a **fused-silica fibre** coated with a polymeric liquid phase, **poly(dimethylsiloxane)** or **polyacrylate**. After equilibration, the fibre containing the absorbed or adsorbed analyte(s) is removed and thermally desorbed in the hot injector of a gas chromatograph. The analyses are then analyzed by gas chromatography (GC) using an appropriate column and detector with or without cryofocusing. Further details of both the theoretical and practical aspects of the technique can be obtained from recent publications by Pawliszyn and coworkers and others.

The SPME method uses a fine **fused-silica fiber** coated with a polymer (i.e., **polydimethylsiloxane or polyacrylate**) to extract organic compounds from their matrix. SPME fiber combines sampling and preconcentration in a single step. After a well-defined adsorption time the fiber is transferred to a standard split/splitless injector, where the organic compounds are thermally desorbed from the polymeric phase. So far, SPME has been used coupled with GC to analyze a wide range of organic compounds in aqueous samples, such as aromatic hydrocarbons, halogenated volatile organic compounds (VOCs), polyaromatic hydrocarbons (PAHs), polychlorinated biphenyls (PCBs), pesticides. nitroaromatic compounds and phenol and its derivatives.

The technique is very simple, fast and does not employ any organic solvents either for sample preparation or cleanup. This makes it highly desirable because, unlike other methods, it does not release environment-polluting organic solvents into the atmosphere. Thus far, the technique has been successfully applied to the determination of a wide variety of organic compounds in water.

5.3   Application For Pyrolysis Liquids

SPME has been used to analyse the specific odour of pyrolysis liquids.

6   CONCLUSIONS

A couple of powerful analytical tools are available for the analysis of pyrolysis liquids. However, the user should be aware of the possibilities and limits of each method, bearing in mind the complexity of the pyrolysis oil. Independent of the objectives of the analysis, proper sample preparation is necessary to facilitate data processing after analysis. This includes for example the determination of the water content which is important for the dilution of samples. It is important to note that neither of the methods is capable of giving a complete analysis of the sample. Obviously, a combination of GC and HPLC methods together with the appropriate detection systems is best suited for the chemical description of the pyrolysis oil sample.

ENVIRONMENT, HEALTH AND SAFETY ASPECTS RELATED TO FAST PYROLYSIS OILS - A GUIDELINE TO NOTIFY A NEW SUBSTANCE

Philippe Girard and **Cristina Diez**
CIRAD-Forêt - Energy Environment Unit
TA 10/16 - 34398 Montpellier Cedex 5 - France

ABSTRACT

Environment, health and safety aspects of liquids from fast pyrolysis process are of very high concern. Pyrolysis being one of the three main thermochemical routes to convert biomass into useful primary energy products, fast pyrolysis has benefited from active research programmes since 1980's. Today, different demonstration plants are set up in Europe as well as in North America and significant quantities of bio-oils are produced for research and development purposes. Bio-oils contain 100s of chemicals and their composition depends on the operating conditions and feedstock. Thus, the question of safety procedures for human health and environment preservation is raised.

The reported composition of fast pyrolysis oil shows some similarities to pyroligneous acids or wood distillate - although they contain different components or different range of concentration - which have been largely used in the food industry for years. They may not thus be dramatically "unfriendly" towards human health. However, the use of bio-oils as fuel oils involves much larger production volumes and industrial units, as well as dispersed locations of storage and transport all over Europe. In this situation, the exposures to the product involve more people (workers on production and use sites and potentially private individuals), longer time and more sites. It is therefore necessary to minimise as far as possible the impacts of the products and to anticipate and foresee the main risks related to this situation in order to define the right ways to handle the product and to be able to quickly and efficiently react in case of an accident, either towards human health or the environment.

Fast pyrolysis oil is a new available complex mixture of chemical compound. The European regulations concerning potentially toxic or toxic products require that all new substances must be notified by the producer (or importer) before being marketed in Europe. Different levels of information have to be supplied, depending on the production quantity. For more than 1 ton/year/producer, complete toxicological and eco-toxicological data are required. For between 1 ton and 100 kg/year/producer, results on acute toxicology and mutagenic effects are required.

The procedure for notification has to be done by the oil producers. The data have to be presented to the legal authority of one of the European country. The legal authority will then statute on the product classification , labelling and inscription in the ELINCS and give authorisation for marketing with the corresponding preventive accompanying measures. This is a mandatory procedure for any companies aiming at commercialising a new chemical substance. In this case, the owners of industrial pilot plants are responsible for initiating the procedure, with the available data, to obtain trading authorisation over Europe. This procedure is a long and complex process. Therefore this paper will attempt to describe the procedure for placing on the EU market a new chemical substance. It has been elaborated in the framework of the EHS PyNe subject group as a guideline for fast pyrolysis independent producers that would produce and sell fast pyrolysis oil as fuel.

TOXICITY OF BIO-OILS

In the long term, biomass will undoubtedly play a significant role in the supply of energy in many countries. Pyrolysis is one of the three main thermochemical routes to convert biomass into useful primary energy products. As a result of the thermal decomposition of the raw material, a gas, a liquid and a solid are formed, which can be used directly or further upgraded to give more value-added fuels.

Fast pyrolysis at a temperature around 500°C, at very high heating rates and short vapour residence times (less than 1 second) gives high liquid yields of up to 80% weight on a dry feed basis. Their composition mainly depends on the feedstock used and the pyrolysis conditions (1). Thus, the toxicology of the oils will also depend on feedstock and process. However, no systematic studies have been made to relate the different parameters.

Bio-oils have been widely studied at laboratory scale for 15 years and are now at a stage to be produced on larger scale in industrial pilot plants and demonstration units. Until now, the attempts to assess the toxicity of these products have been based on bibliographical collection of results concerning close products (pyroligneous acids or wood distillates for smoky food flavours or food preservatives) and concerning individual components of the bio-oils (2). A number of studies exist in historical literature concerning hazards of acute and long term exposure to smoke and to the pyrolysis liquids formed at low temperature.

Diebold (3) has made a review of available toxicological data on individual components that can be found in pyrolysis liquids and smokes. Primary vapours are mainly oxygenated compounds from the partial decomposition of biomass macropolymers (lignin, cellulose, hemicellulose). Secondary vapours are formed by cracking of the primary vapours, at higher temperature or higher residence times, and are phenolic compounds. Tertiary vapours are mainly polyaromatic hydrocarbons (4). In the whole, the pyrolysis oils and smokes are a mixture of primary, secondary and tertiary vapours, one type being predominant depending of process conditions. 74 chemicals have been reported to be either in oils, smokes or both, with a concentration of at least 0.1 wt %. Some of them are reported in table 1. The toxicity of each chemical has been examined. The conclusion is that aldehyde and furan families can lead to acute toxicity of pyrolysis vapours, by ingestion or inhalation. Chronic health effects could also be related to aldehydes unless the severity of the pyrolysis is so high as to create significant quantities of benzene and PAH, potentially leading to tumour development.

It has been assumed until now that similar properties to smokes or wood distillates can be more or less expected for fast pyrolysis bio-oils. Fast pyrolysis bio-oils contain less condensed compounds, which should decrease long term toxicology. On the other hand, they contain different compounds compared to smokes that can give synergetic effects from a toxicological point of view (leading to either increase or decrease in the health effects). One study (5) has been reported concerning tests on bio-oils but contradictory results have been obtained on chronic toxicity, one indicating no tumour promoting ability and the other showing mutagenicity.

Thus, there is still a clear lack of knowledge about the toxicological and eco-toxicological behaviour of bio-oils. The needs are both to relate the process parameters to the toxicity of the final product and to derive new data on bio-oil toxicology, which has never been done before. The relation between process parameters and toxicity will allow to define standard for the future commercialisation of the product, avoiding putting on the market a wide range of products which may have completely different toxicological and ecotoxicological characteristics, though very close from the application point of view. This relation between operating conditions and composition and health and environment impacts of the products may be difficult to define, taken into account the huge number of compounds in the oils. It is expected to identify key parameters allowing the improvement of toxicological properties as well as the determination of the influence of initial feedstock.

Table 1   Concentration of some hazardous compounds identified in fast pyrolysis products
(wt %)

	J.P Diebold from Milne(6) et al 1996 (wet oil)	Meier et al 1999 (7) (water free)
Acids	0.3 – 9.1	
Fornic	0.5 – 1.2	
Acetic	0.1 – 1.8	3.0 – 4.7
Propanoic	0.1 – 0.9	
Hydroxyacetric	0.1 – 0.5	
Butanoic	0.1 – 0.8	
Pentanoic	0.1 – 0.4	
4-oxypentanoic	0.2 – 0.3	
Heptanoic	0.3	
Aldehydes		
Formaldehyde	0.1 – 3.3	
Acetaldehyde	0.1 – 8.5	7.78 – 12.92
2 propenal	0.6 – 0.9	
2 methyl - 2 butenal	0.1 – 0.5	
Pentanal	0.5	
Phenols		
Phenol	0.3 – 3.8	0.03 - .0.2
o-cresol	0.1 – 0.6	0.1
m-cresol	0.1 – 0.2	0.1 – 0.3
p-cresol	0.1 – 0.3	0.02 – 0.1
23 dimethyl phenol	0.1 – 0.5	
2-4 dimethyl phenol	0.1 – 0.3	0.11 – 0.4
2-5 dimethyl phenol	0.2 – 0.4	
2-ethylphenol	0.1 – 0.2	
1.4 diOH benzene	0.1 – 1.9	
Furans		
Furan	0.1 – 0.3	
2 - methyl furan	0.1 – 0.2	
2 -feranone	0.1 – 1.1	0.4 – 0.77
furfural	0.1 – 1.1	
furfural alcohol	0.1 – 5.2	
5-méthylfurfural	0.1 – 0.6	

The growing interest of potential users for bio-oils is leading them to take more and more concern about these aspects of the product. Indeed, fast pyrolysis is the only single stage process that gives a useful liquid at a competitive cost.

The safe handling and transport of bio-oil is a particular problem that reflects on the innate advantage of bio-oil in being a liquid that can be stored and transported. Safe practices for storage and transport need to be defined in relation to UN and other regulations and recommended and required procedures made available to all involved in this activity.

Efforts have been spent within PyNe to assess the toxicity and eco-toxicity data on bio-oils. The work would concern both acute and chronic effects on human health, carcinogenic and mutagenic effects. The environmental impacts of bio-oils liquids would also be measured in water and soil. This would allow a complete MSDS to be defined with the proper preventative and remedial procedures to be adopted during production, transport and use of bio-oils. Unfortunately this activity has not started yet. Nevertheless we already know from Diebold's report (3) that EU notification will be required.

Therefore this guide tries to help and advise about every mandatory administrative step for placing a new chemical substance on the European Community market. It has no legal status so Competent Authorities (C.A.) listed in Annex should be contacted in order to provide particular information.

DEFINITIONS

New substances or preparations cannot be placed on the Community market on their own unless they have been notified to the C.A. of one of the Member States in accordance with Directive 92/32/EEC (8) by means of a notification containing certain information. The procedure of notification was laid down in 1981. Every Member State has designated C.A. which receive and study the dossiers of notification.

Some of the definitions used in the EC directive are given in Table 2.

Table 2   Definitions extracted from the EC directive 92/32/EEC

Substances : "chemical elements and their compounds in the natural state or obtained by any production process, including any additive necessary to preserve the stability of the products and any impurity deriving from the process used, but excluding any solvent which may be separated without affecting the stability of the substance or changing its composition".

New substances : They are not included in EINECS : "European Inventory of Existing Commercial Substances"(10) which contains the definitive list of all substances deemed to be on the Community market on 18 September 1981. Every substance placed on the Community market after 1981 is considered as a new substance and it has to submit a notification process.

Preparations : "mixtures or solutions composed of two or more substances". A preparation is an intentional mixture of substances whereas a substance has been obtained directly as a product formed by some compounds that have not been previously mixed.

Dangerous Substance : Classified depending on its dangers: 15 categories of danger: explosive, oxidizing, extremely flammable, highly flammable, flammable, very toxic, toxic, harmful, corrosive, irritant, sensitizing, carcinogenic, mutagenic, toxic for reproduction and dangerous for the environment defined in accordance to Article 2 of Directive 92/32/EEC (8). New dangerous substances are included, classified and listed in Annex I of Directive 67/548/EEC (9). If substances contain some impurities, additives or individual components they shall be taken into account as far as their concentrations exceed the concentration limits fixed as 0.1 % (if toxic, very toxic, carcinogenic, mutagenic or toxic for reproduction Class 1 or 2) or as 1 % (if harmful, corrosive, irritants, sensitizing, carcinogenic, mutagenic or toxic for reproduction Class 3), except for smaller values specified in Annex I of Directive 67/548/EEC (9).

Dangerous preparation : Preparations which contains at least one classified dangerous substance or which has dangerous properties with a concentration that exceed the concentration limits specified in Annex I of Directive 67/548/EEC (9) (particular concentration limits) or specified in Article 3 of Directive 88/379/EEC (11) (general concentration limits). They are classified as dangerous and listed in Directive 88/379/EEC (11).

In this table, preparations are mentioned as it might concern bio-oil products to be used as fuel for engines (fast pyrolysis oil + ethanol).

EXEMPTIONS FROM NOTIFICATION

There are several cases in which a new substance/preparation does not have to be declared, always in accordance to C.A.

Maximum Quantities

If it is placed on the Community market in quantities of less than 10 kg per year per manufacturer (kg/yr/mf) (provision that can change in some member states).

Exportation - Importation

If it is going to be used out of the Community market, i.e. Exportation. Substances transferred between two establishments making part of the same corporation are not considered as placed on the market. On the other hand, substances imported into the Community customs territory, produced or freely ceded are considered as placed on the market.

If it has been manufactured and placed out of the Community market. It is not mandatory to notify it again in the Community market if the substance has been agreed by a C.A. recognized by the European Community. In any case, C.A. of the member state where the substance will be introduced has to be contacted for national requirements. If a notification has to be submitted, the European C.A will probably accept the precedent classification making the new process of notification easier.

If it has been manufactured outside the Community market and already notified in it, the Representative of the importers shall only inform C.A. about the quantities introduced. On the other hand, if the substances have been notified in the EC by other importers, the new one must submit a new dossier of notification.

Research and Development Stage

Substances which are placed on the market in quantities which are limited for the purpose of "scientific experimentation, analysis or chemical research carried out under controlled conditions which includes the determination of intrinsic properties, performance and efficacy as well as scientific investigation related to product development", and in any case not exceeding 100 kg/yr/mf under controlled conditions, shall not be notified but any manufacturer or importer must maintain written records at the disposal of the C.A. (2 copies for each one), inspectors and company doctors containing the available information concerning:

- identity of the substance (name-purity-impurity),
- spectral and analytical data,
- identity of the manufacturer or importer,
- lists of customers,
- results of the available physico-chemical, toxicological and ecotoxicological tests,
- labelling data. If necessary (if toxic), the label should bear the warning: "Caution - substance not yet fully tested",
- justified necessary quantities for the operation,
- envisaged uses,
- tests already carried out and a summary of the research program.

Substances with a limited number of registered customers in quantities that are limited for the purpose of "the further development of a substance in the course of which pilot plant or production trials are used to test the fields of application of the substance" qualify for an exemption for a period of one year. The notifier should justify the request for an exemption by providing to C.A. similar written records as in the case above with this additional information:

- the research and development (R&D) programme,
- the guarantee that the substance will be handled only by customers' qualified staff in controlled condition and will not be made available to the general public at any time.

In case the substance is toxic, toxicity data, and first aid, fire fighting, accidental release and transport, handling and storage prevention measures, are required.

When no physico-chemical, toxicological and ecotoxicological tests are available, the C.A. ask the notifier to submit some tests depending on the chemical nature of the substance, its quantities and its uses. Furthermore the notifier should communicate to the C.A. in each Member State where supply is to occur any additional information. Consult the national regulations or contact the C.A. for national requirements.

After one year, these substances will normally be subject to notification. The one year exemption period may in exceptional circumstances be extended for a further year if the notifier can demonstrate, to the satisfaction of the C.A., that such an extension is justified.

PRODUCT IDENTIFICATION

Before submitting a dossier of notification to C.A. it is worthwhile to provide them with information concerning the substance to get the particular administrative steps required.

Composition

A detailed composition shall be provided. The main components percentage shall be enough if substances are very complex. The aim is to know if it is a real new substance or if it has been already notified so, both, C.A. and notifier, try to find it registered in one of the "Existing chemical products Inventories" where substances can be found identified with a register number.

If the substance has been notified at European level, it will be registered in one of the following European Inventories:

- **EINECS** "European Inventory of Existing Commercial Substances" (10) : which contains the final and closed list of all substances deemed to be on the Community market between 1st January 1971 and 18th September 1981 (100.195 substances). Substances placed on the market after this date are considered as new and their notification is mandatory. EINECS is published by the Commission (Annex I).

- **ELINCS** "European List of Notified Chemical Substances" (12) : which contains substances introduced after 1981 and until 30th June 1995. They are considered new substances and consequently have to be notified. It is regularly updated and published at the Official Journal of the European Communities.

Substances are considered mono-substances if their principal component is present over 80%, on the other hand they are considered as mixtures. Every component present over 10% must be registered in ELINCS. Impurities shall not be listed unless their contribution shall be very important for the classification of the substance. Commercial substances not appearing on EINECS or ELINCS shall be notified in accordance to this Directive and they will be incorporated in the updated ELINCS version.

For a the substance notified in America, the American Chemical Society publishes the "Chemical Abstracts", a periodical chemical substances registries internationally recognised. It includes information concerning substances: CAS "Chemical Abstracts Service" number, EC number if existing, international nomenclature, trade name, category of danger, danger symbol, R-phrases, S-phrases, concentration limits, European regulations, importation/exportation regulations, etc. To attribute a CAS number to a chemical product, a summary form (55$) shall be completed with: chemical nomenclature, structure and molecular formula of the substance with no particular tests to be carried out. No additional and follow-up information about the product is needed whereas the CAS service completes it with worldwide level related data.

INRS (the French C.A.) has suggested a listed product very similar to bio-oils. Identification numbers and nomenclature are:

EINECS number = 295-321-8
CAS number = 91995-59-4
Nomenclature = distillates, wood tar

Nevertheless, the composition of both products is different as mention by Diebold (2) and they cannot be considered as the same substance. The entry of "distillates, wood tar" has already been used by some manufacturers to transport bio-oils using the established legislation for it. But in the case of bio-oils placed on the market legislation is more severe and a new notification is needed.

Toxicity

The MSDS (Material Safety Data Sheet) of the main components can be provided to give an idea on the substance toxicity if no test has been carried out.

When the substance is thought to be dangerous and no similar substance has been found registered, identification tests have to be carry out by a GPL laboratory (Good Laboratory Practices) (13) (14) to introduce the substance in one of the fifteen categories of danger. These test are relatively costly and had detailed the PyNe activities up to now due to fund constraints.

DOSSIER OF NOTIFICATION

Once a new substance has been identified, a notification has to be submitted for placing them on the market. The notifiers must present or address by mail an identical dossier to each C.A. These dossiers must be provided in a standardised summary form, an electronic one being preferred. The common European summary form software SNIF (Structured Notification Interchange Format), developed by the Commission until June 1992, can be obtained from C.A. A new Windows version will be available soon.

Elements to be provided are:
1. Technical Dossier.
2. Notifier Letter.
3. Risk Assessment.
4. Declaration concerning the Unfavourable Effects.
5. Detailed and full Description of the Tests conducted.
6. Confidentiality of Data.
7. Charges.
8. Proposed Classification and Labelling.
9. Proposal for a Safety Data Sheet.

If it is not technically possible or if it does not appear scientifically necessary to provide some information, the reasons shall be clearly stated and will be subject to acceptance by the C.A.

In some cases, " Comments " has to be included as a separated part since the summary form cannot satisfy the compilation of all data as well as justified reasons for uncompleted sections.

The Technical Dossier

It supplies the information necessary for evaluating the foreseeable risks, whether immediate or delayed, which the substance may entail for human and the environment, and containing all available relevant data for this purpose :

- Identity of manufacturer and of the notifier: Location of the production site;
- Identity of the substance: IUPAC name, trade name, CAS number, molecular and structural formula, composition of the substance, degree of purity, molecular weight. The same data is required for impurities and additives with their percentages and for components if a mixture;
- Spectral and analytical data;
- Description of methods or appropriate bibliographical references;
- Information on the substance: production (technical process), estimated exposure related to production : envisaged uses, desired effects; trade form, waste quantities and their composition; estimated production for each of the envisaged uses;
- Recommended methods and precautions concerning handling, storage and transport;
- Emergency measures in the case of fire, accidental release, reaction with water, explosion, etc;
- Packaging;
- Physico-chemical, toxicological and ecotoxicological properties of the substance. Degradation;
- Possibility of rendering the substance harmless, possibility of recycling, neutralization and destruction.

Detailed explanation to complete the summary form is given in the "Guidance notes to those completing a summary notification dossier of a new substance " written in accordance with Directive 92/32/EEC (8) and available from the C.A.

Before carrying out animal testing for the purpose of collecting test data for the notification of a new substance, the potential notifier must enquire of the C.A. as to whether or not the substance has already been notified (substance notified after 18 September 1981 and introduced on ELINCS). In the case of already notified, substance name and the address of the first notifier, the C.A. has to verify the relevance of the information available. Additional tests can be required if C.A. consider it pertinent.

Depending on the total quantities placed on the market per year per manufacturer (t or kg/yr/mf). The requirements for the technical dossier vary.

- 10kg-100kg/yr/mf, or, 50kg-500kg/mf : **A Reduced Technical Dossier** is necessary. Required information as laid down by C.A. (as fixed by Annex VII C of Directive 92/32/CEE) (8).
- 100kg-1t/yr/mf, or, 500kg-5t/mf : **A Reduced Technical Dossier,** which contains the additional information of Annex VII B of Directive 92/32/CEE (8) is required.
- 1t-10t/yr/mf, or, 5t-50t/mf : **A Technical Dossier** which corresponds to the submission of a full notification as fixed in Annex VII A of Directive 92/32/CEE (8) is required.
- 10t-100t/yr/mf, or, 50t-500t/mf : **A Technical Dossier Level 1** is needed. The C.A. have to be informed of new quantities placed on the market and they may require some or all of the additional tests laid down in Level 1 to be carried out within a time limit which the C.A. will determine.

- 100t-1000t/yr/mf, or, 500T-5000T/mf :Correspond to a **Technical Dossier Level 1**. The additional tests laid down in Level 1 have to be carried out within a time limit which the C.A. will determine, unless the notifier can give good reasons why a given test is not appropriate or an alternative scientific test would be preferable.
- >1000T/yr/mf, or, >5000T/mf : Correspond to a **Technical Dossier Level 2**. The C.A. shall draw up a programme of tests according to Level 2 to be carried out within a time limit which the C.A. will determine.

The tests to be carry out in the case of the Technical Dossier Level 1 and 2 are described in the note for the attention of Competent Authorities for the implementation of Directive 67/548/EEC (9); Guidance to Notifiers on Level 1 and 2 Notifications" ; published by the European Commission on the 28th November 1996.

The Notifier Letter

It is a letter from the person responsible for the notification, who must be the Manufacturer for substances manufactured within the Community.

For substances manufactured outside the Community (Manufacturer's address must be included), the notifier can be the importer who must have a document from the manufacturer assuring the quantities placed on the market, or the representative who is the person established within the Community who need a Statement from the manufacturer for the purpose of submitting a notification for the substance in question that he is designated as the manufacturer's Sole Representative (S.R.).

Risk Assessment

The notifier might submit on voluntarily basis a risk assessment of the notified substance. It will be carried out by the C.A. according to the principles laid down in Directive 93/67/EEC (15). It should include the following information: identification of the hazardous properties of the substance, exposure assessment for the human population and for the environment, comparison between exposure levels and effects, etc. The C.A. shall study the risk assessment and propose some recommendations on the risk reduction before contacting the Commission.

There is a technical guidance document published by the European Commission: EUSES (European Union System for the Evaluation of Substances) available from CEFIC (European Council of the Federation on Chemical Industries), from ECB (European Chemicals Bureau) and from C.A.

Declaration Concerning the Unfavourable Effects

Unfavourable effects of the substance in terms of the envisaged uses can be avoided if the risk assessment has been submitted. It is intended to take into account all potential dangers including those not covered or identified by the classification and labelling. All data should be checked for consistency with the test results.

Detailed and Full Description of Tests Conducted

Tests on chemicals carried out within the framework of this Directive, for the purposes of notification, classification and labelling, shall as a general principle be conducted according to the methods laid down in Annex V of the Directive 67/548/EEC (9):

- the physico-chemicals properties of the substances shall be determined according to the methods specified in its Annex V. A
- their toxicity according with Annex V. B
- their ecotoxicity according with Annex V. C

Most of the tests are internationally recognized OCDE tests; (www.ocde.org : OECD Test Guidelines Programme). They are also published by the standards association .

In case there is no official method published, the notifier can choose the more appropriate or establish one. All detailed explanation of the experimental protocol used has to be described. The main constituents MSDS, bibliographical references or practical works can guide the notifier in finding the correct method. For some of the substances it is possible that test data exist which have been generated by methods other than those laid down in Annex V of 67/548/EEC (9), the need to conduct new test and the modifications have to be indicated.

Laboratory test shall be carried out in compliance with the principles of Good Laboratory Practice. There is no specific summary form for test laid down by the Commission. Laboratories shall send a summary of the tests reports that should contains suitable signed GPL and Quality Assurance Statements.

C.A. may ask for further information, verification and/or confirmatory tests concerning the substances notified. Additionally they may carry out some sampling for control purposes, and require the notifier to supply the quantities of the notified substances.

Confidentiality of Data

The notifier may indicate the information which he considers to be commercially sensitive and disclosure of which might harm him industrially or commercially, and which he therefore wishes to be kept secret from all persons other than the C.A. and the Commission. Full justification must be given in such cases in the summary form or in a separate annex.

The information allowing for confidentiality can be divided in five main categories, namely :
1.   Chemical identity of the substance
2.   Impurities
3.   Uses and desired effects of the substance
4.   Volumes on the market
5.   Names of bodies responsible for the tests

On the other hand, there are some elements that cannot be kept secret, such as:
• the trade name of the substance. If the chemical name is to be kept secret, the name of the chemical family must be provided. Substances that are not classified as dangerous may be included in the list with its trade name for a maximum of three years. However if the C.A. to which the dossier was submitted considers that the publication of the chemical name in the IUPAC nomenclature itself could reveal information concerning commercial exploitation or manufacture, the name of the substance may be recorded under its trade name alone for as long as those C.A. see fit. Dangerous substances may, at the request of the C.A. receiving the notification, be entered on the list in their form of their trade names alone until such time as they are introduced into the list of dangerous substances;
• the name of the manufacturer and the notifier;
• physico chemical properties of the substance;
• the possible ways of rendering the substance harmless;
• the summary results of the toxicological and ecotoxicological tests;
• if essential to classification and labelling, the degree of purity of the substance and the identity of impurities and/or additives which are known to be dangerous;
• the recommended methods, precautions and the emergency measures in the case of an accident;
• the MSDS;

- in the case of dangerous substances, analytical methods that make it possible to detect a dangerous substance when discharged into the environment as well as to determine the direct exposure of humans.

If the notifier should himself later disclose previously confidential information, he shall inform the C.A. accordingly.

Charges

Charges are paid at the same time as the dossier is presented to the C.A. and they are envisaged to study the provided information. They are different in each Member State.

Proposed Classification And Labelling

This is proposed by the notifier. The classification informs about the hazards of the substance in the case of handling, transport, storage, and packaging. The label is established in order to ensure an adequate level of protection for man and the environment.

Labelling is mandatory for dangerous substances and preparations and it is established on their classification in accordance with the List of Dangerous Substances. In the case of dangerous substances not yet listed, their label is designed according with the general principles settle down in Annex VI of Directive 67/548/EEC (9).

- All new substances have to be labelled and packed in so far as the notifier may reasonably be aware of their dangerous properties. If it is not possible to label the substances because the results of tests provided are not all available, the label should bear, in addition to the label deriving from the tests already carried out, the warning "Caution – substance not yet fully tested";
- A dangerous substance which appear on EINECS but which has not yet been introduced in Annex I of Directive 67/548/EEC (9) shall be investigated to know the relevant and accessible data exist concerning the properties of such substances. On the basis of this information they shall be packaged and provisionally labelled according to criteria laid down in Annex VI of Directive 67/548/EEC (9) (Classification Guide).

Every package shall show clearly, indelibly and at least in the official language of the Member State one or more labels with the elements fixed in the List of Dangerous Substances in Annex I of the Directive 67/548/EEC (9) and established with criteria mentioned in Annex VI of this Directive.

In the case of a substance, the name of the substance under one of the designations given in this Directive should be mentioned or, if it is not listed yet, an internationally recognized designation and the EINECS or ELINCS numbers. The label shall also include the words "EC label" and the name and full address including the telephone number of the responsible person established in the Community (manufacturer, importer or distributor).

Peacocke (16) suggested a set of labels, which can be used up to the results of test and the MDS will be available.

Dangerous substances cannot be placed on the market unless their packaging satisfies requirements. Peacocke (16) suggested different types of appropriate packaging depending on the volume of product to be transported or stored.

Proposal for a Safety Data Sheet

The MSDS is mandatory if the substance is classified dangerous. It enables professional users to take the necessary measures as regards the protection of the environment and health and safety at the workplace. The sheet should contain the information necessary for protection of man and the environment. In case of emergency all the procedures will be defined following the content of the MSDS. Although not required it is also recommended that a safety data sheet for non-dangerous substances should be submitted.

All information in MSDS should be checked for consistency with the test results reported in the summary. General rules for the elaboration, distribution, contents and format of the MSDS are referred in Annexe XI of Directive 92/32/EEC (8) and in Directive 91/155/EEC.

Czernik (17) suggested a draft MSDS based on MSDS of related substances and on the information available through Elliott's work (1).

PLACING OF NOTIFIED SUBSTANCES ON THE MARKET
Preliminary Judgement Phase

Substances can be placed on the market no sooner than 15 or 30 days in the case of the Reduced and Special Technical Dossiers are required, 60 days in other cases.

If there is no answer in the time limit, the C.A. consider that the dossier is in conformity with the requirements of the Directive 92/32/EEC (8) and the notifier will receive a notice with the date for placing on the market and the official number which has been allocated to the notification. In Spain the notifier can know the conformity in a time limit of forty days in accordance with Article 44 (law 30/1992, 26 November 1992) but the substance shall be only placed on the market after sixty days.

If the dossier is not in conformity with Directive 92/32/EEC (8), the C.A. advise the notifier of the information required in a limit of other 60/30 days for rendering the notification into conformity. If the notifier does not receive any answer after this period, the substance can be placed on the market.

Deeper Study

After the preliminary judgement phase, the C.A. study the dossiers of notification thoroughly. For example, in France, a Commission of the French Ministry of Environment study the ecotoxicity of the substance and advice the notifier about some recommendations for protecting the environment. INRS, the C.A. transmits to the French Ministry of Work the studies carried out by an interdisciplinary team for protecting the workers.

When a Member State has received the dossier of notification, information on the supplementary testing carried out or follow-up information, it shall as soon as possible send the Commission a copy of the dossier or of the further information or a summary thereof with the test chosen, the reasons for their choice, the results and, if appropriate, an assessment of the results. The assessment of the risks or a summary shall be forwarded to the Commission as soon as it becomes available.

On receipt of the dossiers and information the Commission shall forward copies to the Member States. The other C.A. may consult directly the C.A. which received the original notification, or the Commission, on specific details of the data contained in the dossier and it may, also, suggest that further tests or information be requested.

Follow-Up Information

Any notifier of a substance already notified is responsible on his own initiative for informing in writing the C.A. to which the initial notification was submitted of changes in the annual or total quantities placed on the Community market by him, or, in case of a substance manufactured outside the Community for which the notifier has been designated as S.R., by him and/or others. Any importer of a substance produced by a manufacturer established outside the Community who imports the substance within the framework of a notification previously submitted by a S.R. shall be required to ensure that the S.R. is provided with up-to-date information concerning the quantities of the substance introduced on to the Community market.

C.A. can require complementary tests (name of the laboratory must be provided) to submit a new technical dossier if the total quantity exceed the limits fixed for these dossiers (see above). While waiting for C.A. resolution, notifiers can place on the market the quantities fixed by the precedent technical dossier.

New knowledge of the effects of the substance on man and/or environment of which he may reasonably become aware, new uses, any change in the composition of the substance, any change in the status of the notifier should be notify to the C.A.

TRANSPORT OF DANGEROUS SUBSTANCES

This is covered in the specially commissioned report by Peacocke in Chapter 13.

INDUSTRIAL RISK, PROTECTIVE MEASURES IN THE CASE OF MAJOR-ACCIDENT HAZARDS

Directive 96/82/EC of December the 9 1996 (18) or Directive "Seveso II" on the control of major-accident hazards involving dangerous substances fixes the measures of dangerous establishments to be considered by workers and establishes the control measures taken by the C.A. It replaces from 3 February, 1999, the Directive 82/501/EEC of 24 June 1982 or Directive "Seveso I"(19).

The Directive shall apply to establishments, that is to say "the whole area under the control of an operator where dangerous substances are present in one or more installations, including common or related infrastructures or activities" where dangerous substances are present in quantities equal to or in excess as listed in Annex I, Parts 1 and 2 column 2 and 3 of the sub-mentioned directive. The "presence of dangerous substances" shall mean the actual or anticipated presence of such substances in the establishment, or the presence of those that it is believed may be generated during loss of control of an industrial chemical process.

Member states require the operator to present some information depending on the maximum quantities which are present or are likely to be present at any time (Annex I of Directive 96/82/EC) (18) and on the type of danger. Dangerous substances present at an establishment only in quantities equal to or less than 2% of the relevant qualifying quantity shall be ignored for the purposes of calculating the total quantity present if their location within an establishment is such that it cannot act as an initiator of a major accident elsewhere on the site.

- If they are present in quantities equal to or in excess of the quantities listed in Annex I, Parts 1 and 2 column 2, the operator is required to present to the C.A. only a Notification with the following information: "complete address of the operator and the establishment concerned, information sufficient to identify the dangerous substances or category of substances involved, quantity and physical form of the substance, activities of the installation, immediate environment of the establishment. The Notification concerns the less pollutant and less dangerous activities and it can be established during the utilisation/production of the substance.

- In case they are present in quantities equal to or in excess of the quantities listed in Annex I, Parts 1 and 2 column 3, the operator is required to ask the C.A. for an Authorisation before starting the envisaged activity and which concerns the most pollutant and dangerous activities. The operator is required to produce a Safety Report, available to the general public as fixed in the Annex II of Directive 96/82/EC (18) with:
  - Safety Management System: a major-accident prevention policy (Annex III of Directive 96/82/EC) (18) that includes: organisation and personnel, identification and evaluation of the major-accident hazards, operational control, management of changes, planning for emergencies, monitoring performance and periodic audit and review of the major-accident prevention policy.
  - Identification and accidental risks analysis: An updated inventory of the dangerous substances present in the establishment: chemical name, CAS number, name according to IUPAC nomenclature, the maximum quantity of dangerous substances present, physico-chemical and toxicological characteristics and indication of the hazards.
  - Measures of protection (Annex V of Directive 96/82/EC)(18).
  - Presentation of the environment of the establishment: geographical location, meteorological, geological and hydrographic condition. Identification and description of installations where a major accident may occur.
  - Description of installations: activities, products, and processes.
  - Demonstration that adequate safety and reliability have been incorporated into the design, construction, operation and maintenance of any installation.
  - Internal Emergency Plans have been drawn up (Annex IV of Directive 96/82/EC)(18) including workers' opinion. It is updated every three years.
  - Supplying information to enable the External Plan to be drawn up in order to take the necessary measures in the event of a major accident; providing sufficient information to the C.A. to enable decisions to be made in terms of the siting of new activities or developments around existing establishments. It is updated every three years.

For new establishments, the information should be submitted prior to starting operation.

CONCLUSION AND RECOMMENDATIONS

As Diebold mentioned, wood smoke, wood distillation and fast pyrolysis bio-oil are not the same. The growing interest of potential users for bio-oil as a renewable energy source will increase its market. Therefore production, transport and use of fast pyrolysis bio-oil might become commercial soon. In that case, EU notification will be required for producers.

Toxicity of the fast pyrolysis oil remain unclear and a full MSDS cannot be set up now based on the information available. Due to the presence of certain chemicals, such as formaldehyde, may cause health problem in the long term. Theses uncertainties clearly demonstrate the need for further investigation to determinate real bio-oil toxicity on humans and the environment. If we consider the potential variability of bio-oil composition, related to the process parameter and feedstock, the need for further investigation is even greater.

Taking into account the CA requirement and, the personal protection procedure and management, on site monitoring would be recommended. Indeed, carrying out air sampling during the production on the workspace would contribute to better define workers protection if required in the operating environment.

It would be therefore highly recommend that additional work should be done to fit with EU requirement and to anticipate the market development of fast pyrolysis oil where more people will become involved with this promising renewable fuel.

The relation between process and feedstock on one hand and chemical composition and toxicity for human health and environment on the other hand need to be investigated so as to find the operating conditions to produce bio-oils with the lowest impacts. Then, the optimised and defined compositions of bio-oils have to be submitted to the mandatory tests required by the commission, the objective being the definition of secure handling and storage procedures, in order to control the risks related to the product for the population and the environment. The effects of different ways of exposure (inhalation, ingestion or skin contact) have to be quantified, as well as the effects of long term exposure. The impacts on the environment should also be evaluated by biodegradability.

This should be done in order to:

- Optimise the process for the production of "friendly" fuels towards human health and environment.
- Identify the feedstock which can be used in pyrolysis to obtain low impact bio-oils, and of those which eventually should not be used because of the formation of toxic components,
- Better know the relation between operating conditions and oil composition, which can be used for the optimisation of other end-use properties and further, improve energy efficiency of these fuels.
- Account for health and environment care at an early stage of development of the fuels,
- Allow an official notification to the authorities: the product needs to be recorded in the ELINCS (European List of Notified Chemical Substances) data base, so that all relevant information on health and environment care would be available for any one interested in the production or use of bio-oils such as producers, users, social and medical services, transports, trade associations etc.
- Allow an easier production and circulation of higher quantities of bio-oils as new applications will become available. This will allow larger scale applications of both production processes and end-use processing of bio-oils and thus a better knowledge of bio-oils characteristics and behaviour.

ACKNOWLEDGEMENTS

The support of the European Commission through a "Leonardo grant" to the University of Valladolid (Spain) is gratefully acknowledged.

REFERENCES

1. Elliott, D.C. "Relation of Reaction Time and Temperature to Chemical Composition of Pyrolysis Oils", Pyrolysis Oils from Biomass, E.J. Soltes and T.A. Milne, Eds, ACS Symposium series 376, pp 55-65, 1988
2. Elliot D.C., Hart T.R. "Environmental impacts of thermochemical biomass conversion - Final report" NREL/TP - 433 - 7867 - June 1995
3. J. Diebold, "A review of the toxicity of biomass pyrolysis liquids formed at low temperatures", in Fast Pyrolysis Of Biomass - A Handbook, Eds: A Bridgwater, S Czernik, J Diebold, D Meier, A Oasmaa, C Peacocke, J Piskorz, D Radlein, CPL Press ISBN 1-872-691-072, 1999

4. Evans R.J. and Milne T.A. "Molecular characterisation of the pyrolysis of biomass - Fundamentals" - Energy and Fuels , 1, n° 2 - pp 123 - 138.

5. Elliot D.C. "Analysis and comparison of biomass pyrolysis/gasification condensates - Final report" PNL - 5943/UC - GID - June 1986

6. Milne T.A.; Agblevor F, Davis M., Deutch S. and Johnson D. "A review of the chemical composition of fast pyrolysis oil" in Development in thermochemical conversion of biomass, A.V. Bridgwater and D.G. Boocock eds., Banff, Canada, pp 409-424, 1996

7. Meier D., Oasmaa A., G.V.C. Peacocke - "Properties of fast pyrolysis liquids : states of test methods - Fast pyrolysis of biomass : a handbook - CPL press - pp 75-91 - 1999

8. Directive 92/32/EEC of 30 April 1992 amending for the seven time Directive 67/548/EEC (OJ, No L 154, 5.6.1992, p. 1) which lays down the principles of risk assessment for new substances, introduces "Sole Representative" in the notification system and adds MSDS.

9. Directive 67/548/EEC of 27 June 1967 (OJ, No L 196, 16.8.1967, p.1) on the "Approximation of laws, regulations and administrative provisions relating to the classification, packaging and labelling of dangerous substances". It should ensure the establishment of a common market in the field of dangerous chemical substances and a high level of protection of human health. The directive is permanently amended (9 times until July 1999) and updated (25 times until July 1999) to take account of the scientific and technical progress in the field of dangerous substances.

10. EINECS Inventory: (Official Journal of the European Communities, OJ, No C 146, 15.6.1990, p.1).

11. Directive 88/379/EEC of 7 June 1988 (OJ, No L 187, 16.7.1988, p.14) which fix the general concentration limits of a dangerous substance present into a preparation if no specific concentration limits are laid down in Annex I of Directive 67/548/EEC. Annex II : dangerous preparations list.

12. ELINCS Inventory : 5th amendment (the last one) of the Commission, OJ, No C 72, 11.3.2000. Code (2000/C 72/01).

13. Directive 87/18/EEC which provide the GPL principles (Good Laboratory Practice) (OJ, No L 15, 17.1.1987, p.29). The list of GPL Laboratories is periodically updated by the EC.

14. Directive 87/302/EEC (OJ, Volume 31 No L 133, 30.5.1988) and Directive 92/69/EEC (OJ, Volume 35 No L 383A, 29.12.1992) about the Annex V Methods.

15. Directive 93/67/EEC of 20 July 1993 (OJ, No L 227, 8.9.1993, pp.0009-0018) which establishes the principles of risk assessment for men and the environment of substances notified in accordance to Directive 67/548/EEC.

16. Peacocke G.V.C, Bridgwater A.V. "Transport, Handling and Storage of Biomass Derived Fast Pyrolysis Liquid". Internal text PyNe, Pyrolysis Network for Europe - 2000

17. Czernik S. - Environment, health and safety - "Properties of fast pyrolysis liquids : states of test methods - Fast pyrolysis of biomass : a handbook - CPL press - pp 75-91 -1999

18. Directive 96/82/EC of 9 December 1996 (OJ, L010, 14.01.1997 p.0013 - 0033) on the control of major-accident hazards involving dangerous substances

19. Directive 82/501/EEC of 24 June 1982 on the major-accident hazards of certain industrial activities and with the limitation of their consequences for man and the environment.

1. COMPETENT AUTHORITIES

FRANCE

Notification

I. INRS
Institut National de Recherche et
de Sécurité (Research and Security
National Institute)- e
Service"Contrôle des produits"
30, rue Olivier-Noyer
75680 Paris Cedex 14, France.
Tel. +33 (0)1.40.44.30.00 (standard)
Fax. +33 (0)1.40.44.30.54 (standard)
http://www.inrs.fr

II. Ministère de l'Aménagement du
Territoire et de l'Environnement
(Ministry of Environment)
"Bureau des substances et préparations
chimiques"
20 avenue de Ségur
75302 Paris 07 SP, France
Tel. +33 (0)1.42.19.15.21 (standard)
http://www.environnement.gouv.fr

Transport

Ministère de l'Equipement, des Transports et du Logement
(Ministry of Equipment, Transport and Housing)- Arche Sud
92055 La Défense Cedex 04, France.
Tel. +33 (0)1.40.81.21.22 (standard)
http://www.equipement.gouv.fr

SPAIN

Notification

Ministerio de Sanidad y Consumo
(Ministry of Health and Consumption)
Dirección General de Salud Pública
Subdirección General de Sanidad Ambiental
Departamento de Productos Químicos (Planta 7°)
Paseo del Prado 18,20 - 28014 Madrid, Spain.
Tel. +34.91.596.10.8990/91 (standard)
Fax. +34.91.596.43.15 (standard)
http://www.msc.es

Transport

Ministerio de Fomento (Transportes, Turismo y Comunicaciones)
(Ministry of Transport and Equipment)
Plz. San Juan de la Cruz, 3
28003 Madrid, Spain
Tel. +34.91.533.24.00 (standard)
http://www.mfom.es

UNITED KINGDOM

Health and Safety Executive
Data appraisal unit
Magdalen house, Stanley Precinct
Bootle, Merseyside
L20 3QZ, UK
Tel. +44.054.15.45.500 (standard)
http://www.hse.gov.uk

NETHERLANDS

Rijksinstituut voor Volksgezondheid en milieu(RIVM)
(National Institute for Public Health and the Environment)
Centre for Substances and Risk Assessment (CSR)
Anthonie van Leeuwenhoeklaan 9, PO Box 1
NL – 3720 BA Bilthoven, The Netherlands
Fax +31.30.274.44.01

2. INFORMATION ON NEW SUBSTANCES NOTIFICATION AND CLASSIFICATION, PACKAGING AND LABELLING OF DANGEROUS SUBSTANCES

EUR-LEX: About Européen legislation: http://europa.eu.int/eur-lex/fr/index.html
European Commission: http://europa.eu.int/comm/index.html
Manual of decisions: summary of the meetings between C.A. and the European Commission about the notification of new chemical substances:
http://europa.eu.int/comm/environment/dansub/

ECB (European Chemical Bureau)
JRC Environment Institute
I-21020 Ispra (Varese), Italy
Tel. +39.03.32.78.58.66
Fax +39.03.32.78.58.62
http://ecb.ei.jrc.it

CEFIC(European Council of the Federation
of Chemical Companies)
Avenue E. Van Nieuwenhuyse 4, Box 1
B-1160 Brussels, Belgium
Tel. +32.2.676.72.11
http://www.cefic.be

3. LISTS OF CHEMICAL SUBSTANCES

EINECS and **ELINCS**: CD-Rom available at:

Société Silver Platter Informations
75, av. Parmentier
75544, Paris Cedex 11, France.
Tel. +33 (0)1.40.21.24.59

Silver Platter Information Ltd.
UK Office: 10 Barley Mow Passage,
W4 4PH Chiswick, London, UK
Tel: +44 (0) 181 995 8242

CAS (Chemical Abstracts Service)
2540 Olentangy River Road
P.O. Box 3012
Columbus, Ohio 43210, USA.
Tel. +1.614.447.3600,
Fax +1.614.447.3713
http://www.cas.org

IMPLEMENTATION SUBJECT GROUP REPORT

Maximilian Lauer
Joanneum Research, Institute of Energy Research, Joanneum Research
Elisabethstrasse 5, Graz, A-8010 Austria

The objectives of the Implementation Subject Group were to review the opportunities and problems influencing the successful commercialisation of fast pyrolysis and the application of pyrolysis liquids (bio-oil) and to make recommendations to improve the rate and the success of commercialisation.

In Section 1 a review of opportunities and problems is given, followed by a report on the activities of the Subject Group in Section 2. The results of the Subject Group work are given in the following Sections: workshops on implementation issues (Section 3), competitiveness assessment (Section 4) and recommendations (Section 5).

1. REVIEW OF OPPORTUNITIES AND PROBLEMS

For reviewing the opportunities and problems influencing the successful commercialisation of pyrolysis technology the criteria for the assessment (Section 1.1) and the fields of application to be considered (Section 1.2) are discussed in this section. Different approaches for the assessment are discussed in Section 1.3.

1.1 Criteria for the Assessment of Opportunities and Problems

The criteria for the assessment can be divided into two different groups, the techno-economic criteria and the structural criteria.

1.1.1 Techno-economic criteria

This group of criteria describes the general applicability of a technology from a technical and an economical point of view. For possible investors these criteria have to be met or the disadvantages have to be fully compensated by incentives, otherwise the implementation of the technology will be impossible. In general two techno-economic criteria are to be considered:

- readiness for use
- competitiveness

Readiness for use

A technology has to be ready for use before it can be implemented. This means, that there is no development work left to the investor except perhaps some adaptations to the specific application. The technology has to be proven, reliable, and more or less free from technological risks. A "turn key operation" should be possible.

If the technology is not really ready for use (e.g. in the pilot phase or the demonstration phase) the realisation of an application will only be possible, if there exist incentives that equalise the disadvantages of the technology for the investor. These incentives could be as an example a positive outlook into the future and/or subsidies given by the authorities interested in implementation.

Competitiveness

Competitiveness seems to be a very simple criterium. It describes the attraction of a technology compared to another technology giving the same service. Related to biomass pyrolysis for the production of energy or chemicals two fields for the assessment of competitiveness are to be considered:

Competition with petrochemical products
The use of biomass pyrolysis products competes with petrochemical products. Petrochemical products have the advantage to be very well known, to be integrated on the markets and to be available almost everywhere at very reasonable prices (at least at the moment; the increasing taxation on petrochemical products as in some European countries could cause changes in competitiveness).

Competition with other biomass conversion technologies
The use of biomass as an energy source for thermochemical conversion is well known. Combustion and gasification are the main technologies in this field. Fuel input power range goes from 5 kW for simple stoves up to several hundred MW in power stations. Most of these technologies are highly developed and state of the art (heating boilers, steam boilers with turbines or piston engines etc.).

Also biochemical conversion of biomass for the production of alcohols or other conversion technologies (e.g. "Biodiesel" methylesters made from vegetable oils) are known very well. Flash pyrolysis technology competes with all these biomass conversion technologies depending on the specific use of the pyrolysis liquid.

For the competition with other thermochemical processes using biomass a very important advantage is existing for the implementation of pyrolysis technology. The conversion process is split into two parts, the flash pyrolysis process itself (production of pyrolysis liquid) and the use of the pyrolysis liquid as fuel or chemical feedstock. These two parts can be of different capacity (large pyrolysis unit, small scale utilisation units), can be realised in different places (transport of pyrolysis liquid) with a time shift (storage). Splitting the conversion process of the use of biomass gives the opportunity to locate the pyrolysis facility near the biomass source (e.g. remote areas) and with the optimal capacity (e.g. like big scale paper mills) and to produce independent from the time, when the user will need the service. On the other hand the use of bio-oil is independent from the production in means of capacity limits, place or kinds of use. Another advantage compared to other thermochemical processes is the possibility of the utilisation of pyrolysis liquid as chemical feedstock.

1.1.2 Structural criteria

These criteria describe issues relevant for the implementation of a technology in a market driven and a social environment, as the existence of laws and standards, the possibility to evaluate hazards and risks, the basic motivation of decision makers (image cultivation etc.). These "structural" criteria are "soft" criteria. If some of them cannot be matched, the motivation for investment will decrease, but this will probably not stop implementation of the technology totally. The importance of these "soft" criteria is often underestimated.

Structural criteria are for example:
- laws, standards and political measures
- hazards and risks
- motivation of decision makers

Laws and standards

Laws and legal regulations can influence the implementation of pyrolysis technology in different ways. The activities of companies can be restricted or encouraged in some fields (development plans, economic incentives, political measures etc.). Regulations on environmental issues influence the competitiveness and the possibilities of implementation etc. Laws and legal regulations and other political measures are very different from country to country and can even be different between provinces of the same country. Therefor an overall assessment of the opportunities and problems related to laws and legal regulations is not possible.

Standards are very important for the implementation of new technologies because they contribute reliability in business operations. This is most important for biomass pyrolysis, where production of pyrolysis liquids and the utilisation is likely to be done by different companies, and at different locations. Standards for product quality and for transport and storage technologies will be the basis for a successful implementation of pyrolysis technologies. In PyNe the Subject Group Analysis and Characterisation is preparing specifications as a basis for future standardisation.

Hazards and risks

Knowledge and assessment of hazards and risks is essential for possible investors for being able to identify and manage technical and financial risks. The evaluation of possible hazards and risks may give very different results depending on the particular situation. In PyNe the Subject Group Environment, Health and Safety is discussing these issues and preparing recommendations and specifications for handling and transporting bio-oil.

Motivation of decision makers

Investors have to consider different points for their decision. Even the most economic opportunity can turn out as unfavourable, if the technology has requirements difficult to be met by the companies structure (e.g. restricted management resources). A possible investor has also to consider the company image that e.g. may or may not include also the readiness for taking technological risks.

1.2 Fields of Application to be Considered

Applications of pyrolysis technology can be divided into production (section 1.2.1), shipping and storage (Section 1.2.2) and end use of pyrolysis liquids (Section 1.2.3).

1.2.1 Production of pyrolysis liquids

Production of pyrolysis liquids includes take over and conditioning (drying, grinding) of biomass, the pyrolysis process itself and management (and improvement respectively) of product quality. All this items have already been described very comprehensively (e.g 1), so in this issues can be kept short.

Take over and conditioning of biomass feedstock

This step usually contains storage, drying and grinding of the biomass delivered. Quite similar technologies are used for the preparation of biomass for pellet production, which in the last years has become quite common for the production of high quality biomass fuel in Scandinavia and other regions. Also the capacity of the plants used for pellet production is quite the same as future pyrolysis plants will possibly have (2).

Pyrolysis process

For the pyrolysis process and the reactor design there exist different technologies, that are described by A.V. Bridgwater and G.V.C. Peacocke (3).

Product quality management/improvement

A product quality management is needed to ensure a constant product quality meeting the quality demands given by standards and by users. The need for a quality management/improvement grows with the capacity of the production plant, the number of product users and the demands of the users (depending on the application the product is used for).

The quality of the crude pyrolysis liquids after the pyrolysis process show some characteristics, that will prevent application in some applications. The most important characteristic seems to be instability, char suspended, high water content, etc. (e.g. 1,4). For most of the possibilities of application, especially for the (economically most interesting) small scale CHP applications, there is a need for stabilisation and for upgrading the crude pyrolysis liquid.

1.2.2 Delivery

Concerning the delivery of biomass to the users there is to be considered the delivery to bulk consumers and the delivery to decentralised consumers.

Delivery to bulk consumers

In this case the producer and the user of the pyrolysis product can easily agree on appropriate means of transportation, installations for pumping and storage etc. The use of special materials for installations will be no major barrier.

Delivery to decentralised consumers

Transport will be usually made by trucks, Storage tanks are usually located in buildings (no high peak temperatures). But due to the higher number of consumers, there is a need for simple and reliable installations for tank evacuation and dispensing.

1.2.3 Use of bio-oil

There are different uses intended with bio-oil:

- Energetic use
  - Boiler fuel
    - Large and medium size boilers
    - Small scale boilers
  - CHP-Production
    - Power plants (steam, gas turbine, combined cycle process)
    - Internal combustion engines
      - Large scale, low speed diesel engines
      - Small scale, high speed diesel engines
- Non energetic use
  - Fertiliser
  - Sugars
  - Phenolic glues
  - Other use as feedstock for chemical processes

The PyNe network concentrates on the energetic use of pyrolysis liquids. The international activities on the non-energetic use are observed but without specific effort in terms of research activities.

1.3 Approaches for the Assessment

For the assessment of opportunities and problems different points of view can be taken. Two examples have been considered in the work of the PyNe Subject Group Implementation:

- An assessment of the general competitiveness in order to distinguish between less or more promising future applications of pyrolysis technology compared to other possibilities the investor can find. This assessment can give information on the directions future research and development efforts should focus on, but can't give information on the chances of a specific project.

- An assessment of specific projects, when demonstration scale plants are followed by the first commercial scale plants. Issue of this assessment is the identification of technological and financial risks and of possibilities for minimising these risks for the partners involved in project development, financing and operation.

The assessment of the general competitiveness is tried by a study done by the Subject Group Implementation described in section 4 of this report. The assessment of specific projects is discussed in a study "Evaluation of Bio-Energy Projects" by Patricia Thornley , PB Power, UK. This study is integrated as Chapter 13 in this handbook.

2. ACTIVITIES OF THE SUBJECT GROUP IMPLEMENTATION

The Subject Group Implementation started its work with two workshops at the PyNe Meeting in De Lutte, Netherlands, 28th Nov. – 1st December 1998. This meeting was dedicated to the work of Subject Group Implementation and the Subject Group Environment health and Safety. The Subject Group Implementation Meeting took 1,5 days and offered 8 presentations by speakers involved actively in the commercialisation of Pyrolysis technology and two workshops focussed on the topics:

- What are the technical and non-technical barriers on implementation?
- What should be done to improve the rate and the extend of implementation?

The results of these workshops are summarised in Section 3. The discussions at these workshops resulted in a draft working plan for the further work of the Subject group:

(1) General overview on
  - technical and non technical barriers
  - what should be done to overcome these barriers?
  These overview is the result of the workshops in De Lutte (see Section 3)
(2) Evaluation of Bio Energy Projects, Study done by PB Power on behalf of PyNe (see Chapter 7 of this handbook).
(3) Assessment of the competitiveness of specific bio-oil applications (see section 4)
(4) Preparation of recommendations (see section 5)

At the PyNe Meeting in Semmering, Austria, 28th –31st January 2000 the study "Evaluation of Bio. Energy Projects" by Patricia Thornley was presented and discussed. It is integrated in this handbook as Chapter 13. At the same meeting the working plan for the Subject Group and the approach to the competitiveness assessment study were discussed and agreed. A questionnaire for the data collection was discussed and circulated to the PyNe members in August 2000.

The results of the competitiveness assessment study were presented and discussed at the PyNe Meeting at Birmingham, UK, $2^{nd} - 4^{th}$ December 2000 (see Section 4).

Based on the work of the PyNe Subject Group Implementation a proposal was submitted to the EC – ALTENER programme to continue and develop the competitiveness assessment study. The proposal was approved by the European Commission, the work will begin during spring 2001. This project includes sub-contracts to all PyNe members for providing data.

3. WORKSHOPS OF THE SUBJECT GROUP IMPLEMENTATION

At the PyNe Meeting in De Lutte, Netherlands, 28th Nov. – 1st December 1998 two workshops took place focussing the barriers on implementation of pyrolysis technology and how to overcome these barriers.

3.1 Workshop 1: What Are the Technical and Non-Technical Barriers on Implementation?

For the assessment of possibilities of a new technology and chances for the implementation as a first step it is necessary to identify possible barriers. Result of the workshop is a prioritised list of barriers identified by the PyNe members (mostly seen from the RTD point of view). The discussion was divided into three parts, the production of bio-oil, the energetic use and the non-energetic use of bio-oil.

3.1.1 Barriers to implementation of bio-oil production

Barriers identified as very important:

- The investment cost for plants in an industrial scale are too high.
- There is not sufficient experience in the reliability and availability of pyrolysis plants.
- No proven methods for industrial scale application of upgrading and stabilisation of the bio-oil are known.
- Product standards are missing.

Barriers identified as less important:

- The costs of biomass are very high
- The availability of biomass with defined quality is not given in practice.

An additional result of the discussion was, that the interaction of feedstock quality and product quality is not known sufficiently.

3.1.2 Barriers to energetic use of bio-oil

The discussion on barriers to energetic use was divided into three sections, Boiler operation, CHP with internal combustion engines and big scale CHP (power plants)

The overall results of the discussion in the three sections were, that there is nearly no experience with the utilisation of bio-oil. As the reasons the practically non existing availability of bio-oil and the high cost for extensive tests with bio-oil coming from the existing small scale pyrolysis units were identified.

3.1.3 Barriers to non-energetic use of bio-oil

The barrier identified as most important is, the poor availability of bio-oil in sufficient quantities for extensive testing (no market existing).

Barriers identified as less important are the lack of product development and the limited resources for RTD.

An additional result of the discussion was, that there exist substantial information lacks on possibilities for bio-oil applications. It was suggested, that a number promising applications are unidentified up to now.

3.2 Workshop 2: What Should Be Done to Improve the Rate and the Extend of Implementation

After having identified and prioritised the barriers on implementation of pyrolysis technology (section 3.1) the next step was to answer the question on what to do to overcome these barriers. This was tried in the second workshop. The summarised result of this discussion was:

- Information on cost structures of production and use of the bio-oil is needed in order to put the right focus on future research and development work.
- There is a need for a number of demonstration plants in order to get some experience concerning reliability and to produce sufficient bio-oil (at reasonable cost) to do further research on energetic and non-energetic use of bio-oil.
- Standards for bio-oil as well as handling and shipping procedures should be discussed and agreed.
- Further RTD is particularly necessary for:

  - upgrading and stabilisation of bio-oil
  - the interaction between feedstock quality and product quality.

It was agreed, that the investigation of cost structures should be subject to the ongoing activities of the Subject Group Implementation.

4. COMPETITIVENESS ASSESSMENT

The competitiveness assessment done by PyNe tries to give information on the cost structures and on the future situation of different applications of pyrolysis technologies taking into account the specific economic boundary conditions given in different countries by energy prices and energy taxation. The method of assessment (Section 4.1), the data used (Section 4.2) and the results (Section 4.3) are discussed.

4.1 Method of Assessment

Competitiveness as one of the techno-economic criteria is the most important issue for assessing the opportunities and chances for the implementation of a technology. In this study competitiveness is seen as relative economic attractiveness of a specific application compared to other opportunities an investor has. This attractiveness normally is an economic issue. Among different opportunities the one with the least overall cost is in general the most attractive to the investor.

There are different possibilities used to assess the competitiveness of technologies and their applications without going into details of financing and company specific conditions. The comparison of technology data and the comparison of specific product cost are discussed below and a third possibility, the comparison of energy supply cost is introduced as a third possibility.

Comparison of technology data

Often technologies are compared to each other using their characteristic data (efficiencies, specific cost etc.). Comparison of technologies is quite useful, if they are similar and used for the same purpose and in a similar size. In this case a comparison of technology related data can give good information.

But it seems to be impossible to compare the characteristics of different technologies and different applications in order to assess their competitive situation, because competitiveness depends very much on the situation technologies are used for and the technical, economical and social environment they are integrated in. As the bio-oil applications are quite different in technology, size and purpose the comparison of technologies is not applicable for this competitiveness assessment.

Comparison of specific product cost

Comparison of the specific cost for heat and power production is a widely used method for the assessment of different applications. It is very useful for applications of the same size and for a similar purpose, if there is only one product or if the heat/power ratio is similar. If sizes, purposes or technologies are different, the comparison of product cost is not practical. As an example a big power station can produce electricity at relatively low specific cost, but will earn only low prices for feeding power to the high voltage grid. If a small company produces electricity with a e.g. fixed bed biomass gasifier for its own needs, specific power generating cost will be much higher as with the power station. But as the company has to pay retail prices for the electricity, the small gasifier in the example could be competitive from the company's point of view.

Comparison of energy supply cost

This approach for the assessment of competitiveness compares the overall cost for energy supply with a commonly used technology to the overall cost of an alternative possibility (here the application of biomass pyrolysis technology) from the investor's point of view. In most industrialised countries the investor can choose between the conventional energy supply (heat/power) and the application of an alternative (here the application of biomass pyrolysis technology).

The comparison of energy supply cost is used for this competitiveness assessment. It is assumed, that pyrolysis technology and its application is ready for industrial use, and that the technological risks for the investor are small and well known. So the results of this competitiveness assessment are not applicable to pilot and demonstration plants in the next few years, but give an information on the anticipated competitive situation of the "tenth" commercial plant realised.

The conventional energy supply normally is producing heat with a fossil oil/gas boiler and/or buying electricity from utilities. This is specified as the standard situation for conventional energy supply in this paper. The term application is used to describe a specific plant by technology, it's purpose, size, the fuel used, the annual operation time, the labour demand and the cost for the conventional alternatives for the investor.

Competitiveness in this paper is seen from the investors point of view as the relation of annual cost of two possible applications using different technologies but providing the same service. For

94

easy comparison between a variety of applications a "competitiveness factor" CF is introduced using the relation:

$$CF = \frac{\text{annual cost of conventional energy supply}}{\text{annual cost of biomass pyrolysis application}}$$

As CF is a non-dimensional factor describing a cost relation, the economic competitiveness of different applications of different biomass conversion technologies can be compared to each other.

The annual cost are calculated as an overall sum of cost related to investment, to operation and to consumption following the guideline VDI 2067 edited by the association of German Engineers. Cost related to investment are calculated as annuity (constant annual settlement of the investment including the interest rates over a time period corresponding to the technical lifetime of the plant). Cost related to operation include personnel, service and maintenance cost. Cost related to investment and to operation are fixed cost. Cost related to consumption are variable cost and include cost for fuel, for transportation and all other cost related to the production intensity.

The annual cost of biomass application is generated using :
  (1) Technology related data (power range, efficiencies, specific investment, specific operation and maintenance cost, lifetime etc.).
  (2) Application related data (annual operation time, power output, labour cost, interest rate etc.).
  (3) Fuel related data (biomass availability and price, specific bio-oil production cost)

The annual cost of the conventional energy supply for every application is calculated by using specific cost data for fossil fuels and for electric power purchased from utilities.

In the discussion with the PyNe members a list of five applications for bio-oil to be investigated was generated:

- 2 MWth bio-oil boiler for industrial use (process heat)
- 8 MWth bio-oil boiler for industrial use (process heat)
- 20 MWth bio-oil boiler for industrial use (process heat)
- 20 MWe CHP power plant with gas turbine
- 5 MWe CHP Diesel (low speed) engine, industrial use

4.2 Data Used for the Assessment

4.2.1 Technology related data

The data for pyrolysis technology are assumptions based on data for other biomass technologies (5) and for fossil fuel based technologies. A selection of the assumptions for the technology related data are shown in Table 1.

Table 1   Selection of technology related data

Technology	Size range (bio-oil input)		electrical efficiency		thermal efficiency		life-time	spec. Investment (related to bio-oil input)	
	Pmin [kW]	Pmax [kW]	eta el min	eta el max	eta th min	eta th max	[a]	min [EUR/kW]	max [EUR/kW]
Bio-oil process heat boiler 1to 10 MWth	1000	10000	0.00	0.00	0.85	0.92	20	120	220
Bio-oil process heat boiler 10 to 30 MWht	10000	30000	0.00	0.00	0.85	0.92	20	90	120
Bio-oil Combined Cycle CHP Plant	0	110000	0.45	0.45	0.40	0.40	20	530	530
Bio-oil Gas Turbine 20 -50 Mwel	60000	150000	0.32	0.34	0.40	0.45	20	500	650
Bio-oil Diesel Engine (LS) 2-5 MWe	5000	15000	0.37	0.39	0.45	0.45	15	550	650

4.2.2 Application related data

The application related data describe the specifications of the application as size (input power), annual operation time, labour cost etc. These data are assumptions discussed with the PyNe members in order to use applications that are possible and common in most of the countries of the PyNe members.

In Table 2 a selection of the most important application related data are shown.

Table 2   Selection of application related data

Application	bio-oil input [kW]	operation time [hrs/yr]	personal cost [EUR/hr]	labour demand [hrs/yr]
Bio-Oil Boiler 2MW	2319	4000	14.5	200
Bio-oil boiler 8 MW	9427	5000	19.0	100
Bio-oil boiler 20 MW	24797	5000	19.0	100
Bio-oil gas turbine CHP Plant 20 MWe	62500	5000	23.3	2400
Bio-oil Diesel (ls) CHP plant 5 MWe	13265	5000	19.0	700

4.2.3 Fuel related data

Fuel related data were collected by a questionnaire sent to all PyNe members. The data returned by the PyNe members reflect the actual situation of sources, condition, quantities available and prices for biomass that could be used for a Pyrolysis plant in the different countries. For some of the countries no data were available or some of the data are lacking. As in the next two years a refined assessment will be the subject to an ALTENER project with all the PyNe members involved, no excessive effort was laid in completing the data for every country. The data returned are shown in Table 3. A special case is the information given for Canada. This was given for a plant a Canadian Company intends to construct for a sugar industry in Guatemala using the sugar cane bagasse as fuel. As this could be an interesting option for biomass pyrolysis, the calculation was also done for this special case.

All the prices given refer to the situation given in August 2000.

Table 3 Biomass sources and cost in different countries.
(The numbers after the Country code refer to different biomass sources available in the country)

Country	Source	Water cont. % w,b,	Quantity dt/a	Price EUR /dt
UK	forestry	50	182,500	50.0
AUT	industry	50	200,000	26.0
BEL 1	chips forestry	45	110,000	74.5
BEL 2	bark industry	50	50,000	8.0
BEL 3	chips industry	45	16,500	36.4
BEL 4	offcuts ind.	45	55,000	38.4
BEL 5	sawdust ind	45	110,000	21.8
CAN (Guat)	industry (bagasse)	50	320,000	0.0
DK 1	agriculture	15	850,000	70.6
DK 2	forestry	40	84,000	70.0
DK 3	industry	15	12,750	76.5
FIN	forestry	50	200,000	18.0
GER 1	forestry	50	15,000	80.0
GER 2	industry	15	25,500	-47.0
GER 3	agriculture	20	24,000	37.5
GRE 1	agriculture	20	22,400	25.8
GRE 2	industry	20	22,400	18.4
GRE 3	agroindustry	20	67,200	14.8
IRL 1	forestry	35	60,450	23.1
IRL 2	industry	60	39,600	37.5
IRL 3	industry	60	92,400	100.0
NOR 1	forestry	50	40,000	74.0
NOR 2	industry	10	9,000	56.0
ESP 1	olive stones	33	20,000	27.3
ESP 2	bark	45	55,000	27.3

For the application of biomass pyrolysis technology the cost of the bio-oil are relevant. For this first step of an competitiveness assessment a formula was used to calculate the bio-oil cost using the fuel related data given by the PyNe members. This formula was taken from A.V. Bridgwater (6). It gives a relation of bio-oil cost versus production capacity and feed cost.

$$c_{bo} = 8.87 * (cap)^{-0,3407} + fc * (0.625 * LHV_{bm})^{-1}$$

c_{bo} = specific cost for bio-oil [EUR/GJ $_{bio-oil}$]

cap = capacity of the bio-oil production plant [dry t $_{biomass}$ /hr]

fc = Biomass feed cost [EUR/dry ton)

LHV_{bm} = Lower heating value of Biomass [GJ/ t)

Using this relation for the bio-oil cost and the information on biomass availability and prices in different countries the cost for bio-oil were calculated. It was assumed, that the biggest plant capacity will be 200,000 dry tons per year (respectively 40 dt/hr assuming a operation time of 5000 hrs/yr). Table 4 shows the results of this calculation.

Table 4 Calculation of bio-oil cost for the different countries.
The numbers after the Country code refer to different biomass sources available in the country. The items in bold are the results for all biomass sources in the country.

Country	Quantity dt/a	Plant capacity (5000 h/y) dt/h	Price EUR /dt	Fixed Cost EUR/MWh	Var. Cost EUR/MWh	bio-oil cost total EUR/MWH
UK	**182,500**	**36.5**	**50.0**	**10.7**	**15.2**	**25.8**
AUT	**200,000**	**40.0**	**26.0**	**10.4**	**7.9**	**18.3**
BEL 1	110,000	22.0	74.5	12.5	22.6	35.0
BEL 2	50,000	10.0	8.0	15.8	2.4	18.3
BEL 3	16,500	3.3	36.4	22.2	11.0	33.2
BEL 4	55,000	11.0	38.4	15.4	11.6	27.0
BEL 5	110,000	22.0	21.8	12.5	6.6	19.1
BEL Total	**341,500**	**40.0**	**17.0**	**10.4**	**5.2**	**15.5**
CAN (GUAT)	**320,000**	**40.0**	**0.0**	**10.4**	**0.0**	**10.4**
DK 1	850,000	40.0	70.6	10.4	21.4	31.8
DK 2	84,000	16.8	70.0	13.5	21.2	34.7
DK 3	12,750	2.6	76.5	24.0	23.2	47.2
DK Total	**946,750**	**40.0**	**72.0**	**10.4**	**21.8**	**32.2**
FIN	**200,000**	**40.0**	**18.0**	**10.4**	**5,5**	**15.8**
GER 1	15,000	3.0	80.0	22.8	24,2	47.1
GER 2	25,500	5.1	-47.0	19.4	-14,2	5.2
GER 3	24,000	4.8	37.5	19.8	11,4	31.2
GER total	**64,500**	**12.9**	**37.5**	**14.7**	**11,4**	**26.0**
GRE 1	22,400	4.5	25.8	20.2	7,8	28.0
GRE 2	22,400	4.5	18.4	20.2	5,6	25.8
GRE 3	67,200	13.4	14.8	14.5	4,5	19.0
GRE total	**112,000**	**22.4**	**14.8**	**12.4**	**4,5**	**16.9**
IRL 1	60,450	12.1	23.1	15.0	7,0	22.0
IRL 2	39,600	7.9	37.5	17.0	11,4	28.4
IRL 3	92,400	18.5	100.0	13.1	30,3	43.4
IRL total	**192,450**	**38.5**	**100.0**	**10.5**	**30,3**	**40.8**
NOR 1	40,000	8.0	74.0	17.0	22,4	39.4
NOR 2	9,000	1.8	56.0	26.7	17,0	43.6
NOR total	**49,000**	**9.8**	**56.0**	**15.9**	**17,0**	**32.9**
ESP 1	20,000	4.0	27.3	20.9	8,3	29.2
ESP 2	55,000	11.0	27.3	15.4	8,3	23.7
ESP total	**75,000**	**15.0**	**27.3**	**14.0**	**8,3**	**22.3**

The results show very different prices for bio-oil in the countries. The price differences are result of the biomass price in the countries and the size of the plant (depending on the biomass quantity available).

The lowest price was obtained for German industrial dry wood chips that - at least at the moment - are seen as a waste material with negative prices (There will be paid for taking away the waste material). As the biomass price is –47 EUR per dry ton in this case, the cost of the bio-oil made with this material will be as low as 5.2 EUR/MWh. A second special case is Canada with the information on a plant in Guatemala. As the biomass price is zero, bio-oil cost would be as low as 10.4 EUR/MWh. Also in Finland bio-oil would be relatively cheap with cost of 15,8 EUR/MWh. In most of the countries the cost for bio-oil will be between 20 and 30 EUR per MWh. This price in the same range as heavy fuel oil in Europe depending on quality and taxation. Relatively high prices for bio-oil are to be expected in Denmark (31.8 up to 47,.) and in Norway (32.9 to 43.6) and in Ireland (28.4 to 43.4).

4.2.4 Cost of the conventional energy supply

The cost of the energy supply was subject to the questionnaire sent to all PyNe. The cost for the heat production and the cost for buying electricity respectively selling "green electricity to the grid are calculated separately.

An overview on the answers for the heat production are shown in Table 5. The answers of some of the members were incomplete or unsure due to the specific national situation, so this data were excluded from the evaluation. For every country the fuel quality is taken, that commonly is used for heat production in the indicated power range. In the UK, Finland, Ireland and Spain natural gas is reported to be usually the fuel for industrial boilers of the indicated size. A special situation is reported from Greece: In the urban areas LFO has to be used whereas in the nonurban areas HFO may be used. For this calculation LFO was assumed as fuel for the smaller boilers, HFO for the bigger ones.

All the information is based on the situation in August 2000.

Table 5   Calculation of the cost for the conventional heat production

Country	heat cost (including 20 % fix cost)			
	2MWth EUR/MWh	8MWth EUR/MWh	20 MWth EUR/MWh	80 MWth EUR/MWh
UK	10.80	10.80	10.80	10.80
AUT	3490	34.90	27.90	27.90
BEL	45.03	45.03	45.03	45.03
FIN	15.60	15.60	15.60	15.60
GER	18.29	18.29	18.29	18.29
GRE	84.39	84.39	19.43	19.43
IRL	21.60	21.60	21.60	21.60
NOR	51.28	48.21	45.03	32.69
ESP	9.60	9.60	9.60	9.60

It can be seen from Table 5, that the cost for conventional heat production is very different in Europe at the moment. The reason for this are quite different taxation systems and other political measures influencing the market. Another big influence have the hectic price increases on the markets for HFO and LFO in 2000 whereas the prices for natural gas are following more slowly.

The cost for the investor for buying electricity from an utility or the prices he will get for selling "green" electricity are shown in Table 6. In countries, where at the moment no special prices for selling "green" electricity are reported, the space in Table 6 is kept clear. In these countries at the moment selling electricity to the grid normally is not profitable for industrial producers. I can be seen, that the cost for buying electricity in the countries investigated ranges from 37 EUR/MWh in Belgium to 60 EUR/MWh in Ireland; Guatemala with electricity cost of 69 EUR/MWh can be seen as a special case. In Germany very high prices are paid for "green" electricity. In Ireland the price for selling "green" electricity is quite low.

Table 6. Cost for buying electricity and prices for selling "green" electricity to the grid.

Country	buy electricity 5 MW [EUR/MWh]	sell green electricity 5 MW [EUR/MWh]	buy electricity 20 MW [EUR/MWh]	sell green electricity 20 MW [EUR/MWh]
UK	58.5		58.5	
AUT	47.3		47.3	
BEL	37.0	62.0	37.0	62.0
CAN (GUAT)	69.0		69.0	
FIN	50.0		50.0	
GER	46.5	87.0	46.5	87.0
GRE	50.3	56.9	43.2	56.9
IRL	60.0	44.0	57.0	44.0
NOR	48.8	38.3	48,8	38.3
ESP	54.0	73.5	54.0	73.5

For the calculation the higher value (buying electricity respectively selling "green" electricity) was taken

4.3 Results of the Assessment

Table 7 shows the results for the competitiveness assessment for all countries with complete information available.

For better interpretation of these results in Figure 1 as an example for a sensitivity analysis is shown the graph for the bio-oil diesel CHP in Austria. It can easily be seen, that the most important parameter for the competitiveness is the cost of the conventional energy supply followed by the fuel cost and the operation time per year. The specific investment cost and the other parameters investigated are of minor importance. This situation is not only seen for this particular application but is typical for the cost structure of all the applications investigated.

Table 7   Competitiveness factor CF for bio-oil applications in different countries.

Country	Boiler 2 MWth	Boiler 8 MWth	Boiler 20 MWth	Gas turbine 20 MWe	Diesel engine 5 MWe
UK	0.25	0.30	0.31	0,. 5	0.60
AUT	1.10	1.31	1.33	0.99	0.90
BEL	1.59	1.93	1.97	1.65	1.34
CAN (Guat)	0.82	1.04	1.07	1.57	1.17
FIN	0.54	0.66	0.67	0.96	0.75
GER	0.44	0.51	0.52	1.14	1.30
GRE	2.80	3.37	0.79	1.07	1.64
IRL	0.36	0.41	0.41	0.61	0.54
NOR	0.99	1.14	1.03	0.83	0.76
ESP	0.26	0.31	0.31	0.83	0.90

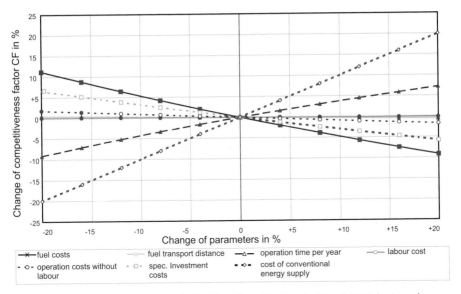

Figure 1 Sensitivity analysis for bio-oil application in Diesel CHP in Austria

The results for CF show big differences between the countries and the applications investigated. The reason is the different situation in:

- biomass availability and prices
- cost for fuels used for heat production
- cost for buying electricity or prices for selling "green" electricity respectively.

The different situations are depending on specific national energy politics (prices for "green" electricity, prices and taxation for fossil fuels, restrictions for the use of specific fuels) and by the different specific biomass prices and biomass availability.

But as important as the specific national situation the international market for energy is influencing the result. At the moment of data acquisition (August 2000) the prices for heating oil was increasing rapidly going on even more until the end of 2000, whereas the prices for gas are following only moderately and with a time delay. The prices for electricity in contrary have been fallen in most countries substantially in succession to the liberalisation of the electricity markets. In some countries the liberalisation is completed, in other countries the liberalisation is just beginning. The same situation can be stated for the prices for selling "green" electricity and taxation of fuels or other energy related items. So the results of the investigation presented here mirrors partly also the energy markets going through a time of upheaval.

As an overall result it can be stated, that pyrolysis technology after being developed to readiness for use could be quite competitive depending on the boundary conditions in the specific countries. A difference of competitiveness between boilers and CHP application cannot be stated as a general result. Depending on the development of national prices for electricity the competitive situation of bio-oil CHP applications can be good or bad compared to bio-oil application in boilers.

It is the general mid term expectation, that fossil fuel prices and the prices for electric energy will increase moderately compared to the situation in August 2000. If this expectation comes true, the competitive situation of pyrolysis technology compared with conventional energy supply will improve in most of the countries. In some countries with low biomass availability and high cost of biomass feedstock the competitive situation of pyrolysis technology will be problematic also in the future.

As shown in another investigation with the same approach (but different input data) the competitive situation of pyrolysis technology compared to other biomass to energy conversion technologies is encouraging as well (7). This will be investigated in the next two years in an ALTENER project.

5. RECOMMENDATIONS

Based on the work of the PyNe Subject Group Implementation the following recommendations for improving the opportunities and chances of pyrolysis technology are given:

The industries involved should be supported especially by improving the common pre-competitive basis needed for their business:
1. Standards for bio-oil quality and handling and shipping procedures should be discussed and agreed.
2. Ongoing and increased efforts should be made to find economically viable, standardised methods for quality improvement of bio-oil.
3. Pilot and demonstration plants should be realised and operated in order to get substantial long term experience and to proof the reliability of the pyrolysis technology.
4. Bio-oil should be made available more easily and in the quantities needed for developing bio-oil applications and in order to enable better experience with bio-oil applications.
5. The investigation of cost structures and the competitiveness evaluation should be refined.

REFERENCES

1 Bridgwater, A.V.: "Fast Pyrolysis of Biomass in Europe", in Biomass Gasification & Pyrolysis, State of the Art and Future Prospects, Kaltschmitt, M., Bridgwater, A.V., eds., p.53 ff., CPL Press, Newbury, UK, 1997
2 . Rensfelt, E.: "Pyrolysis Technologies"; presentation at the 5th PyNe Meeting 28th March to 3rd February 1998, Salzburg, Austria, Minutes by Aston University, 1998.
3 Bridgwater, A.V., Peacocke, G.V.C.: "Fast Pyrolysis Processes for Biomass", Renewable and Sustainable Energy Reviews 4 (2000), p 1-73, Pergamon press.
4 Oasmaa, A., et al.: "Physical Characterisation of Biomass- Based Pyrolysis Liquids", VTT Publications 306, Espoo, Finland, 1997.
5 Schaller, W. et al.: Sustainable Bioenergy Strategy for Austria ("Nachhaltige Bioenergiestrategie für Österreich"). Energy Research Association of the Assosciation of Austrian Utility Companies (Energieforschungsgemeinschaft des Verbands der Elektrizitätswerke Österreichs, VEÖ). Report on EFG project No 1.26, Vienna, (to be published in Spring 2001.
6 A.V. Bridgwater et al: in "Fast Pyrolysis of Biomass, a Handbook"; pp 12, 13; CPL Press, Newbury, UK, 1999.
7 Lauer, M., Pogoreutz, M.: "Competitiveness assessment of Applications of Thermochemical Biomass Conversion Technologies"; Paper presented to the Conference "Progress in Thermochemical Biomass Conversion", 17th – 22nd September 2000, Tyrol, Austria; (Proceedings to be published by Aston University, UK in Spring 2001).

FUNDAMENTALS, MECHANISMS AND SCIENCE OF PYROLYSIS

Jan Piskorz
RTI - Resource Transforms International Ltd, 110 Baffin Place, Unit 5, Waterloo,
Ontario, N2V 1Z7, CANADA

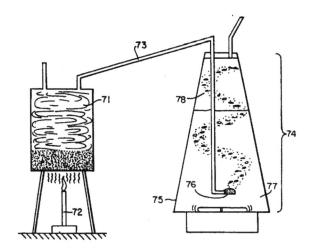

Figure 1   Biomass pyrolysis
(from U.S. Patent # 4,594,251, issued June 10, 1986, to Nicholson)

ABSTRACT

Many prominent speakers have attended and contributed to PyNe organized meetings and workshops. The participation of industrialists and decision/policy makers was also considerable. Interaction between members of PyNe and other experts in the field was maintained on an ad-hoc basis and directed toward determination of what knowledge would be useful to the pyrolysis community at large. Developments in pyrolysis processing are proceeding at a rapid pace. The future of the thermal conversion of biomass seems assured.

INTRODUCTION

Renewable carbon is produced at a huge annual rate in the biosphere. Progressive utilization of renewable feedstocks can lead toward a CO_2-neutral system of energy consumption. The concept of biomass *fast pyrolysis* promises to convert renewable carbon into useful chemicals, intermediates and liquid fuels.

Mankind's fascination with pyrolysis started with discovery of fire; we all know the magic of sitting in front of a good-fire. It is pyrolysis that generates the visible cloud of a persistent smoke-aerosol. Indeed the most ancient and well-known application of biomass pyrolysis is in the production of smokes for food aromas and colouring. It takes around 1 hour to burn (pyrolysis + combustion at ~1000 °C) a typical 0.5 x 0.1m wood log to completion. The crackling heard during the log fire attests to the difficulty with which volatile products escape. In comparison, the process of fast pyrolysis of wood accomplishes generation of combustible aerosols at much lower temperature, ca. 500 °C, in a few seconds or less. Such a fast thermal conversion rate is even more remarkable considering that the wood biopolymer is an excellent thermal insulator; see Table 1.

The science of pyrolysis tries to elucidate all the complex interrelated aspects of pyrolysis, for example, heat and mass transfer phenomena, sample-particle size effects, particle and gas residence times, reactor configuration, fluid dynamics, and so forth.

Table 1 Thermal conductivities of biomaterials in relation to other common materials. Typical values.

	Steel	Water	Air	Paper	Wood	Wood char
Thermal conductivity, W/mK	45-16	1.1	0.03	0.13	0.10-0.34	0.08-0.05

It was felt that a forum was needed where all available know-how could be compiled, collated and disseminated.

DEFINITIONS

According to Encyclopaedia of Energy Technology and the Environment (1):

> *"Pyrolysis is a process of the thermal decomposition to produce gases, liquids (tar) and char (solid residue). Pyrolysis is usually understood to be thermal decomposition, which occurs, in an oxygen-free atmosphere, but oxidative pyrolysis is nearly always an inherent part of combustion processes. Gaseous, liquid and solid pyrolysis products can all be used as fuels, with or without upgrading, or they can be utilized as feedstock for chemical and material industries. The types of materials which are candidates for pyrolysis processes include coal, biomass, plastics, rubber, and the cellulosic fraction (50%) of municipal waste".*

Thus in addition to interest in pyrolysis *per se*, it may be seen that the pyrolysis step is of critical importance in any **combustion** or **gasification** scheme.

The pyrolysis step, particularly in so-called "fast" pyrolysis happens in a time interval of a few seconds or less. In just such a short period of time, chemical reaction kinetics, mass transfer processes, phase transitions, and heat transfer phenomena all play important roles and can influence the overall outcome. However, due to these same short time intervals, fundamental aspects of these chemical engineering processes are not easily elucidated with the consequence that considerable controversy exists in the open literature.

The classical model of cellulose pyrolysis is shown in Figure 2 as developed by Broido and Shafizadeh (2) and further expanded by many other researchers as indicated in Figure 2.

The complexity of pyrolysis is illustrated in Figure 3 as presented by Radlein (3).

To derive a more comprehensive understanding of all phenomena involved is a goal of basic research in pyrolysis science. With the growing activities focused on renewable fuels, new chemicals and materials, the benefits from such fundamental work will in time become manifest in technological developments and innovations.

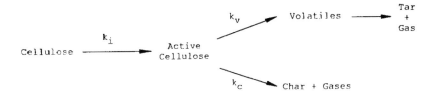

$$k_i = 2.8 \times 10^{19} e^{-(242,000/RT)} \text{ s-1}$$

$$k_v = 3.2 \times 10^{14} e^{-(198,000/RT)} \text{ s-1}$$

$$k_c = 1.3 \times 10^{10} e^{-(151,000/RT)} \text{ s-1}$$

**Figure 2   Classical "Kinetic Model of Pure Cellulose Pyrolysis under Vacuum",
Broido-Shafizadeh's Model of Cellulose Pyrolysis (but also presented/developed by
others including Kilzer (4), Bradbury (5), Diebold (6), Arseneau (7)).**

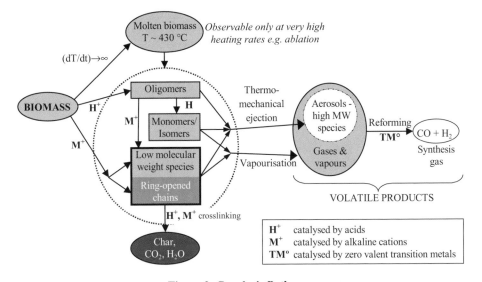

Figure 3   Pyrolysis Pathways

HISTORICAL BACKGROUND

Historically, pyrolysis of biomass has been mostly concerned with the ancient art of charcoal
production. While the 20th century, which is behind us, brought some novel developments, some
of the early milestones in improving the production of char, tar and gases are often forgotten and
no longer reported in contemporary literature. Some facts or examples relevant to present day
activities are worth recalling:

Charcoal

- Palmer, in 1914, reported increased charcoal yields, up to 43.3 wt% from birch and maple sawdust and chips (8). He used an externally heated autoclave heated in the bath of sodium and potassium nitrates. The processing was carried out under internal pressure of 150 psig.
- A similar concept of charcoal production under increased pressure is presently under development (9).

Liquid

- Regarding **liquid product** of pyrolysis, Klason reported a very early attempt at vacuum pyrolysis in 1914. As described in the English translation of his work (10), this was accomplished by *"distilling wood under the lowest possible vacuum (cathode-light vacuum) and in a short time"*. In this vacuum distillation of wood, the tar amounted to 43.6% and charcoal to 19.4%. As a point of interest to present-day modelers of pyrolysis – Klason was convinced that the sometimes observable exothermicity of wood pyrolysis reactions was due to the decomposition of evolved volatiles into secondary tar and char and not to the primary step of pyrolysis,
- The present-day development of the vacuum pyrolysis of wood was instigated by Chornet et al. (11) and is currently approaching commercial status (12).
- The earliest "fast" process for high liquid yield is probably the Garrett process of the 1970's (13), which used a circulating fluid bed of char as heat transfer medium to produce liquid fuel. However apparently economic considerations and/or technical problems related to MSW conversion led to its abandonment.
- Other players who appeared on the scene in the 1980's demonstrated a variety of technologies specially optimized for high liquid yields. These notably include the entrainment reactor from Georgia Tech, the NREL Vortex reactor, Vacuum Pyrolysis of Sherbrooke and WFPP (University of Waterloo) among others (14). Liquid yields were raised to 70%. The NREL technology was licensed to Interchem Inc. Parts of the University of Waterloo Intellectual Property have since been licensed to Encon, Fenosa, Red Arrow and Ensyn. It is important to note that the primary pyrolysis liquids are essentially the same whether they are destined for the fuel or the chemicals or for food colouring and preservation.

Gas

- **Gases** of pyrolysis: The possibility of ethylene and acetylene production by pyrolytic gasification of lignocellulosics was described not so long ago. Thus Hearon et al., (15) proposed the thermal decomposition of organic materials shredded to sizes below 1 mm at temperatures close to 1000°C in a *substantially inert gaseous environment*. Ethylene, acetylene combined yield reached 17% from Douglas fir powder. On the other hand, the collaborative effort of two Canadian Universities (Waterloo and Western Ontario, Scott et al. (16)) showed that at a temperature of 800°C, conversion of wood to carbon monoxide exceeded 35 %.
- Presently, high temperature pyrolytic conditions (with some steam addition) are adopted in the wood gasifier under development in Burlington, Vermont (17).

Chemicals

Regarding particular chemicals or chemical classes the following are some important and topical examples:

- Early in this century, and even at the present time in a few locations, methanol (wood alcohol), acetic acid, preservative tars etc. have been produced commercially by the so-called dry distillation of wood. Wood smoke aromas applicable in food processing are manufactured around the world in around a dozen of "calciners". These are rather slow, low temperature processes in which the liquid yield is low.

- The discovery of **levoglucosan** by Pictet and Sarasin and patented in 1918 (18) later led to some patent activity. See, for example US Patents # 3,298,928 to Esterer, (19); #3,374,222 to Peniston (20); # 4,880,473 to Scott (21); #5,355,782 to Moens (22); and #5,395,455 to Scott (23).

- **Hydroxyacetaldehyde**, initially a curiosity chemical (Bobrov, (24)), (Goos, (25)), (Hodge, (26)), (Byrne, 1966), was found in concentration exceeding 10% in the University of Waterloo **fast** pyrolysis oil (Elliott, (27). As stated by Antal (28):

  "For many in the community the biggest surprise in recent years was the discovery by Piskorz, Radlein and Scott (1986) of a major pyrolysis pathway which leads to the selective formation of glycolaldehyde from cellulose".

- Piskorz (1986) reported yields of hydroxyacetaldehyde around 17% from cellulose (29). Similarly high concentrations of hydroxyacetaldehyde were found in tars from pyrolysis of corn starch (30). Even higher yields of this compound, circa 22% from powdered dextrose and from corn syrups, were claimed by Stradal and Underwood, (31).

- Recently, hydroxyacetaldehyde also known under the name of glycolaldehyde became a chemical celebrity. As reported by NASA, June 2000, glycolaldehyde was discovered in a giant cloud of gas and dust near the center of our own Milky Way Galaxy by scientists using the National Science Foundation's 12 Meter radio telescope on Kitt Peak, Arizona. It was the first discovery of a simple sugar molecule in space.

- **Pyrolytic lignin**, the term used by Radlein (32) to describe the whole water insoluble fraction of wood derived bio-oil, was successfully tested as a partial replacement, up to 50%, of phenol in phenol-formaldehyde resins (novolaks and resols); see ref. (33). A very specific recipe for a similar accomplishment can be found in the PCT international application WO 99/38935 to Freel and Graham. In another, already granted patent one can read:

  "Lignins are a natural product and as such there may be natural chemical variations in the lignin macro-molecule. These variations may include different functional groups or different concentrations of these functional groups. These variations become rather significant when you are trying to formulate a consistent product day in and day out. This is a major reason why lignins have not been extensively utilized as replacement for phenolic resins".

Opposite to that:

  "A synthetic phenolic resin can be made to exact and reproducible specifications". (34)

GENESIS OF THE FUNDAMENTALS SUBJECT GROUP

During the PyNe Inaugural Meeting in Salzburg, February 1998, the general assembly of PyNe participants decided to create a working group dedicated to the fundamental science of pyrolysis.

Science and fundamentals are essential features of successful process development. The understanding of the underlying basic phenomena avoids semi-cooked technological, premature and apparent innovation and permits process optimization.

The intention was to develop the work carried out in the IEA Bio-Energy PYRA programme from 1995 to 1997 (35, 36, 37).

OBJECTIVES AND GOALS

In the original (1998) statement of targets, goals and undertakings. the working group was perceived to be dedicated to discussion, analysis, dissemination and constructive review in the area of fundamental science of pyrolysis, to extend and develop the work carried out in the IEA Bioenergy PYRA program of 1995-1997 (35, 36).

The group was expected to attempt to determine what knowledge would be of use to it and the pyrolysis community at large to assist in the development and implementation of fast pyrolysis. The requirements for engineering design would need to be considered and areas where further work is required highlighted.

Members of the group expressed a desire to consider topics of interest that included not only the science of primary pyrolysis but also dealt with the secondary steps of liquid collection and focused on the nature of products.

Examples of possible topics of interest (as suggested by participants).
1. *Thermogravimetry*: Literally hundreds of papers have been published in the past describing biomass pyrolysis kinetics derived from thermogravimetric measurements. Very often the kinetic parameters obtained varied from paper to paper and controversies persist about the validity of the values reported in application to the fast pyrolysis. It was hoped the Group would reach a consensus on the issue so that a defining paper could be published in a scientific journal by experts in the field of thermogravimetry.
2. *Time and temperature history of particles undergoing pyrolysis:* Some of the questions raised included whether it was possible to pyrolyze wood particles of 6 mm dimensions in 20 millisecond as reported in (38), and what are the actual bio-polymer temperatures attainable in any fast thermal processing i.e. is it possible to have wood particles at 600°C?
3. *Lignin thermal decomposition:* Lignin degradation studies are obscured by the difficulties in preparation of lignin samples in its "native" form. Macromolecular lignin chemistry as applied to bio-oil characterization was emphasized. Novel and unusual techniques like MALD-MS, MALDI-TOF and similar hyphenated methods were to be considered.
4. *Single particle modeling :*The extant variety of published models for single particle decomposition as well as overall reactor models should be reviewed, explained and validated if possible with the emphasis on whether they could be used to illustrate the engineering aspects of specific reactor design with different heat transfer requirements?
5. *Liquids collection:* Engineering fundamentals, theory and mathematics of devices like electrostatic precipitators or venturi quenchers with specific application to bio-oil aerosol recovery should be examined.
6. *Charcoal combustion, utilization, and activation:* The importance of this topic was underlined.
7. *Micro-emulsions:* Fundamental aspects of macro-molecular self-organization as relevant to flocculation, aging and stability behaviour of highly oxidized oligomeric molecular fragments in bio-oils.

The following is a quotation from the original statement:

> At least one workshop will be organized in the tradition of very well received
> previous IEA-PYRA meetings in La Coruna, Banff, Denver and Montreal.

Scientific papers will be prepared for publication by PyNe, presented in conferences or submitted to the scientific journals when warranted.

Scientific experts and movers in the field of biomass pyrolysis should be contacted, approached and invited.

Interaction between members of PyNe and others shall be maintained on an ad-hoc basis and perceived deficiencies in pyrolysis knowledge should be inquired into.

Science, basic research and fundamentals in general are essential features of successful process development. Improvement in understanding of the basic phenomena permits process optimization. Information and views on ongoing work to be maintained and recommendations made for a more coordinated and interactive programme to resolve major uncertainties.

An additional contribution was made by Lédé (39);

The PyNe Group on the Science of Pyrolysis should give the researchers involved the opportunity to prove how much the fundamental research can be useful for:
- *improving the pyrolysis processes*
- *inducing the (often-unpredictable) spin-offs that can be expected in other domains.*

The Group should focus its efforts on the pure aspects of chemistry (kinetic mechanisms, analysis...) but also on the problems connected to the process itself which belong to the science of chemical and process engineering:
- *choice of the reactor*
- *reactor optimization*
- *reactor couplings*
- *heat and mass transfer considerations*
- *hydrodynamics*
- *mathematical modeling*

ACHIEVEMENTS

Two major workshops were held which are described and summarised below.

WORKSHOP 1: THE SCIENCE OF BIOMASS PYROLYSIS
Stratford-Upon–Avon Workshop, 22-24th July 1998

"The Science of Biomass Pyrolysis" workshop held in Stratford-upon-Avon, U.K. 22-24th July 1998 was organised by a scientific committee consisting of J Piskorz, M Antal, S Czernik, M Gronli, J Lédé, C Peacocke and T Bridgwater,

This workshop examined and explored the fundamental processes occurring in biomass pyrolysis and featured international experts in the field of mathematical modeling and computer simulation (Di Blasi, Wojtowicz, Suuberg, Gronli, Arauzo, Peacocke, Antal, Solantausta). The current status of know-how, recent advances and potential of future developments were discussed.

Apart from the formal, oral contributions by participants, two informal workshops were held. The presentations given have been submitted to a peer review process, some have already been published in the scientific journals. Some are included in this report.

For years, many researchers have evaluated the kinetic parameters of cellulose pyrolysis by means of thermogravimetric analysis. The historical importance of the meeting lay in bringing together scientists on different sides of controversial issues related to these kinetics. As stated by M.R. Gray (40):

> *While an understanding of reaction mechanism permits the development of rigorous kinetics, kinetics analysis also has value in the opposite sense of testing hypotheses with respect to mechanism. Predictions from a hypothetical kinetic model can be tested by experiment to help to define the underlying reaction mechanisms.*
>
> *At its best, chemical kinetics has value as a critical tool for understanding reactions, both for scale-up and testing of hypotheses. At its worst, when done badly or trivially, kinetics provides superficial regression of experimental data that neither informs nor predicts.*

Introduction to the Workshop

The introductory opening of the Meeting was in the form of two of quotations:

Man of the century Albert Einstein in "The Evolution of Physics" had this to say about notion of questions and answers:

> *The formulation of a problem is often more essential than a solution, which may be merely a matter of mathematical or experimental skills. To raise new questions, new possibilities, to regard old questions from a new angle, requires creative imagination and marks real advances.*

One of the most cited chemists in the period of 1981-1997, F.A. Cotton in C&EN, March 30, 1998, pp. 43-46 wrote:

> *The concept of curiosity-driven research is now openly attacked. The refrain is – 'why should the public pay unless they know what they are going to get out of it'. Scientists are generally inept if not downright uninterested in trying to explain to the public what they do and why they do it. They are even naïve enough to admit that they enjoy what the do! Thus we have the confrontation so aptly captured in the title of an editorial in Science last May: -'(The) scientifically illiterate vs. (the) politically clueless'.....*

Programme

J. Piskorz, D. Radlein, D.S. Scott, "Pyrolysis of Cellulose - From Oligosaccharides to Synthesis Gas".

O. Boutin, J. Lédé, "Radiant Flash Pyrolysis of Cellulose: Evidence for the Formation of Short Life Time Species. Experimental Determination of Mass Balances".

L. Garcia, M.L. Salvador, R. Bilbao, J. Arauzo, "Gas Production from Catalytic Pyrolysis of Biomass".

E.M. Suuberg, V. Oja, "Tar Vaporization in Biomass Pyrolysis".

H. Pakdel, J.N. Murwanashyaka, C. Roy, "New Developments in Vacuum Pyrolysis".

G. Dobele, G. Rossinskaja, G. Telysheva, D. Meier, O. Faix, "Cellulose Dehydration and Depolymerization during Pyrolysis in the Presence of Phosphoric Acid".

M.J. Antal Jnr, X. Xu, "Total, Catalytic, Supercritical Steam Reforming of Biomass".

J.M.L. Penninger, "Chemistry of Diphenyl-Ether in Supercritical Water".

C. Lahousse, X. Delavallee, R. Maggi, B. Delmon, "The Role of Coke (or Char) in Biomass Pyrolysis".

C. Di Blasi, "Formulation and Application of Biomass Pyrolysis Models for Process Design and Development".

M. Gronli, M. Melaaen, "A Mathematical Model for Wood Pyrolysis. Comparison of Experimental Measurements with Theoretical Predictions".

N. Robinson, T. Bridgwater, "Adiabatic Modeling of Ablative Pyrolysis".

Y. Chen, S. Charpenay, A. Jensen, M.A. Serio, M.A. Wojtowicz, "Pyrolysis Behavior of Different Classes of Biomass".

Y. Solantausta, "Estimating the Performance of Industrial Fast Pyrolysis Processes".

R. Bilbao, A. Millera, M.B. Murillo, M.L. Salvador, J. Arauzo, "Kinetic Studies about the Thermal Decomposition of Lignocellulosic Residues". (Delivered in the form of a poster.)

Some Conclusions from the Presentations:

In general, there was an agreement that bio-oil application as a liquid fuel remains no more as the single driving force behind interest in biomass pyrolysis. It has become apparent that higher value products/chemicals would be of significant importance in early commercialisation stages of biomass conversion.

Some specific outcomes from the presentations and discussions are summarised below. References to names are those of the speakers at the workshops.

- Supercritical water appears to be an effective medium for the conversion of wet biomass material into liquid and gaseous fuels components. Formation of char typical for fast pyrolysis can be avoided. Unfortunately, technical issues related to high temperature/pressure combination remain to be solved, (Antal, Penninger). See also (41).

- There are several varieties of research grade ("pure") and technical quality cellulose (micro-crystalline and fibrous) available to researchers. These can be easily differentiated and "finger-printed" by routine thermogravimetric analyses, often coupled with other modern analytical methods. It is possible to derive rigorous chemical kinetics from carefully obtained thermogravimetric data, but such kinetic parameters should be reported with a detailed characterization of the feed used and products formed. Clearly, the important effects of mass and heat transfer limitations in acquiring and evaluating thermogravimetric data have to be addressed by researchers (Piskorz). See also references (41, 42, 43).

- Under high heating fluxes cellulose particles degrade and pass through a melted phase before liberation of vapours and aerosols into the gas phase. This "intermediate" phase is liquid at pyrolysis temperature but solid at room conditions and is soluble in water. Although sometimes speculatively referred to as "active cellulose", this intermediate phase is not cellulose anymore, but rather is composed of anhydro-oligosaccharides of DP 2-10 (Lédé). See ref. (44, 45, 46, 47, 48).

- The actual concentrations of volatile phenols in bio-oils could be augmented in fractional vacuum pyrolysis, by distillation with steam or in vacuo (Pakdel). It was reported that those phenolic compounds posses anti-diarrheal and anti-motility properties, exhibit germicidal activity, herbicidal effects, and antiseptic values. They have also been used in the tanning industry and as food aromas in a meat processing. The resin acids are another abundant class of chemicals found in bio-oils.

- MAB (Metastable Atom Bombardment) analysis of vacuum pyrolysis oil detected molecular ions of up to 700 atomic mass units. Although the chemical complexity of bio-oils is an obstacle for commercial applications, it seems to be of no deterrent in the field of adhesive formulation. See references (49, 50, 51, 52).

- Dehydration agents like mineral acids can drastically change the direction of cellulose pyrolysis. Phosphoric acid catalysed water removal from glucopyranose unit upon pyrolysis leads to a generation of cyclic levoglucosenone, a doubly dehydrated anhydrosugar, a potential, versatile synthon for the organic chemistry (53, 54, 55, 56).

111

- The continuous removal of char from the pyrolysis zone limits the extent of heterogeneous gas phase reactions (57). The presence of char in the freeboard space of fluidised bed reactors of any kind (from a bubbling fluid bed to a transport fluid bed) catalyses cracking reactions leading to secondary products. Similar observations were reported during the workshop in the case of pyrolysis of model compounds like guaiacol (Lahousse). Often, the role that char plays interacting with pyrolysis volatiles is improperly neglected.

- The volatile products of cellulose pyrolysis (and several related materials) contain components of molecular weight approaching a few hundreds of Daltons. Those components are not only thermally labile but have limited volatility inside and outside the solid matrix of decomposed bio-polymer, (Suuberg). They are manifested in the creation of persistent aerosols so evident in practice of pyrolysis. How limited the volatility of these materials really is, it can be estimated by careful extrapolation of experimental measurements of the vapour pressure of sugars.

- There have been many recent attempts to model both the intra- and extra particle processes of biomass pyrolysis, (Di Blasi). Most are limited to a single particle of cellulose pyrolysis. Other models, semi-global in character by lumping products of pyrolysis to 3 streams of char, tar and gases, are empirically validated for specific conditions only. Usually, the detailed products distribution is not taken into consideration. There is a real need of modellers for the "real conditions" pyrolysis data (58, 59) that could be used in model simulations. As of now, we do not have fully adequate models, which could be adopted for practical process engineering including scale up. Models with predictive technological powers are needed.

- As proven by Gronli, it takes 4 minutes to heat a wood sample to a temperature of ~400 °C to the depth of 8 mm under the significant heat flux of 130 kW/m^2. When the substrate changes from wood to a more porous char, the local thermal conductivity decreases. The increased char layer thickness has the consequence that longer pyrolysis times are required for a given conversion. Such findings seem to indicate that some promotional statements of "fast" and complete rapid pyrolysis of 6 mm particles of sawdust in ~300ms under typical fluid bed conditions are exaggerated if not impossible. The confirmed laboratory fast pyrolysis results reporting ~70%yield of bio-oil were obtained using 0.5 mm as an upper limit of the size of feed particles.

- On a practical, commercial level there is a need for a robust performance model of a pyrolysis plant. So far, overall thermal efficiencies of between 55 and 78 percent for a variety of process configurations have been reported. At present there are no industrial scale fast pyrolysis plants producing alternative fuel oil in operation, (Solantausta). No reliable mass and energy balances are available in the public domain. Because of that, reported product cost of 0.11 US$/kg, (equivalent to about 7.5 US$/GJ), when using wood cost of 44 $/dry tonne, are so far, rather speculative. Nevertheless these numbers are close to those reported by Black in the early days of fast pyrolysis technology development (60). Black assessed oil production cost at 5 US$/GJ for a 1000 tpd plant (50 % wet feed at 15 US $/t).

- A limitation of fast pyrolysis is the requirement for pulverized feeds. Preferentially, feeds of choice are those passing through ~1mm sieve opening. Ablative pyrolysis is an exciting prospect, still at the laboratory level of development, due to possibility of using larger chip sizes of biomass (Robinson). Modelling of ablative pyrolysis requires an additional input concerning the dimension of the ablated liquid layer, its physical properties, "melting" temperatures and so forth. A so called "ablation velocity" was measured experimentally by Lédé (61) for wood pyrolysis at 1073 K, and under a pressure of 3.7×10^5 Pa , as approaching the value of 3 mm per second.

- The stoichiometry of biomass (cellulosics in particularly) can be represented roughly by equal molar portions of $CO + H_2$. Fast pyrolysis of biomass can be conducted in a pyro-gasification mode. Then, the main products are gases: $CO + H_2$; (carbon monoxide and hydrogen). Pyro-gasification conversion requires temperatures circa 800°C. At such temperatures water is no longer a totally inert component. Heterogeneous catalysis is an option in pyro-gasification to improve tar conversion to gas and to lower the temperature of processing. The main problem of catalytic pyro-gasification is the loss of catalyst activity caused by carbon deposition (Garcia, Bilbao).

- Biomass is a tremendously diversified feedstock, which makes it difficult to develop comprehensive and robust models as a function of feedstock characteristics. A suggestion was made (Wojtowicz) to adopt and adapt previously developed models to be found in existing coal pyrolysis science.

In addition to the formal, oral presentations, two discussions were held to address the following questions:

- Fundamentals – how do we help and who do they help?
- How much more do we know now than 20 years ago?
- What models should we be using, developing or researching?

Discussion On
Fundamentals – How Do They Help and Who Do They Help?

This first workshop was led by M. Antal and D. Meier and the conclusions are summarised in Table 2 below followed by a report on the discussion.

Table 2   Conclusions from Workshop

Pay-off
• Have a look at coal industry, improved process efficiencies, clean coal technology
• Fundamental ideas now required from industry
• Science comes after application
Funding for fundamentals
• -a common language is needed; patent activities
• -define clear cut goals
• -driving forces, e.g. carbon dioxide emission
Requirements from the Scientific Community
• -define targets for modelling and data needed
• -improvement of the predictive level of mathematical models.

Re: "define clear cut goals" – this is an open statement which is not good enough, we have to be more specific than that. For example – predictive powers of modelling should be brought into focus. Present models of pyrolysis are not as yet adequate or generic enough for concrete cases of mass-heat transfers. Thermal efficiencies, vapour pressures and thermodynamic data in general are still question marks in view of the many assumptions and guesses. Verification of models with experimental data on a significant scale has yet to be accomplished.

One of the reasons to consider fundamental questions of pyrolysis science is a strong conviction that somewhere potential, practical applications do exist. Our focus should not be narrowly defined to modelling or mathematical simulation but the practicality of specific product commercialisation should be stressed. Examples of charcoal and ethanol were mentioned.

Participants were reminded here that apart from the "fundamentals" sub-group of PyNe there are four other sub-groups dealing exactly with problems of the chemistry of products, feasibility, health and environment concerns. The "fundamentals" sub-group remains one fifth of the overall pie in the plethora of PyNe activities.

Still PyNe should try to raise outside interests (industry) in the Network work. New technological developments in ethanol, charcoal, cetane improvers and gasoline additives areas should be highlighted (although, at present, there is a very little support for an ethanol fuel progressing in Europe).

The defined focus of PyNe up to now has been clearly directed towards liquid fuels and chemicals but consideration should be given to the inclusion of charcoal production in the agenda. However, as of now, topics like bio-oils stability, biodegradability, toxicology of bio-fuels come more naturally to the existing work of PyNe members.

For many in attendance "fundamentals" meant reaction mechanisms and pathways rather than mathematical modelling.

There was also general agreement that bio-oil as fuel no longer remains as the single technological objective behind interest in biomass pyrolysis. It is becoming apparent that higher value products and chemicals may be of significant importance in early commercialisation stages.

The fundamental science of biomass pyrolysis has already achieved augmented understanding of the roles played by the main factors, including physical parameters (heat transfer properties like thermal capacity and thermal conductivity), in the overall picture of biomass pyrolysis. The compilation and publication of those physical data particularly of relevance at typical pyrolysis temperatures (above 300°C) should be one of the goals of this PyNe Sub-group. It was recommended that a starting point in such an undertaking should be distribution between interested participants (Suuberg, Di Blasi, Arauzo, Wojtowicz) of the recently published PhD thesis of MN Gronli entitled "A Theoretical and Experimental Study of the Thermal Degradation of Biomass" kindly supplied by the author.

Discussion On
What Models Should We Be Using, Developing or Researching?
This second workshop was led by Y. Solantausta and E. Suuberg and is summarised below:

- It was agreed that fundamental work on pyrolysis is needed to advance the art of those aspects of reactor modeling related to chemistry, kinetics, and physical properties. It was agreed that Piskorz will co-ordinate an effort to summarize existing data. The emphasis in the summary was to be on the physical and, particularly, the thermal properties of cellulose and biomass, relevant to modelling their thermal conversion.
- The analogue to coal science was emphasized by several speakers. It would be quite useful for the biomass community to be aware of how similar kinds of problems were solved in coal science. For example, what approaches should be taken in modelling of pyrolysis phenomenon, what data is most urgent, and how the data available should be utilized.
- It is recognized that some aspects of the modeling done by the coal pyrolysis community cannot be used without major modification (e.g. coal pyrolysis models are basically random scission models, with no allowance for ionic or unzipping types of processes). The hand-in-glove development of robust pyrolysis models together with performance of validation experiments over a wide range of heating conditions will be a key to success. The question of how to use the lessons learned in biomass applications might be posed in a manner, which is more specific to the application. For example, the coal community has been most concerned with developing models for pulverized combustion and gasification. What new issues will arise in dealing with pulverized biomass? Large particle coal pyrolysis models are considerably less developed. What efforts have been made, what obstacles have been encountered and how important will these be in applications involving biomass, which will often be converted in larger particle form?

114

- A more close interaction between coal scientists and the biomass community would be beneficial. Perhaps suitable keynote speakers should be invited to bio-energy conferences.
- It was also suggested that a certain reactor system could be selected as the driver, and then models would be designed for this system. However, even though there are some sound arguments for this approach, generally speaking, it did not receive wide support. The feeling was that even if we aim to understand biomass pyrolysis on a fundamental level, we will necessarily be initially driven to explore conditions defined by certain applications. Perhaps rather than trying to define a single application that drives all of the research, we should recognize that several will exist, in different places. There should, however, be a continual concerted effort to reconcile the results obtained at low and high heating rates. This led to the discussion of making sure that some attempt is made to use well-characterized, common samples, so as to permit this reconciliation to take place (see below).
- Several speakers emphasized that, always when generating fundamental data, the applications for which the data is generated, should be presented. It is also necessary to present the applicability of data.
- The need for standardized samples (again an analogy to coal science is obvious) was recognized. However, it was felt that at least in Europe it would be the right forum, where a summary of past and existing standardized samples could be carried out.
- The goal was to offer a formal publication describing the history of sample standardization, and some guidelines for those seeking to work with relatively standard samples.

Discussion On
How Much More Do We Know Now Than 20 Years Ago?

To answer those questions we may go back in time to the writings of Doug Hayes, the "chief" and past manager of Bioenergy Development Program in Canada. (Canada, in the middle of eighties, was a "hot bed" of basic research and innovation in biomass pyrolysis.) In his programme review in 1988 Hayes wrote (62):

University of Waterloo.
D.S. Scott, J. Piskorz, D. Radlein and co-workers are well known for their fluidized-bed flash pyrolysis development, also known as the WFPP. The WFPP includes – direct thermal processing at 450-550°C, atmospheric pressure, and about 500 ms vapour residence time. They report high liquid yield (80% including water, based on input wood) that is a suitable fuel for conventional boilers.

University of Western Ontario.
M. Bergougnou, R. Graham and co-workers have developed an ultra-rapid pyrolysis or ultrapyrolysis process. Although there are similarities in this work and the research at the University of Waterloo, there are important differences. Whereas the Waterloo process utilizes fluidized-bed heat transfer, Bergougnou employees a very rapid (30 ms) mixing and heat transfer in a vortical contactor or vortactor followed by a plug-flow entrained-bed downflow reactor (50-900ms) and quenching (30 ms) with cryogenic nitrogen in a cryovortactor. Also dissimilar to the Waterloo process are the process conditions (650-1000°C, 50-900 ms residence time), and the main product at these temperatures is gas rather than liquid. The current objective, in collaboration with Ensyn Engineering (see below), is the production of chemicals and, in particular, olefins.

It is interesting to note here that the Universities of Waterloo and Western Ontario conducted an extensive data comparison from each of their reactor systems. Using selected data from both groups at around 500ms residence time, liquid and gas production data were plotted versus temperature. The temperature ranges were as follows: Waterloo at 400-750°C, and Western Ontario at 650-900°C. With combined data for each of the gas yield versus temperature, and liquid yield versus temperature data, there was, as expected considerable concurrence of overlapping data.. A simple first-order kinetic model is able to describe the oil yield over the temperature range of both experiments.

Ensyn Engineering.
Ensyn Engineering is a recently formed company whose principal investigator, R. Graham, has scaled-up the University of Western Ontario ultrapyrolysis reactor by a factor of 20 to a 5-10 kg/hr capacity RTP (Rapid Thermal Processor).

Laval University.
The multi-stage vacuum pyrolysis was developed by C. Roy and co-workers, initially at the Universite de Sherbrooke and, currently, at the Universite Laval. The technology is based on a multiple hearth vacuum pyrolysis process development unit operating at a 30 kg/hr biomass capacity.

And where are we today?

As reported elsewhere in a variety of sources including the PyNe newsletters, the fluid bed pyrolysis as originally extensively investigated at the University of Waterloo is at different stages of development at dozens of research centres and universities around the world from Malaysia to England, Spain, and the USA.

Ensyn Engineering Canada became Ensyn Group USA whose main achievement has been the supply of several commercial units to Red Arrow, Wisconsin for smoke aromas production.

The Laval University spin-off - Pyrovac Inc., has recently erected (2000) what is so far the biggest pyrolysis demonstration unit (vacuum) of 3500 kg/hr in Jonquiere, Quebec.

Although Solantausta stated (63) in his IEA Bioenergy Task 22; Final Report, Espoo 2000, pp. 28, stated:

There are no commercial biofuel liquid production plants on the energy market today.

from the above developments and from the ongoing activities at ENEL, VTT, BTG, Pyrovac, Dynamotive, Wellman, Fortum, we are very close to that goal. Without the progress and help from the science of pyrolysis this would be impossible.

Biomass is the only source of renewable carbon available and as stated by Bridgwater in his proposal to the European Commission for continuation of the PyNe Network (PyNe Meeting, Birmingham 2000):

Thermal processing of biomass has the potential to offer a major contribution to meeting the increasing demands of the bio-energy and renewable energy sectors and to meet the targets set by the E.C. and members countries for CO_2 mitigation.

Scope:

A review of the role that modeling can play in understanding, developing, improving and implementing pyrolysis technologies.

Oral presentations:

Pathways and mechanisms

T. Bridgwater, Jan Piskorz, "What is bio-oil"?

Kinetics and design

C. Di Blasi, "Numerical simulation of biomass fast pyrolysis. -Influence of external heat transfer resistances and reaction constants".

W. Prins, "Modeling the rotating cone pyrolyser".

J. Naber, "Priorities in modeling the HTU process".

A. Burnham, J. Reynolds, "The Use of Nucleation-Growth Kinetics Models for Polymers and Biopolymers".

E. Suuberg, "Char porosity".

Scale-up and systems

J. Bellan, "Validated biomass particle kinetics and its application to vortex reactor yield prediction and scaling".

C. Roy, "Scaling vacuum pyrolysis".

H. Pakdel, "Oil stability".

A. Himmelblau, "Making phenolics – industrial point of view".

M. Lauer, "Modeling bio-energy demand and predicting the opportunities for fast pyrolysis".

Open Forum

"What do we know and where should we be going"?

Record of the Meeting by N. Ahrendt

"What is bio-oil"? by T. Bridgwater, J. Piskorz

A short introduction was made by T. Bridgwater and J. Piskorz to explain the nature, substance and problems of fast pyrolysis liquid. While progress is being made in deriving a better understanding of the liquid, there is still much more work needed to improve the pyrolysis process and to develop bio-oil to suit the end users and hence improve the rate of commercialization.

"Numerical simulation of biomass fast pyrolysis. – Influence of external heat transfer resistances and reaction constants" by C. Di Blasi

Columba Di Blasi (University of Napoli) is currently working on convective/radiative pyrolysis for a single particle model to study events outside the particle, particularly the influence of external heat transfer resistance.

Simulations have been made with four different models representing different external heat transfer rates. Two sets of conditions were studied:
1. Particle half thickness from 0.1 to 5mm for a reactor temperature of 600K
2. Reactor temperature variations from 600 to 1100K for a particle with a half thickness of 1mm.

117

The conclusions are that external heat transfer resistances significantly reduce the particle heating rate and there are many other factors which can have an important effect. There is a continued need for generation of high quality data to derive relationships and to validate models.

"Modelling the rotating cone pyrolyzer", by W. Prins

Single particle analysis is very important as the design of a reactor is only possible with a good knowledge of the kinetics. However there are large variations in published primary kinetics. The results of particle modeling are useful but very variable due to unknown elements in particle kinetics.

Wolter Prins (University of Twente) has been looking at permeability and comparing anisotropic and isotropic reactions, analyzing the movement of gas, pressure, velocity vectors and conversion contours. The Biot number, Bo, is defined as the internal heat transfer resistance over the external heat transfer resistance. A high Bo value means the rate of heat transfer inside the particle is minimal. The cooling effect in the anisotropic case is considerably less than in the isotropic case.

"Priorities in modeling for the HTU process", by J. Naber

Jaap Naber (Biofuels bv) started his presentation by giving a basic description of the types of modeling, which have been important through the technical development path for the HTU process. This is a low temperature-high pressure liquid phase thermal conversion process which gives 50% yield of bio-crude with a heating value of around 35 MJ/kg. Biofuels bv are presently setting up a pilot plant to process 20-100kg/hr dry biomass.

In the process development the main modeling activities were:
- Supportive process scouting
- Reaction engineering
- Updating of flow sheets
- Continuous pilot plant
- Product studies.

For the commercial prototype the following elements are of major importance for its development:
- Fully developed flow sheeting
- Dynamic modeling for process control, operation support and automation
- Identification of improvement potential.

Jaap Naber concluded that models are extremely important in technology development and that it is essential to apply the right model at the right time with an appropriate level complexity.

"Use of nucleation-growth kinetic models for polymers and biopolymers", by A. Burnham, J. Reynolds

Alan Burnham and John Reynolds (Lawrence Livermore Laboratory) have been working on global kinetic analysis of complex systems. Numerous models are available for this purpose but their use is often confused/hindered by a lack of computer capability.

The three main kinds of models available are:
- First order
- Narrow profile reactions
- Distributed reactivity

A range of models have been applied to many polymeric materials including cellulose and coal. The conclusion is that cellulose and other linear polymers require an nth order nucleation model while newsprint and coal require a distributed energy model.

"The role of porosity and surface area in carbon combustion and gasification reactions", by E. Suuberg

The first part of the presentation was concerned with the nature of char reactivity. The transport process and resistance are complex and there is evidence that there are mass transfer limitations in gasification. Analysis of data suggests that reactivity is best represented as a function of surface area rather than mass.

The second part of Eric Suuberg's (Brown University) presentation was about the vapour pressure of tars and related materials and how these can be measured. A comparison of the vapour pressures of different saccharides was made but as these are not very volatile, the tests were not easy. Coal tars tend to have lower vapour pressures than cellulose tars.

"Validated biomass pyrolysis kinetics model and its application to vortex tar yield prediction and scaling", by J. Bellan

Josette Bellan (JPL) listed a number of motivations for the Jet Propulsion Laboratory (JPL) biomass kinetic model development. Porosity is a key feature of the model. The mechanism adopted was that of Bradbury.

Three sets of experimental data were used to derive the model which was then tested against 11 sets of published data.

From experimental data available particles have been divided into Class I and Class II. Class I contains particles that are so small that they are denominated by kinetics. Class II contains larger particles.

The model was applied to the NREL vortex reactor with good agreement with experimental results. The model was used to derive optimum temperatures for maximum liquid yield (~900K) and also for scale-up with recommendation that multiple reactors should be used with low aspect rations and multiple concentric walls.

"Scaling-up vacuum pyrolysis reactors", by C. Roy

Christian Roy of Laval University described the steps in the evolution of the current 3.5 t/hr vacuum pyrolysis plant in Jonquiere. When scaling up vacuum pyrolysis reactors, some elements are not possible to predict and experimental studies are needed to understand the conditions for scaling, particularly related to heat transfer. Agitation in the vacuum pyrolysis reactor is very important and Pyrovac Inc. is working on finding the optimum agitation for different feedstocks and also finding the relationship between feedstock, optimum reactor size and process, and products.

"Stability of vacuum pyrolysis oil", by H. Pakdel

Hooshang Pakdel (Laval University has been working on the ageing of pyrolysis oil by examining the variation in molecular weight distribution from fresh oil to aged oil and also the effect of aqueous phase addition on the ageing of bio-oil. The stability of the oil is improved by increasing the aqueous phase but in more than 15% of the cases the problem of phase separation arises. More research is needed to find more effective stabilizing agents

"Making phenolics – industrial point of view", by A. Himmelblau

For his experiments Andrew Himmelblau (Biocarbons Co.) is using a bubbling fluidized bed with air addition to provide process heat. This gives lower oil yields than conventional fast pyrolysis. The optimum temperature for oil production under these conditions is 450°C but a higher temperature of 550-600°C might be considered for more interesting compounds.

The commercialization issue is not just technical as market readiness and market needs are very important. There has been a great deal of discussion about how to enhance the value of pyrolysis oil especially outside North America and many researchers believe that it is essential to look into other chemical products.

"Modelling bio-energy demand and predicting the opportunities for fast pyrolysis", by M. Lauer

Max Lauer (Joanneum Research) is presently working on an assessment of the opportunities for and the barriers against successful commercialization of fast pyrolysis of biomass and applications for pyrolysis liquids.

The information for his assessment will be collected from PyNe members via questionnaire and a competitiveness assessment tool will be used to compare applications for different technologies (conventional and renewable), at different sizes, and for different purposes by using a non-dimensional competitiveness factor based on the ratio of conventional cost to renewable cost for the same service provided.

This review will both identify the most promising opportunities and provide a list of recommendations to improve the rate and success of commercialization.

"Closing summary", by J. Piskorz, T. Bridgwater

The importance of the fundamental science in the development and successful implementation of fast pyrolysis is poorly appreciated by funding sources. These PyNe workshops, of which this is the second, are intended to provide both a forum and a focus for those who are active to help all involved to keep up to date with recent and current developments. This will ensure that the research community makes the most effective contribution to commercial developers by providing clear messages and robust evidence of the value of fundamental research to successful industrial development of these emerging processes.

Another benefit of such international workshops is the collaborations and exchanges that can result from meeting other workers. The meetings are not only valuable for PyNe members to meet with researchers outside the network to learn of complementary work, but also to provide a forum for the whole community to share results and experience and develop better links and ideas and procedures.

Contributed Comments on the Oakland PyNe Workshop by Colomba Di Blasi

The presentations at the Workshop have confirmed that several laboratories are interested in the modeling of the "particle system". The models, at least from the mathematical point of view, include a good description of physical and chemical processes, in some cases integrated by proper external (reactor) conditions. However, there are two critical points, which still deserve further developments/improvements, before such models can be applied with confidence:
1)   determination of correct input data (especially property values, kinetic constants, heat and mass transfer coefficients)
2)   extensive model validation.

As for the point 1) reliable kinetic data are available only for cellulose pyrolysis. Published data on wood degradation kinetics do not give acceptable results for fast pyrolysis, even from the qualitative point of view. I think that it is necessary that such information is obtained through laboratory experimentation. As pointed out in this workshop (Burnham and Reynolds), differences between the different experimental devices and the subsequent analyses still exist and further efforts are needed in this direction. The approach based on the use of guessed values for the degradation kinetics of wood components, without any experimental basis, (Bellan) has limited validity. A posteriori (successive) validation through comparison with literature data (Bellan) cannot be considered conclusive because sample properties are not reported in the majority of these studies and, mainly, the actual heating conditions are largely unknown. Simulations carried out for fluid-bed pyrolysis (Di Blasi) by means of two widely used correlations for the external heat transfer coefficient show that conversion times can be different up to a factor of about 2. Hence temperature profiles and product distribution also present very large variations. Consequently "ad hoc" selections of the percentage of the three wood components (not to say about the influence of ash content and composition on reaction activity and selectivity) and the heat transfer coefficients can easily lead to agreement between theory and experiments. However, this does not mean that we can apply that model (equation + input data) to other situations and assume that the model predictions are quantitatively correct.

My view is that independent experiments are needed for
a)   the determination of input parameters (in particular, chemical kinetics should be determined in the absence of heat and mass transfer limitations) and
b)   the production of data for model validation (conditions where both chemical kinetics and transport phenomena are important).

It is important that experimental data for model validation are produced under exactly known conditions (properties of feedstock and heating conditions), which should be implemented/included in the model.

From the Workshop, interest from different laboratories has also been evidenced in the modeling of the "reactor system". The complexity of this system requires the introduction of simplifications. These should be introduced, keeping in mind the motivations for model development, that is, the aspects of the pyrolysis process that should be understood/predicted. The considerations given above also apply for this system.

In conclusion, future developments are needed on the following points:
-     Semi-global primary reaction mechanisms (and kinetic constants) for wood and biomass under "real conditions", determined through different experimental systems
-     Secondary reactions mechanisms and constants
-     Heat and mass transfer coefficients
-     Properties, -data for model validation
-     Reactor models, - applications (industrial scale).

Contributed Comments on the Oakland PyNe Workshop by Josette Bellan

Regarding the conclusions from the Workshop, I think that by now we have reasonably good biomass kinetics for the fast pyrolysis regime at atmospheric pressure. This is proven by our very extensive validation both for small and large particles. What we do not have is kinetics at higher pressures (see Philippe Girard's work) and low pressure close to vacuum (see Christian Roy's work). And our kinetics is not acceptable in the low temperature regime desirable for producing char. So, I suggest that TGA and isothermal studies should be conducted at different pressures than atmospheric. We, at JPL, have developed a high-pressure flow model that has been validated with micro-gravity experiments, and if we get funding we could use the high-pressure data to first obtain a kinetic model and then develop a particle flow model based upon the flow model that we have already validated. A "tall order", but certainly worthwhile!

ACKNOWLEDGMENTS AND THANKS

Those go to all contributors, co-authors and participants in the PyNe workshops held in Stratford and Oakland:

Joseph Spitzer, Tony Bridgwater, Colomba Di Blasi, Eric Suuberg, Stefan Czernik, Yrjo Solantausta, Dietrich Meier, Nina Ahrendt, Josette Bellan, Jaques Lédé, Michael Jerry Antal, Desmond Radlein, Donald S. Scott, Olivier Boutin, R. Bilbao, A. Millera, M.B. Murillo, M.L. Salvador, Jesus Arauzo, Vahur Oja, Hooshang Pakdel, J.N. Murwanashyaka, Christian Roy, Oskar Faix, Galina Dobele, Galina Rossinskaja, Galina Telysheva, Xiaodong Xu, Johannes M.L. Penninger, C. Lahousse, X. Delavallee, Rosanna Maggi, B. Delmon, Carmen Branca, Gabriella Signorelli, Federico Buonanno, Morten G. Gronli, Morten C. Melaaen, Nick Robinson, Yonggang Chen, Sylvie Charpenay, Anker Jensen, Michael A. Serio, Marek Wojtowicz, L. Garcia, Alan K. Burnham, Andrew Himmelblau, Wolter Prins, John G. Reynolds, Sreekumar Ramakrishnan, Ed Hogan, Jody Barclay, Angela Garcia, Keith Morris, Rolf Strenziok, Bob Hague, Max Lauer, Don Stevens, Barry Freel, Don Huffman, Cordner Peacocke, Claire Humphreys, Pearse Buckley, Paula Costa, Philippe Girard, Erik Rensfelt, Ann Segerborg-Fick, Carlos Amen-Chen, Iannis Boukis, John Brammer, Ibrahim Gulyurtlu, Blaise Labrecque, Jaap naber, Peter Wickboldt, Douglas Elliott, David Beckman, Anja Oasmaa, Yrjo Solantausta, Anders Ostman, Erich Podessor.

BIBLIOGRAPHY

The required reading list of pyrolysis fundamental aspects and know-how of the last 25 years should include the following compendia:

1. Proceedings. Specialists' Workshop on Fast Pyrolysis of Biomass, Copper Mountain, Colorado, (J. Diebold) 1980.
2. Proceedings of IEA sponsored Thermochemical Biomass Conversion Conferences (TBC) Eds Bridgwater AV et al., (Phoenix 1988, Interlaken 1992, Banff 1996, Tyrol 2000).
3. Proceedings of Biomass Conferences of Americas (Burlington 1993, Portland 1995, Montreal 1997, Oakland 1999).
4. Special issues of Journal of Analytical and Applied Pyrolysis like Vol. 51, Nos. 1-2, Pyrolysis in Resource Recovery, July 1999.
5. Proceedings of EU-Canada Workshops on Biomass Thermal Processing, Ottawa, 1990 and 1996.
6. Proceedings of IGT Energy from Biomass and Wastes, D. Klass (Ed.), 1976-1990.
7. ACS Symposium Series 376, "Pyrolysis Oils from Biomass" Eds. J. Soltes, T.A. Milne, ACS 1988.
8. "Fast Pyrolysis of Biomass: A Handbook", Ed. A Bridgwater et al. CPL Press 1999.

ANNEXES

There are two annexes:
1   Abstracts from Stratford on Avon Workshop
2   Bibliography and Patents, 1997-2000

REFERENCES

1   M. Serio, M. Wojtowicz, S. Charpenay, "Pyrolysis", Chapter in Encyclopedia of Energy Technology and the Environment, John Wiley & Sons, Inc., pp. 1181-2308, 1995.

2   A. Broido, "A Kinetics of Solid Phase Cellulose Pyrolysis", Thermal Uses and Properties of Carbohydrates and Lignins, F. Shafizadeh et al (Eds.), Academic Press, 1976.

3   D. Radlein, "The Production of Chemicals from Fast Pyrolysis Bio-oils", in A. Bridgwater et al. Fast Pyrolysis of Biomass: A Handbook, pp. 165, CPL Press 1999.

4   F.J. Kilzer, A. Broido, "Speculation on the Nature of Cellulose Pyrolysis", Pyrodynamics, Vol. 2, pp. 151-163, 1965.

5   A. Bradbury, Y. Sakai, F. Shafizadeh, "A Kinetic Model for Pyrolysis of Cellulose", J. Appl. Polym., Sci., 23, 3271, 1979.

6   J. Diebold, "A Unified Global Model for the Pyrolysis of Cellulose", Biomass and Bioenergy, Vol. 7, Nos. 1-6, pp. 75-85, 1994.

7   D.F. Arseneau, "Competitive Reactions in the Thermal Decomposition of Cellulose", Can. J. Chem., Eng., 49, 632-638, 1971.

8   R.C. Palmer, "Effect of Pressure on Yields of Products in the Destructive Distillation of Hardwood", The Journal of Industrial and Engineering Chemistry, Vol. 6, No. 11. 1914.

9   M.J. Antal, "Proces for Charcoal Production from Woody and Herbaceous Plant Material", U.S. Patent # 5,435,983 , 1995.

10   L.F. Hawley, "Wood Distillation", ACS The Chemical Catalog Co. Inc., pp. 55, 1923.

11   E. Chornet, C. Roy, "Organic Products and Liquid Fuels from Lignocellulosic Materials by Vacuum Pyrolysis", Canadian Patent # 1,163,595, 1984.

12   C. Roy et al., "Industrial Scale Demonstration of the Pyrocycling Process for the Conversion to Biofuels and Chemical", Biomass for Energy and Industry, Seville 2000.

13   D.E. Garrett, G.M. Mallan, "Pyrolysis Process for Solid Wastes", U.S. Patent # 4,153,514, 1979.

14   A.V. Bridgwater, G.V.C. Peacocke, "Fast pyrolysis processes for biomass", Renewable and Sustainable Energy Reviews, 4, 1-73, 2000.

15   W.M. Hearon, D.W. Goheen, J.T. Henderson, "Preparation of Unsaturated Hydrocarbons from Oxygen – Containing Organic Materials", U.S. Patent # 3,148,227, 1964.

16   D.S. Scott at al., "The Role of Temperature in the Fast Pyrolysis of Cellulose and Wood", Ind. Eng. Chem. Research, 27, 8-15, 1988.

17   M.A. Paisley, M.C. Farris, J. Black, J.M. Irving, R.P. Overend, "Commercial Demonstration of the Battelle/FERCO Biomass Gasification Process: Startup and Initial Operating Experience", Proc. of the Fourth Biomass Conference of the Americas, R.P Overend, E. Chornet (Eds), pp. 1061-1066, Oakland, 1999.

18   A. Pictet, I. Sarasin, Helv. Chim. Acta 1, 87, 1918.

19   A.K. Esterer, "Pyrolysis of Cellulosic Material in Concurrent Gaseous Flow", U.S. Patent # 3,298, 928, 1967.

20   Q.P. Peniston, "Separating Levoglucosan and Carbohydrate Derived Acids from Aqueous Mixtures Containing the same by Treatment with Metal Compounds", U.S. Patent # 3,374,222, 1968.

21   D.S. Scott, J. Piskorz , "Process for the Production of Fermentable Sugars from Biomass", U.S. Patent # 4,880,473, 1989.

22   L. Moens, "Isolation of Levoglucosan from Pyrolysis Oil derived from Cellulose", U.S. Patent # 5,371,212, 1994.

23   D.S. Scott, J. Piskorz, D. Radlein, P. Majerski, "Process for the Production of Anhydrosugars from Lignin-and Cellulose-Containing Biomass by Pyrolysis", U.S. Patent # 5,395,455, 1995.

24   P.A. Bobrov, Destructive Distn. of Wood, 5, 3-41,Trudui Tzentral Nauch. Issledovatel. Lesokhim. Inst. Narkomlesa S.S.S.R., 1934.

25   A.W. Goos, "The Thermal Decomposition of Wood", Wood Chemistry, L.E. Wise, E.C. Jahn (Eds.), Vol. 2, pp. 846, Reinhold Publishing 1952.

26   J.E. Hodge, "Browning Reaction Theories Integrated in Review", Agricultural and Food Chemistry, Vol. 1, 1953.

27   D.C. Elliott, IEA Co-operative Project D-1, Biomass Liquefaction Test Facility, Stockholm, Vol. 4, 1983.

28   M.J. Antal Jnr, I&EC Research, 34, pp. 703-717, 1995

29   J. Piskorz, D. Radlein, D.S. Scott, "On the Mechanism of the Rapid Pyrolysis of Cellulose", J. Anal. Appl. Pyrol., 9, pp. 121-137, 1986.

30   D.S. Scott, J. Piskorz, D. Radlein, S. Czernik, P. Majerski, "Production of Hydrocarbons from Biomass Using the Waterloo Fast Pyrolysis Process", Waterloo Research Institute, 1991.

31   US Patent # 5,393,542, 1995

32   D. Radlein, J. Piskorz, D.S. Scott, "Lignin Derived Oils from the Fast Pyrolysis of Poplar Wood", J. Anal. Appl. Pyrol., 12, pp. 51-59, 1987.

33   J. Piskorz, D.S. Scott, D. Radlein, S. Czernik, "New Applications of the Waterloo Fast Pyrolysis Process", pp. 64-73, Biomass Thermal Processing, E. Hogan, J. Robert, G. Grassi, A.V. Bridgwater (Eds), CPL Press 1992.

34   K.R. Kurple, "Modified Lignins", US Patent # 6,054,562, Apr. 2000

35   IEA Bioenergy. Pyrolysis Activity. Final Report, 1998.

36   A. Bridgwater et al. "Fast Pyrolysis of Biomass: A Handbook", CPL Press, 1999.

37   J. Lédé, J.P. Diebold, G.V.C. Peacocke, J. Piskorz, "The nature and properties of intermediate and unvaporised biomass pyrolysis materials", in A.V. Bridgwater, D.G.B. Boocock (Eds) Development in Thermo-Chemical Biomass Conversion, pp. 27-42, Blackie Academic & Professional, 1997.

38   C. Rossi, R. Graham, in Biomass Gasification & Pyrolysis, M. Kaltschmitt, A.V. Bridgwater (Eds), CPL Press, pp. 303, 1997.

39   J. Lédé, Private communication 1998.

40   M.R. Gray, Can. J. of Chem. Eng., Vol. 75, pp. 482, June 1997.

41   M.J. Antal, S.G. Allen, D. Schulman, X. Xu, "Biomass Gasification in Supercritical Water", Ind. Eng. Chem. Res., Vol. 39, No. 11, 2000, 4040-4053.

42   M.J. Antal, G. Varhegyi, E. Jakab, "Cellulose Pyrolysis Kinetics: Revisited", Ind. Eng. Chem. Res., 37, 1998, 1267-1275.

43   M. Gronli, M.J. Antal, G. Varhegyi, "A Round-Robin Study of Cellulose Pyrolysis Kinetics by Thermogravimetry", Ind. Eng. Chem. Res., Vol. 38, No. 6, 1999, 2238-2243.

44   S.B. Nordin et al. "An indication of Molten Cellulose Produced in a Laser Beam", Textile Research Journal, p. 152, 1974.

45   A. Vladars-Usas, "Thermal Decomposition of Cellulose", M.A.Sc. Thesis, University of Waterloo, Ont., Canada, 1993.

46   J. Piskorz, D.S. Scott, D. Radlein, "Mechanisms of the Fast Pyrolysis of Biomass: Comments on Some Sources of Confusion", in Minutes of the 2nd PYRA Meeting, Appendix 19, La Coruna, Nov. 1995.

47   O. Boutin, M. Ferrer, J. Lédé, "Radiant flash pyrolysis of cellulose – Evidence for the formation of short life time intermediate liquid species", J.A.A.P. 47, 1998, 25-40.

48   J. Piskorz, P. Majerski, D. Radlein, A. Vladars-Usas, D.S. Scott, "Flash pyrolysis of cellulose for production of anhydro-oligomers", J.A.A.P., 56, 2000, 145-166.

49   C. Roy, H. Pakdel, "Process for the Production of Phenolic-rich Pyrolysis Oils for Use in Making Phenol-Formaldehyde Resole Resins", U.S. Patent # 6,143,856, 2000.

50   K.R. Kurple, "Modified Lignins", U.S. Patent # 6,054,562, 2000.

51   B. Freel, R. Graham, "Natural resin formulation" patent application PCT/CA99/00051.

52   H.L. Chum, S.K. Black, "Process for Fractionating Fast Pyrolysis Oils and Products Derived there from", U.S. Patent # 4,942,269, 1990.

53   Z.J. Witczak, "Levoglucosenone and Levoglucosans", ATL Press, 1994.

54   G. Dobele, G. Rossinskaya, G. Domburg, "Monomeric Products of Catalytic Thermolysis of Cellulose and Lignin", pp. 482-489, in Biomass, Gasification and Pyrolysis: State of the Art and Future Prospects, M. Kaltschmitt and A.V. Bridgwater (Eds), CPL Press, 1997.

55   Y. Halpern, R. Riffer, A. Broido, "Levoglucosenone – A Major Product of the Acid-Catalyzed Pyrolysis of Cellulose and Related Carbohydrates", J. Org. Chem., 38, 204-209, 1973.

56   V. Boelsen, "Levoglucosenone – A High Value Chemical from the Waterloo Fast Pyrolysis Process", University of Waterloo, 1994.

57   M.L. Boroson, J.B. Howard, J.P. Longwell, W.A. Peters, "Heterogeneous Cracking of Wood Pyrolysis Tars over Fresh Wood Surfaces", Energy & Fuels, Vol. 3, pp. 736-740, 1989.

58   R.S. Miller, J. Bellan, "Analysis of Reaction Products and Conversion Time in the Pyrolysis of Cellulose and Wood Particles", Combust. Sci. and Tech., Vol. 119, pp. 331-373, 1996.

59   J. Bellan, private communication, 1999.

60   J. Black, "Preliminary Evaluation of the Waterloo Fast Pyrolysis Process", in D.S. Scott "Technical Evaluation of the Waterloo Fast Pyrolysis Process", Report for BDP Canada, 1986.

61   J. Lédé, H.Z. Li, J. Villermaux, "Fusion-like Behavior of Wood Pyrolysis", J. of Anal. and Appl. Pyrol., 10, 291-308, 1987.

62   D. Hayes, "Biomass Pyrolysis Technology and Products", in Pyrolysis Oils from Biomass J. Soltes, T.A. Milne, Eds., ACS Symposium Series 376, pp. 8-15, 1988

63   Y. Solantausta, IEA Bioenergy Task 22: Final Report, Espoo 2000.

PyNe
Subject Group: Fundamentals, Mechanisms and Science of Pyrolysis.
The Science of Biomass Pyrolysis
Stratford-upon-Avon Workshop, 22-24th July 1998.

ABSTRACTS

PAPERS are reproduced in Chapters 14-18

TOTAL, CATALYTIC, SUPERCRITICAL STEAM REFORMING OF BIOMASS
Michael Jerry Antal, Jr. and Xiaodong Xu

Hawaii Natural Energy Institute and the Department of Mechanical Engineering, University of Hawaii at Manoa, Honolulu, HI 96822, USA

By mixing wood sawdust with a corn starch gel, a viscous paste can be produced that is easily delivered to a supercritical flow reactor by means of a cement pump. Mixtures of about 10 wt % wood sawdust with 3.65 wt % starch are employed in this work, which we estimate to cost about $0.043 per lb. Significant reductions in feed cost can be achieved by increasing the wood sawdust loading, but such and increase may require a more complex pump. When this feed is rapidly heated in a tubular flow reactor at pressures above the critical pressures of water (22 Mpa), the sawdust paste vaporizes without the formation of char. A packed bed of carbon catalyst in the reactor operating at about 650°C causes the tarry vapours to react with water, producing hydrogen, carbon dioxide, and some methane with a trace of carbon monoxide. The temperature and history of the reactor's wall influences the hydrogen-methane product equilibrium by catalysing the methane steam reforming reaction. The water effluent from the reactor is clean. Other biomass feedstocks, such as the waste product of bio-diesel production, behave similarly. Unfortunately, sewage sludge does not evidence favourable gasification characteristics and is not a promising feedstock for supercritical water gasification.

CHEMISTRY OF DIPHENYL-ETHER IN SUPERCRITICAL WATER
Johannes M.L. Penninger

SPARQLE International B.V., Hengelo, The Netherlands, e-mail: Spql@introweb.nl

Supercritical water appears as an effective medium for the conversion of wet biomass material into liquid and gaseous fuel components. Drying of the wet feed stock is not needed and a substantial improvement in useful energy results. Formation of coke that is typical for thermal pyrolysis doesnot occur in a supercritical water environment.

The chemistry and mechanisms of these transformations are not clear yet and conceptual process studies must use "black-box"approaches when it comes to reactor design.

Raw biomass is a complex material and it is therefore difficult to delineate specific chemical reactions. But with model compounds with structural features that also prevail in biomass the detailed chemistry can be revealed.

The present experimental study deals with diphenyl-ether as a chemical model structure for the lignin constituents of biomass. The lignins contain structural oxygen functionalities of which the aryl-aryl oxygen bonds are most refractory.

From previous work it became clear that chemistry with supercritical water specifically occurs at hetero-atom functional groups; therefore the preference for Diphenyl-ether as a model compound.

Experiments were carried out by contacting the model compound with supercritical water for a specific length of time in a high pressure bomb at temperatures of $450 - 480^\circ C$ and bomb water densities from zero to 0.5 gr/ml. (Water critical temp is $374^\circ C$, critical density is 0.33 gr/ml).

The results show a gradual change in product selectivity with bomb water density ; this indicates a gradual transition from radical-type pyrolysis to ionic hydrolysis.

A new and unexpected catalytic effect of low concentrations of NaCl on the hydrolysis chemistry was observed. This effect doesnot obey classic Debye-Huckel theorem for reaction rates of substances in aqueous solutions with varying ionic strength.

A new theory is postulated that accounts for the observed first-order rate dependence on NaCl concentration of Diphenyl-ether hydrolysis in supercritical water.

PYROLYSIS OF CELLULOSE - FROM OLIGOSACCHARIDES TO SYNTHESIS GAS

Jan Piskorz [1], Desmond Radlein [1], Donald S. Scott [2]

[1] Resource Transforms International Ltd. Waterloo, Ontario, Canada
[2] University of Waterloo

Cellulose Pyrolysis Products

From a commercial and industrial point of view the driving force of all the activity is the existence of products. The first challenge is therefore to define a target product.

- What are the possible products from cellulose pyrolysis? Five principal methods or pathways of decomposition can be identified, each characterized by one of the following specific products: levoglucosan, cellobiosan and other anhydro-oligosaccharides, hydroxyacetaldehyde, levoglucosenone and synthesis gas.
- What are the determining factors for the dominant pathway for cellulose decomposition?
- What are the required process conditions for their optimization?

Thermogravimetric data suggest that differently prepared cellulose show different pyrolysis characteristics. Therefore in discussions of cellulose pyrolysis kinetics it should be stated which of the possible decomposition modes is being referred to, i.e. which is dominant under the given experimental conditions.

RADIANT FLASH PYROLYSIS OF CELLULOSE : EVIDENCE FOR THE FORMATION OF SHORT LIFE TIME SPECIES. EXPERIMENTAL DETERMINATION OF MASS BALANCES.

O. Boutin and J. Lédé

Laboratoire des Sciences du Génie Chimique, CNRS-ENSIC-INPL, 1, rue Grandville - BP451, 54001 NANCY Cedex, France

All the processes of biomass thermochemical conversion (pyrolysis, gasification, carbonisation, ...) begin with elementary steps of decomposition of each component of the starting material (cellulose, hemicellulose and lignin). In the case of cellulose, they occur in very short times (a few milliseconds) and are hence in competitions with transfer phenomena. At the present time, no consensus is reached on the corresponding kinetic pathways. The main reasons are that the studies are usually performed in systems that are incompatible with the rapidity of the reactions. The heat and mass transfer efficiencies are often bad and unknown. Moreover, the systems are such that secondary reactions (cracking, re-condensation, etc) are possible. They are often unable to detect reactions occurring without mass losses. An original experimental approach consists in using a concentrated radiation as a heat source.

This paper first presents an experimental device relying on the use of a discharge lamp (5 kW Xenon lamp) associated with two concentrating elliptical mirrors. This imaging furnace provides, in very clean conditions, high heat flux densities (more than 10^7 W m^{-2}) that can be accurately measured. The system is also able to deliver flashes of light as short as 0,01 s. Associated with a quench of the primary products, it is hence possible to study the first steps of fast thermal reactions. Cellulose powder is pressed in order to prepare small pellets. These pellets are placed at the focus of the imagining furnace, inside a transparent reactor fed by Argon. The jet of inert gas flows close to the pellets, in order to immediately cool down the products liberated from the reaction zone, before trapping in a glass wool filter. The experiments are made with different fluxes and irradiation times in conditions of true heating rates up to 5000 K s^{-1}.

First visual and microscopic observations of the surface of the pellets reveal that cellulose passes through a melted phase (producing agglomeration of the powder), before liberation of vapours and aerosols into the gas phase. These intermediate species are liquid at the reaction temperature, solid at room temperature and water soluble. These results bring evidence for the formation of the so called "Active Cellulose" much debated in the literature since about 20 years. Finally the paper reports mass balances relying on the values of the masses of unreacted cellulose, of the intermediate liquid species and of the products trapped in the filter. The first results show that no noticeable quantities of gases (H_2, CO, CO_2, CH_4) and char are formed in these conditions.

NEW DEVELOPMENTS IN VACUUM PYROLYSIS

H. Pakdel, J. N. Murwanashyaka and C. Roy

Pyrovac Institute Inc., Parc technologique du Quebec metropolitain, 333, rue Franquet, Sainte-Foy (Quebec), Canada, G1P 4C7

A 3.5 t/h vacuum pyrolysis pilot plant is under construction in the city of Jonquiére, Quebec, for the conversion of bark residues to biofuels and chemicals. A smaller but otherwise similar unit with a throughput capacity of about 100 kg/h is operating since 1994 in Quebec city. The researchers at Pyrovac Institute are actively involved with various aspects of pyrolysis products upgrading to biofuel, specialty chemicals and activated carbon. Another area of interest involves membrane technology for on-line separation of the aqueous phase from the oil phase.

The production of chemicals by vacuum pyrolysis of biomass offers significant economic and technological opportunities. An extensive effort has been made for the characterization of the complex oxygenated pyrolysis oil. Amongst the newest analytical techniques, the Metastable Atom Bombardment source (MAB) has gained some interest for the analysis of the vacuum pyrolysis oil. The pyrolysis oil obtained from a bark mixture of spruce, pine and fir species was analysed with MAB N_2^* with EI 70 eV on an Autospec-Q spectrometer. The MAB source was developed by Dephy Technology (Quebec, Canada) and was further modified at the Université de Montréal. The pyrolytic oil was introduced by using a direct insertion probe heated from 35°C to 500°C. Due to its high sensitivity, the exact mass measurement and MS/MS experiments were carried out on the typical product ions. At a probe temperature of 150°C, the resin acids were the most abundant species in the MAB N2* spectrum. At a probe temperature of 300°C, the molecular ion distribution of the pyrolysis oil was in the range of 90 to 700 mass units which was in close agreement with that obtained by GPC. This oil was found interesting for adhesive formulations due to the absence of low molecular weight volatile compounds which are detrimental to phenol formaldehyde resin. A phenol formaldehyde resin was formulated by replacing 50% by weight of commercial phenol with the pyrolysis oil. The performance properties of the surface resin product compared favourably with that of a commercial phenol formaldehyde resin.

Phenolic compounds are the major pyrolysis products of wood lignin and are produced at relatively high pyrolysis temperatures. These compounds are valuable products with potential industrial applications. However, the complexity of the pyrolysis oil is an obstacle for the separation and purification of targeted compounds. Fractional pyrolysis of a mixture of birch bark (54%) and birch wood (46%) was carried out at the temperature range of 25-550°C in a laboratory batch vacuum pyrolysis reactor. The active zone of decomposition leading to phenolic compounds was found to be between 275 and 350°C. This fraction contained 4.8 % by wt. of phenols on an anhydrous wood basis which was the highest phenol concentration ever reported by this laboratory for a whole pyrolysis oil, before any further separation process had been carried out.

Several methods have been studied to recover phenols from pyrolysis oils. Amongst these methods, steam distillation followed by a vacuum distillation yielded a number of sub-fractions with 50-70 % phenol concentration. It was found that steam distillation can separate a volatile fraction at a low steam temperature with only minor modification of the chemical composition of the pyrolysis oil. Further work on the separation and purification of phenols is underway in the authors' laboratories.

CELLULOSE DEHYDRATION AND DEPOLYMERIZATION REACTIONS DURING PYROLYSIS IN THE PRESENCE OF PHOSPHORIC ACID

Galina Dobele [1], Galina Rossinskaja [1], Galina Telysheva [1], Dietrich Meier [2]
and Oskar Faix [2]

[1] Latvian State Institute of Wood Chemistry, 27 Dzerbenes St., LV-1006, Riga, Latvia
e-mail: ligno@edi.lv
[2] Federal Research Centre for Forestry and Forest Products, Institute for Wood Chemistry and Chemical Technology of Wood, Leuschnerstr. 91, D-21031 Hamburg
Federal Republic of Germany

The low-temperature pyrolysis process and dehydration reactions of cellulose impregnated with phosphoric acid (1-10 wt. %) have been studied. Based on the yield of pyrolytic water and alterations of glycopyranose units of cellulose it was found that dehydrated cyclic structures are formed in the solid product. Phosphoric acid also catalyzes the subsequent depolymerisation process of the dehydrated polymer giving rise to 20% (b. o. cellulose) of levoglucosenone. The yield of levoglucosenone depends on the amount of phosphoric acid and the time of thermal treatment. Thus, the degree of dehydration reactions can be controlled by variation of these two parameters.

THE ROLE OF COKE (OR CHAR) IN BIOMASS PYROLYSIS

C. Lahousse, X. Delavallée, R.Maggi, B. Delmon

Laboratoire de Catalyse et Chimie des Matériaux Divisés, Université catholique de Louvain , 2/17 place Croix du Sud, B-1348 Louvain-la-Neuve, Belgium

The aim of this communication is to present results obtained while searching for a more efficient cracking catalyst for pyrolytic vapours. These results turned out to shed light on some little investigated and poorly understood aspects of pyrolysis itself.

Using guaiacol as a model compound, the cracking activity of various catalysts has been tested. Vapours of the model compound were produced by an injection system using He as carrier. To avoid condensation without excessive heating in an attempt not to favour the decomposition of the reactant before the reactor, the reactant partial pressure chosen was 4 Torr. In an effort to approach the actual pyrolysis conditions, the residence time of the gas mixture was fixed at 1 s. The reactor temperature was 400°C. The tests were carried out until a stable level of conversion and selectivity was obtained, namely for one or two hours.

At this temperature all the catalysts tested showed a very low activity. The activity was rapidly masked by an independent process which occurred with all catalysts. The results which will be presented during the workshop show that the conversion rapidly grows and that this growth is associated with the appearance of a mass balance defect. (The guaiacol converted is not completely converted to GC detectable products.) Shortly after the appearance of this phenomenon, products that are not structurally linked to guaiacol are detected. Whatever the catalyst, a similar final state is reached. This final state is characterised by a conversion close to 100%, a mass balance defect of 80% and by the presence of a large number of products, namely both the guaiacol decomposition products like phenol and many other compounds. In all case, the catalytic bed is black after test; coke is formed.

Coke formation is obviously responsible for the missing part of the mass balance. The association of a balance defect to the growth of conversion suggests that coke causes the increase in conversion. Coke seems to convert guaiacol, but this mainly into coke with only a small proportion of volatile products. The catalyst role in coke formation thus appears as very limited. Coke formation seems to be an auto-catalytic process.

The products, which are not derived from guaiacol, certainly must result from a partial decomposition of this coke. Their presence indicates that this char-like material gets continuously formed and is simultaneously undergoing a decomposition process. While releasing GC-detectable oxygenated compound, the original guaiacol polymer certainly transforms itself into something more graphite-like.

The use a model compound (which enables to accurately measure conversion) makes possible to show that the coke formed from guaiacol possesses a surprising activity. It converts guaiacol far more efficiently than industrial catalysts can do, and actually produces by its decomposition most of the cracked products. It may also possess some catalytic activity on its own, namely be able to directly convert guaiacol into refractory compounds like phenol.

Some of our results nicely confirm the data collected on real pyrolytic compounds. Indeed a recent UCL study of a ZSM-5 sample, which had been used for bio-oil upgrading, showed that the coke it contained was formed from bio-oil independently from the action of the catalyst. Big coke particles were found. They did not correspond in their size or aspect to the agglomerates of small catalyst crystallites. Moreover, the multiple analyses performed in the corresponding work have shown that the coke formed from bio-oil was rich in oxygen. Like the coke formed from guaiacol, this coke produced GC-detectable oxygenates upon heating. The formation process and the properties of bio-oil and guaiacol cokes seem thus very similar. The chemical reactivity and possible catalytic activity of the coke formed from pyrolytic vapours is similar to those of the coke formed from guaiacol. This suggests that, when dealing with the catalytic upgrading of pyrolytic vapours, a non-negligible portion of the "cracking" product could come from coke decomposition (or coke catalytic activity) rather than from the catalyst itself.

In addition, this study showed that the presence of the catalyst actually only has a very limited influence on the properties and on the evolution of coke. Like the other solid produce in biomass pyrolysis called "char", coke properties are exclusively determined by the nature of pyrolytic vapours. The chemical composition of these two products, as reported in the literature, is very similar. There is thus a strong possibility that the pyrolytic "char" could be as active as coke and therefore generate a significant amount of the compounds produced during pyrolysis.

The results shortly described above highlight one aspect of biomass pyrolysis, which seems to have been improperly neglected. This study clearly demonstrates that the role of char during the pyrolysis process and its contribution to the composition (and properties) of the oil produced should be examined. This is feasible, especially in the frame of a cooperative research. Indeed, along with test on real biomass, this study shows that the use of model compound is a complementary method which allows to obtain fundamental information that are not easily accessible otherwise. Our laboratory at UCL would be happy to cooperate to a research on this aspect, namely a study on "char" role in pyrolysis, and has the knowledge and the experience to perform the model compounds part of the study.

TAR VAPORIZATION IN BIOMASS PYROLYSIS

Eric M. Suuberg and Vahur Oja

Division of Engineering, Brown University, Providence, RI 02912 USA
e-mail: Eric_Suuberg@brown.edu

This presentation reviews evidence suggesting that mass transfer limitations might play a key role in cellulose and biomass pyrolysis. The main focus of the presentation concerns the problems inherent in characterization of the vaporization behaviour of pyrolysis tars in general and biomass pyrolysis tars in particular. The nature of cellulose pyrolysis tars is discussed, and some new results are presented on the measurement of vapour pressures of various saccharides. These materials are difficult to characterize in this manner because they are exceedingly thermally labile. The techniques for performing the characterization are discussed.

The results on cellulose pyrolysis tars are compared with the results on low-rank coal pyrolysis tars. The difficulties in application to biomass of various correlations which have been derived for coal tars are noted. The need for more careful experimental work, specifically directed at characterizing the vapour pressures of pyrolysis tars from biomass components, is highlighted. The special issues involved in developing correlations suitable for mathematical modeling of the pyrolysis process will be noted.

131

FORMULATION AND APPLICATION OF BIOMASS PYROLYSIS MODELS FOR PROCESS DESIGN AND DEVELOPMENT

Colomba Di Blasi, Carmen Branca, Gabriella Signorelli and Federico Buonanno

Dipartimento di Ingegneria Chimica, Universita degli Studi di Napoli Federico II Piazzale V. Tecchio, 80125 Napoli, Italy
Tel: +39-081-7682232; Fax: +39-081-2391800; e-mail: diblasi@unina.it

Pyrolysis of biomass is the result of complex interactions among numerous chemical and physical processes, which take place across the particle and the reaction environment. The relative importance of the different processes is highly dependent on particle characteristics (thermo-chemical properties, size, moisture and ash content) and the type of conversion unit (heat and mass transfer between particle and reactor, extra-particle activity of secondary reactions).

In this paper, the different approaches chosen to model intra- and extra particle processes of biomass pyrolysis are illustrated in relation to fixed-bed, fluid-bed and ablative reactors. The experimental measurements needed to produce data for process simulation and model validation are listed. Simulation results are presented for several conversion regimes and, when appropriate, predictions are compared with experiments. Finally, suggestions are made on how modeling can be used for process development and optimization.

A MATHEMATICAL MODEL FOR WOOD PYROLYSIS. COMPARISON OF EXPERIMENTAL MEASUREMENTS WITH THEORETICAL PREDICTIONS

Morten G. Grønli [1] and Morten C. Melaaen [2]

[1] SINTEF Energy Research, Thermal Energy and Hydropower, N-7034 Trondheim, Norway
Phone: + 47 73 59 37 25, Fax +47 73 59 28 89, e-mail: Morten.Gronli@energy.sintef.no
[2] Telemark Institute of Technology (HiT-TF), N-3914 Porsgrunn, Norway

Experimental and modelling work on the pyrolysis of biomass under regimes controlled by heat and mass transfer are presented. In a single-particle, bell-shaped Pyrex reactor, one face of a uniform and well-characterized cylinder (D=20 mm, L=30 mm) prepared from Norwegian spruce has been one-dimensionally heated by using a Xenon-arc lamp as a radiative heat source. The effects of heating conditions, i.e. heat flux and grain orientation on the product yields distribution and reacted fraction have been investigated. The experiments show that heat flux alters the pyrolysis products as well as the intra-particle temperatures to the greatest extent. A comprehensive mathematical model, which can simulate pyrolysis of wood is presented. The thermal degradation of wood involves the interaction in a porous media of heat, mass and momentum transfer with chemical reactions. Heat is transported by conduction, convection and radiation and mass transfer is driven by pressure and concentration gradients. The simulation of these processes involves the simultaneous solution of the partial differential, conservation equations for mass, energy and momentum with kinetic expressions describing the rate of reaction. By using three parallel competitive reactions to account for primary production of gas, tar and char, and a consecutive reaction for the secondary cracking of tar, the predicted intraparticle temperature profiles, ultimate product yields distribution and reacted fraction agreed well with the experimental results.

ADIABATIC MODELLING OF ABLATIVE PYROLYSIS

Nick Robinson and Tony Bridgwater

Bio-Energy Research Group, Aston University, Birmingham B4 7ET, UK

Ablative pyrolysis is an exciting prospect due to the possibility of employing small intensive reactors with high throughputs and low inert gas requirements. Consequently the product collection train is much simplified and significantly smaller and more cost effective.

A new type of ablative fast pyrolysis reactor and associated product collection system with a nominal throughput of 5 kg/h has been designed and is currently under construction. In order to improve design procedures, extensive modelling of the reactor and product collection system was undertaken to enable optimisation and prediction of performance. As an integral part of this model fundamental work on the adiabatic ablation of a wood chip has been investigated. This involves solving the conservation of mass, energy and momentum equations for the ablation process. As an input the model requires the physical properties of the wood chip being ablated and the associated liquid layer, its dimensions, the operating and melting temperatures, the applied pressure and the applied normal velocity. As an output the model calculates the wood chip ablation velocity, the friction force; and the relative motion gap, temperature and pressure distribution in the produced liquid film beneath the wood chip.

This enables various factors to be investigated and their effect on the ablation rate established.

PYROLYSIS BEHAVIOR OF DIFFERENT CLASSES OF BIOMASS

Yonggang Chen [1], Sylvie Charpenay [1], Anker Jensen [2], Michael A. Serio[1], and Marek A. Wójtowicz [1]

[1] Advanced Fuel Research, Inc., 87 Church Street, East Hartford, CT 06108-3742, USA
[2] Department of Chemical Engineering, Technical University of Denmark, 2800 Lyngby, Denmark

Over the next decade there will be a renewed emphasis on the production of chemicals and liquid fuels from biomass, the use of agricultural wastes as feedstocks, and the co-firing of coal and biomass materials. In view of the tremendous diversity of biomass feedstocks, a great need exists for a robust, comprehensive model that could be utilized to predict the composition and properties of pyrolysis products as a function of feedstock characteristics and process conditions. The objective of this work was to adapt an existing coal pyrolysis model and make it suitable for the pyrolysis of biomass. The soundness of this approach is based on numerous similarities between biomass and coal. There are important differences, however, which preclude direct application of the coal model. This work involved: 1) selection of a set of materials representing the main types of biomass; 2) development of a biomass classification scheme; 3) development of a modeling approach based on modifications of a coal pyrolysis model; 4) calibration of the model for a set of standard materials against pyrolysis data taken over a range of heating rates; 5) validation of the model against pyrolysis data taken under other (higher) heating rate conditions.

ESTIMATING THE PERFORMANCE OF INDUSTRIAL FAST PYROLYSIS PROCESSES

Yrjö Solantausta

VTT Energy, Combustion & Conversion Technology, Biologinkuja 5, PO Box 1601, Espoo, FIN-02044 VTT, Finland; Tel: +358 9 456 5517, Fax: +358 9 460 493, e-mail: Yrjo.Solantausta@vtt.fi

One of the basic unknown issues related to the production of fast pyrolysis liquids from biomass is the process efficiency. No truly industrial scale plants are in operation, and no performance data is available in public domain concerning the largest plants in operation. This makes performance analysis difficult, and only few serious analysis on the overall process performance have been published.

An estimate for the thermal efficiency of pyrolysis liquid production as an alternative fuel oil from wood is presented. AspenPlus™ is employed as the modelling tool. Some challenges and uncertainties in modelling include: presentation of the pyrolysis liquid, heat of pyrolysis, suspension density in the reactor, temperature profile of the reacting wood, oil recovery, role of aerosols.

There is a considerable variation in process efficiencies published for relatively similar process concepts, between about 65 to 78 % based on lower heating values. An efficiency of 73 % is estimated in this work. However, it is accepted that because of major process uncertainties and deficiency in data, considerable uncertainty remains related to these values.

ANNEX 2

To illustrate what was happening and to show how wide the frontier of basic research in biomass conversion and biomass pyrolysis has been in the last 4 years, a chronological selection of references is put together below. (Presentations at major conferences and symposia are omitted). Also a list of recent, relevant U.S. patents is compiled.

References, 1997-2000
1997

J.A. Conesa, R. Font, A. Marcilla, "Comparison between the Pyrolysis of Two Types of Polyethylene in a Fluidized Bed Reactor", Energy & Fuels, 11, 126-136, 1997.

D. Wang, S. Czernik, D. Montane, M. Mann, E. Chornet, "Biomass to Hydrogen via Fast Pyrolysis and Catalytic Steam Reforming of the Pyrolysis Oils or Its Fractions", Ind. Eng. Chem. Res., 36, 1507-1518, 1997.

M.J. Antal, G. Varhegyi, "Impact of Systematic Errors on the Determination of Cellulose Pyrolysis Kinetics", Energy & Fuels, 11, 1309-1310, 1997.

R.V.Pindoria, J-Y Lim, J.E. Hawkes, M-J. Lazaro, A.A. Herod, R Kandiyoti, "Structural characterization of biomass pyrolysis tars/oil from eucalyptus wood waste: effect of H_2 pressure and sample configuration", Fuel, Vol. 76, No. 11, 1013-1023, 1997.

M. Serio, M.A. Wojtowicz, Y. Chen, S. Charpenay, A. Jensen, R. Bassilakis, K. Riek, "A Comprehensive Model of Biomass Pyrolysis", Final Report, United States Department of Agriculture, 1997.

M.Y. Wey, S.C. Huang, C.L.Shi, "Oxidative pyrolysis of mixed solid wastes by sand bed and freeboard reaction in a fluidized bed", Fuel, Vol. 76, No. 2, 115-121, 1997.

Y. Chen, S. Charpenay, A. Jensen, M.A. Serio, M.A. Wojtowicz, "Modeling Biomass Pyrolysis Kinetics and Mechanisms", ACS Div. of Fuel Prepr. 42 (1), 96, 1997.

M. Kelsey, R. Truttmann, "Complete thermogravimetric analysis", American Laboratory, 17-20, January 1997.

H. Pakdel, C. Amen-Chen, C. Roy, "Phenolic Compounds from Vacuum Pyrolysis of Wood Wastes", The Canadian J. of Chem. Eng., Vol. 75, 121-126, 1997.

C. Amen-Chen, H. Pakdel, C. Roy, "Separation of Phenols from Eucalyptus Wood Tar", Biomass and Bioenergy Vol. 13, Nos ½, 25-37, 1997.

J.G. Reynolds, A.K. Burnham, "Pyrolysis Decomposition Kinetics of Cellulose-Based Materials by Constant Heating Rate Micropyrolysis", Energy & fuels, Vol. 11, No. 1, 88-97, 1997.

C. Di Blasi, M. Lazetta, "Intrinsic Kinetics of Isothermal Xylan Degradation in Inert Atmospere", JAAP, 40-41, 287-303, 1997.

M. Lazetta, C. Di Blasi, F. Buonanno, "An Experimental Investigation of Heat Transfer Limitations in the Flash Pyrolysis of Cellulose", Ind. Eng. Chem. Res. ,36, 542-552, 1997.

J.A. Caballero, J.A. Conesa, R. Font, A. Marcilla, "Pyrolysis kinetics of almond shells and olive stones considering their organic fractions", JAAP 42, 159-175, 1977.

E. Bjorkman, B. Stromberg, "Release of Chlorine from Biomass at Pyrolysis and Gasification Conditions", Energy & Fuels, 11, 1026-1032, 1997.

J.G. Olsson, U. Jaglid, J. Pettersson, "Alkali Metal Emission during Pyrolysis of Biomass", Energy & Fuels, 11, 779-784, 1997.

J.P. Diebold, S. Czernik, "Additives to Lower and Stabilize the Viscosity of Pyrolysis Oils during Storage", Energy & Fuels, 11, 1081-1091, 1997.

G.J. Suppes, Y.Rul, E.V. Regehr, "Hydrophilic Diesel Fuels – Ignition Delay Times of Several Different Blends", SAE Paper 971686, 1997.

E. Jakab, K. Liu,, H.L.C. Meuzelaar, "Thermal decomposition of wood and cellulose in the presence of solvent vapors", Ind. Eng. Chem. Res., 36, 2087-2095, 1997.

J. Arauzo, D. Radlein, J. Piskorz, D.S. Scott, "Catalytic Pyrogasification of Biomass. Evaluation of Modified Nickel Catalysts", Ind.. Eng. Chem. Res., 36, 67-75, 1997.

J. Delgado, M.P. Aznar, J. Corella, "Biomass Gasification with Steam in Fluidized Bed: Effectiveness of CaO, MgO, and Cao-MgO for Hot Gas Cleaning", Ind. Eng. Chem. Res. 36, 1535, 1997.

Y. Solantausta, T. Makinen, T. Koljonen, 'Modelling of a flash pyrolysis process'. EU AIR Contract AIR2-CT94-1162, VTT Energy, 1997.

1998

K. Miura, T. Maki, "A Simple Method for Estimating f(E) and $k_0(E)$ in the Distributed Activation Energy Model", Energy & Fuels, 12, No 5, 864-869, 1998.

S. Charpenay, M. Wojtowicz, M.A. Serio, "Pyrolysis Kinetics of the Waste-Tire Constituents: Extender Oil, Natural Rubber, and Styrene-Butadiene Rubber", ACS Div. of Fuel Chem. Prepr , .43 (1) 185-191, 1998.

M.J. Antal, G.Varhegyi, E. Jakab, "Cellulose Pyrolysis Kinetics: Revisited", Ind. Eng. Chem. Res. 37, 1267-1275, 1998.

K. Sipila E. Kuoppala, L. Fagernas, A. Oasmaa, "Characterization of Biomass-Based Flash Pyrolysis Oils", Biomass and Bioenergy, Vol. 14, No. 2, 103-113, 1998.

R. Blaine, "A faster approach to obtaining kinetic parameters", American Laboratory, 21-23, January 1998.

G. Varhegyi, P. Szabo, F. Till, B. Zelei, M.J. Antal, X. Dai, "TG, TG-MS and FTIR Characterization of High-Yield Biomass Charcoals", Energy & Fuels, 12, 969-974, 1998.

Y. Chen, S. Charpenay, A. Jensen, M.A. Wojtowicz, M.A. Serio, "Modeling Biomass Pyrolysis Kinetics Under Combustion Conditions", 27th International Symposium on Combustion, Boulder, 1998.

R.S. Miller, J. Bellan, "Numerical Simulation of Vortex Pyrolysis Reactors for Condensable Tar Production from Biomass", Energy & Fuels, 12, 25-40, 1998.

O. Boutin, M. Ferrer, J. Lédé, "Radiant flash pyrolysis of cellulose – Evidence for the formation of short life time intermediate liquid species", JAAP 47, 13-31, 1998.

M.D. Guillen, M.L. Ibargoitia, "New Components with Potential Antioxidant and Organoleptic Properties, Detected for the First Time in Liquid Smoke Flavoring Preparations", J. Agric. Food Chem., 46, 1276-1285, 1998.

A. Jensen, K. Dam-Johansen, M. Wojtowicz, M. A. Serio, "TG-FTIR Study of the Influence of Potassium Chloride on Wheat Straw Pyrolysis", Energy & Fuels, 12, No 5, 929-938, 1998.

P.A. Della Rocca, "Study on biomass thermal conversion processes", D. Phil. Thesis, Facultad de Ciencias Exactas y Naturales, Universidad de Buenos Aires, 1998.

T. Hsisheng, Y.C. Wei, "Thermogravimetric Study on the Kinetics of Rice Hull Pyrolysis and Influence of Water Treatment", Ind. Eng. Chem. Res,. 37, 3806-3811, 1998.

A.M.C. Janse, H.G. de Jonge, W. Prins, W.P.M. van Swaaij, "Combustion Kinetics of Char Obtained by Flash Pyrolysis of Pine Wood", Ind. Eng. Chem. Res. , 37, 3909-3918, 1998.

A. Shihadeh, "Rural electrification from local resources: Biomass pyrolysis oil combustion in a direct injection diesel engine", ScD Thesis, MIT, Dep. Of Mechanical Engineering, Sep., 1998.

C.R. Shaddix, P.J. Tennison, ""Effects of Char Content and Simple Additives on Biomass Pyrolysis Oil Droplet Combustion", 27th Symposium (International) on Combustion, The Combustion Institute, 1998.

D. Wang, S. Czernik, E. Chornet, "Production of Hydrogen from Biomass by Catalytic Steam Reforming of Fast Pyrolysis Oils", Energy & Fuels, 12, 19-24, 1998.

S. Czernik, D. Wang, E. Chornet, "Production of Hydrogen from Biomass by Catalytic Steam Reforming of Fast Pyrolysis Oil", Proc. U.S. DOE Hydrogen Program Review, Vol. II, NREL/CP-570-25315, pp557-576, 1998.

L. Garcia, M.L. Salvador, J. Arauzo, R. Bilbao, "Influence of Catalyst Weight/Biomass Flow Rate on Gas Production in the Catalytic Pyrolysis of Pine Sawdust at Low Temperatures", Ind.Emg. Chem. Res. 37, 3812. 1998.

1999

A.K. Burnham, R.L. Braun, "Global Kinetic Analysis of Complex Materials", Energy & Fuels, 13, No 1, 1-22, 1999

A. Borys, T. Platek, J. Wegrowski, "Antioxidative properties of smoke preparations obtained from ash and beech wood", Warsaw Meat and Fat Research Institute T.XXXVI, 197-206, 1999.

C. Roy, A. Chaala, H. Darmstadt, JAAP 51, 201-221, 1999.

M. Gronli, M..J. Antal, G. Varhegyi, "A Round-Robin Study of Cellulose Pyrolysis Kinetics by Thermogravimetry", Ind. Eng. Chem. Res., Vol. 38, No. 6, 2238-2243, 1999.

A. Oasmaa, S. Czernik, "Fuel Oil Quality of Biomass Pyrolysis Oils-State of the Art for End Users", Energy & Fuels 13, 914-921, 1999.

A.K. Burnham, R.L. Braun, "Global Kinetic Analysis of Complex Materials", Energy & fuels, Vol. 13, No. 1, 1-22, 1999.

A. Bridgwater, S. Czernik, J. Diebold, D. Meier, A. Oasmaa, C. Peacocke, J. Piskorz, D. Radlein, "Fast Pyrolysis of Biomass: A Handbook", CPL Press, 1999.

M. Marquevich, S. Czernik, E. Chornet, D. Montane, "Hydrogen from Biomass Reforming of Model Compounds of Fast-Pyrolysis Oil", Energy & Fuels, 13, 1160-1166, 1999.

M.D. Guillen, M.J. Manzanos, "Extractable Components of the Aerial Parts of Salvia lavandulifolia and Composition of the Liquid Smoke Flavoring Obtained from Them", J. Agric. Food Chem., 47, 3016-3027, 1999.

C. Di Blasi, C. Branca, "Global Degradation Kinetics of Wood and Agricultural Residues in Air", The Canadian J. of Chem. Eng., 77, 555-561, June 1999.

L. Helsen, E. Van den Bulck, S. Mullens, J. Mullens, "Low-temperature pyrolysis of CAA treated wood: thermogravimetric analysis", JAAP, 52, 65-86, 1999.

A.V. Bridgwater, D.Meier, D. Radlein, "An overview of fast pyrolysis of biomass", Organic Geochemistry 30, 1479-1493, 1999.

C. Di Blasi, G. Signorelli, C. Di Russo, G. Rea, "Product distribution from pyrolysis of wood and agricultural residues", Ind. Eng. Chem. Res. 38, 2216-2224, 1999.

X. Dai, M.J. Antal, "Synthesis of a High-Yield Activated Carbon by Air Gasification of Macadamia Nut Shell Charcoal", Ind. Eng. Chem. Res., 38, pp. 3386-3395, 1999.

M.S. Tam, M..J. Antal, "Preparation of Activated Carbons from Macadamia Nut Shell and Coconut Shell by Air Activation", Ind. Eng. Chem. Res. 38, pp. 4268-4276, 1999.

M.D. Guillen, M.L. Ibargoitia, "Influence of the Moisture Content on the Composition of the Liquid Smoke Produced in the Pyrolysis Process of Fagus sylvatica L. Wood", J. Agric. Food Chem. 47, 4126-4136, 1999.

R. Aguado, "Combustion and Pyrolysis of Wood Residues in Conical Spouted Beds", Ph.D Thesis, University of the Basque Country, Bilbao, Spain, 1999.

2000

Y. Solantausta, E. Podesser, D. Beckman, A. Ostman, R.P. Overend, "IEA Bioenergy Task 22: Techno-economic assessment for bioenergy applications", 1998-1999, Final Report, 2024. VTT Research Notes.

A.V. Bridgwater, G.V.C. Peacocke, "Fast pyrolysis process for biomass", Renewable and Sustainable Energy Reviews, 4, 1-73, 2000.

S. Chebil, A. Chaala, C. Roy, Fuel, 798, 671-683, 2000.

C. Amen-Chen, B. Riedl, C. Roy, 34th International Particleboard/Composite Materials Symposium, Pullman, Washington, April 3-6, 2000.

F. Chan, B. Riedl, X.M. Wang, C. Roy, X. Lu, C. Amen-Chen, Wood Adhesives 2000, Lake Tahoe, Nevada, June 2000.

L. Helsen, "Low-temperature pyrolysis of CCA treated wood waste", Ph.D. Thesis, Katholieke Universiteit Leuven, Belgium, 2000.

L. Helsen, E. Vanden Bulck, "Kinetics of the low-temperature pyrolysis of chromated copper arsenate-treated wood", JAAP, 53, 51-79, 2000.

D. Xianwen, W. Chuangzhi, L. Haibin, C. Yong, "The Fast Pyrolysis of Biomass in CFB Reactor", Energy & Fuels, 14, 552-557, 2000.

S.T. Srinivas, A.K. Dalai, N.N. Bakhshi, "Thermal and Catalytic Upgrading of a Biomass-Derived Oil in a Dual Reaction System", The Canadian Journal of Chemical Engineering, Vol. 78, pp.343-354, April 2000.

M. G. Gronli, M. C. Melaaen, "Mathematical Model for Wood Pyrolysis-Comparison of Experimental Measurements with Model Predictions", Energy & Fuel, 14, 791-800, 2000.

G.L. Juste, J.J.S. Monfort, "Preliminary test on combustion of wood derived fast pyrolysis oils in a gas turbine combustor", Biomass and Bioenergy, Vol. 19, Issue 2, pp. 119-128, 2000.

A. Sihadeh, S. Hochgreb, "Diesel Engine Combustion of Biomass Pyrolysis Oils", Energy & Fuels, 14, 260-274, 2000.

R. Aguado, M. Olazar, M.J. San Jose, G. Aguirre, J. Bilbao, "Pyrolysis of Sawdust in a Conical Spouted Bed Reactor. Yields and Product Composition", Ind. Eng. Chem. Res., 39, 1925-1933, 2000.

D. Radlein, "Study of Levoglucosan Production. A Review", RTI, Final Report, March 2000.

J.Piskorz, P. Majerski, D. Radlein, A. Vladars-Usas, D.S. Scott, "Flash pyrolysis of cellulose for production of anhydro-oligomers", J.A.A.P., 56, 145-166, 2000.

M.E. Boucher, A. Chaala, H. Pakdel, C. Roy, "Bio-oils obtained by vacuum pyrolysis of softwood bark as a liquid fuel for gas turbines. Part I: Properties of bio-oil and its blends with methanol and a pyrolytic aqueous phase", Biomass and Bioenergy, Vol. 19, Issue 5, pp. 337-350, 2000.

M.E. Boucher, A. Chaala, H. Pakdel, C. Roy, "Bio-oils obtained by vacuum pyrolysis of softwood bark as a liquid fuel for gas turbines. Part II: Stability and ageing of bio-oil and its blends with methanol and a pyrolysis aqueous phase", Biomass and Bioenergy, Vol. 19, Issue 5, pp. 351-361, 2000.

Y.O. Solantausta, "Cost and Performance Analysis of New Wood-Fuelled Power Plant Concepts", PhD Thesis, Aston, University, May, 2000.

G.L. Juste, J.J.S. Monfort, "Preliminary test on combustion of wood derived fast pyrolysis oils in a gas turbine combustor", Biomass and Bioenergy 19, 119-128, 2000.

J. Lédé, "The Cyclone: A Multifunctional Reactor for the Fast Pyrolysis Biomass", Ind. Eng. Chem. Res., 39, 893-903, 2000.

M.J. Antal, S.G. Allen, D. Schulman, X. Xu, "Biomass Gasification in Supercritical Water", Ind.Eng. Chem. Res., 39, 4040-4053, 2000.

M.C. Samolada, A. Papafotica, I.A. Vasalos, "Catalyst Evaluation for Catalytic Biomass Pyrolysis", Energy & Fuels, 14, 1161-1167, 2000.

US Patents issued in 1997-2000

D.S. Scott, J. Piskorz, D. Radlein, P. Majerski, "Process for the Thermal Conversion of Biomass to Liquids", U.S. Patent # 5,605,551 issued Feb. 25, 1997, priority data Nov. 26, 1992, assignee: University of Waterloo, Waterloo, Canada, European Patent 0 670 873, 12, 03, 1997.

S. Dumitriu, P.F. Vidal, E. Chornet, " Polyionic insoluble hydrogels from chitosan and xanthan", U.S. Patent # 5,620,706, issued April 15, 1997, assigned to University of Sherbrooke, exclusive license to Kemestrie Inc.

M. Mansour, K. Durai-Swamy, D. Warren, "Endothermic spent liquor recovery process", U.S. Patent # 5,637,192 issued June 10, 1997, assigned to Manufacturing and Technology Conversion International, Columbia, MD.

K.H. Oehr, G.A. Simons, J. Zhou, "Reduction of Acid Rain and Ozone Depletion Precursors",

U.S. Patent # 5,645,805, filed June 19, 1995, issued July 8, 1997, assigned to Dynamotive Corporation, Vancouver, Canada.

P.W. Moeller, "Method of making a tar-depleted liquid smoke", U.S. Patent # 5,637,339, appl. Apr. 27, 1995, issued Jun. 10, 1997, assigned to Hickory Specialties, Inc., Brentwood, Tenn.

R. Arsenault, M. Trottier, E. Chornet, P. Jollez, "Process of Aqueous Extraction of Maltol", U.S. Patent # 5,646,312 (July 9, 1997), assigned to Florasynth Inc. Teterboro, N.J.

D. Radlein, J. Piskorz, P. Majerski, " Method of producing slow-release nitrogenous organic fertilizer from biomass", U.S. Patent # 5,676,727, filed Dec. 7, 1995, issued Oct. 14, 1997, (priority data - Dec. 9, 1994).

D.N. Bangala, E. Chornet, "Steam Reforming Catalyst and Method of Preparation", U.S. Patent # 5,679,614 issued on Oct. 21, 1997, exclusive license to Kemestrie Inc.

G.L. Underwood, "Method of preparing a smoke composition from a tar solution", U.S. Patent # 5,681,603, filed Sep. 29, 1995, issued Oct. 28, 1997, assigned to Red Arrow Products Co., Inc., Manitowoc, Wis.

R.R. Suchanec, "Asphalt Emulsion with Lignin-Containing Emulsifier", U.S. Patent # 5,683,497, Nov. 4, 1997, assignee Hercules Inc. Wilmington, Del.

J. Monnier, G. Tourigny, D.W. Soveran, A. Wong, E. Hogan, M. Stumborg, "Conversion of biomass feedstock to diesel fuel additive", U.S. Patent # 5,705,722 issued Jan. 6, 1998, assigned to Natural Resources Canada, Ottawa.

K. Fischer, J. Katzur, R. Schiene, "Organic Fertilizer and Method of Manufacturing It", U.S. Patent # 5,720,792 issued Feb. 24, 1998, assigned to Technische Universitaet Dresden, Germany.

O. Brioni, D. Buizza, "Method and Apparatus for Producing Wood Charcoal by Pyrolysis of Wood-like Products or Vegetable Biomasses in General", U.S. Patent # 5,725,738, issued Mar. 10, 1998.

J. Piskorz, P. Majerski, D. Radlein, "Energy Efficient Liquefaction of Biomaterials by Thermolysis", U.S. Patent # 5,728,271, issued Mar. 17, 1998, assigned to RTI Resource Transforms International Ltd. Waterloo, Canada.

J. Shoop, G. L. Underwood, "Browning Composition and Method of Browning Dough-Based Foodstuffs", U.S. Patent # 5,756,140 issued May 26, 1998, assignee Red Arrow products Company Inc., Manitowoc, Wis.

D.B. Brown, J. Black, "Method for Ablative Heat Transfer", U.S. Patent # 5,770,017, issued Jun. 23, 1998, assigned to Ireton International, Inc. Nova Scotia, Canada, filed June 7, 1995.

B.A. Freel, R.G. Graham, "Method and Apparatus for a Circulating Bed Transport Fast Pyrolysis Reactor System", U.S. Patent # 5,792,340 issued Aug. 11, 1998, filed Apr. 7, 1995 assigned to Ensyn Technologies, Inc. Ontario.

F.A. Agblevor, "Process for Producing Phenolic Compounds from Lignin", U.S. Patent # 5,807,952 issued Sep. 15, 1998, filed May 2, 1996, assigned to Midwest Research Institute, Kansas City, Mo.

H. Ishitoku, T. Sugiwaki, M. Kawamura, T. Nakamoto, "Lignin Composition Method of Producing the Same and Dispersing Agent for Cement Used the Same", U.S. Patent # 5,811,527, issued Sep. 22, 1998, assigned to Nippon Paper Industries Co. Ltd. Tokyo.

D. Radlein, G. Simons, K.H. Oehr, J. Zhou, "Reduction of Nitrogen Oxides", U.S Patent # 5,817,282, issued Oct. 6, 1998, assigned to Dynamotive Technologies Co., Vancouver, Canada.

M. Ikura, S. Mirmiran, M. Stanclulescu, H. Sawatzky, "Pyrolysis Liquid-In-Diesel Oil Microemulsions", U.S. Patent # 5,820,640 issued Oct. 13, 1998, assigned to Natural Resources Canada, Ottawa, Canada.

R.J. Evans, H.L. Chum, "Pyrolysis and Hydrolysis of Mixed Polymer Waste Comprising Polyethyleneterephthalate and Polyethylene to Sequentially Recover", U.S. Patent # 5,821,553, issued Oct. 13, 1998, assigned to Midwest Research Institute, Kansas, MO.

G.L. Underwood, J.J. Rozum, "Method of removing hydrocarbons from liquid smoke compositions", U.S. Patent # 5,840,362 issued Nov. 24, 1998, assignee - Red Arrow Products, Inc., Manitowoc, Wis.

R. Holighaus, K. Niemann, M. Rupp, "Process for the processing of salvaged or waste plastic materials", U.S. Patent # 5,849,964, issued Dec. 15, 1998, assigned to Veba Oel Aktiengesellschaft, Gelsenkirchen, Germany.

J. Piskorz, P. Majerski, D. Radlein, "Energy efficient liquefaction of biomaterials by thermolysis", U.S. Patent # 5,853,548 issued Dec. 29, 1998, assignee: RTI Ltd, Waterloo, Canada.

T.M. McVay, F. Baxter, F.C. Dupre, "Reactive Phenolic Resin Modifier", U.S. Patent # 5,866,642 issued Feb. 2, 1999, assigned to Georgia-Pacific Resins, Inc. Atlanta, Ga.

P.S. Singh, "Method for Browning Precooked, Whole Muscle Meat Products", U.S. Patent # 5,952,027 issued Sep. 14, 1999, asssigned to Swift-Eckrich, Inc., Downers Grove, Ill.

B.A Freel, "Process to Produce Grilled Flavor Composition", U.S. Patent # 5,952,029, issued Sep. 14, 1999, assigned to Ensyn Technologies, Inc. Ontario, Canada.

B.A. Freel, R.G. Graham, "Apparatus for a circulating bed transport fast pyrolysis reactor system", U.S. Patent # 5,961,786 issued Oct. 5, 1999, Foreign Application Priority Data – Jan 31, 1990 [CA].

K.R. Kurple, "Modified Lignins", U.S. Patent # 6,054,562 issued Apr. 25, 2000.

C. Roy, H. Pakdel, "Process for the Production of Phenolic-rich Pyrolysis Oils for Use in Making Phenol-Formaldehyde Resole Resins", U.S. Patent # 6,143,856 issued Nov. 7/2000.

The recent application: B. Freel, R. Graham, "Natural resin formulation", PCT/CA99/00051; priority data: 30 Jan. 1998.

REVIEW OF METHODS FOR UPGRADING BIOMASS-DERIVED FAST PYROLYSIS OILS

S. Czernik
National Renewable Energy Laboratory, USA
R. Maggi
Université Catholique de Louvain, Belgium
G.V.C. Peacocke
Conversion and Resource Evaluation Ltd., UK.

INTRODUCTION

Fast pyrolysis has been developing for the last twenty years as a technology for producing liquid product, bio-oil, from biomass. The bio-oil has the potential to become an important liquid fuel and a source of chemicals. Yields of bio-oil as high as 60-75% based on the biomass weight can be obtained using different reactor types including bubbling fluidized beds[1], circulating and transported beds[2], cyclone reactors[3], rotating cone[4], and vacuum reactors[5]. At present, several fast pyrolysis technologies have reached near commercial status. Two circulating fluidized bed plants are operated by Red Arrow Products in Wisconsin, with a nominal capacity of 50 t/day. A 15 t/day installation based on the same technology has been constructed by ENSYN Technologies in Umbria, Italy. A bubbling fluidized process developed at the University of Waterloo (Canada) has been demonstrated at 3 t/day scale by Union Fenosa, a Spanish Electric Company, in Galicia, Spain. Dynamotive operates a 1.5 t/h fluidized bed unit in Vancouver, Canada and Wellman recently opened a 5 t/h plant in U.K. based on the same technology. Pyrovac constructed and started operation of a 100 t/h vacuum pyrolysis plant in Québec, Canada.

Over the years, bio-oils generated in different pyrolysis units from various biomass feedstocks have been analyzed, characterized, and tested for fuel applications. Combustion tests were performed by research institutions (Sandia National Laboratories, VTT, MIT, University of Kansas) and utility companies (Neste Oy, Wärtsilä, Orendra, Ormrod Diesels) using different scale burners and engines. These tests have demonstrated that the bio-oils could be used as a conventional fuel oil substitute. However, they also revealed several challenges in such applications resulting from the unique properties of bio-oils. This paper will review the work that has been done up to date on improving these undesired properties and upgrading quality of bio-oils as fuels.

PROPERTIES OF BIO-OILS

Bio-oils are usually dark brown, viscous liquids. They are comprised of different size molecules derived from depolymerization/fragmentation of cellulose, hemicellulose, and lignin. Therefore, the elemental composition of bio-oils resembles that of biomass. In contrast to petroleum fuels, bio-oils contain a large amount of oxygen, usually 45-50 wt. %, which is present in more than 200 compounds that have been identified in the oils[6]. The single most abundant component is water. The other major groups of compounds are carboxylic acids, hydroxyaldehydes, hydroxyketones, sugars, and phenolics. Most of the phenolic compounds are present as oligomers having a molecular weight in the range of 900 to 2500[7]. Typical properties of bio-oils important for fuel applications were compared to those for conventional petroleum fuel in a previous review[8]. The lower heating value of bio-oils is in the range of 14-18 MJ/kg, which is less than half of that for standard fuel oils. Such a low value results from the high water (15-25 wt. %) and high oxygen content (35-40 wt. % on water-free basis) of the bio-oils. Because of higher density, the heating value of bio-oils based on volumetric basis is about 60% of that for diesel fuel. The presence of water and oxygenated compounds is also a cause of a high polarity and, consequently, of the immiscibility of bio-oils with hydrocarbon fuels. A beneficial effect of water on bio-oil is reducing viscosity, thus improving its flow characteristics. Another undesired property of bio-oils is corrosiveness caused by carboxylic acids (pH of 2-3, acid number of 50-100 mg KOH/g). The corrosiveness to carbon steel and aluminium is especially severe at elevated temperatures[9,10]. Fortunately, the oils are non-corrosive to stainless steel and to polymeric materials.

Bio-oils usually contain solid particles, mostly sand and char entrained from the reactor. The particle content is in the range of 0.1 to 3% depending on the pyrolysis technology and on the efficiency of char removal from pyrolysis vapors. These solids are highly undesirable because they tend to settle at the bottom of the vessel in a form of sludge, can obstruct the flow and cause erosion of injection nozzles. Char is also a source of alkali present in bio-oil at the level too high for certain applications (turbines). In addition, char seems to catalyze reactions leading to the increase in bio-oil viscosit[11].

Probably the most challenging property of bio-oils is their chemical instability. The oils contain compounds that can polymerize. The main chemical reactions observed are polymerization of unsaturated hydrocarbon chains[12], esterification occurring between carboxyl and hydroxyl groups, and etherification, mostly formation of acetals by carbonyl and hydroxyl functionalities[13]. These reactions result in an unwanted increase of viscosity and a decrease of volatility.

PHYSICAL UPGRADING OF BIO-OILS FOR FUEL APPLICATIONS

The properties that negatively affect bio-oil fuel quality are foremost low heating value, incompatibility with conventional fuels, solids content, high viscosity, and chemical instability. The heating value can be significantly increased, but it requires extensive changes to chemical structure of bio-oils that will be discussed in the next section of this paper. The other undesired characteristics can be improved using simpler methods. For example, solids can be removed from bio-oil by filtration, viscosity can be reduced by addition of a suitable solvent, which also improves bio-oil chemical stability, and at least partial compatibility with conventional fuels can be achieved through emulsification. Below, we will review the state of the art of the relevant processes and methodologies.

Hot-Gas Filtration For Char Removal

In most of the existing pyrolysis plants, char is separated from gases and vapors using cyclones. However, the efficiency of cyclones to remove solids from gases is low for small particles, especially those below 10μm. For this reason a wide range of char content has been observed in the oils generated in different plants. More efficient char removal technology is essential to increase the quality of biomass pyrolysis oil. Filtration of the liquid oil has not been very successful. The oil tends to agglomerate around the particles and to form sludge, which results in plugging the filters and in losses of the oil adsorbed on char particles. Therefore, gas filtration prior to condensation was proposed as a more efficient method for char removal from the oil. Such filtration has been performed using sintered metal and flexible ceramic fabric elements. The filters were operated at about 400-420°C, the range suitable for avoiding oil condensation and to minimize the oil yield losses caused by thermal cracking[14]. Hot-gas filtration resulted in an oil yield of 50-55%, which is 10% less than that obtained in the same reactor system using cyclones. However, the ash content of the hot-filtered oil was less than 0.01% and the alkali content was less than 10 ppm - much lower than reported for biomass oils produced in systems using only cyclones. The solids removal also had a positive effect on lowering bio-oil viscosity and reducing its rate of increase during storage.

The thermal cracking of pyrolysis vapors during hot filtration, although decreasing the oil yield, had a beneficial impact on the oil quality. Cracking caused size reduction of oligomeric molecules, which not only decreased the oil viscosity but also affected its combustion chemistry. Diesel engine tests performed on crude and on hot-filtered oil showed a substantial increase in burning rate and a lower ignition delay for the latter, due to the lower average molecular weight for the filtered oil[15].

Hot gas filtration has not yet been demonstrated over a long-term process operation. Experiments performed to date revealed difficulties with effective removal of char cake from the filter surface when using traditional back-flush methods. New developments are needed to implement this technology in a commercial process.

Emulsification

Pyrolysis oils are not miscible with hydrocarbon fuels but with the aid of surfactants they can be emulsified with diesel oil. A process for producing stable micro-emulsions with 5-30% of bio-oil in diesel has been developed at CANMET[16]. The resultant emulsions showed promising ignition characteristics. A drawback of this approach is the cost of surfactants and the high energy required for emulsification.

Solvent Addition

Polar solvents have been used for many years to homogenize and to reduce viscosity of biomass oils. The addition of solvents, especially methanol, also showed a significant effect on the oil stability. Diebold[17] observed that the rate of viscosity increase ("aging") for the oil with 10 wt. % of methanol was almost 20 times less than for the oil without additives. This effect is due to physical dilution and to chemical reactions between the solvent and the oil components that prevent further chain growth. The chemical reactions that can occur between the bio-oil and methanol or ethanol are esterification and acetalization. Though thermodynamically non-favored, they can proceed to a significant extent if appropriate conditions are applied. For example, in the presence of an acid catalyst and molecular sieves (to adsorb water and to shift the reaction equilibria), bio-oil effectively reacted with ethanol forming ethyl acetate, ethyl formate, and diethoxyacetal of hydroxyacetaldehyde at the expense of formic acid, acetic acid, and hydroxyacetaldehyde[18]. Eventually, in addition to the decrease in viscosity and in the aging rate, other desirable changes like reduced acidity, improved volatility, heating value, and miscibility with diesel fuels were also achieved. The simplicity, the low cost, and beneficial effects of solvents, especially methanol and ethanol, on bio-oil properties, favor the application of solvent addition as a practical upgrading method.

CHEMICAL/CATALYTIC UPGRADING OF BIO-OILS

The chemical/catalytic upgrading processes aim at the removal of oxygen, which is the main cause of instability and other unwanted characteristics of bio-oils. They are more complex and expensive than physical methods, but offer significant improvements ranging from simple stabilization to high-quality fuel products[8].

Full deoxygenation to high-grade products such as transportation fuels was proposed in the 1980s. Two main routes explored were hydrotreating and catalytic vapor cracking. Hydrotreating of bio-oil carried out at high temperature, high hydrogen pressure, and in the presence of catalysts results in elimination of oxygen as water and in hydrogenation-hydrocracking of large molecules. The catalysts (sulphided CoMo or NiMo supported on alumina) and the process conditions are similar to those used in the refining of petroleum cuts. Because of the tendency to polymerization, a low temperature (250°C) pretreatment of bio-oil precedes the standard hydrotreating processing at 400°C[19,20]. Catalytic vapor cracking makes deoxygenation possible through simultaneous dehydration-decarboxylation over acidic zeolite catalysts. At 450°C and the atmospheric pressure oxygen is rejected as H_2O, CO_2, and CO producing mostly aromatics[21]. The low H/C ratio in the bio-oils imposes a relatively low limit on the hydrocarbon yield and, in addition, the technical feasibility is not yet completely proven. The catalyst deactivation still raises many concerns for both routes. The processing costs are high and the products are not competitive with fossil fuels[22].

The second approach is a partial upgrading aiming at the chemical stabilization that improves storage and handling of bio-oils for the use in internal combustion engines. Mild hydrotreating process seems to be a very effective method for this purpose. It requires much less hydrogen because only double bonds are hydrogenated and only the unstable oxygen (carbonyl and carboxyls) is eliminated. New developments in the field are activated carbon supported hydrotreating catalysts[23] that selectively reduce coke deposition. However, their activity needs more improvement. Promising performance showed Pd catalysts used at very mild hydrotreating conditions (hydrogen pressures as low as 2-3 bar)[24]. Vapor cracking using basic oxides instead of acidic zeolites and steam addition allowed for partial deoxygenation and an increase in hydrocarbon yield[25,26].

Similarly, using ZSM-5 or Y zeolite during pyrolysis in the fluidized bed or co-fed with the biomass results in significant changes in the bio-oil composition[27]. Other possible catalytic upgrading is the production of methyl aryl ethers (MAEs) from bio-oil in a fixed-bed reactor packed with alumina supported K_2SO_4[28]. These MAEs could be used as gasoline octane enhancers.

CONCLUSIONS

1. Pyrolysis liquids have unique properties that are being addressed by a number of research activities, aimed at improving their acceptance in boilers, engines and turbines.
2. It seems that at present only physical methods such as filtration and solvent addition can be economically viable. These methods improve certain important bio-oil properties, especially viscosity and stability.
3. Chemical processes of hydrogenation and catalytic cracking that can convert bio-oil into a high-grade hydrocarbon fuel, though technically feasible are too expensive and their products are not competitive with fossil fuels.
4. Emulsification of bio-oil with diesel fuel seems to be a more promising approach. Also reacting of bio-oil with alcohols can prove to be an attractive option but still requires more research and development effort.

REFERENCES

1. Scott, D.S.; Piskorz, J.; Radlein, D., Ind. Eng. Chem. Process Des. Dev. 1985, 24, 581-586.
2. Graham, R.G.; Freel, B.A.; Bergougnou, M.A., In Research in Thermochemical Biomass Conversion; Bridgwater, A.V., Kuester, J.L., Eds.; Elsevier Applied Science, London 1988; pp. 629-641.
3. Diebold, J. and Scahill, J., In Pyrolysis Oils from Biomass: Producing, Analyzing, and Upgrading; Soltes, E.J, Milne, T.A., Eds.; ACS Symposium Series 376, ACS, Washington, D.C. 1988; pp. 31-40.

4. Janse, A.M.C., Prins, W., van Swaaij, W.P.M., In Developments in Thermochemical Biomass Conversion, Bridgwater, A. and Boocock, D, Eds., Blackie Ac. and Prof., 1997, pp.368-377.

5. Roy, C.; de Caumia, B.; Menard, H, In Fundamentals of Thermochemical biomass Conversion, Overend, R.P., Milne, T.A., and Mudge, L.K., Eds, Elsevier Applied Science Publishers, London, 1988, pp. 237-245.

6. Elliott, D.C., 1986, Analysis and Comparison of Biomass Pyrolysis/Gasification Condesates – Final Report. PNL-5943, Contract DE-AC06-76RLO 1830.

7. Meier, D.; Scholtze, B., In Biomass Gasification and Pyrolysis, Kaltschmitt, M. and Bridgwater, A.V., Eds., CPL Press, Newbury, 1997, pp. 431-441.

8. Maggi, R. and Elliott, D., In Developments in Thermochemical Biomass Conversion, Bridgwater, A. and Boocock, D, Eds., Blackie Ac. and Prof., 1997, pp. 575-588.

9. Soltes, E.J.; Lin, J.-C.K., Hydroprocessing of Biomass tars for Liquid engine Fuels. In *Progress in Biomass Conversion*, Tillman, D.A. and Jahn, E.C., Eds., Academic Press, New York, 1984, pp. 1-69.

10. Jay, D.C.; Sipilä, K.H.; Rantanen, O.A.; Nylund, N.-O., Wood pyrolysis oil for diesel engines. In: International Combustion Engine Division of the ASME, 1995 Fall Technical Conference, September 24-27, 1995.

11. Agblevor, F.A.; Besler, S.; Montané, D.; Evans, R., Presented at ACS 209th National Meeting, Anaheim, CA, April 2-5,1995.

12. Polk, M.B. and Phingbodhippakkiya, M., Development of Methods for the Stabilization of Pyrolytic Oils, EPA-600/2-81-201, September 1981.

13. Czernik, S.; Johnson, D.K., and Black, S., Biomass & Bioenergy, 1994, 7, 187-192.

14. Scahill, J.W.; Diebold, J.P.: Feik, C.J, In Developments in Thermochemical Biomass Conversion; Bridgwater, A.V. and Boocock, D.G.B., Eds.; Blackie Academic & Professional, London 1997; pp. 253-266.

15. Shihadeh, A.L., Ph.D. Thesis, 1998, Massachussetts Institute of Technology, USA.

16. Ikura, M.; Slamak, M.; Sawatzky, H., US Patent 5,820,640, 1998

17. Diebold, J.P. and Czernik, S., Energy & Fuels, 1997, 11, 1081-1091.

18. Radlein, D.; Piskorz, J.; Majerski, P., European Patent # EP 0718392, 1999.

19. Laurent, E. and Delmon, B., In Biomass for Energy and Industry, Hall, D., Grassi, G. and Scheer, H., Eds., Ponte Press, Bochum, 1994, pp. 177-181.

20. Elliot, D. and Baker, E., In Energy from Biomass and Wastes X, Klass, D., Ed., 1983, pp. 765-782.

21. Chang, C. and Silvestri, A., J. of Catalysis, 1977, 47, p. 249.

22. Bridgwater, A., In Biomass for Energy and Industry, Kopetz, H., Weber, T. et al., Eds., CARMEN, 1998, p. 268-271.

23. Centeno, A. David, O., Vanbellinghen, C., Maggi, R., and Delmon, B., In Developments in Thermochemical Biomass Conversion, Bridgwater, A.V. and Boocock, D.G.B., Eds., Blackie Academic & Professional, London 1997, pp. 589-601.

24. Meier, D., Bridgwater, A.V., Di Blasi, C., and Prins, W., In Biomass Gasification and Pyrolysis, Kaltschmitt, M. and Bridgwater, A.V. Eds., CPL Press, Newbury, 1997, pp. 516-527.

25. Lahousse Ch., 1997, Private communication.

26. Williams, P. and Nugranad, N., In Biomass for Energy and Industry, Kopetz, H., Weber, T. et al., Eds., CARMEN, 1998, pp. 1589-1592.

27. Salter, E. and Bridgewater, A., ibid, pp. 1773-1776.

28. Samolada M. and Vasalos, J., ibid, pp. 1716-1719.

PRODUCTION, PROPERTIES AND USE OF WOOD PYROLYSIS OIL - A BRIEF REVIEW OF THE WORK CARRIED OUT AT RESEARCH AND PRODUCTION CENTRES OF THE FORMER USSR FROM 1960 TO 1990

G Dobele
Latvian State Institute of Wood Chemistry, 27 Dzerbenes Strett, Riga, LV-1006, Latvia

1. INTRODUCTION

Wood pyrolysis has a long history in Russia. In previous centuries, pine tar and birch bark tar formed a major Russian export. Since the 18th century, acetic acid has been obtained from the volatile products of wood pyrolysis as calcium salt. Methanol, ethyl acetate, acetone and formalin were all produced by pyrolysis in Russia in the 19th century. In the 1930s, the first two large-scale wood pyrolysis plants were built on the territory of the former USSR, in the towns of Sjava and Ashinsk. At present, Russian wood pyrolysis plants are equipped with vertical continuous-action retorts. The capacity of each retort reaches 70,000 m^3 of wood per year. All wood pyrolysis plants in the former Soviet Union have included full condensation of liquid products, with local automatic control systems at separate stages.

Considerable attention has been paid to studies on the obtaining of the liquid products of pyrolysis. A whole range of studies has been devoted to the investigation of the mechanism of thermal transformation of wood and its components, the improvement of pyrolysis technologies, obtaining of individual chemicals, as well as devising of new processes and equipment. The present review focuses mainly on the work of the following scientific schools: the Forestry Technical Academy (St. Petersburg); the Scientific Research Institute of the Wood Chemical Industry (Nizhnij Novgorod); the Institute of Chemistry and Chemical Technology of the Siberian Department of the Academy of Sciences of USSR and Technological Institute (Krasnojarsk); and the Institute of Wood Chemistry of the Latvian Academy of Sciences (Riga).

In the USSR, as world-wide, an interest in wood as an energy source arose mainly in connection with the inevitable depletion of natural resources of gas, oil and coal. The development of pyrolysis technologies in Russia was associated with the production of a solid energy raw material - char - and with wood gasification processes. Since the 1960s, considerable attention has been paid to the development of a production technology for individual chemicals, in particular, furfural and levoglucosan. The research in the field of fast pyrolysis and liquefaction of wood, aimed at oil production, has not received wide recognition.

1.1 Formation of Liquid Products During Pyrolysis of Wood

1.1.1 Obtaining and classification of oils

Useful liquid products are obtained from the vapour gases of pyrolysis by means of cooling or adsorption. Components are distilled over a wide temperature range, from below 40°C to above 300°C, depending on their boiling points. In the process of separating liquid products, some components tend to condense and to form oligomers and high-molecular compounds. As a consequence about half of the products are obtained as non-volatile pitch, due to the polymerisation and condensation reactions proceeding during repeated heating in the distillation process in the presence of organic acids.

During storage, the liquid condensate is commonly separated into three layers, namely, the upper layer - floating oils (less than 1% from dry wood, hereinafter referred to as oils), the middle layer

- oil and water, and the lower layer - heavy or settled oil. The yield and composition of the oil, oil-water and heavy oil depend on the wood species and pyrolysis regime (Figure 1.1).

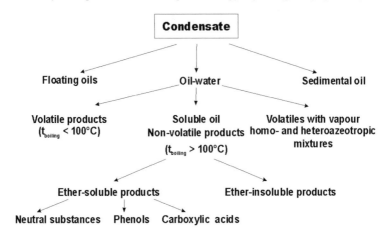

Figure 1.1   Classification of liquid products of pyrolysis

The composition of the water-soluble products (oil-water) of pyrolysis includes about 180 components. All the compounds may be divided into volatile ($t_{boiling}$ < 100 °C), volatile-with-vapour (jointly boiling homo- and hetero-azeotropic mixtures), and non-volatile (soluble oil, $t_{boiling}$ > 100°C) substances. In terms of the group composition, volatile products include: alcohols (methanol etc.), aldehydes (formaldehyde, acetaldehyde, etc.), ketones (acetones, methyl-ethyl ketone, etc.), ethers (dimethyl ether) and esters (methyl acetate, ethyl acetate, methyl formate, etc.). The boiling point of volatile-with-vapour organic substances is commonly above 100°C, while the boiling point for those distilled from oil-water as a result of the formation of azeotropes is 100°C and below. Volatile-with-vapour products comprise carbonic acids (acetic, formic, propionic, butyric and valerian acids and their isomers), alcohols (allyl, crotyl, furfuryl alcohols, etc.), aldehydes (furfural, hydroxy- methyl furfural, etc.) and some phenols.

The group composition of non-volatile products of oil-water (water-soluble oil) is rather complicated and is not completely understood. All oil-water components are divided into two groups, i.e. those soluble in organic solvents (diethyl ether) and those insoluble. Admittedly, the latter include the products of incomplete thermal destruction of the carbohydrate complex of wood, namely, hydroxyacids and lactones. Ether-soluble products are generally divided into neutral substances (non-interacting with alkalis), phenols (reacting with strong alkalis such as NaOH, KOH, etc.) and carbonic acids (entering into exchange reactions with all alkaline reagents).

Heavy oil, in contrast to soluble oil, does not contain any substances that are insoluble in ether. The group composition of heavy oil is analogous to that of the ether-soluble substances of soluble oil (neutral, phenols, carbonic acids). However, the components of heavy oil have a higher molecular mass, and the phenolic hydroxyls of the components are replaced mainly by methoxyl groups.

1.2   Contribution of Separate Components of Wood to the Formation of Oil

The yield and composition of liquid products obtained from pyrolysis of the separate components of wood are not equal, and are determined by their chemical structure.

Hemicelluloses are one of three main components of wood cell wall. Their content in wood reaches 43%. The hemicelluloses of coniferous species are less thermally stable than deciduous ones. Compared with other components of wood, hemicelluloses are less thermally stable in any raw material, and tend to be degraded during heating to temperatures of 175°C.

It is accepted that acetic acid is formed from the acetyl groups of hemicelluloses, and methanol mainly from methoxyl groups. The xylan content in wood determines the furfural yield from pyrolysis. In the products of the pyrolysis of xylan pre-treated with hydrogen peroxide, mono-atomic phenols, mainly o-cresol, are also found [1].

Cellulose, another main component of wood, has a higher thermal stability in contrast to hemicelluloses. The composition of cellulose pyrolysis products is complex, comprising more than 150 individual substances [2].

One group of substances retains the carbon skeleton of the elementary link of cellulose consisting of six atoms, namely, glucose, 1,6 anhydroglucopyranose (levoglucosan), etc.; furan derivatives (5-oxymethyl furfural and its derivatives; maltol, etc.). The remainder of the pyrolysis products result from the destruction of the elementary cellulose unit. Among these there are substances with three (glyceric aldehyde, dioxyacetone, hydroxymethyl glyoxal, methyl glyoxal, pyroracemic acid, acetone, acrolein), two (glyoxal, glycolic aldehyde, acetaldehyde, acetic acid) and one (formaldehyde) carbon atom. Compounds with a higher number of carbon atoms than in the cellulose elementary unit are also found, e.g. cyclo-octatriene, toluene, cresols, xylenols.

The main product of pyrolysis of cellulose under vacuum conditions and fast pyrolysis is *levoglucosan (LG)* and products of its dehydration.

Lignin is the third main component of wood. The aromatic nature of lignin predetermines the composition of heavy oil, which consists of 40-50% phenolic substances [3]. The main group of substances is comprised of phenols, their methyl ethers and alkyl derivatives (in the site of the cleavage of the propane chain). These can be divided into the following groups:

- monoatomic phenols such as phenol; o-, m-, p-cresol, xylenols (dimethyl phenols); trimethyl phenols; ethyl phenol; propyl phenol, etc.
- biatomic ortho-phenols such as pyrocatechol; 4-methyl and ethyl-pyrocatechols; monomethyl ether of pyrocatechol and, mainly, its para-substituted derivatives: methyl-, ethyl-, propyl and allylguaiacol, cis- and trans-isoengenols, vanillin, acetovanillin and many others.
- triatomic phenols such as pyrogallic acid, monomethyl ether of pyrogallic acid, dimethyl ether of pyrogallic acid and their alkyl derivatives, analogous derivatives of guaiacol.
- methadioxibenzenes-resorcin and its alkyl derivatives, p-dioxibenzene-hydroquinone, are present in the lignin pyrolysis oil in negligible amounts.

Another large group of heavy oil components comprises neutral substances. A whole range of benzene row aromatic hydrocarbons have been identified. The content of oxygen-containing neutral substances in heavy oil is 20-30%.

In addition to phenols and neutral substances, resin, higher fatty, aromatic and aliphatic acids are present in heavy oil (10-20%). Resin and higher fatty acids are formed in the oil distillation process, since these are present in all woody species in certain amounts, representing the extractives class. Such acids are not formed as a result of pyrolysis of isolated native lignin.

In oil-water, whose yield is much higher than that in the case of cellulose and hemicelluloses, substances such as acids, alcohols and aldehydes, ketones, ethers, etc. are present.

1.3   Effect of Pyrolysis Conditions

The yield of products depends on the characteristics of the raw material (raw material factors) and the pyrolysis regime (regime factors).

The raw material factors affecting the course of the pyrolysis process and the yield of products include the chemical composition of wood (dependant on the species); the presence of bark and rot; the size of wood chips; and moisture content. The acetyl and methoxyl groups of hemicelluloses are responsible for the formation of acetic acid and methanol, respectively, during pyrolysis. An increase in the lignin content in wood results in an increased heavy oil yield. In heavy oil from coniferous species, the phenols and pyrogallic acid derivatives prevail. Deciduous wood oil contains derivatives of both pyrogallic acid and pyrocatechol [4].

In the case of pyrolysis of wood affected by brown stem rot, the yield of methanol and acetic acid decreases, while the yield of heavy oil increases [5].

As pyrolysed wood chips increase in size, the yield of almost all liquid products, particularly soluble oil, acetic acid and methanol, tends to decrease [6]. As the sizes of fine particles of wood increase from 0.05 to 0.50 mm, the yield of liquid products changes moderately.

An increase in the moisture content of wood from 0% to 26% results in a 9.4% fall in the oil yield, mainly owing to an increase in the char yield [7,8]. In this case, the yield of methanol falls by approximately 28% and that for formic and propionic acid by almost 50%. The yield of furfural and furfuryl alcohol decreases moderately.

1.4   Effect of the Mode of Heating and Retention Time of Liquid Products in the Reaction Zone

Depending on the method of heat supply into the reaction zone, pyrolysis processes can be characterised as externally and internally heated. In apparatus with exterior heating, thermal degradation begins near the walls of the apparatus and, owing to weak circulation of the steam-gas mixture and the thermal conductivity of wood, proceeds unevenly. In apparatus with internal heating, the heat from the gas heat carrier is transferred directly to the wood, the latter is degraded evenly throughout the whole volume, and good conditions for the removal of pyrolysis products are obtained.

The effect of the gas-carrier (inert gases, hydrogen, carbon oxides) on the yield and composition of heavy oil from pyrolysis of birch wood at 700°C has been studied [9]. It has been shown that, as the gases are introduced, i.e. the residence time of volatile products in the reaction zone decreases, the yield of oil (especially high boiling and thermally unstable components) tends to increase. The nature of gas (inert, reductive, oxidative) does not have any effect on the oil yield. In the pyrolysis of birch wood in the flow of recirculating gases, specific capacity can reach 16.7 t of wood per m^3 of the reaction zone, oil yield can reach 37%, and char yield can fall to 12% [10].

The results of an investigation of the pyrolysis of wood in low-temperature argon plasma are of particular interest [11]. It is shown that, from pyrolysis of birch and spruce wood, it is possible to obtain soot at a yield of 19-26%. The soot has an optical density of less than 0.03 and is obtained at a specific power consumption of about 7 kWt*h/kg.

Investigations into the effect of heating rate on the yield of pyrolysis products have shown that, during slow pyrolysis, the release of the main part of the products is completed below 500°C. A comparison of the yield of pyrolysis products from birch wood shows that, as the heating rate increases from 3 to 20°C/min, the yield of char and non-condensable gases decreases by a factor of 1.5, and the yield of phenols grows by a similar factor [12]. Data presented in Table 1 show that an increase in the heating time of large-chip wood up to a temperature of 500°C from 3 hours to 14 days has a minor effect on the yield of methanol and acetic acid, but has a major effect on the yield of char and oil, i.e. oil yield decreases 10 times, and char yield, owing to cracking of resinous substances, tends to increase approximately 1.5 times [13].

Table 1.1. Effect of pyrolysis time on products' yield from birch wood

Heating time up to 500°C, hours	Yield, % from o.d. wood			
	Char	Volatile acids	Methanol	Oil
3	25.5	7.8	1.5	18.0
8	30.8	7.9	1.5	17.0
16	33.2	7.1	1.5	10.1
336	39.1	6.8	1.4	1.8

At high heating rates, the yield of liquid products increases and depends on the final temperature of pyrolysis. Thus, as the heating rate grows from 20 to 320°C/min, the yield of oil-water and heavy oil, including phenols, passes through a maximum at a temperature of 500°C [14]. Both a reduction and an increase of temperature by 100°C from this maximum point results in a more than 2-fold decrease in phenols yield (Figure 1.2) [15].

An increase in the residence time of pyrolysis products in the reaction zone results in decreased yields of oil owing to secondary cracking reactions. In this case, the most thermally stable components, i.e. those of lower molecular mass and thermo-reactive capacity, accumulate in the oil [15].

The alteration of pressure in the reaction zone has a similar effect to change in residence time [16].

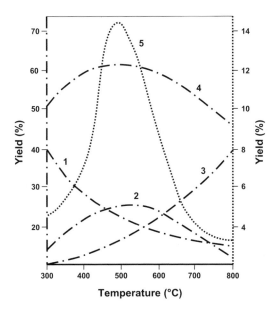

Figure 1.2 Yield of char, liquid, gas, condensate and phenols at different temperatures for pyrolysis of birch wood

151

1.5 References

1. A.Lopatin. Synopsis of Thesis for a Candidate's Degree, 1969, 20 pp. (in Russian).
2. I.Kirshbaum,, G.Domburg, G.Dobele, I.Berzina. Gas-chromatography analysis and identification of wood and its components thermolysis products. Khimija drevesini (Wood Chemistry), 1979, No. 2, pp. 80-90 (In Russian).
3. G.Domburg, G.Dobele, T.Sharapova, V.Sergeeva. Lignins structure influence on their thermodestruction in vacuum. Khimija drevesini (Wood Chemistry), 1976, No.5, pp. 64-72 (In Russian).
4. S.Sukhanovskij, E.Akhmina, and L.Stepanova. Changes in the physico- mechanical properties and capillary-porous structure of spruce, pine and birch wood and its basic components in hydrolysis and pyrolysis processes. In: The Chemistry and Use of Lignin, Zinatne, Riga, 1974, pp.164-174 (in Russian).
5. A.Rychkova. On the chemical composition of birch and aspen rotten wood. Zhurnal prikladnoj khimiji (Journal of Applied Chemistry), 1958, v. 31, No. 2, pp. 31-34 (in Russian).
6. L.Abduraghimov, A.Androsov and B.Tanchenko. Investigation of the regularities of thermal decomposition of wood under the action of exterior thermal flows. Fizika gorenija i vzryva (The Chemistry of Burning and Explosion), 1980, v. 16, No. 6, pp. 119-121 (in Russian).
7. V.Korjakin. Drying of Technological Wood in the Wood Chemistry Industry. Goslesbumizdat, Moscow, 1961, 293 p. (in Russian).
8. V.Korjakin and V.Akodus On the problem of the effect of the moisture content of wood on the yield of products of its dry distillation. In: Proceedings of Scientific Research Institute of the Wood Chemical Industry, Gorki', 1959, v. 13, pp. 23-27 (in Russian).
9. V.Pijalkin, S.Gurilev, E.Tsiganov and A.Slavjanskij. Improvement of methods for wood pyrolysis and new trends in utilising pyrolysis products. In: Proceedings of Wood Forestry Akademie, Leningrad, 1969, v. 132, pp. 74-78 (in Russian).
10. V.Fjodorov, A.Slavjanskij and V.Pijalkin. Electromagnetic pyrolysis of ground wood. Izvestija. Vuzov, Lesnoj zhurnal (Proceedings of Higher Schools. Wood Journal), 1972, v. 52, pp. 52-58 (in Russian).
11. V.Pijalkin, A.Nikinen, V.Zaitsev and V.Fjodorov. Pyrolysis of wood in low-temperature plasm. Wood Chemistry and Tapping. Proceedings of Scientific and Technical Papers, 1973, No. 8, pp.11-127 (in Russian).
12. Z.Matvejeva. Investigation of the Composition and Origin of Phenols of Low-Temperature Pyrolysis of Wood. Synopsis of Thesis, Leningrad, 1970, 20 p. (In Russian).
13. A.Slavjanskij and F.Mednikov. Technology of Wood Chemistry Productions, Lesnaja promyshlennost', Moscow, 1970, 392 p. (in Russian).
14. A.Slavjanskij. New Methods for Pyrolysis. Lesnaja promyshlennost', Moscow, 1965, 253 p. (In Russian).
15. A.Kislitsin. Pyrolysis of Wood: Chemical Activity, Kinetics, Products, New Processes. Lesnaja Promyshlennost', Moscow, 1990, 313 p. (in Russian).
16. Technology of Wood Chemistry Productions. Ed. V.Virodov, A.Kislitsin, V.Gluhapeva. Lesnaja promyshlennost', Moscow, 1987, 350 p. (in Russian).

2. SEPARATION AND COMPOSITION OF PYROLYTIC OILS, PRODUCTION OF END PRODUCTS

2.1 The Condensation of Volatile Products of Pyrolysis

The vapour-gas mixture formed during pyrolysis of wood contain a wide variety of components which differ in their physical and chemical properties. The liquid products of pyrolysis are represented by the range of components in the vapour-gas mixture which can be condensed at atmospheric pressure using a cooling medium at a temperature no less than 5°C.

Two device types for liquid products condensation are used in Russia's wood chemistry laboratories; scrubbers (direct-contact condensers) and surface-type heat-exchangers.

Cooled oil-water is commonly used as a cooling liquid for the direct-contact condenser, in which liquid products are condensed as a result of a decrease in temperature and are absorbed by the oil-water. The efficiency of recovery of liquid products reaches 99%.

Surface-type heat exchangers are of a shell-and-tube design. Here, the volatile products are led through pipes, and the cooling liquid (water) is supplied to the inter-tube space. To improve the separation of light boiling liquid products (methanol, acetone, methyl acetate, etc.), the tube space is also sprinkled with oil-water. Up to 98% of all liquid products of wood pyrolysis are condensed in surface-type heat-exchangers.

2.2 Chemical Composition and Use of Oils

As has been mentioned above, the total condensate released from the pyrolysis vapour-gas mixture separates into three layers, namely, floating oils, oil-water and heavy oil (see Figure 1.1). An approximate composition for pyrolysis oils is given in Table 2.1. Altering the pyrolysis regimes, especially the heating rate and residence time of the vapour-gas mixture in the reaction zone, can cause the content of organic substances in the condensate to vary substantially, mainly at the expense of altering the content of soluble and heavy oils (Figure 2.1).

Table 2.1. Main components of pyrolysis oils.

Component	Yield,%			
	Birch moisture content,%		Spruce moisture content,%	
	8.8	33.7	7.2	33.0
Floating oils	1.0	0.7	1.8	1.2
Acids	9.8	5.6	6.4	2.9
Alcohols	4.0	2.3	2.6	1.0
Esters	4.0	3.6	3.2	2.1
Aldehydes	0.9	0.6	0.5	0.2
Ketones	1.9	1.1	1.2	0.8
Heavy oil	16.5	6.7	20.4	5.6
Soluble oil	17.1	4.2	12.5	5.0
Other organic compounds	0.3	0.2	0.2	0.1
Water	44.5	75.0	51.2	81.1

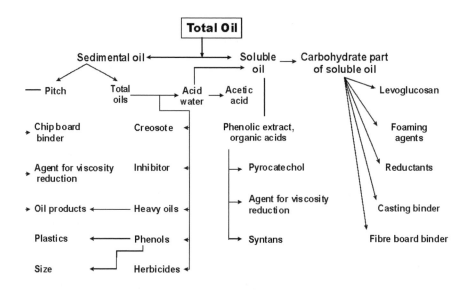

Figure 2.1 Scheme of wood oil processing

Oil-water contains (in descending order): acids (acetic, formic, propionic, butyric, valerianic etc.); alcohols (methanol, propanol, allyl alcohol, crotyl alcohol, furfuryl etc.); ketones (acetone, methyl ketone, methyl-propyl ketone etc.); aldehydes (acetaldehyde, formaldehyde, furfural etc.), methyl ethers of acetic, formic, propionic etc. acids; phenols, methyl ethers of phenols, lactones, hydroxy-acids and a whole range of other compounds.

Heavy oil contains 45-60% of *phenols*, 25-30% of *neutral substances* and 10-15% of *carbonic acids*. **Heavy oil** *phenols* are represented mainly by their incomplete methyl ethers, at levels no less than 40% in the case of the pyrolysis of coniferous species and no less than 60% in the case of the pyrolysis of deciduous species. Pyrogallic acid derivatives and pyrocatechol derivatives prevail in the oil *phenols* of deciduous and coniferous species respectively.

Neutral substances of **heavy oil** consist of oxygen-containing derivatives (75%) and hydrocarbons (25%). The hydrocarbon part contains aromatic compounds such as naphthalene, dimethyl naphthalene, dehydroretene and paraffins. The oxygen-containing part consists of alcohols (about 40%), aldehydes and ketones (about 30%).

The acidic part of the **heavy oil** includes aliphatic, aromatic and resin carbonic acids. The elemental composition of heavy oil is almost constant at C - 64-68%, H - 7-9%, O - 26-28%.

The yield of **soluble oil** from pyrolysis with exterior heating is 7-8%, and that from pyrolysis with interior heating and processing of ground wood reaches 16-25%. The pyrolysis oil of coniferous species chips processed by the gasification method consists of 15-20% of organic volatile acids, 30-35% of oxyacids and oxyacid lactones, up to 30% of levoglucosan and 15-20% of phenols. The phenols content in the soluble oil of deciduous wood species is 40-45%.

The scientific literature contains a whole range of studies on the various uses of pyrolysis oils. In the present review, only a few are considered [1-10].

The **light floating oils** obtained from pyrolysis of birch wood are used for production of tar or a veterinary preparation (barbs-control).

The **soluble oil** can be separated by extraction with organic solvent (diethyl ether) into the carbohydrate and phenolic parts. The phenolic extracts are used for obtaining *technical pyrocatechol*, for synthesis of *tanning agents* (syntans) and decreasing the clay drilling mud viscosity. The carbohydrate part of soluble oil-water, evaporated to a moisture content of 25-30%, has been used as a phenol-free *casting binder,* and also as a binder for producing fibreboards by the dry method.

One of the pathways for application of the residue formed after separation of levoglucosan is the production of *dyes,* since the water-soluble compounds as well the oil-insoluble part contain chromophoric groups [11]. In the presence of mordants, the components of oil residue derived from both birch wood and corn can dye the wool and synthetic fibres into colours ranging from yellow to dark-brown, depending on the solution concentration and time of treatment. The dying is even and resistant to physico-mechanical actions. Using these oil residues, qualitative dyes were obtained by reaction with diazosulphonyl or diazoantranyl acids. The product of diazo coupling can be used in the form of a solution or a powder (in the composition of dry dyes) obtained by evaporation of the solution under a vacuum.

The method for obtaining dyes is as follows. Pyrolysis oil is dissolved with 0.1 N NaOH and cooled to 5°C. Simultaneously, a solution of diazosulphonyl or diazoantranyl acid is prepared. A potassium carbonate solution is added such that the solution would retain an acidic reaction towards litmus, and mixed for 15 min. A portion of the acid solution is poured to the cooled solution of pyrolysis oil at a rate such that the temperature of the reaction medium remains at 5°C, and 2N NaOH is added to ensure the constant alkalinity of the medium (pH 10-11). A further 20 minutes mixing completes the reaction, and then the product is ready for use.

The soluble oil obtained after the separation of levoglucosan from corn cob pyrolysis oil and the distillation of acetone from it, was used for obtaining *a tanning agent.* For this purpose, the oil was diluted with water to a 40% moisture content and oxidised with hydrogen peroxide in a reactor at constant mixing and cooling [12]. The obtained solution, containing 14% of substances determined as tannins, was prepared for experimental lots of tanned leather. The tanning agent may be obtained in the form of a powder.

The residue obtained after separation of levoglucosan was tested and displayed good results as *a plasticiser* for concrete mixtures [13].

For obtaining *the agent for viscosity reduction* in drilling oil and gas wells, **soluble oil phenols** are condensed with formaldehyde in an acidic medium at a temperature of 80°C. The formed novolak, after washing with hot water, is dissolved in alkali and sulphamethylated. The obtained product is evaporated and dried. The reagents consumption (t) per 1 t is as follows:

Soluble oil		0.9-1.1
Formalin, 40%		0.22
Water-free sodium sulphite	0.25	
NaOH, 92%		0.11
Technical H_2SO_4	0.02	

For obtaining of a *casting binder* the extracted **oil-water** or **soluble oil** is evaporated to a density of 1270-1300 kg/m^3. Sometimes, oil-water is evaporated directly and a non-phenol-free binder is obtained, although the phenol content and rather a high acidity restricts its application. In this case, to improve the binder quality, it is neutralised with lime (lime milk). An additional treatment of the obtained product with alkali solution (NaOH), or its mixing with moulding clay and sulphite liquor taken after fermentation for ethanol production, enables the obtaining of a binder with new properties.

Soluble oil is used to obtain a *cold curing wood oil* for foundry production. For this purpose, the soluble oil is oxidized with oxygen from air, with simultaneous distillation of water to a final moisture content of no more than 15%. A curing agent, manganese ore or chromium anhydride, is introduced directly prior to use. The mixture is low-toxic and ensures a notable improvement of labour conditions in foundries as compared to the commonly used mixtures based on synthetic resins.

For obtaining *hydroconcrete plasticisers* used in building for hydroinsulation, **the soluble oil** is neutralised with a 40% NaOH solution to pH 10-12, then lime milk is added at a rate of 2% CaO from oil. The final product contains up to 50% of water.

Smoking agents for hot and cold smoking of meat, dairy and fishery products, have been obtained from **pyrolysis oil**. For this purpose, the oil is washed with water and the wash water is evaporated to a density of 1270-1300 kg/m^3. One of the smoking agents is the residue from evaporation diluted 10 times with water prior to use. The distillate from evaporation has also been used for smoking. An analogous smoking agent can be obtained by evaporating the soluble oil in the manufacture of the non-phenol-free casting binder.

Saponified wood oil, synthetic wood varnish, *oil for the regeneration of rubber* and other products have been obtained from **heavy oil**.

Saponified wood oil is a product of saponification with alkali. It is used in building as an air involving admixture (0.1-0.3% from the cement mass) for mortars or concrete mixtures.

The *wood oil varnish* is a solution of oxidised pyrolysis oil, modified with formaldehyde. The **heavy oil** is oxidised with air oxygen at temperatures of 80-90°C for 2 hours. The oxidised oil is condensed with formaldehyde for 3-4 hours without a catalyst. A curing agent, e.g. chromium anhydride, is introduced in an amount 4-5% to the condensation product. An equal (by volume) amount of organic solvent is added to the mixture obtained. The products obtained are used as anti-burn coatings, improving the quality of the surface of steel castings and reducing the labour input for the cleaning of castings.

The *oil for rubber regeneration* is produced from coniferous wood pyrolysis **heavy oil**. The quality of the rubber softener is conditioned by the content of resin and higher fatty acids. More often, the **heavy oil** is used for rubber regeneration immediately after washing and distillation under a vacuum (the removal of volatile acids, water and light boiling substances).

The major pathway for processing **heavy oil** is *distillation* with utilisation of the obtained oil fractions and pitch.

Wood-tar oils and *pitch* are produced by batch or continuous distillation under a vacuum. In the case of the batch method, the oil is distilled in a still supplied with a small rectification column. Live steam is simultaneously supplied into the still. In this case, three fractions are extracted: *light oils* with a boiling point up to 180°C; *creosote oils* with a boiling point in the range 180-240°C; *inhibiting oils* (antioxidants) with a boiling point in the range 240-310°C.

In the case of continuous distillation in tube furnaces, two oil fractions are extracted, namely creosote and inhibiting oils. In both cases, pitch remains in the residue at levels of 45-55% of the mass of organic matter in the oil.

Wood-tar *creosote oils* display antiseptic properties and are used for preservative treatment of leather in tanning plants, and sometimes also for preservative treatment of wood.

The wood-oil *inhibitor* is used as an admixture to cracking-benzene (0.1% by mass) for the purpose of stabilisation.

The wood-resin *inhibitor* is used also as an interrupter of polymerisation processes in the production of synthetic rubber.

Wood-oil pitch is used mainly for obtaining a *wood-pitch binder* and a *oil-binder* in the production of granulated active carbons for manufacturing pig iron and steel casting moulds, as well as an agent preventing the formation of burnt-on sand in casting moulds.

For preparation of *a bi-component binder*, a pitch with a softening temperature $\geq 70^{\circ}$C is ground and mixed with dry moulding clay in the ratio 7:3. A more efficient *three-component binder* is obtained from pitch and clay, and dry sulphite-yeast mash is added in the ratio 50:25:25.

The oil evaporated to a moisture content of 3-4% is used for *char granulation*. In another method for the same purpose, the pitch is dissolved in different oils. For example, 55 to 60 parts of pitch are mixed with 15 to 20 parts of creosol oils and 20 to 30 parts of light oil (or green petroleum oil) as a thinner. If necessary, the formulation is modified according to the required viscosity of the finished product.

There are various other proposed uses for wood oils and their distillation products. However, they are not competitive with their counterparts based on natural fossil resources.

2.3   References

1.   A.Kiprianov. Investigation of Wood Resin Processing Processes. Synopsis of Thesis, Leningrad, 1970, 31 p. (in Russian).
2.   V.Koryakin. Thermal Degradation of Wood. Goslesbumizdat, Moscow, 1962, 294 p. (in Russian).
3.   A.Kislitsin and A. Chashchin. Prospects for use of dendrochemical products as a raw material for organic synthesis. Prospects for Utilisation of Wood as a Raw Material. Zinatne, Riga, 1982, pp. 67-78 (in Russian).
4.   L.Azhar and E.Levin. Utilisation of tar-water of pyrolysis of hydrolysis lignin for wood preservation. Derevoobr. promyshlennost' (Woodworking Industry), 1971, No. 3, pp. 17-19 (in Russian).
5.   J.Juriyev, A.Kiprianov and J.Judkevich. Separation of phenols from soluble tar with acetone solvents. Ghidroliznaja i lesokhimicheskaja promyshlennost' (Hydrolysis and Wood Chemistry Industry), 1977,. No. 8, pp. 21-22 (in Russian).
6.   J.Goldshmit, Z.Rodionova and G.Petrovicheva. New avenues of the processing   and use of wood tars. Thermal Processing of Wood and Its Components, Krasnojarsk, 1988, p. 66 (in Russian).
7.   M.Shirokova and S.Smetanina. Obtaining of new products on the basis of components of thermal processing of wood. Ibid., p. 66 (in Russian).
8.   S.Smetanina and E.Chuprov. Prospects for use of surface-active additive. Ibid., p.   70 (in Russian).
9.   L.Gusher, J.Goldshmit and L.Kasilova. A binder for production of granulated active carbons. *Ibid.*, p. 71 (in Russian).
10.   V.Kozlov and A.Smolenskij. Floating reagents - foaming agents from wood tar," Problems of Pyrolysis of Wood and Extractives of Coniferous Species. Proceedings of the Institute of Forestry Problems", Riga, 1958, pp. 127-139 (In Russian).
11.   A.c. USSR, No. 340672 (In Russian)
12.   A.Rijkuris. Composition   and Properties of Oil   Formed at Vacuum-Pyrolysis of Lignocellulose with Superheated Steam. PhD Thesis, 190 p. (In Russian).
13.   A.c. USSR, No. 833713 (In Russian)

3. CONVERSION OF WOOD AND LIGNIN IN THE PRESENCE OF LIQUID MEDIA.

3.1 Process of the Liquefaction of Hydrolysis Lignin in Solvent Medium

A process of the liquefaction of hydrolysis lignin in a medium of low-boiling solvents, isopropanol, ethanol and methanol [1,2], was studied in the 1980s as part of more general studies on the obtaining of liquid fuel from biomass. These studies were preceded by investigations into the obtaining of a liquid fuel from coal, shale and peat by the method of supercritical dissolution in lower aliphatic alcohol, which serves as a proton donor.

Commercial hydrolysis lignin used as the raw material for liquefaction was obtained by percolation hydrolysis, with diluted sulphuric acid.

Experiments were carried out in an autoclave (3 l) with an electromagnetic stirrer at a ratio of solvent to organic mass of lignin of 10:1. The heating rate was $3^{o}C/min$. As the required temperature was reached, the furnace was switched off. The formed liquid products were distilled from the solid residue under a vacuum at $60^{o}C$.

A decrease in the conversion degree in the order of: isopropyl alcohol > ethyl alcohol > methyl alcohol is attributed by the authors [1] to a decrease in the donor capability of the alcohols. The conversion degree of lignin depends essentially on temperature and pressure, i.e. an increase in pressure from 11 to 35.5 MPa at a temperature of $400^{o}C$ increases the conversion degree of lignin in isopropyl alcohol from 59% to 83%. The conversion degree of lignin is increased by a decrease in moisture content (from 70% to 11%) and an increase in the alcohol:lignin ratio.

The liquid product after distillation represents a viscous dark liquid flowing at room temperature. Its yield depends on the process conditions and reaches 60 wt% of the organic mass of lignin.

A comparison of the physical and chemical properties of the liquid product of the supercritical solution of lignin and fuel oil (Table 3.1) shows that these are close, and the product obtained from lignin may be used as a low-sulphur boiler fuel.

The lignin liquefaction gas contained: CO_2 8-22%; CO 8-15%; C_nH_{2n} 35-50%; C_nH_{2n+2} 30-40%; H_2 and C_2, each 0.5-1.5% (volumetric basis). The chromatographic analysis shows that the hydrocarbon part of gases consists of C_1-C_4 compounds, among which propane (35.2% mol) and propylene (38.3%) prevail. This testifies that the major contribution in gas formation is made by the destruction of alcohol [1]. The solid carbonised residue has a heat of combustion of about 25 MJ/kg and may be recommended for use as a solid fuel [1].

As isopropyl alcohol is rather expensive, the possibility was studied of activating the process of liquefaction of hydrolysis lignin in methanol using catalysts of Fe-Zn-Cr oxide composition with alkaline additives. High efficiencies have been demonstrated [2].

The catalyst was introduced by way of impregnation. The impregnated hydrolysis lignin (15 g) and methanol (30 ml) were loaded into the autoclave (250 ml), heated to $380-400^{o}C$ and stirred, and held at this temperature for 2 h. The heating and cooling times were 1 h and 2 h, respectively. The reaction mass from the autoclave was freed from methanol, water and easy-boiling compounds, by drying it under a vacuum. Maltenes and asphaltenes, whose yield was 41% from hydrolysis lignin, were separated by hexane and benzene extraction. Taking into account the gas formation, the conversion degree of the organic mass of hydrolysis lignin was close to 90%.

Table 3.1. Physical and chemical properties of furnace fuel oil and liquid product of the supercritical dissolution of hydrolysis lignin in isopropyl alcohol[1]

Characteristics	Furnace fuel oil	Liquid product of supercritical dissolution
Kinematic viscosity at 80°C, centistokes, no more than	118.0	100.0
Ash content, % no more than	0.14	0.1
Sulphur content, % no more than	0.5-3.5	0.2-0.6
Flash point in an open crucible, °C, no lower than	110.0	105.0
Solidification temperature, °C, no lower than	25.0	15.0
Heat of combustion, kJ/kg, no lower than	$39.7*10^3$	$33.0*10^3$-$35.5*10^3$

In earlier studies on the catalytic liquefaction of brown coal in methanol [3] and its catalytic hydrogenation with molecular hydrogen, the catalytic process of the liquefaction of hydrolysis lignin was described by the following scheme:

$$Zn\text{-}Cr$$
$$CH_3OH+H_2O \leftrightarrow CO_2+3H_2$$

$$Fe\text{-}Zn\text{-}Cr$$
$$lignin + H_2 \leftrightarrow product\ Ar\text{-}H$$

$$product\ Ar\text{-}H + CH_3OH \leftrightarrow product\ Ar\text{-}CH_3 + H_2O$$

It is noted that the Zn-Cr composition accelerates the methanol decomposition reaction, while the Fe oxide strengthens the hydrogenation activity of the catalytic system.

The maximum yield of maltenes (hexane soluble fraction), the most valuable part of the liquid product, reached 37.6% from lignin with the use of the Fe-Zn-Cr catalyst (taken in the quantities 10, 1.3, 0.67% of lignin) and 47.3% with the use of caustic potash (15% of lignin), at a process temperature of 380°C. The total yield of maltenes and asphaltenes was 67.0% and 53.1% respectively.

An increase in the process temperature to 410°C in the presence of the Fe-Zn-Cr catalyst decreased the maltenes yield to 17.6%, while the total yield of maltenes and asphaltenes increased to 82.6%.

Maltenes obtained from the liquefaction of hydrolysis lignin and brown coal have a very similar elemental composition, i.e. 83.7% C, 8.6% H, 7.4% O for maltenes from lignin (380°, the catalyst Fe-Zn-Cr taken in the amount 10; 1.3 and 0.67%, respectively) and 83.4% C, 9.1% H, 7.2% O% for maltenes from brown coal (380°C, the Zn-Cr catalyst), despite the distinctions in the elementary composition of the source material (hydrolysis lignin: 61.8% C, 5.83% H, 31% O; brown coal: 70.8% C, 4.98% H, 23% O).

The possibility of joint processing of the products of the liquefaction of hydrolysis lignin and brown coal has been suggested [2].

3.2 Liquid Fuel on the Basis of Gasification Products

A process for obtaining liquid hydrocarbons and olefins by the catalytic pyrolysis of the gasification products of raw materials (wood processing wastes, logging debris, agricultural wastes) has been developed. Gasification can be carried out with steam or air [4].

In contrast to steam gasification, 50% of the products of air gasification consists of nitrogen. Therefore, the synthesis of engine fuels of C_5-C_{22} composition from such a gas was a new concept [5]. However, air gasification results in a lower yield of products suitable for the synthesis of paraffin hydrocarbons found in the composition of diesel fuels and olefins. Thus, the feedstock is used less effectively in the case of air gasification than in the case of steam gasification (Table 3.2.).

Table 3.2. Basic indices of the process of hydrocarbon synthesis from plant biomass gasification products [5]

The catalyst 32% Co - 2% Mg - 3% ZnO_2 - kieselguhr; pressure 1 MPa, volume rate of raw materials supply $100 m^3$/h.

Composition of gasification products				Optimum temperature of synthesis, °C	Yield of synthesised hydrocarbons, g/m^3				Composition of liquid hydrocarbons, %		
CO	CO_2	H_2	N_2		C_1	C_2-C_4	C_5-C_{22}	total	olefin	paraffins	
										branched	normal
Steam gasification											
33	33	3	-	190	22	2	1	161	21	17	22
20	35	4	-	220	45	2	1	193	5	13	82
45	20	4	-	250	24	3	7	159	13	36	51
		5				4	1				
		3				2	1				
		5				3	4				
							1				
							0				
							7				
Air gasification											
30	5	1	5	230	3	9	5	65	4	16	80
30	5	5	0	230	2	4	2	48	13	27	60
15	15	1	2	210	12	1	4	66	3	22	75
		5	5			3	1				
		2	5				4				
		0	0				1				

The summary yield of liquid hydrocarbons, boiling above C_4, from gases obtained by wood waste steam gasification was 107-117 g/m^3, and that for C_2-C_{22} fractions was 139 g/m^3. Taking into account the fact that, in the case of gasification with steam, 1 kg of raw material yields about 2 m^3 of gas, the yield of liquid oil will be 200-250 g per 1 kg of plant raw material, and that of the C_2-C_{22} fraction 250-300 g. In the case of air gasification, 3 m^3 of gas is formed from 1 kg of the raw material, and 40-50 g of hydrocarbons from 1 m^3 of gas, i.e. 1 kg of the raw material yields 120 to 150 g of hydrocarbons, from which different products can be obtained, analogous to petroleum products.

The process of obtaining C_2-C_4 olefins from plant biomass comprises the following stages: gasification of the raw material with steam at 1000-1200°C; synthesis of liquid hydrocarbons on a Co catalyst (200°C, 1 MPa) from gasification gases; pyrolysis of the C_2-C_{22} synthesis products with a KVO_4 catalyst [6] on a solid carrier such as kieselguhr or corundum at 790°C (time of

160

contact 0.2 s or less). A part of the heat obtained by gasification can be used for warming up the raw material for catalytic pyrolysis.

The catalytic pyrolysis of liquid hydrocarbons was realised under semi-industrial conditions in furnaces consisting of 24 pipes (an inner diameter of 100 mm), containing 900 L of vanadium catalyst (granules with a diameter of 10-15 mm). A functional check of the installation has been performed in a continuous regime for 2 months.

Table 3.3. Olefins (C_2-C_4) yield upon catalytic pyrolysis of liquid hydrocarbons synthesised on the basis of plant biomass gasification products

Olefines	Yield (kg) per 10 t biomass	
	with air	with steam
Ethene	350-450	1500
Total C_2-C_4	550-600	1000
Butadiene	40	80

3.3 Liquefaction Of Biomass In The Presence Of Compounds Of Alkaline Metals And Iron.

To increase the efficiency of wood conversion to liquid products, melts of formates of potassium and sodium as well as complex catalysts of reduced iron were investigated in the late 1980s [7, 8].

Studies showed [7] that liquefaction of wood (spruce sawdust), affected by formates, proceeded under atmospheric pressure. Formates are reductants of wood carbonyl groups at temperatures below the decomposition point. It has been established that the reduction of carbonyl groups to carbinol ones by formate occurs at a temperature of 300°C. Upon further heating to 450°C, liquid products with a low oxygen content (8-9%) are formed. The formate ions formed during the degradation of carbohydrate components can also reduce oxygen containing products of wood alkaline pyrolysis. Yield of oil after wood liquefaction at 450°C depends on the formate-alkali ratio and reaction time at 300-320°C and varies from 12 to 15% from dry wood.

The conversion of wood in the presence of systems based on reduced iron was carried out under a working pressure of 1 MPa [8]. A mixture of sawdust, iron oxide (Fe_2O_3) and carbonates of alkali metals was loaded into the reactor. Argon was passed through the reactor, and the mixture was heated from 1.5 h up to 850°C, then held at this temperature until the production of CO stopped (0.5-1 h). The solid residue containing sodium carbonate, metallic iron and a small admixture of Fe_2O_4 (data of X-ray analysis) was mixed with a new portion of wood, and the cycle was repeated.

In the presence of reduced iron, the oil yield increases 1.5 times, and the proportion of the wood carbon passing into the oil increases from 16 to 24%. The yield of water-soluble products decreases by about 30% and their carbon content decreases from 1.84% to 0.85%.

As the number of liquefaction cycles increases, the oil yield gradually decreases, due to the accumulation of charcoal and, possibly, non-active iron carbides in the solid residue.

The replacement of high-melting Na_2CO_3 (melting point 852°C) for a low-melting eutectic mixture of sodium, potassium and lithium carbonates (melting point 400°C) ensured the decrease of the oxygen content in oils from 13.5 to 10.8%. The addition of Fe in the eutectic mixture did not affect the oxygen content, in contrast to its action in the composition with Na_2CO_3. In this process, oxygen is removed in the form of CO and not H_2O, as it is commonly in the pyrolysis process.

3.4   Production of Levulinic Acid

Levulinic acid, which has application in plastics production, was also an object of investigation for researchers engaged in the area of thermo-catalytic transformation of the plant raw material.

A method was proposed for the obtaining of levulinic acid from technical pulp [9]. The pulp was first dissolved in concentrated acids, then diluted with water. In the case of hydrochloric acid, the process was carried out in an autoclave heated to 180-190°C with superheated steam for 5 min. Then the aqueous solution of hydrochloric and formic acids was distilled, and the crude levulinic acid was separated from humic substances on a nutch-filter and distilled under a vacuum. The yield of levulinic acid was 44% from dry pulp, and the degree of its purity was 99.4%.

It was also shown that the treatment of aspen wood in the presence of sulphuric acid and its salts in an autoclave at 240°C resulted in the formation of 16% (on o.d. wood) of levulinic acid [10]. Under these conditions, 34% of levulinic acid can be obtained from wood cellulose [11].

A method for the obtaining of levulinic acid from lignocellulose residue after pre-hydrolysis of sulphuric acid upon furfural production has also been investigated [12]. The method included lignocellulose treatment in a vortex flow of superheated steam at a raw material - heat carrier ratio of 1:3-1:10. The levulinic acid yield was 7-10% and the char yield was 40-50%.

3.5   References

1. F. Teghay, V. Menshov, V. Ryzhkov, V. Korniets and E Korniets. Liquefaction of hydrolysis lignin by the method of supercritical dissolution in lower aliphatic alcohols. Khimija tverdogo topliva (Chemistry of Solid Fuel), 1984, No. 5, pp.91-96 (in Russian).
2. V. Taraban'ko, G. Gul'bis, A. Kudryashev, M. Shipyakova, E. Shevtsov, B.Kuznetsov, A. Rumailo, and O.Bondarenko. The alkali-metal formates affected liquefaction of wood under atmospheric pressure. Khimiya drevesiny (Wood Chemistry), 1989, No. 1, pp. 95-99 (Russian).
3. V. Taraban'ko, N. Beregovceva, N. Ivancenko, E.Korniets and P. Kuznetsov. Investigation of influence of methanol on the process of coal catalytic hydrogenation. Khimija tverdogo topliva (Chemistry of Solid Fuel),1985, №4, pp.76-81.
4. J..Pushkin, A..Lapidus and S..Adelson. Plant biomass as a raw material for obtaining olefins and motor fuels. The Chemistry and Technology of Fuels and Oils, 1994, No. 6, pp. 3-5 (in Russian).
5. J. Pushkin, G. Golovin, A. Lapidus, A..Krylova, V.Gorlov and V. Kovach Obtaining of motor fuels from plant biomass gasification gases. Khimija tverdogo topliva (Chemistry of Solid Fuel), 1994, No. 3, pp.62-72 (in Russian).
6. A.c.USSR No. 277743 (in Russian).
7. V.Taraban'ko, G. Gul'bis, A. Kudryashev, M. Shipyakova, E. Shevtsov, B.Kuznetsov, A. Rumailo and O.Bondarenko. The alkali metals formates affected liquefaction of wood under atmospheric pressure. Khimiya drevesiny (Wood Chemistry), 1989, No. 1, pp.95-99 (in Russian).
8. V.Taraban'ko, B. Kuznetsov, G. Gul'bis, A. Kudryashev, E. Shevtsov. (1989) Study of Wood Pyrolysis Process in the Presence of Iron. Khimiya drevesiny (Wood Chemistry), 1989, No. 5, pp. 76-79 (in Russian).
9. A.c. USSR No. 249364 (in Russian).
10. A.Efremov, G. Slashchinin, E. Korniets and B.Kuznetsov. Catalytic thermolysis of aspen wood in the presence of sulphuric acid and ferric, cobalt and aluminium sulphates applying high pressures. Khimiya drevesiny (Wood Chemistry), 1990, No. 5, pp. 57-60 (in Russian).

11. A.Efremov, V. Kuznetsov, A. Konstantinov, S. Kuznetsova and V.Krotova. New thermocatalytic methods of chemicals producing from lignocellulose materials in the presence of acid-type catalysts. Proceedings of 8th International Symposium on Wood and Pulping Chemistry, 1995, 6-9 June, Helsinki, Finland, Vol. 1, pp. 689-696.
12. A.c. USSR No 947177 (in Russian).

4. UTILIZATION OF AGRICULTURE WASTES FOR PYROLYTIC PROCESSING

Agriculture wastes (corn cobs, sunflower husk, rice husk) were tested as a raw materials for the obtaining of chemicals by pyrolysis.

Chemicals such as furfural can be obtained from the pyrolysis of sunflower husk, pre-treated with a catalyst solution, with a yield of approximately 8% from the raw material and germanium catalyst (0.015 g GeO_2 per 1 t of raw material) (Table 4.1).

Table 4.1   Yield of chemicals obtained by pyrolysis of sunflower husk impregnated with acid or salt solutions [1]

Chemicals	Yield, % in terms of dry husk	
	at 8% H_2SO_4	At 10% $ZnCl_2$
Furfural	8.9	7.4
Acetoacetic sodium	9.2	8.8
Total phenols	5.8	6.1
Neutral oils	5.2	5.3
Active carbon	10.8	12.0
Germanium dioxide	15 mg/kg	was not isolated

For obtaining germanium, by-products of the processing of non-ferrous metal ore, tar-water from the cake production process, and charcoal combustion ash are commonly used. Germanium is isolated from ash by treatment with 31% HCl at 105-110°C. A similar technique was used for the isolation of germanium from the carbonised residue of the pyrolysis of sunflower husk, from which, after washing and calcination, activated carbon was obtained [1].

Prior to pyrolysis, sunflower husk was crushed to particles 1-3 mm in size, in order to disturb the surface wax-like film and provide conditions for more homogeneous impregnation. Pyrolysis was carried out in a retort with a temperature gradient of 130-150°C in the upper part to 420-460°C in the lower part. The residence time of the raw material in the retort was 90 min.

Phenols and neutral oils were removed from the retort with furfural as a part of the liquid condensate and separated from the furfural by vacuum distillation. About 40% of the phenolic fraction (i.e. 3% from the initial husk mass) was represented by cresol isomers.

Rice hulls, comprising 20% from the weight of the commercial rice produced, contain 16-18% of silicon oxide. Therefore, the pyrolysis process was used as a pretreatment stage for manufacturing silicon of semiconductor purity for solar batteries (photoelectric transducers). In this case, novolak resins were synthesised from the obtained liquid products, for use at other stages of the pure silicon production process [2].

The resins synthesized from the liquid products were also tested as binders for different pressed and granulated materials. It is shown that the carbonised residues can be used in adsorption chromatography as well as for a commercial sorbent [2,3].

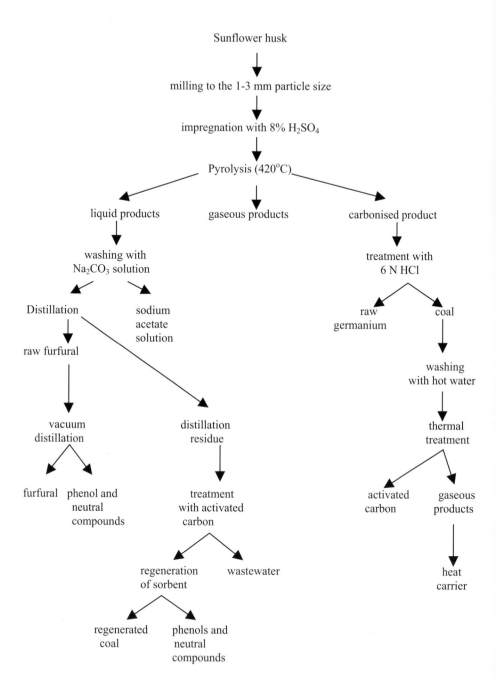

Figure 4.1 Scheme of a process for obtaining chemicals by pyrolysis of sunflower husk [1]

The product obtained upon pyrolysis in an air atmosphere (white ash), after crushing and screening the fractions with particle size 0.20-0.25 mm, had a bulk density of 0.264 g/ml, a pore specific volume 1.6 ml/g, and a pH (5% aqueous suspension) of 7.1. In terms of its properties, this product is comparable with the known carrier, "Chromaton" (produced in the Czech Republic), but its surface is more inert, which makes it possible to exclude the preliminary silanisation of the carrier, commonly used for chromatography of extremely polar substances (methanol, acetone, acetic acid etc.).

The solid residue obtained upon the thermal pyrolysis of rice hulls and hydrolysis lignin of rice hulls in an inert atmosphere (black ash), according to data from chemical and rentgenography analyses, consists of amorphous silicon dioxide and carbon. This product can be used as a commercial sorbent, analogous to active carbon, while the narrow fraction screened after the crushing of "black ash" is suitable for gas-adsorption chromatography. The efficiency of a column filled with "black ash" is higher than that of a column of active carbon.

4.1   References

1.   L.Panasyuk. Assessment of the Possibility of Obtaining Phenols and other Chemical Products from Plant Raw Materials. 1966, Dnepropetrovsk, 23 pp. (in Russian).
2.   L. Saprykin.Basic regularities of the thermal degradation of rice hulls", PhD Thesis, 1989, Riga, 156 pp. (in Russian).
3.   A.c. USSR No. 1352350.

5.   PYROLYSIS TECHNOLOGIES FOR OBTAINING OF INDIVIDUAL CHEMICAL COMPOUNDS

5.1   Levoglucosan

5.1.1   Mechanism of levoglucosan formation upon thermal degradation of cellulose.

The main product of pyrolysis of cellulose under vacuum conditions and fast pyrolysis is levoglucosan (LG) and the products of its dehydration.

There are several theories explaining the process of thermal depolymerisation of cellulose (Shafizadeh, Halpern and Patai; Levin and Basch; Hardiner; Wodley et al). The two foremost mechanisms for the cleavage of cellulose molecule are by the homolytic process with the formation of free radicals, or by the heterolytic process with the formation of intermediate carbonium ions, then anhydrosaccharides. In the former USSR, one of the basic hypotheses for levoglucosan formation was put forward by Golova and co-authors when studying the thermal degradation of cellulose under a vacuum; later this theory was developed by Sergeyeva, Kalnins, Kislitsin [1-3].

From data on the change in the degree of polymerisation of cellulose and levoglucosan yield upon heating at 305°C under a vacuum, Golova and co-authors have established that the process of thermal degradation of cellulose proceeds in two stages. It was concluded that the first stage of degradation proceeds basically in less ordered regions, mainly with the cleavage of bonds in the elementary unit. At this stage, the total cellulose mass loss is minor, i.e. about 4-8%. LG is formed with a low yield, and degree of polymerisation falls to the value 200-300. At the second stage, a chain process of depolymerisation proceeds in more ordered regions by way of successive splitting off of the elementary units and their isomerising with the formation of LG and new active centres. At this stage, the degree of polymerisation does not change, degree of decomposition is directly proportional to heating time, and the levoglucosan yield rapidly reaches a constant value, i.e. about 75% from the mass of decomposed cellulose.

The authors assumed that thermal depolymerisation is a free-radical process with an intramolecular transfer of the chain, whose initial act is the homolytic breaking of the 1,4-glucoside bond between the C(1) atom and oxygen [4-6]. In the authors' opinion, the presence

165

of hydrogen bonds increases the thermal stability of crystallites, which ensures their retaining up to temperatures sufficient for the homolytic dissociation of the glycoside bond. Owing to the proceeding of depolymerisation in the solid phase and the absence of conditions for diffusion, the radical process is localised in the "cage" and does not transfer into the volume. Thereby, the direction of the levoglucosan formation process is ensured.

5.1.2   Factors affecting the levoglucosan obtaining process

Different factors affecting the depolymerisation process and LG yield were established. For example, it has been shown that the increase of the ash content of cotton cellulose from 0.001 to 0.07% decreases the levoglucosan yield from 65-70 to 40% [1].

Inorganic salts, being introduced into cellulose, promote under the thermal action the cleavage of cellulose via C-C bonds with the formation of products of deep degradation.

In terms of the efficiency of the inhibition of the LG formation reaction, chlorides of alkaline metals are arranged in the following series: lithium < sodium < potassium < caesium according to the inverse dependence on the ion radii value [7].

Prokhorov and co-authors assumed that the degree of structural order of cellulose after alkaline and acidic pretreatments tends to increase, and the standard entropy of cellulose tends to diminish, which is responsible for the increase in the LG yield [5,8].

The effect of the additives of organic substances such as glucose as well as naphthylphenyldiamine, a known inhibitor of free-radical reactions, has been studied. The LG yield upon the introduction of these organic additives also decreased [4].

No reagents accelerating the process of cellulose depolymerisation and promoting the increase of LG were found. The positive effect of some reagents upon pyrolysis of wood was attributed to the catalysis of degradation of hemicelluloses (not affecting the cellulose structure) which led to the development of a material porous structure, thereby providing a more intensive evacuation of the LG formed. The increase in the LG yield upon birch wood pyrolysis observed upon the introduction of monochloroacetic acid, manganese chloride and iodic acid was explained by this mechanism [9,10].

It has been established [11,12] that the relationship between two basic directions of thermodegradation of cellulose, namely, dehydration and depolymerisation, determines the yield of the corresponding products and depends on the pyrolysis temperature, heating rate and other factors (Figure 5.1).

The summarization of data relating to the mechanism of thermal degradation of cellulose was carried out by many researchers (Shafizadeh, Philipp, Levin, Bash, Franklin, Golova, Domburg, Kislitsin et al.). However, a whole range of issues are still under discussion, since the studies were performed under non-identical conditions and on different kinds of celluloses. It is objectively pointed out in the scientific literature that the thermo-degradation of cellulose represents a complex process consisting of a whole range of parallel and successively proceeding reactions [13]. The thermo-degradation of cellulose depend on the material properties and pyrolysis conditions. Therefore, when discussing the results of the studies obtained under different conditions, the following should be taken into account [14]:

1.   The cellulose degradation process includes the stage of degradation in disordered regions and subsequent degradation of crystallites according to two competing reactions, i.e. dehydration and depolymerisation. The ratio of these reactions is affected by the temperature-time parameters of the process.

2. The supramolecular structure of cellulose as well as its degree of polymerisation and admixtures strongly affect its chemical transformations.
3. The kinetic parameters of the process are dependent on the change in the physical and chemical characteristics of the system proceeding upon thermo-destruction.
4. It is correct to compare the kinetic parameters of the thermal degradation of cellulose obtained under conditions of isothermal and non-isothermal heating in the temperature range of the maximum destruction rate.

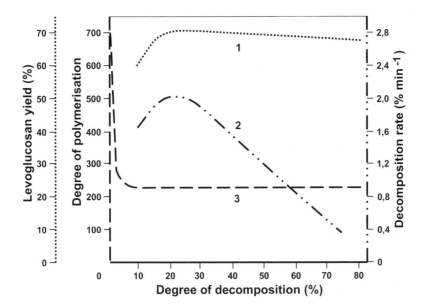

1 Levoglucosan yield, % from cellulose degraded at the present moment
2 Degradation rate
3 Degree of polymerisation of cellulose

Figure 5.1 Dynamics of the process of cellulose pyrolysis indices at 300°C and the pressure 7*10^{-3} Pa.

Studies of the mechanism of thermal depolymerisation of cellulose formed the basis of the commercial method for levoglucosan production from plant raw materials with full utilisation of wastes formed. Based on the regularities of levoglucosan formation, it was proposed to carry out a process of cellulose pyrolysis under a high vacuum in a flow of superheated steam or inert gases in reactors with exterior heating [15]. The above-mentioned measures made it possible:

- irrespective of the residual pressure value, to displace the air from the reactor, thereby to prevent the oxidative processes;
- to perform a direct contact of the heat carrier with the raw material and provide an even heating;
- to use a heat carrier for speeding up the evacuation of thermo-destruction products from a high temperature zone.

This method was realised under pilot and industrial plant conditions.

5.1.3 Levoglucosan production technology at the Krasnodar Hydrolysis Plant

In late 1950's, a research programme "Comprehensive Processing of Plant Materials" was conducted by research workers at a range of scientific research institutes of the Academy of Sciences of the former USSR and the Latvian Institute of Wood Chemistry. The programme objective was the realisation of a pyrolysis technology for obtaining valuable products, including technical levoglucosan, at the Krasnodar Hydrolysis Plant [16].

In the first stage of the work, two pilot pyrolysis reactors, one continuous with exterior heating and a raw material capacity of 5-10 kg/h and one batch with interior heating and a raw material loading of 2-5 kg, were designed. Pyrolysis was carried out under a vacuum and in a flow of superheated steam. Corn cobs were the main raw material.

The experience on the pilot units provided a wealth of data on the process of obtaining levoglucosan. In 1968, the construction of a larger unit was accomplished.

When formulating a new project (Institute of Wood Chemistry, Latvia), the possibility of using different types of raw material was taken into account, including corn cobs, deciduous wood, sun-flower husks, etc. obtained after pre-hydrolysis in different regimes directed to obtaining of xylite or furfural. The pre-hydrolysis units were not specially designed, since it was already intended to use the corn cob lignocellulosic residue obtained upon xylite production (treatment of corn with diluted sulphuric acid) at the Krasnodar Hydrolysis Plant, and the deciduous wood lignocellulosic residue obtained upon furfural production (treatment of wood with concentrated sulphuric acid) on a unit manufactured by the Processing Design Office of the Institute of Wood Chemistry, Latvia. Two reactors for pyrolysis were proposed; a continuous action reactor with a capacity of 100 kg/h raw material, and a batch reactor with a loading of 270 kg raw material.

In the technology for levoglucosan production realised at the Krasnodar Hydrolysis Plant (Figure 5.2), the lignocellulosic residue (moisture content 75%), after pre-hydrolysis of corn cobs, was heated to $90^{\circ}C$ in a solution of 2% sulphuric acid. Lignocellulose was separated and twice washed with water, then was washed three times with condensates of vapour and vapour units for removal of sulphuric acid and sodium salts.

The washed lignocellulosic residue was filtered and dried in a drum dryer up to a moisture content of 20%, then sorted and supplied to the pyrolysis reactors.

Pyrolysis was carried out in a flow of superheated steam under a vacuum (80-100 mm of mercury column) in the temperature range $280^{\circ}C$ to $500^{\circ}C$. The volatile products of pyrolysis were supplied to the condenser system. The condensate obtained from the batch reactor was directed to a vacuum evaporation unit, where it was evaporated to a paste-like state (residual moisture content 10-12%). The paste was supplied to the levoglucosan crystallisation unit, and the condensate was supplied to the washing of the lignocellulosic residue.

When operating the continuous reactor, the fractional condensation of products was envisaged. In this case, the concentrated solution obtained at the first step of condensation contained the majority of levoglucosan. The water condensate obtained at the second and third steps was collected and used for washing the lignocellulosic raw material.

In the crystalliser, the concentrated LG solution was mixed with acetone for cooling. Then the mixture was supplied to a pressure filter. The mother liquor was removed under the pressure of carbon dioxide, and the LG crystals were washed with acetone. Then LG was supplied to a purification station. Acetone was distilled from the mother liquor for recycling the solvent to production. The residual tar was loaded to a container for subsequent use, mainly as a binder for casting forms and as a tanning agent.

168

Figure 5.2 is broken down into 6 sections for clarity. These are linked as shown in each part of the figure.

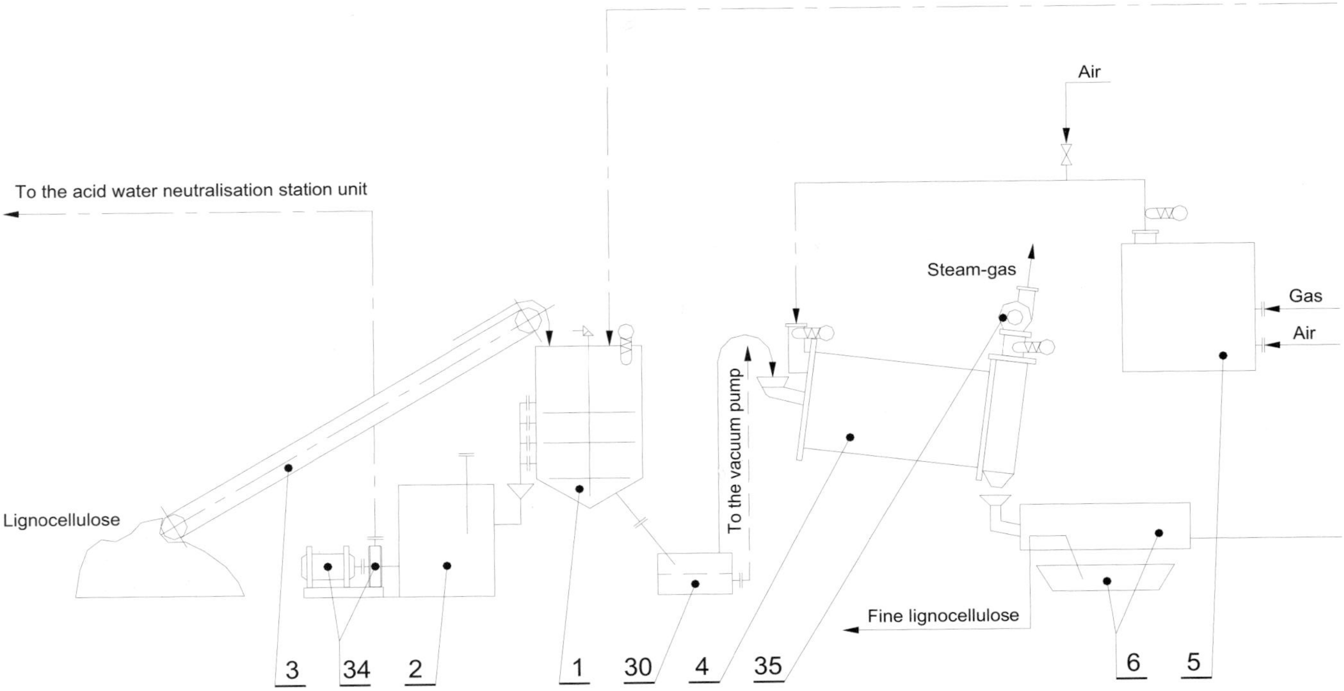

Figure 5.2A

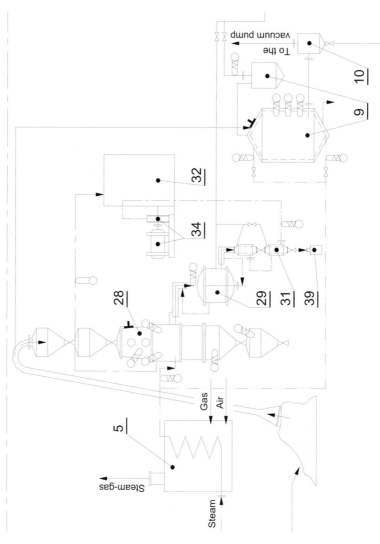

Figure 5.2B

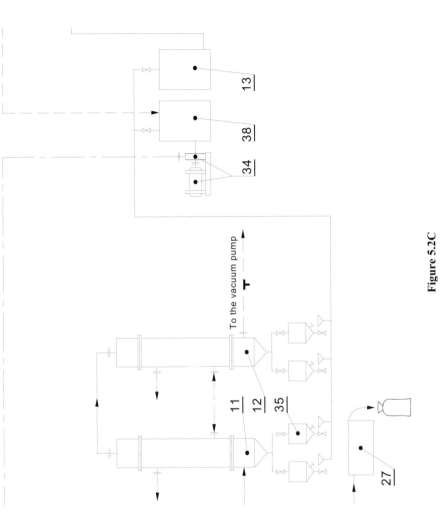

Figure 5.2C

To the vacuum pump

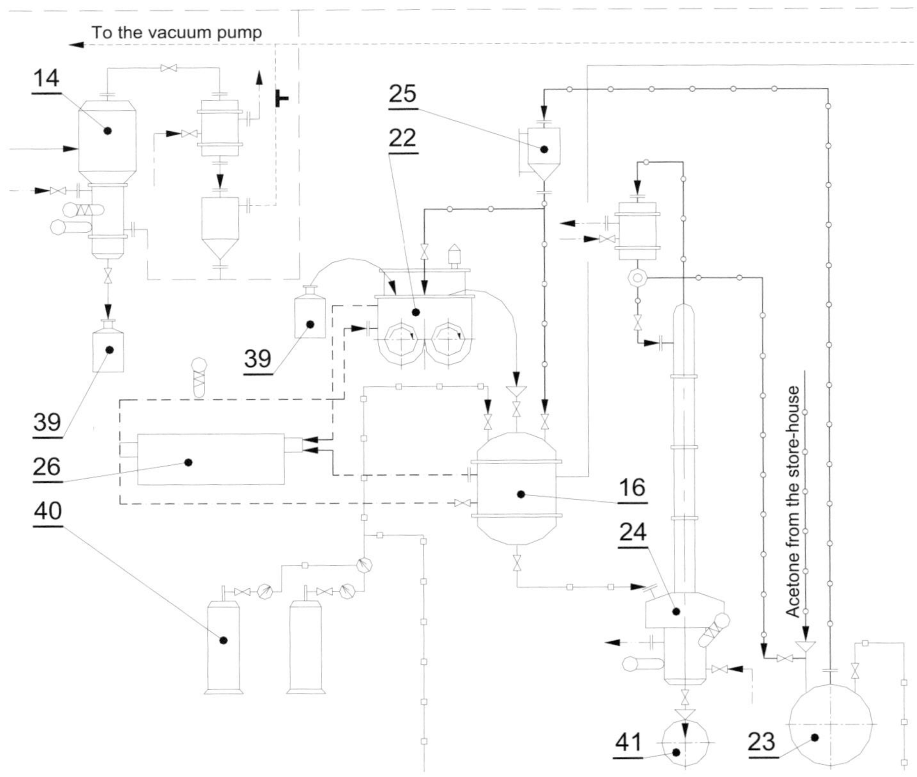

To the vacuum pump

Acetone from the store-house

14

25
22

39

39
26
40

16
24

23

41

Figure 5.2D

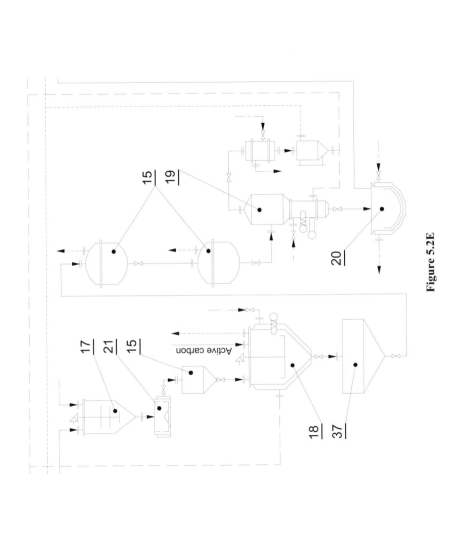

Figure 5.2E

15
19

17
21
15

Active carbon

18
37

20

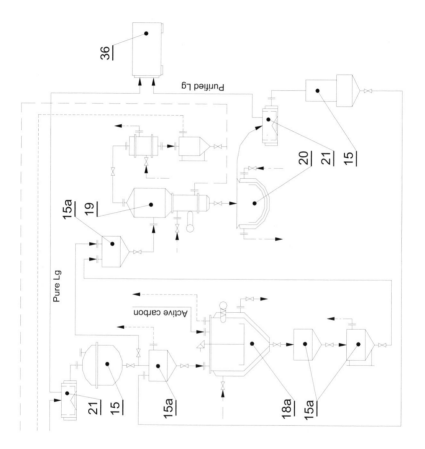

Figure 5.2F

Specifications

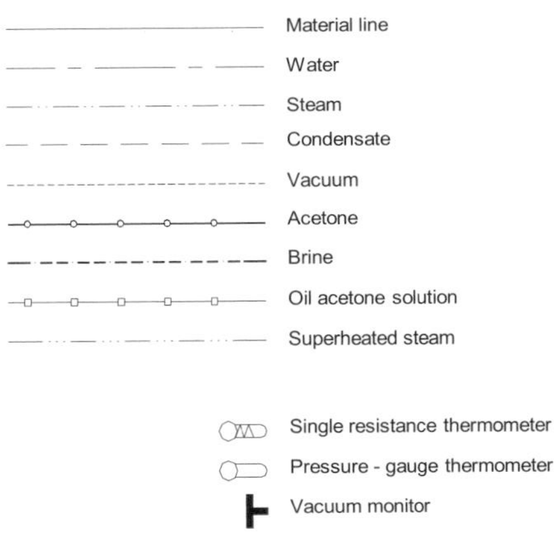

———	Material line
— — —	Water
— ·· —	Steam
— · —	Condensate
‑‑‑‑‑	Vacuum
—◇—◇—	Acetone
‑ ‑ ‑ ‑	Brine
—▫—▫—	Oil acetone solution
— · ▫ · —	Superheated steam

⊶⊷ Single resistance thermometer

⊂⊃ Pressure - gauge thermometer

⊥ Vacuum monitor

1	Souring and washing reactor
2	Acidic water tank
3	Conveyer
4	Drum dryer
5	Gas furnace
6	Sorting
7	Vacuum pump
8	Vacuum pump
9	Thermoreactor of periodical action
10	Carbonised residue cyclone
11	Tubular condenser
12	Tubular condenser
13	Thermolysis condensate collector
14	Vacuum evaporation
15	Collectors
16	Pressure filter
17	Solvent
18	Hot clarification mixer
19	Vacuum evaporation
20	Crystallizer - bowl

21	Centrifuge
22	Mixer
23	Acetone collector
24	Acetone distillation still
25	Acetone measuring tank
26	Cooling appliance
27	Ball mill
28	Thermoreactor of continuous action
29	Spiral heat exchanger
30	Vacuum filter
31	Paste collector
32	Hot water tank
33	Condensate collector
34	Centrifugal pump
35	Fan
36	Levoglucosan dryer
37	Vacuum filter
38	Condensate collector
39	Cans
40	Carbon dioxide container
41	Oil barrel

Figure 5.2 Flow diagram of a unit for producing levoglucosan, carbonized residue and oils from lignocellulose [16].

At the purification station, the LG was dissolved until a solution with a concentration of 20% was obtained, mixed with active carbon and cooled for 1 hour. Then the carbon was filtered off and the LG centrifuged. The LG solution was subsequently mixed with clarifying carbon for 1 hour at 80°C and filtered on a Nutch-filter. The LG solution obtained was supplied to a vacuum evaporation unit, where it was evaporated to 70% of dry matter and supplied to a crystalliser. The obtained solution was supplied to a centrifuge and filtered. The mother liquors obtained at the purification station were collected, evaporated to a concentration of 70% and poured out to the crystalliser, and treated as described above.

The material balance (in terms of raw material) on the unit of the Krasnodar Hydrolysis Plant (capacity in terms of levoglucosan 75 kg/day) was:

   Raw material, corn cob lignocellulosic residue - 5000 kg/day
   Moisture content of the raw material - 75%
   Content of the fine (less than 2 mm) fraction* - 36% -1250 kg/day
   Levoglucosan yield - 9.4% (on o.d. raw material)
   Oil yield - 15.6% - 125 kg/day
   Yield of carbonised residue -25% - 2000 kg/day;

*The fine fraction was separated and used for briquettes production - 500 kg/day (taking into account the moisture content 20% and losses 4.0%).

About 3 t of levoglucosan was produced on pilot and industrial units at the Krasnodar Hydrolysis Plant during 1958-1970, and was passed on to different research centres in the former USSR to study the possibility of its utilisation as a raw material in organic synthesis and polymer production (see below). Later the pyrolysis units were dismantled and, in succeeding years, work on improving the levoglucosan production technology was carried out at the Latvian Institute of Wood Chemistry.

5.1.4 Levoglucosan production technology developed at the Latvian Institute of Wood Chemistry.

Since 1971, to improve the levoglucosan production technology, researchers at the Institute of Wood Chemistry [17] have worked on the following main problems:
   –   increasing of the LG yield when employing high-capacity reactors;
   –   reduction in the maintenance time of the condensation system heat exchange equipment.

By 1975, a continuous reactor with a capacity of 50 kg/day of raw material [18] and a cyclone condenser with automatic thermostatting had been developed and a closed water-management system devised. This unit was designed for utilising as the raw material the lignocellulosic residue after prehydrolysis of birch wood for the production of furfural and acetic acid.

The following data characterised the operation of the unit (products yield from the mass of o.d. lignocellulose):

LG in thermolysis condensate	20%
Purified LG (96% basic substance)	10%
Residual tar	15%

The content of sulphuric acid in the lignocellulosic raw material used for obtaining LG must be in the range 0.10-0.15% of the o.d. raw material mass. Therefore, washing is carried out until the pH of the wash water reaches 2-2.5. To attain such a pH level, the washing is performed with water at a temperature of 60-70°C in the ratio (in terms of o.d. wood) of around 1:8.

Moist fine-disperse lignocellulose was dried in a cyclone with flue gases to a moisture content of 5-10% (rel.).

Table 5.1   Change in the composition of the lignocellulosic raw material after washings

Substances	Content before washing, %	Content after washing, %
Lignocellulose	93.3	99.2
Reducing agents	4.5	0.5
Organic acids (in terms of acetic acid)	1.0	0.15
Sulphuric acid	1.0	0.15
Free furfural	0.2	-

The main technical parameters of the products of biomass pyrolysis aimed at obtaining LG are listed in the Table 5.2.

Table 5.2 Main technical characteristics of the products

Characteristics	Standards		
Levoglucosan			
	pure	purified	technical
Melting point in the range, $^{\circ}$C	177-179	167-172	130-140
Specific rotation of 2.5% water solution, $^{\circ}$C	67	66	-
LG content in terms of the increase of reducing agents after inversion, %	98.0	97.0	92.0
Light transmission of 0.4% aqueous solution, %, no less than	96.0	84.0	14.0
Mass share of losses upon drying, %, no more than	0.07	0.07	0.67
Explosive range (in terms of sugar), g/m^3	8.9		
Residual tar			
Appearance	Dark-brown viscous liquid without mechanical impurities		
Moisture content, % from tar mass	10-15		
Acid value, mg KOH/g	30-60		
OH groups content, % from the mass of o.d. tar	6-9		
LG content, % from the mass of o.d. tar	15-25		
Density at the temperature 50°C, g/m^3	1.225-1.361		
Viscosity at the temperature 50°C, $*10^3$ Nsm^{-2}	84-398		

Fast thermolysis of lignocellulose was performed in a cyclone reactor in a flow of superheated steam at normal pressure and temperatures of 370 to 390°C. The mass ratio of lignocellulose to heat carrier was 1:4.

The formed vapour-gas mixture containing LG is condensed in a scrubber and a heat-exchanger, then evaporated under a vacuum until the dry matter content is 90-92%. The evaporated condensate (thermolysis paste) is diluted with acetone (5 l of acetone per 10 kg of paste) and, after crystallisation, separation of the mother liquor and washing, a technical LG is obtained equivalent to 13% of the mass of the dry lignocellulosic raw material.

The technical LG containing 10-12% of admixtures (water, tar) is purified by a selective extraction method [19]. After crystallisation, the purified LG is dried under a vacuum to remove the solvent and water. The yield of purified LG is 10% of the mass of o.d. lignocellulose.

The mother liquor after separation of technical LG contains LG, tar, water and acetone. Acetone recovery is performed in a still with a reflux column.

To prevent environment pollution, it was proposed to burn the carbonised residue and non-condensing gases jointly with liquid fuel in a superheater for the production of process steam.

5.1.5 Description of the process

Please note that the flowsheet (Figure 5.3) could not be redrawn or reproduced, but copies can be provided on request to the editor.

Moist lignocellulose is immersed in a collector (1). From the collector (1), the raw material is loaded in a loading bunker (5) using a screw transport and a bucket elevator (2). From the loading bunker (5), it is supplied through a sector unloader to a continuous cyclone dryer (6), where it is dried to a moisture content of 5-10% with flue gases. The dried lignocellulose from the lower part of the cyclone dryer (6) is supplied to a collector of dry lignocellulose (9) through a sector unloader with the help of a screw feeder (7) and a bucket elevator (8). The lignocellulose dust from the spent heat carrier is trapped in a cyclone (10), while the dust-free spent heat carrier is ejected into the atmosphere with the help of a centrifugal fan (11).

The dried raw material, with the help of a screw feeder (12) and a bucket elevator (13), is loaded periodically into a transfer bunker (19). From there the lignocellulose is agitated in the transfer bunker and poured into the loading bunker (20), from where it is continuously supplied to the channel of a fast thermolysis thermo-reactor (21) with a heated jacket. The superheated steam (pressure 0.30 MPa), is supplied via a steam trap (17) to a superheater (15), where it is heated to 500-520°C. Raw material particles are entrained by the flow of superheated steam in the spiral chamber of the fast thermolysis reactor, and pyrolysis occurs in 6-10 sec at 370-390°C. The carbonised residue from the lower part of the thermo-reactor (21), is collected in the transfer bunker through the sector feeder, and is thence supplied to a discharge hopper (23).

The vapour-gas mixture from the fast thermolysis reactor (21) is cooled in a scrubber (24) to 105-110°C and in a heat exchanger (27) to 30-40°C.

The non-condensable gases are supplied by a fan (28) to the furnace of the superheater (15), where in passing, they entrain the carbonised lignocellulosic residue which is continuously supplied to the pipe-line with the help of the sector feeder (23). The condensate from the scrubber (24) and the heat exchanger (27) is supplied to the collectors (25, 29) which, in parallel, serve as hydrogates. The scrubber (24) is sprayed by the thermolysis condensate which is supplied with a pump (30) *via* a settler (26) through injectors to the scrubber (24). The condensate is gathered in the settler (26) from the collectors (25, 29) and is allowed to settle for about 8 hours.

The settled condensate from the settler (26) is pumped in batches to the condensate collector through a filter (32, 33).

The thermolysis condensate is evaporated in an evaporation apparatus (35) at a residual pressure of 150-200 mmHg (64). The evaporated condensate (thermolysis paste) with 90-92% solids is gathered in intermediate collectors (36) and is supplied periodically to a mixer-crystalliser (37), where it is dissolved with acetone from an acetone batcher (38 or 39) and cooled to 10-15°C.

The vapour from the evaporation apparatus (35) is condensed in a heat exchanger (44) and gathered in condensate collectors (45).

The crystalline mass of levoglucosan from the mixer-crystalliser (37) is transferred into a filtering centrifuge (41), where the technical LG is separated from the mother liquor, which is gathered in a collector (43). Then the technical LG is washed with acetone from a batcher (39), and the acetone from the washings is gathered in a collector (42). The technical LG is gathered and purified (by the method according to application No. 3534298/04) using a batcher (46), a reactor (47) a heat exchanger (48), a pressure filter (49) and a collector (50).

The solution of purified LG from the collector (50) is fed by a proportioning pump (52) for evaporation in a continuous action rotor film evaporator (53) through a heat exchanger (51). The evaporated solution is gathered in intermediate collectors (54). The solution vapours are cooled in a heat exchanger (55) and gathered in collectors (57).

The evaporated solution of purified LG from the intermediate collector (54) is supplied periodically to a crystalliser (56), where it is cooled to 15-20°C and crystallised. Then the mass is transferred to a filtering centrifuge (58), where the crystals are separated from the mother liquor, which is gathered in a collector (60). The purified LG is washed with the solution from the collectors (57) on a filtering centrifuge. The washing solution is gathered in a collector (59), and then fed as needed to the batcher (46). The purified LG is dried in a shelf vacuum dryer (62). For catching the solution vapours, a heat exchanger (63) is installed, which is cooled by the brine from a cooler (67) with the help of a circulating pump (66). The solvents are gathered in a collector (65). To recover the solvents and obtain the residual tar, a reflux condenser is used. The acetone mother liquor from the collector (43) is pressed out to the still of the rectifying apparatus (69). The acetone is distilled, by gradually heating the solution with heating steam. Acetone vapours are caught in a dephlegmator (70) and a cooler (71). The solvent is gathered in a collector (72) and returned to the measuring batcher (39). The residual tar is poured into a container.

5.1.6   Experimental installations, batch and continuous reactors

Designs of several pyrolytic units with thermo-reactors of different types, such as the reactor with coaxial fixing grid, the fluid bed vibro-reactor and the directing plate cyclone reactor, were devised and installed by the Processing Design Office of the Latvian State Institute of Wood Chemistry.

In the **reactor with a coaxial fixation grid** (Figure 5.4), the material is moved through a perforated tube from the top down, superheated steam is introduced into the material flow, and its movement is directed along the radius to the coaxial grid. The use of a such a reactor prevents the ejection of the material into the condensation system, excludes the necessity of an insulating steam jacket around the reactor and simplifies the transportation of the material.

The reactor was run in a **continuous regime** with a periodical supply of the lignocellulosic raw material of 200-700 kg through a shaker feeder. In the pyrolysis zone, the lignocellulose was subjected to a thermal shock by superheated steam with a temperature of 370°C. The formed steam-gas mixture was removed from the pyrolysis zone through a coaxial grid, was rapidly cooled to 180-200°C and removed through a jet screen to a cyclone condenser. A conical steam distributor rotated jointly with a discharge tray, ensuring an even movement of the material.

The design of the **fluid bed vibro-reactor** was based on the experience of operating dryers of a similar principle. The attractions of this unit are an intensive heat exchange and a minimum consumption of heat carrier. The reactor was run on a continuous basis. The pyrolysis of the wood lignocellulosic raw material was performed at a lowered pressure in a flow of superheated

179

steam. The reactor's capacity in terms of raw material was 20 kg/h at a moisture content of 10%. The reactor had two zones: a heating zone and a pyrolysis zone. The perforated bottom of the thermo-chamber was set on a spring hanger and connected with the vibrator through a core. The movement rate of the material was controlled by varying the slope of the vibrator.

1 Raw material bunker

2 Thermoreactor

3 Fixation grid

4 Steam distribution cone

5 Discharge tray

6 Built-in tray-pressure regulator

7 Carbonised residue bunker

8 Extinguishing bunker

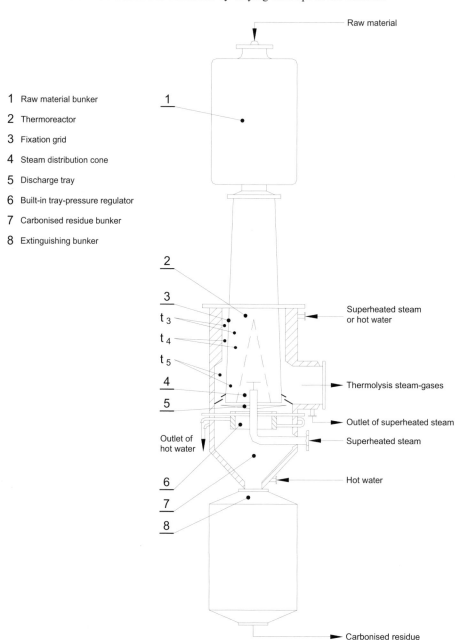

Figure 5.4   Flow chart of a continuous action thermoreactor with batch loading

Lignocellulose was loaded from the bunker to the heating zone, on the vibrating bottom. The bed of lignocellulose was fluidised with superheated steam, which ensured good conditions for heat mass exchange. Under the action of vibrations, the lignocellulose was moved from the heating zone to the pyrolysis zone. The temperature of the heat carrier was 450°C, and the residual pressure was 0.1 atm in the pyrolysis zone. The formed vapour-gas mixture was directed from the pyrolysis zone to the condensation system through a cyclone. To prevent the condensation of the products, the cyclone was heated with superheated steam. The cooling of the vapour-gas mixture supplied from the pyrolysis zone at a temperature of around 300°C occurs in a scrubber sprayed with the condensate or the water from the condenser. In the scrubber, the major part of the LG and accompanying products are condensed.

A fast pyrolysis unit based on **the cyclone-type reactor** with a directing plate had a capacity of 30 kg/h of raw material [20]. The reactor consisted of a spiral thermo-chamber with a heated jacket with a cross section of 100x110 mm and a length of 19 m. Pyrolysis was performed at a normal pressure.

The raw material (lignocellulosic residue after acid prehydrolysis) was transported through a screw channel by a heat-carrier flow in a pulsing mode. The heat-carrier was superheated steam. The raw material to superheated steam mass ratio was 1:4.5, pyrolysis time 8-10 sec, and temperature 370-415°C (Figure 5.5).

1	Drip pan
2	Steam superheater
3	Bunker
4	Reactor
5	Carbonised residue receiver
6	Container
7	Scrubber
8	Waterlock collector
9	Cooler
10	Waterlock collector
11	Condensate collector
12	Fan
13	Evaporator system
14	Cooler condenser
15	Vacuum - pump
16	Cristalliser
17	Centrifuge
18	Filtrate collector
19	Pump

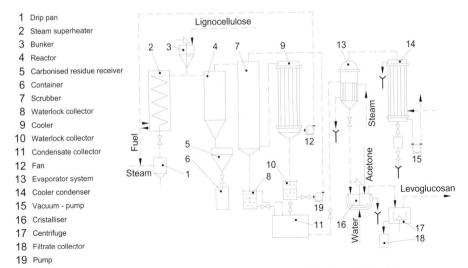

Figure 5.5 Flow diagram of a levoglucosan production pilot unit

Studies carried out earlier at the Institute of Wood Chemistry have shown that the ash content of the raw material considerably affects the levoglucosan yield. It has been established that the presence in the lignocellulose of non-organic salts as well as the organic compounds (e.g. monosaccharides) formed during prehydrolysis decreases the LG yield [1]. Since the cation-exchange capacity of lignocellulose is practically equal to zero at pH 2.3-2.8, it was recommended to carry out the washing of lignocellulose at this acidity; at a higher pH, the LG yield decreased [21,22].

It was established that the LG yield was 20-25% of the mass of o.d. lignocellulose, or 47.5-63% of that of cellulose, in a cyclone type reactor in a flow of superheated steam, at atmospheric pressure and a temperature of 370-385°C. It was shown that an increase in pyrolysis temperature to 415°C not only decreases the LG content in oil produced, but also adversely affects the process of its isolation from oil, decreasing the yield of pure LG. Upon isolating of the LG from the acetone solution of oil, up to 57% of LG (from the content in the oil) or 65.6% (from the content in the paste) was obtained [20].

5.1.7 Methods for separation and purification of levoglucosan

One of the problems of the technology of LG obtained by pyrolysis of cellulose-containing materials is the isolation of levoglucosan from the products of thermo-destruction and obtaining LG in crystalline form with the content of the base substance above 95%.

The technical LG obtained upon the fast pyrolysis of lignocellulose at a temperature of 375-400°C has a brown-grey colour and a melting point of no more than 155°C [23].

It was suggested that ionites (ion exchange resins) could be used for purification of LG [24, 25]. A 5-10% aqueous solution of technical LG was successively passed through cationites and anionites, and a colourless solution was obtained from which a crystalline levoglucosan was isolated after evaporation. However, the method was too laborious because of the complexity of the ionites regeneration, and levoglucosan losses were too high [26].

According to another method, LG was precipitated from an n-butanol solution, adding a 3-5-fold surplus of diethyl ether [25]. The purification degree did not exceed 90%. This method proved impractical due to a low-efficiency.

In the laboratories of the Institute of Wood Chemistry, a method was used for the recrystallisation of technical LG from an ethanol and acetone mixture, then from water or n-butanol, without the use of active carbon. Clarification on active carbon decreases the targeted product yield to 40% of the content in the technical product. The method was tested under pilot plant conditions at the Krasnodar Hydrolysis Plant, using acid, neutral and alkaline active carbons. In this case, the consumption of active carbons was more than 200 g per 1 kg of purified LG. According to the laboratory acetone method developed later at the Institute of Wood Chemistry [27] it was suggested that LG be separated using 2.5 ml of acetone per 5 g of pyrolysis oil containing 45-70% of LG (dry mass of oil). The acetone method ensured the isolation of 91-92% of technical LG. For separating the colouring substances, a selective solution of LG in ethanol was used [28].

Work was carried out on obtaining a purified LG product directly from the pyrolysis paste containing 78% of LG, without the preliminary isolation of technical LG by extraction [29]. In this case, the yield of the purified LG product was 50%, with a pure LG content of 96%. Upon a repeated evaporation of mother liquors and the isolation of purified LG, its yield increased (Table 5.3).

The LG purification technology was tested under pilot plant conditions. 15.2 kg of technical LG (moisture content 3%, pure substance content 83.2%) was dissolved in 160 l of ethanol at 20°C in an enamelled steel (glass lined) reactor and mixed for 45 min. Then the solution obtained was filtered on a filtering centrifuge. 1.3 kg (8.8% of the mass of technical LG) of a dark water-soluble substance containing 64.4% of LG was collected on a filter. The filtrate was evaporated to a volume of 25 l, crystallised at 20°C, filtered and washed with 10 l of ethanol and dried under a vacuum at 105-110°C. As a result, 6.34 kg of purified LG was obtained. A repeated evaporation of the mother liquor to 50% dry matter content gave an additional 2.14 kg of purified product. The total yield of purified LG was 8.48 kg or 69.8% of its content in the technical LG. The purified product contained 95.2% of LG and had a melting point of 174-175°C. The further evaporation of the mother liquor and washing solution to a dry matter content of 50%, after crystallisation, gave an additional 0.58 kg of the product containing 93.9% of the LG, with a lower melting point of 167-169°C. During the subsequent operation, another 0.42 kg of LG of the same quality was obtained.

Table 5.3. Yield of purified LG upon selective extraction and crystallization from ethanol.

Raw materials		1st crystallization		2nd crystallization		Purified LG		
Amount of technical LG, g	Amount of ethanol, mL	Amount of solution, mL	Amount of isolated cryst LG, % from its content in technical LG	Amount of solution, mL	Amount of isolated cryst. LG, % from its content in technical LG	Total amount, g	Amount of isolated cryst. LG, % from its content in technical LG	Melting point, °C
100	1400	270	51,5	120	15,6	59,2	67,1	...
200	3000	375	71,5	100	10,1	144,1	81,7	...
100	1600	200	80,0	50	9,7	72,4	89,7	171-173
200	3000	300	72,0	100	3,5	133,2	75,5	174
2500	35500	4500	59,9	1500	17,9	1715,0	77,8	175

5.1.8 Products and materials on the basis of levoglucosan.

In parallel with developing the levoglucosan production technology, research into the properties of 1,6-anhydrosaccharide was carried out. The main research centres were:

- Institute of Wood Chemistry (Riga, Latvia),
- Institute of Organic Chemistry and
- Institute of Element Organic Chemistry of the Academy of Sciences of the USSR (Moscow)
- a range of laboratories engaged in the study of the consumer properties of the synthesis products.

The results of these research activities are reflected in numerous scientific publications and authorship certificates [30-33]. Main findings are summarized in a monograph by Pernikis [34]. The monograph addresses not only the chemical properties of levoglucosan under different conditions of synthesis, but also the properties of commercial products based on it. A significant contribution to the development of the area of application was made by the studies performed by Golova and Ponomarenko [35-37].

LG as an initial raw material was used for synthesis of oligoethers and oligoesters. The synthesis of polyether on the basis of LG and propylene oxide has been developed for obtaining a diversity of polyurethane materials. The synthesis of modified polyethers containing chlorine and phosphorus has been accomplished. Polyurethanes produced from LG displayed a low flammability and a high heat stability, and were proposed for heat-insulation. LG oligoethers and LG modified oligoesters were used for obtaining rigid polyurethane foams.

The synthesised 2,3,4-trimethacryl levoglucosan was used for production of high-module rubbers applicable for manufacturing parts for mobile seal units. The new material improves on commercial rubbers in rupture strength value, hardness, static modulus of dilatation and heat resistance in the stressed state.

Reactive oligocarbonate acrylates of levoglucosan have been synthesised and, on their basis, a photopolymerising composition for galvanoplasty models has been developed. The models have an enhanced thermal sensitivity, are readily covered with silver and are chemically stable to electrolytes in galvanoplastic vats.

Jointly with the Irkutsk Institute of Organic Chemistry of the Siberian Division of the Academy of Sciences of the USSR, epoxide resins based on LG have been developed with an increased elasticity and high dielectric properties.

A complex compound of LG, iron and zinc, arbitrarily called zinc ferrilevoglucosan, has been synthesised. This compound has been successfully tested to control the chlorosis of fruit trees.

A method for obtaining a new LG-containing carbamide-p-toluene sulphamide formaldehyde resin has been developed. As LG is added to the resin composition, its softening temperature tends to increase to 200°C, rigidity tends to increase and the solubility in organic solvents decreases significantly. Based on the new resin, fluoropigments of all spectral colours have been obtained and are used in the printing industry.

LG was used for stereo-orientated synthesis of chiral natural compounds of different classes. For example, the synthesis of some antibiotics has been carried out.

Technical LG has been added to core sand mixtures in the water glass production process instead of food-stuffs such as glucose, starch, etc.

Overall, the studies carried out at the Latvian Institute of Wood Chemistry can be represented by the scheme of Figure 5.6.

5.2   Furfural

5.2.1   Areas of furfural application

Furfural (α-furalaldehyde) is really the only starting monomer for organic synthesis obtained from renewable plant raw materials rather than oil. Due the unique properties of furfural, its price on the world market has increased 5 times (up to 1800 USD per ton) over the last 20 years (Figure 5.7), and the number of countries producing furfural has increased from 9 to 27. The industrial production of furfural was started in 1922, and it is currently being produced in more than 100 enterprises world-wide.

The importance of furfural has increased substantially with the expansion of its fields of application and those of its derivatives, resulting from the selectivity of furfural as a solvent and its high reactivity. Furfural and its derivatives are widely used in the chemical industry, medicine, metallurgy, agriculture and machine-building. Furfural is widely used as a solvent for resins and waxes, purification of animal fats and vegetable oils, extraction of mineral oils, and separation of vitamins A and D from fish liver fats. It has long been applied in the petroleum industry as a selective solvent. Furfural is used also as an initial compound for obtaining furan compounds, medicines, polymer materials and chemicals for agriculture (Figure 5.8).

The production of polymer fural resins, which are stable against solvents, acids and alkalis, is still important. In the USA, furan resins are currently produced at 18 plants. Thermoreactive resins synthesised on the basis of furfural are characterised by a special plasticity, which is vital for production of large plastic articles with elaborate configurations. Furfural-containing resins are used for moulding of powders with different fillers (asbestos, graphite, etc.) featuring anti-frictional properties. Furfural resins are used as a basis for high-quality water-resistant and heat-stable lacquers and coatings, which can be used as electric and thermal insulation.

LG OLIGOMERS

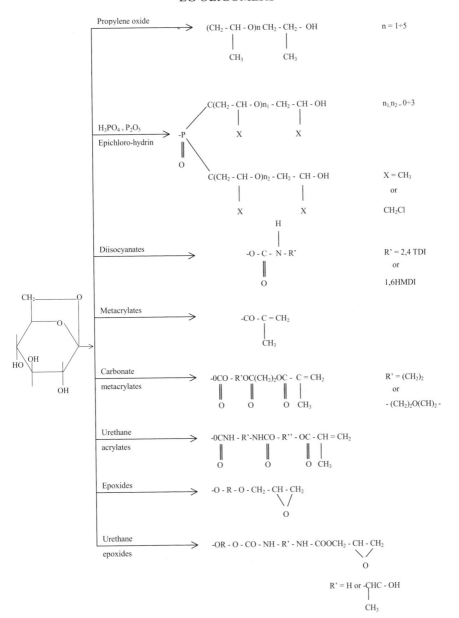

Figure 5.6 Major directions of the syntheses on the basis of levoglucosan

5.2.2 Homogeneous and heterogeneous catalysis of furfural formation

Extensive studies were carried out in the former USSR into the development of methods for obtaining furfural from the thermochemical decomposition of wood and other plant materials.

Different methods for the thermal processing of wood were tested, such as pyrolysis, pre-pyrolysis in the liquid heat carrier, and gasification.

It is known that furfural from plant biomass is formed from pentose hemicelluloses according to the scheme:

$$pentosans \rightarrow pentoses \rightarrow furfural$$

Pentosans are present in all types of lignified plant tissue. However, only raw materials rich in pentosans, including deciduous wood (18-32% content), are of commercial importance.

In the case of a common (slow, without catalysts) pyrolysis of birch wood, the furfural yield does not exceed 1% of the mass of o.d. wood, or no more than 10% of the theoretical yield. At such yields, its separation from the total mass of liquid products is not economically viable. Wood containing about 25% of pentosans can produce furfural at a theoretical yield of around 10% of the mass of o.d. wood [14], together with the main products of pyrolysis, namely, charcoal and acetic acid.

The mechanism of furfural formation from pentoses was widely studied [38-42]. In studies on the kinetics of the furfural formation under homogeneous conditions in the temperature range 120-180°C carried out by the Institute of Wood Chemistry [43], it has been shown that the opening of the pyranose cycle with the splitting-out of a water molecule is the limiting stage of the process.

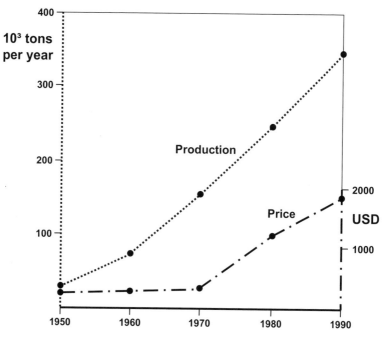

Figure 5.7 Dynamics of furfural production and price worldwide

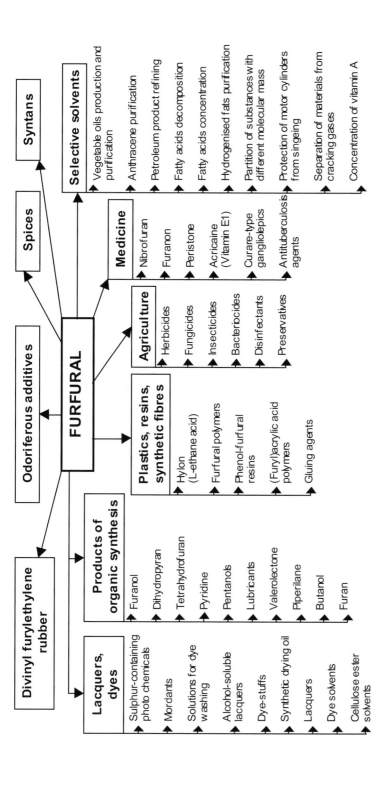

Figure 5.8 Areas of furfural application

The furfural formation process is catalysed mainly by acids and salts of weak base [14]. The catalytic activity of salts was explained by the formation of hydrogen ions upon hydrolysis, although a range of experimental data cannot be explained in this way. Thus, it has been established that the catalytic activity of chromium and aluminium sulphates is 10-15 times as high as the activity of sulphuric acid used at the same concentration. It has been shown that catalytic activity of zinc salts in the pentoses dehydration reaction is much higher than the activity rated according to the concentration of hydrogen ions formed as a result of hydrolysis [44]. The reaction rate constant of xylose decomposition in the presence of aluminium, gallium, indium, scandium, tin (III) and titanium is about three or more times (in the case of chromium chloride 23 times) the reaction rate constant in the presence of hydrochloric acid of the same concentration. A dependence of the catalytic activity of cations in the composition of chlorides in the process of furfural formation from birch wood on their ionisation potential has been established [45]. The obtained dependence is approximated by a polynomial of the third degree:

$$K = -1.6*10^{-3} + 3.89*10^{-4}E - 1.66*10^{-5}E^2 + 5.0*10^{-7}E^3 \qquad (1)$$

where K - catalytic activity, E - ionisation potential, eV.

Based on the kinetic characteristics of furfural formation from birch wood in the presence of different metal chlorides (Table 5.4), it is assumed that, in the presence of metal cations, the process proceeds through the formation of an intermediate of ion type between hydrated cations and pentoses. This is confirmed by high negative entropy activation values, which is indicative of the decrease in the degree of freedom of the activated complex in comparison with the initial state of the reagents [46].

Table 5.4. Kinetic parameters of furfural formation in the presence of metal chlorides

Cation	Kinetic parameters				
	$k*10^3$, min^{-1}	E_a, kJ/mol	A, min^{-1}	ΔH^*, kJ/mol	ΔS^*, J/K.mol
Cr^{3+}	17.27	113.5	$3.29*10^{12}$	110.0	-50.4
Al^{3+}	15.75	105.4	$2.91*10^{11}$	102.0	-70.6
Fe^{3+}	6.92	116.8	$3.50*10^{12}$	113.4	-49.9
Zn^{2+}	0.350	106.5	$8.93*10^9$	103.1	-99.5
Ni^{2+}	0.207	98.6	$5.34*10^8$	95.2	-123.0
Fe^{2+}	0.091	120.2	$1.22*10^{11}$	116.7	-77.8
Mn^{2+}	0.081	107.4	$2.69*10^9$	104.7	-109.5
Mg^{2+}	0.079	101.6	$4.86*10+8$	98.21	-123.7
Na^+	0.026	104.0	$3.16*10^8$	100.5	-127.3

The production of furfural through pre-pyrolysis of wood in the medium of a liquid heat carrier was investigated in the 1960's [47]. In the case of pre-pyrolysis of birch chips in a kerosene medium, pentosans degradation was observed; however the furfural yield did not exceed 2% of the mass of o.d. wood. Pre-pyrolysis of birch chips impregnated with 6.0% hydrochloric acid under the same conditions gave a furfural yield of 8-10%. However, in this case, in parallel with a complete decomposition of pentosans, a considerable degradation of cellulose occurred; hence, the main target posed by the authors, i.e. the consecutive pyrolysis of wood components aimed at obtaining a series of individual products, was not reached. The use of different salts instead of hydrochloric acid as catalysts did not produce positive results; the furfural yield did not exceed 6% even at a high consumption of reagents.

It has been shown that, in the presence of an acid-peroxide catalyst (sulphuric acid and hydrogen peroxide), the furfural yield in the water vapour medium reaches about 9% of the mass of o.d. wood (birch chips for pulp production) [48].

In the search for conditions providing a high conversion of pentoses into furfural with a minimum effect on cellulose, it has been established [49] that the introduction of a salt catalyst in a solid state is the most efficient. Thus, using ammonium chloride, ammonium nitrate or monosubstituted calcium phosphate at a consumption rate of 2.5% of the mass of o. d. birch wood, the furfural yield made up 70-80% of the theoretical one. The application of salts in the solid state excludes the introduction of surplus moisture into the reaction zone.

Based on theoretical studies, a new approach has been developed at the Latvian State Institute of Wood Chemistry to prevent secondary reactions of the furfural formed upon the thermal treatment of the raw material in the presence of catalyst. According to this approach, pentose dehydration proceeds on the porous surfaces of the carrier material. Silica gel and, later on, phosphates were studied as a carrier material, and their applicability for industrial scale production of furfural was demonstrated [50, 51]. It has been shown that by modifying the silica gel with salts such as aluminium sulphate, a 22-25 times increase in catalytic activity is obtained, owing to the formation of Brönsted acidic centres.

In the literature, information is available on so-called specific catalysts of pentose dehydration reactions with the formation of furfural [52-54]. These are primarily phosphorus-containing salts and superphosphates.

Owing to a whole range of advantages of catalytic processes carried out under heterogeneous conditions (a high selectivity of solid-phase catalysts, an increase in the economic of the process, decreasing consumption of the raw material and volume of wastes), studies were carried out in the late 1980s to develop so-called micro-heterogeneous systems for the catalytic dehydration of pentoses. These were catalytic centres, which were covalently immobilised in a cross-linked polymer network [55-58]. It has been found that the conversion of pentoses into furfural is highly efficiently catalysed by Cr, Ti, Zr-porphyrin complexes, immobilised in a cross-linked N,N-methyl-bis-acrylamide copolymer. In this case, the furfural yield reached 85-87% of the theoretical yield. Krupenskij and co-authors [56] have also demonstrated the possibility of increased furfural yield with simultaneously reduced process time (from 62 to 3 hours) by using the catalytic system titanyl ammonium sulphate (10-20 mass %) - polyacrylamide (carrier) - phenolformaldehyde oligomer (13 mass %).

Despite the positive aspects of the application of the above-mentioned heterogeneous catalytic systems, the cost and labour inputs of the synthesis of catalysts hampers their application on an industrial scale. These systems, as with mineral acids commonly used as catalysts of pentose dehydration, do not exclude the resinification of pentoses and furfural in the dehydration process.

A comparative analysis of pentose dehydration kinetics under conditions of homogeneous and heterogeneous catalysis using acid-type catalysts has shown that under homogeneous catalysis conditions, regardless of the various compositions and activities of the catalysts used by various researchers, the mechanism of furfural formation reaction is the same in all cases [59]. The process may be described with the mono-molecular reaction first order equation and that of Arrhenius in the 120-180°C temperature interval. The activity of cations increased along with their valence, and the activation energy decreased. The analysis of activation parameters shows that the rate-limiting stage is the opening of the pyran cycle splitting a molecule of water.

Under heterogeneous conditions at 110-150°C the reaction occurs in the kinetic region, but on increasing temperature to 180°C the process occurs in the diffusion region. It has been shown that through modification with salts and acids the catalytic activity may be increased 22-25 times for silica gels and 1.6-1.8 times for alumosilica gels. This is caused by formation of Brönsted acid centres and changes in the porous structure of heterogeneous catalysts [59].

At present, furfural is produced at commercial scale through the processing of different types of raw material rich in pentosans (residues of annual agricultural plants such as corn cobs, sunflower husk, deciduous wood), with catalyst solutions at elevated temperatures. The furfural yield reaches 55% of theoretical yield [14].

The majority of these furfural production processes are based on the thermal treatment of a raw material impregnated with a catalyst solution. A completely novel furfural production technology [60-67] has been developed by the researchers of the Latvian State Institute of Wood Chemistry, excluding the raw materials soaking-impregnation stage and, at the same time, ensuring a high furfural yield and the production of a lignocellulosic residue (lignocellulose) suitable both for processing to hexose sugars by a conventional hydrolysis method and for pyrolytic conversion to levoglucosan.

The technology is based on the differentiation catalysis theory developed by scientists from the Latvian State Institute of Wood Chemistry. According to the theory [68], the concentrated sulphuric acid covers only some of the sample surface, tends to chemisorb firmly and does not penetrate into the wood tissue. The acetic acid released during the treatment of wood with steam catalyses the hydrolysis of hemicellulose components. Pentoses diffuse towards the material particle surface and are degraded there into furfural under the action of concentrated sulphuric acid. As a result, firstly, no surplus quantity of pentoses is accumulated in the cell walls and, secondly, the pentoses formed during diffusion from the cell walls to the particles surface cannot undergo side reactions with intermediate products of dehydration. Furfural formation at the surface of the particles of the raw material favours its rapid transition into a vapour phase; therefore, the possibility of proceeding secondary reactions is excluded, and furfural losses are negligible. The yield of furfural is significantly higher than that for other production methods and reaches 80% of the theoretical yield. Owing to the surface localisation of sulphuric acid, cellulose destruction tends to decrease considerably, and lignocellulosic residue is a valuable raw material for further chemical processing in the hydrolysis industry, in composite materials production and in obtaining of levoglucosan [69].

The technology has been implemented in 6 plants in the former USSR. The licence has been sold to Yugoslavia. Seven contracts with firms in Yugoslavia, Hungary, Czechoslovakia, and Finland were successfully completed. At the plants built by Escher Wyss and Rosenlew, in Hungary, the furfural yield increased by 27-29%. In 1997 the improved furfural and bioethanol production technology from hardwood was implemented for the first time in the Kirov Biochemical Plant, Russia [67].

5.2.3   Macrokinetics of furfural formation

The transformation of pentosans into furfural is a multi-stage process and, during the catalytic thermolysis, the chemical reactions of hydrolysis and dehydration are superimposed by processes such as catalyst diffusion into the cell wall, diffusion of the formed furfural towards the material particles surface, etc. Drying of the material occurs simultaneously with furfural formation. Three main steps can be considered in the furfural production process: chemical conversion of pentosans to furfural; diffusion of furfural out of the raw material particles; and furfural removal from reactor with the vapour-gas flow [70]. The development of each step depends on the size of raw material particles, the diffusion of catalysts, moisture, the concentrations of intermediates and furfural, internal and external heat change, hydrodynamic conditions etc. Study of the process macrokinetics provided the regularities necessary for the calculation of technological parameters.

Unlike studies carried out by various authors [71-73], who had studied the pentoses dehydration process under static conditions without furfural removal from the reaction zone, a series of investigations on furfural formation macrokinetics was performed at the Latvian State Institute of Wood Chemistry using a specially designed reactor where pentoses dehydration occurred in the fluidised state in the heat-carrier flow [74-80].

The cyclone chamber designed for this purpose (Figure 5.9) had the heat carrier inlet at the bottom and products collection at the top (the ratio of the cross-sectional area of the air inlet to the chamber was 0.035). This ensured:
1.   a rapid loading of the material in the heat carrier flow;
2.   a stable rotating state of the raw material without sedimentation from the flow or significant premature carry-out of particles;
3.   rapid unloading and cooling of the solid residue;
4.   continuous removal of the gases formed.

Superheated steam was used as the heat carrier.

1   Steam line
2   Mixing chamber
3   Bunker
4   Lock valve
5   Loader lever
6   Tangential branch pipe
7   Cyclone reaction chamber
8   Upper outlet branch pipe
9   Shell- and- tube cooler
10 Regulation compensation electric heater
11 Timer
12 Electromagnetic valves
13 Conic valve
14 Conic valve spring
15 Lower cone
16 Cooled receiver

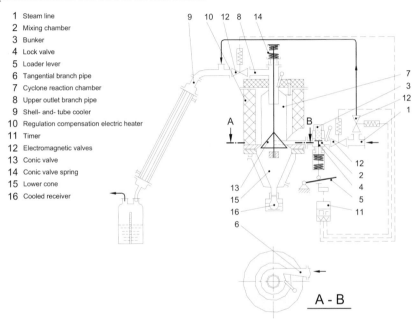

Figure 5.9   Scheme of a laboratory fast thermolysis installation.

To prevent sedimentation of the material particles, the bottom of the chamber is made in the form of a cone with the top inside the unit. The tangentially introduced gaseous jet forms a vortex inside the unit, which holds the raw material particles near the vertical side wall in a rotating pseudo-fluid state. A stable rotation without sedimentation to the bottom is obtained at an air velocity at the inlet cross-section of 14-18 m/sec for dry sawdust (size fraction 1.0-1.6 mm, relative moisture content 55%), 17-20 m/sec for moist sawdust (relative moisture content 55%) and 14-18 m/sec for lignocellulose (relative moisture content 4.5%). In the case of superheated steam, velocity at the inlet cross-section of the apparatus varies from 25 to 30 m/sec and ensures not only the holding of particles in the fluidised state, but also rapid removal of the formed furfural from the reaction zone.

The investigations carried out on the above-mentioned laboratory installation have shown that fast pyrolysis (heating rate > 100°C/min) of birch sawdust (size fraction 0.63-1.00 mm) in a fluidised layer in a superheated steam flow at 220-230°C yields up to 8% of furfural from dry wood. The total rate of furfural formation passes through a maximum at a sulphuric acid consumption of 2.4% of the wood dry mass. The main losses of furfural have been observed during the heating period, when pentose dehydration proceeds with the formation of not only furfural but also other products. In the isothermal heating period, the balance of the formed furfural and its theoretical content in the residue converges for the whole temperature range under study (150-240°C).

The trends of the change in the kinetic parameters of the furfural formation process (apparent effective activation E, effective rate constant, activation entropy) indicate that there are three regions of reaction in the temperature range 150-240°C:
1. a kinetic region in the temperature range 150-170°C, $E_a^* = 139.3$ kJ/mol;
2. a transition region at 180-210°C, E_a^* tends to decrease to 70 kJ/mol
3. a diffusion region at 220-240°C, where the total reaction rate is slowed down by internal diffusion retardation, and the activation energy falls to zero.

Table 5.5 lists the kinetic characteristics and activation parameters (effective rate constant C, apparent effective activation energy E_a^*, Arrhenius preexponent A, activation enthalpy ΔH^*, activation entropy ΔS^*, free activation enthalpy ΔG^* and steric factor P) of the furfural formation from xylose in solution, in a thin layer on a heat carrier, i.e. silicagel, and a pentosane-containing material. In all the cases under study, furfural formation from xylose is a homogenous reaction proceeding under homophase conditions which is satisfactorily described by a first order irreversible reaction equation and the Arrhenius equation.

Table 5.5. Kinetic characteristics and activation parameters of furfural formation from xylose and pentosanes containing raw materials

Raw material	Process conditions	t, °C	K, min^{-1}	A, min^{-1}	E, kJ/mol	ΔH^*, kJ/mol	ΔS^*, J/K.mol	ΔG^*, kJ/mol	P
Xylose solution, 0.1.n H_2SO_4	Statistic	150	0.05	$1.3 \cdot 10^{-13}$	117	113	-39	129	$0.9 \cdot 10^{-2}$
		180	0.50	$5.1 \cdot 10^{-14}$	130	126	-9	130	$3.3 \cdot 10^{-1}$
		210	5.80	$1.2 \cdot 10^{-14}$	137	133	7	129	2.4
Xylose, 1.5 n H_2SO_4 on silicagel	Fast pyrolysis	150	0.40	$3.4 \cdot 10^{-6}$	56	53	-165	122	$2.3 \cdot 10^{-9}$
		210	2.30	$7.7 \cdot 10^{-5}$	51	47	-178	133	$4.7 \cdot 10^{-10}$
Birch wood impreg-nated with H_2SO_4	Fast pyrolysis	150	1.20	$2.0 \cdot 10^{-17}$	139	136	41	117	14.2
		210	27.00	$1.2 \cdot 10^{-10}$	70	66	-117	123	$7.1 \cdot 10^{-7}$
Birch wood	Fast pyrolysis	230	0.14	$9.3 \cdot 10^{-15}$	162	144	14	137	5.5

The high negative values of entropy in the case of xylose dehydration on silica gel (Table 5.5) can be attributed to the conformational transformations of xylose as well as a low mobility of the transitional state in comparison with the initial state of the reagents, according to the scheme suggested by Krupenskij [81].

On the other hand, in the case of xylose dehydration in solution, the entropy tends to increase when the temperature of the process is raised, changing its sign (Table 5.5). This could be connected with the change in the limiting stage of the process, i.e. the formation of an activated complex at negative values of entropy (150 and 180°C), and its cleavage at their positive value (210°C).

Conditions for fast pyrolysis of birch wood sawdust ensuring a furfural yield of 60% of the theoretical yield were determined to be: moisture content of sawdust 30-45% rel.; consumption of catalyst (sulphuric acid) 2.4-3% of oven dry sawdust; temperature of the heat carrier in the reaction zone 220-230°C; thermolysis time 60-70 sec. A process for obtaining furfural from logging residues by low-temperature pyrolysis has been developed at pilot scale in Latvia (a 20 m^3 reactor) and Belarus. The maximum yield of furfural from birch sawdust on a cyclone-type fast thermolysis continuous pilot unit was 39% of the theoretical yield [79].

5.2.4 Experimental installations for furfural production, batch and continuous reactors

Technologies for furfural production based upon catalytic pyrolysis using acids (mainly sulphuric acid) as catalysts have been developed at the Latvian Institute of Wood Chemistry over a number of years. Processes were developed of furfural production by the low-temperature pyrolysis of fine dispersed wood particles treated with acidic catalysts in a flow of gas generated from lignocellulose, the solid residue of low-temperature wood pyrolysis, [82, 83].

An installation for performing a two-stage wood gasification was designed, i.e. low-temperature pyrolysis of the material treated with a catalyst for furfural production, and gasification of the solid residue from low-temperature pyrolysis. Based on the data obtained during the laboratory studies, an industrial batch reactor has been designed (Figure 5.10). The feature of the retort design (2) is that its body acts as a calorifer, whose external side is heated by combustion products (CP) of the gas generated from lignocellulose. The heat is supplied directly to the material by a vapour-gas mixture inside the retort, heating up the material uniformly throughout the retort cross-section. The circulation of the vapour-gas mixture is achieved by an axial fan mounted in the bottom of the retort.

The excess of the vapour-gas mixture in the material during drying and the combustion products in the blow-through period are removed from the bottom of the retort into refrigerator-condensers (3, 4). The non-condensable gases are washed with water in a foamer (6).

Three stages may be identified in the low-temperature pyrolysis, namely, warming up of the raw material up to 100°C, drying of the material, and its subsequent warming to and residence at the final temperature. The low-temperature pyrolysis process may be intensified by the acceleration of the stages of raw material heat-up and drying. The intensification of these stages does not have any noticeable action on the furfural yield, which is very important for devising continuous high-capacity units.

At the stage of raw material drying, only 8-12% of the total amount of the obtained furfural is separated. The onset of an intensive separation of furfural is observed upon reaching a mean temperature of 135-145°C in the retort (point t_3). The average time when the greatest amount of furfural is yielded in processes of different duration is around 1.5 hours.

As a result of the industrial tests carried out, the following main parameters of the process regime were established and specified:
1. The supply of CP directly into the retort is required only after the drying of the raw material and its heating to a mean temperature of 170-180°C in the retort;

2. 210-215°C is the optimal final temperature of the process; upon the further increasing of temperature, although the furfural yield increases from 6.5-7.0% to 7.5-7.8% of the dry raw material, tar-like products tend to be formed which hamper the furfural yield.
3. The optimal consumption of sulphuric acid is 1.5% of the dry raw material.

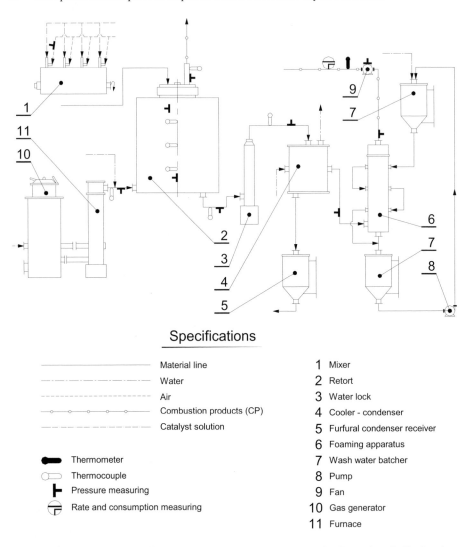

Specifications

———————————————	Material line
—·—·—·—·—·—·—·—·	Water
-------------------------------	Air
—o—o—o—o—o—	Combustion products (CP)
— — — — — — —	Catalyst solution

⬤━	Thermometer	
⊂⊃	Thermocouple	
⊢	Pressure measuring	
⊖	Rate and consumption measuring	

1	Mixer
2	Retort
3	Water lock
4	Cooler - condenser
5	Furfural condenser receiver
6	Foaming apparatus
7	Wash water batcher
8	Pump
9	Fan
10	Gas generator
11	Furnace

Figure 5.10 Scheme of an industrial low temperature pyrolysis plant of periodical action

Based on the analysis of the results of experiments carried out on the batch reactor, a continuous retort with a capacity of 25 kg/h of o. d. mass was designed (Figure 5.11) and tested at industrial scale. The sequence of operation of the pyrolysis process on the retort is as follows.

The raw material loaded into the retort is preheated in a feeding screw (7) to a temperature of 80-85°C using the CP removed from the retort jacket. The supply of the heated material into the retort prevents a partial condensation of furfural from the vapour-gas mixture circulating inside the retort on the raw material particle surface.

194

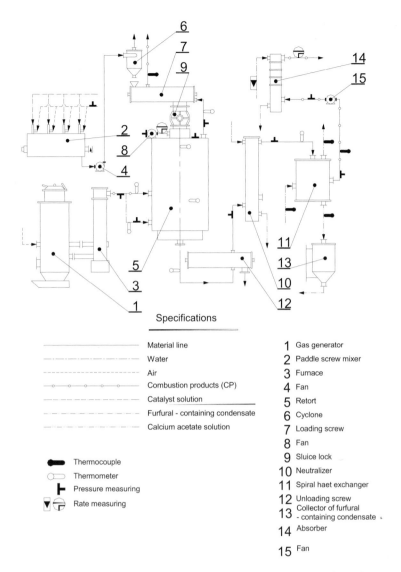

———————————————— Material line	**1** Gas generator
—·—·—·—·—·—·—·— Water	**2** Paddle screw mixer
– – – – – – – – Air	**3** Furnace
—∘—∘—∘—∘—∘—∘— Combustion products (CP)	**4** Fan
—————————— Catalyst solution ———	**5** Retort
— — — — — — Furfural - containing condensate	**6** Cyclone
— · · — · · — · · — Calcium acetate solution	**7** Loading screw
	8 Fan
▬ Thermocouple	**9** Sluice lock
⬭ Thermometer	**10** Neutralizer
⊢ Pressure measuring	**11** Spiral haet exchanger
▼ Rate measuring	**12** Unloading screw
	13 Collector of furfural - containing condensate ⸽
	14 Absorber
	15 Fan

Figure 5.11 Scheme of an industrial low temperature pyrolysis plant of continuous action

The retort is divided into two zones. In the upper zone, drying of the material and its heating to the final temperature of the process occurs. The supply of the heat to the material in this zone is performed by circulation of the vapour-gas mixture inside the retort and along the retort body heated by the CP. To improve the convective heat exchange from the CP to the vapour-gas mixture, the body of the retort is covered with spiral edges on both sides. Under these conditions, the total heat transfer rate reaches values of the order of 35-45 $W/m^2 °C$.

In the lower zone of the retort, the material is blown through with the CP cooled to 210-215°C to remove furfural from this zone more rapidly. The furfural-containing vapour-gas mixture is collected from the lower part of the retort. The surplus of the steam from the drying of the material in the upper zone also blows through the material in the lower zone of the retort.

The optimum process conditions for low-temperature pyrolysis on the continuous action plant were established as follows:

1. The maximum furfural yields (8.8-9.9% of the o. d. mass) were obtained at a final process temperature of 210-215°C. Increase in the final temperature of the process did not increase the furfural yield. Obviously, the performing of the process in a flow of vapour-gas mixture rich in steam (76%) directs the thermal decomposition of the furfural-forming substances towards the formation of furfural at a lower temperature.

2. The rate CP supply to the blow-through zone was 1.4 m^3 per kg of the converted raw material. The furfural yield with blow-through was almost 2% higher than that in experiments without blow-through.

3. The raw material layer in the first zone of the retort (2-2.5 m in height) ensured cooling of the heat carrier (the vapour-gas mixture) by 100-110°C (t_4-t_5).

4. The duration of the thermal treatment of the raw material in the upper zone had a direct effect on the furfural yield. The furfural yield grew as the time of treatment was increased.

Although the maximum furfural yield was obtained at a treatment time in the upper zone of 3.0-3.5 h, 2.4-2.5 h should be considered the optimum treatment time. Under these conditions, the furfural yield was decreased only by 0.1-0.2% of the o.d. mass, while the capacity of the retort increases by 25%.

The total time of the low-temperature pyrolysis process (in both zones) under optimum conditions is 3-3.2 h.

Techno-economic calculations have shown that the continuous process is already commercial at small volume of production, specifically 400 tons furfural per year, and could be implemented for furfural production using small wastes from wood sanitary felling as the raw material.

5.3 Pyrocatechol

In the late 1970s, a technology for obtaining pyrocatechol from products of the vapour-phase pyrolysis of oils was developed. Wood tar pyrolysates are good inhibitors of the oxidation of cracking-benzenes and monomers used for the production of synthetic rubber. The yield of pyrolysates is twice as high as that of an analogous inhibitor obtained from a batch process of tar distillation [84].

The pyrocatechol content in pyrolysate is about 5 times higher than in the tar. To obtain pyrocatechol, pyrolysates of wood tar of different origins were used, such as summary wood tar from pyrolysis plants, settling gas generation tar, and beech wood pyrolysis heavy tar. Pyrolysis was carried out at a temperature of 550°C, and the contact time with the moving layer was 1 sec. The installation's productivity in terms of raw material was 7 kg/h.

For isolation of pyrocatechol from pyrolysate, washing with water at 80°C was used. It is a known technique employed for the isolation of pyrocatechol from heavy gas generator tar. At a pyrolysate-water ratio of 1:2, an average of 10% of organic substances was washed, and the yield of phenols from this quantity of organic substances was 50%. A method for purification of pyrocatechol concentrate by way of repeated crystallisation from solvents (chlorobenzene or trichloroethylene) has been developed. It was shown that, at a pyrocatechol concentrate-solvent ratio of 2:1, the yield of pyrolysed pyrocatechol is 70-75%, with a content of pure pyrocatechol of 94-96%, and the yield of the residue after distillation of the solvent is 15-20%.

The water washing of pyrolysate allows the separation of up to 30% pyrocatechol from its content in pyrolysate and, at the same time, the preservation of the anti-oxidative property of the residual product, which can be used as a commercial inhibitor of oxidative processes.

The possibility has been demonstrated of using the residue after the separation of pyrocatechol from pyrolysate for synthesis of sulphomethylated novolaks, which are used in the production of pigment pastes for light tones in the dye-topping of natural leathers.

5.4 References

1. O.Golova. Thermal dpolymerisation of cellulose. Doklady AN SSSR (Proceedings of the Academy of Sciences of USSR), 1960, No. 6, pp. 1391-1394 (in Russian).
2. O.Golova, J.Epshtein, V.Sergejeva, A. Kalninsh and P.Odintsov. New method for comprehensive chemical processing of plant tissues. Ghidroliznaja i lesokhimicheskaja promyshlennost' (Hydrolysis and Wood Chemistry Industry), 1961, No.7, pp. 4-8 (in Russian).
3. O.Golova. Chemical transformations of cellulose upon the thermal action. Uspekhi khimiji (Advances in Chemistry), 1975, v. 14, No. 8, pp. 1454-1474 (in Russian).
4. A.Kislitsin, Z.Rodionova, V.Savinykh, A.Guseva. Investigation of the mechanism of thermal degradation of cellulose. Zhurnal prikladnoj khimiji (Journal of Applied Chemistry), 1971, v. 44, No.11, pp. 2518-2524 (in Russian).
5. A.Prokhorov, V.Sergeyeva, A.Kalnins. On the theory of directed thermolysis of cellulose-containing materials resulted in obtaining levoglucosan and other products. I. Levoglucosan production at thermolysis of cellulose-containing materials and basic reactions of this process. Khimiya drevesiny (Wood Chemistry), 1977, No. 6, pp.101-106 (in Russian).
6. A.Prokhorov, V.Sergeyeva, E.Fainberg, A.Kalnins On the theory of directed pyrolysis of cellulose-containing materials for obtaining levoglucosan and other products. III Thermodynamic analysis of cellulose pyrolysis, Khmiya drevesiny (Wood Chemistry), 1978, No. 3 pp. 18-25 (in Russian).
7. O.Golova, J.Epshtein, L.Durynina. Pyrolysis of cellulose in the presence of inorganic salts. Khimiya vysokom. soed. (Chemistry of High-Molecular Compounds), 1961, No. 3, pp. 536-541 (in Russian).
8. A.Prokhorov. Calculation of thermodynamic characteristics of some cellulose modifications to predict its processing conditions. Khimiya drevesiny (Wood Chemistry), 1981, No. 4, pp. 73-80 (in Russian).
9. A.c. USSR, No. 1133279 (in Russian).
10. V.Savinykh, A.Kislitsin and Z.Rodionova. Directed thermal destruction of the carbohydrate components of the wood complex in the presence of chemical reagents. Depositor No. 457, Gorky, 1985, 7 p. (in Russian).
11. Technology of Wood Chemistry Industry. Et. al. V.Vyrodov, A.Kislitsin, L.Glukharjova, Lesnaja promyshlennost', 1987, 352 p. (in Russian).
12. G.Domburg. Direction of the thermal reactions of cellulose under fire retardant conditions. In: Theoretical and Practical Aspects of Fire Protection of Wood Materials, Zinatne, Riga, 1985, pp. 54-66 (in Russian).
13. I.Ermolenko, I.Ljubliner and N.Gulko. Element-Containing Carbon Fibrous Materials, Nauka i Tekhnica, Minsk, 1982, 272 p.(in Russian).
14. A.Kislitsin. Pyrolysis of Wood: Chemical Activity, Kinetics, Products, New Processes. Lesnaja Promyshlennost', Moscow, 1990, 313 p. (in Russian).
15. O.Golova, A.Pakhomov, E.Andrijevskaya, R.Krylova. New data on the correlation between the structure of polysaccharides with the direction of chemical reactions proceeding upon thermal degradation. Doklady AN SSSR (Proceedings of the Academy of Sciences of USSR), 1957, v.115, pp.1122-1127 (in Russian).
16. Experimental Levoglucosan Production unit at the Krasnodar Hydrolysis Plant, Report IWCh, Riga, 1968, 86 p.(in Russian).
17. Preprojected Studying of a Levoglucosan Production Unit, Report IWCh, Riga, 1984, 177 p. (in Russian).
18. A.c. USSR, No. 562959 (in Russian).
19. A.c. USSR, No. 1155604 (in Russian).

20. M.Pluminsh and J.Zandersons. Production of levoglucosan from fast pyrolysis of lignocellulose. Khimiya drevesiny (Wood Chemistry), 1984, No. 5, pp. 89-92 (in Russian).
21. A.Prokhorov, I.Alsups and V.Sergeyeva. Preparation of lignocellulose vacuum thermolysis to obtain levoglucosan. Khimiya drevesiny (Wood Chemistry), 1976, No. 6, pp. 91-96 (in Russian).
22. A.c. USSR, No. 503906 (in Russian).
23. J.Zandersons and M.Plumins. The effect of technologic parameters on the yield of levoglucosan during the rapid pyrolysis of lignocellulose. Khimiya drevesiny (Wood Chemistry), 1989, No. 2, pp. 74-79 (in Russian).
24. I.Sorokin and D.Tishenko. Separation of levoglucosan from soluble tar. Ghidroliznaja i lesokhimicheskaja promyshlennost' (Hydrolysis and Wood Chemistry Industry), 1962, No. 3, pp. 8-9 (in Russian).
25. G.Mikhailov and I.Sorokin. Separation of Admixtures from Levoglucosan Concentrate. Proceedings of LTA, Leningrad, 1957, pp. 90-94.
26. N.Merlis, O.Golova, K.Saldadze and I.Nikolajeva. On the application of ionites for the removal of substances accompanying levoglucosan from products of cellulose pyrolysis under a vacuum. Izvestija AN SSR, Ser. Khim. (Proc. of AS of USSR, Ser. Chem.),1957, No.7, pp.880-881 (in Russian).
27. I.Alsups, B.Medne, and I.Berzinya. On isolation of levoglucosan from lignocellulose thermolysis products. Khimiya drevesiny (Wood Chemistry), 1980, No. 4, pp. 88-92 (in Russian).
28. A.c. USSR, No. 1155604 (in Russian).
29. J.Zandersons, M.Plumins and R.Vitolina. Purification of levoglucosan, Khimiya drevesiny (Wood Chemistry), 1990, No. 6, pp. 93-96 (in Russian).
30. R. Pernikis, I. Zaks. The synthesis, properties and polymerization of levoglucosane oligomers curred the action of UV radioactive irradiation. In book: Cellulosic, Chemical, Biochemical and Material Aspects., Ellis-Horwood, N.-J.,L.,P., 1993, pp.189-194.
31. R. Pernikis, J. Zanderson, B.Lazdina. Obtaining of levoglucosan by fast pyrolysis of lignocellulose. Pathways of levoglucosan use. In book: Developments in Thermochemical Biomass Conversation, ed. A.V. Bridgwater, L., UK, 1997, pp. 536-549.
32. A.c. USSR, No. 583134 (in Russian).
33. A.c. USSR, No. 1038344 (in Russian).
34. R.Pernikis. Oligomers and Polymers on the Bases of Sugar Anhydrides. Riga, Zinatne, 1976, 178 p. (in Russian).
35. A.c. USSR, No. 255259 (in Russian).
36. A.c. USSR, No. 303324 (in Russian).
37. E.Ponomarenko, V.Lapenko and G.Markova.Vinyl derivatives of levoglucosan. Monomers and High-Molecular Copmpounds. Proceedings of the Voronjezh University, 1972, v. 95, No. 2, pp. 72-74 (in Russian).
38. N.Kochetkov et al. Chemistry of Carbohydrates. Moscow, Nauka, 1967, 671 p. (In Russian).
39. E.Morozov. Furfural Production. Moscow, Khymiya, 1979, 200 p. (In Russian)
40. V.Sergeeva, G.Domburg, Furfural Formation and Methods for Its Production.Riga. Latv.SSR Academy of Science, 1962, 84 p. (In Russian).
41. Y.Holkin. Technology of Hydrolysis Industry. Moscow, Lesnaya promislennost, 1989, 496 p. (In Russian).
42. M.Dudkin, V.Gromov, N.Vedernikov et al. Hemicelluloses, Riga, Zinatne, 1991, 488 p. (In Russian).
43. N.Vedernikov, S.Popovs, A.Bucena, I.Kruma, V.Zakharov, D.Baldzens, The kinetics of the furfural formation from xylose in the solution of metal sulphates of various valence. 2.The furfural formation at the temperature 120-180°C. Khimya drevesiny, (Wood Chemistry) 1993, No. 6, pp. 53-59 (In Russian).

44. T.Lopatina, I..Korolkov. Investigation of salts catalytic action on pentoses dehydration. Hidroliznoe proizvodstvo (Hydrolysis Industry), Moscow, Lesnaya promislennost, 1970, No.7, pp. 3-6 (In Russian).

45. N.Vedernikov, A.Kalninsh. Regularities of cations catalytic activity in the furfurol formation process. Khimya drevesiny (Wood Chemistry), 1972, No.11, pp.11-114 (In Russian).

46. N.Vedernikov, On the analogy between Arrhenius law and dependence of cations catalytic activity at pentoses dehydration on respective atoms ionisation energy. Khimya drevesiny (Wood Chemistry), 1980, No. 1, pp. 114-115 (In Russian).

47. A.Zav'alov, V.Moroz, L.Petrovicheva. Furfural obtaining from directed catalytic wood pyrolysis. Hidroliznaya i Lesokhimicheskaya Promislennost' (Hydrolysis and Wood Chemistry Industry), 1979, No. 3, pp. 7-8 (In Russian).

48. V.Pialkin, S.Gurilev, E.Tsiganov, A. Slavyanskij, Wood pyrolysis medium influence on yield and quality of products obtained. In: Improving Methods of Wood Pyrolysis and Novel Directions of Pyrolysis products Application: LTA Scientific papers, Leningrad, Wood-Technical Academy, 1969, No.132, pp. 74-78 (In Russian).

49. E.Morozov. Theory and technology of heterophase processes for furfural obtaining. Doctor Thesis. Leningrad, Wood-Technical Academy, 1981, 30 p.(In Russian).

50. N.Vedernikov, S.Popov, A.Bucena. The kinetics of the xylose dehydration in thin layer on the surface of inert carrier. Khimya drevesiny (Wood Chemistry), 1993, No.5, pp.36-41; No.6, pp. 41-45 (In Russian).

51. N.Vedernikov, A.Bucena, V.Egle. The kinetics of furfural formation from pentose monosaccharides under heterogeneous catalysis conditions. Khimya drevesiny (Wood Chemistry), 1993, No. 6, pp. 19-26 (In Russian).

52. E.Morozov. Theoretic and technological aspects of improving furfural production. Hydrolysis and Wood Chemistry Industry. Hidroliznaya i lesokhimicheskaya promislennost' (Hydrolysis and Wood Chemistry Industry), 1986, No.5, pp.11-12 (In Russian).

53. E.Morozov, V.Shkut, M.Kebich. Penotosans conversion to furfural under salts action in solid phase. Celuloza , Hartia, 1980, v.11, No. 5, pp.6-8.

54. E.Morozov. Superphosphate application in continuous process of furfural obtaining. Scientific-Technical Information Microbioprom, 1964, No.1, pp.18-20 (In Russian).

55. G.Potapov, V.Krupenskij, M.Alieva. Aldoses dehydration catalyzed by gel- immobilized Zr-porphyrin complex. Kinetic and Katalysis, 1987, v. 28, No. 1, pp.205-207 (In Russian)

56. A.c. USSR, No.1155600 (In Russian).

57. A.c. USSR, No.1351649 (In Russian).

58. G.Potapov, V.Krupenskij. Dehydration of aldoses catalyzed by gel-immobilized circonium porphyrin complex. React. Kinet. Catal. Lett., 1985, v. 28, No. 2, pp.331-337.

59. N.Vedernikov, A.Bucena, D.Baldzens. Kinetics of furfural formation from pentose monosaccharides under conditions of heterogeneous catalysis. 4. Comparative analysis of pentose interaction with homogeneous and heterogeneous catalysts. Latvijas Khimijas Zurnals, 1996, No.1-2, pp. 118-127 (In Russian).

60. A.c. USSR, No.151683 (In Russian).

61. France pat., No.1421046.

62. Italy pat., No. 752841.

63. Japan pat., No. 512090.

64. Sweden pat., No. 322785.

65. Germany pat., No. 1493864.

66. A.c. USSR, No. 391140 (In Russian).

67. Russian pat., No.1365674; Russian pat. No.2123497 (In Russian).

68. V.Beinart, N.Vedernikov, V.Kalnina. Pentosans and polyuronids conversion to furfural. In: Wood Cell wall and its Transformations under Chemical Action. Riga, "Zinatne", 1972, pp. 484-491 (In Russian).

69. I.Roze, V.Kalnina, N.Vedernikov. Study of products from lignocellulose destruction under influence of sulphuric acid small amounts. Khimya Drevesiny (Wood Chemistry), 1976, No. 4, pp. 66-75 (In Russian).

70. E.Morozov, T.Tsedrik. Furfural formation macrokinetic in the presence of diluted sulphuric acid. Hydrolysis Industry in USSR, 1982, No.5, pp. 8-11 (In Russian)

71. A.Kislitsin. Investigation of chemical mechanisms of wood components thermodestruction. Doctor Thesis, Leningrad, Wood-Technical Academy, 1974, 35 p. (In Russian).

72. V.Kostenko, B.Levitin. Mechanism of monoses dehydration in the presence of acidic catalysts. Hydrolysis Industry in USSR, 1972, No.6, pp. 13-16 (In Russian).

73. V.Kulnevich, Y.Falkovich, B.Ershov. On the kinetics of pentoses high-temperature dehydration in the closed system. Izvestija VUZov, Pischevaja technologia, 1966, No. 3, pp.72-77 (In Russian).

74. U.Ziemelis, A.Kulkevits, A.Bucena. Laboratory installation for study of low-temperature fast thermolysis. Khimya Drevesiny (Wood Chemistry), 1976, No. 1, pp. 90-92 (In Russian).

75. A.Bucena, A.Kulkevits. Furfural formation study during fast law-temperature thermolysis. Khimya Drevesiny (Wood Chemistry), 1980, No. 5, pp.93-98 (In Russian).

76. A.Bucena, A.Kulkevits. Furfural production by fast law-temperature thermolysis. 1. Formation of furfural from xylose and the effect on furfural yield of the process temperature, sulphuric acid and xylose concentration. Khimya Drevesiny (Wood Chemistry), 1985, No. 6, pp.75-79 (In Russian).

77. A.Bucena, A.Kulkevits. Furfural production by fast law-temperature thermolysis. 2. The yield of furfural from xylose in different stages of furfural formation. (Khimya Drevesiny (Wood Chemistry), 1985, No.6, pp. 80-83 (In Russian)

78. A.Bucena, A.Kulkevits. Furfural production by fast law-temperature thermolysis. 3. Macrokinetic data for the total process of furfural formation from xylose. Khimya Drevesiny (Wood Chemistry), 1986, No. 1, pp. 69-72 (In Russian).

79. A.Bucena, A.Kulkevits. Furfural production by fast law-temperature thermolysis. 4. The effect of different technological parameters on the yield of furfural from birch sawdust. Khimya Drevesiny (Wood Chemistry), 1986, No. 3, pp. 75-78 (In Russian).

80. A.Bucena, A.Kulkevits. Furfural production by fast law-temperature thermolysis. 5. Macrokinetics of the furfural from birch sawdust. Khimya Drevesiny (Wood Chemistry), 1986, No.3, pp. 79-84 (In Russian).

81. V.Krupenskij. The nature of cations katalytic effect on monosaccharides destruction. Khimya Drevesiny (Wood Chemistry), 1978, No. 1, pp. 72-75 (In Russian).

82. A.c. USSR, No. 215231.

83. A.c. USSR, No. 454206.

84. Yu. Goldshmidt. Technology for pyrocatechol obtaining from wood-tar oils pyrolyzates, PhD Thesis, Leningrad, Wood-Technical Academy (Lesotechnicheskaya Akademia), 1978, 19 p. (In Russian)

6. CATALYTIC PYROLYSIS OF LIGNOCELLULOSIC RAW MATERIALS WITH PRODUCTION OF LEVOGLUCOSENONE

The effect of chemical reagents on the wood pyrolysis process manifests itself mainly in the change of the temperature regions of active degradation, acceleration or inhibition of the formation of some volatile products, as well as in the change in the kinetic parameters of the process [1].

The studies in this area, carried out in the former USSR, were connected mainly with obtaining a solid product of pyrolysis, char, as well as the regulation of its properties and the devising of new methods for its production. As regards the liquid products of pyrolysis, the studies were aimed mainly at the search for conditions increasing the yield of furfural, levoglucosan (see Section 5) and levoglucosenone (LG – none).

Since the 1970's, in parallel with investigations of Shafizadeh, Broido, Halpern, Fung and Lipska, research into the change of the wood pyrolysis mechanism and the composition of liquid products under the action of different acidic additives were carried out in the former USSR [2].

At the Latvian State Institute of Wood Chemistry, intensive studies on the basic mechanism of acid catalysed pyrolysis were conducted. Dehydrated 1,6-anhydrosaccharide, levoglucosenone (LG-none), was identified as the main product of pyrolysis; its conformation was established by NMR spectroscopy, and its mass spectrum was obtained for the first time [3,4]. A method for obtaining levoglucosenone using the lignocellulosic residue after furfural production as the raw material was proposed [5]. Aspen wood prehydrolysis was carried out by two methods, i.e. using sulphuric acid or aluminium sulphate. Then the lignocellulosic residue was pyrolysed without additional treatment with acid catalysts under vacuum conditions at a temperature of 300-400°C, or at atmospheric pressure at 475-500°C. The levoglucosenone yield under these conditions reached 4.5-6% of cellulose. Catalytic pyrolysis was investigated for increasing the LG-none yield. The mechanism of the action of different dehydrating additives such as boric and phosphoric acids and ammonium chloride was studied [6]. It has been shown that additives inhibit the LG formation during fast pyrolysis and increase the LG-none yield (Table 6.1). The highest LG-none yield is reached by adding phosphoric acid.

Table 6.1 Levoglucosenone formation upon pyrolysis of "Taircell" pulp in the presence of dehydrating additives (temperature 350oC)

Additive 5% in terms of o. d. cellulose	Yield, % in terms of o. d. cellulose
-	0.6
NH_4Cl	4.7
H_3BO_3	5.8
H_3PO_4	16.8

A different mechanism of the action of chemical additives has been established, that reveals itself mainly in the development of dehydration reactions, which are primary processes and determine the further destruction process (Fig 6.1).

It is shown that dehydration of cellulose under the action of phosphoric acid proceeds in a more low-temperature region. Ester bonds are formed according to multiple step mechanisms followed by a β -cis - elimination of phosphoric acid, and double bonds are formed. In contrast to other dehydrating additives, phosphoric acid catalyses not only the dehydration reactions, but also the reactions of the thermal depolymerisation of cellulose.

To elucidate the mechanism of the action of phosphoric acid, the process of the catalytic thermal decomposition of model compounds, i.e. cellobiose and glucose, was investigated. Based on the data obtained on the amount of unchanged glucopyranose units, the development of intra- and intermolecular dehydration reactions in the process of the thermal decomposition of glucose was determined [7-10].

The effect of the conditions of impregnation (amount of phosphoric acid, temperature pretreatment) and pyrolysis (heating rate, temperature) on the LG-none yield was investigated [11]. According to the findings of the Institute of Wood Chemistry, the highest amount of levoglucosenone was obtained upon the introduction of 5% phosphoric acid into "Taircell" pulp under slow pyrolysis conditions, through heating to a temperature of 350°C (5°C/min) in an inert gas flow [12].

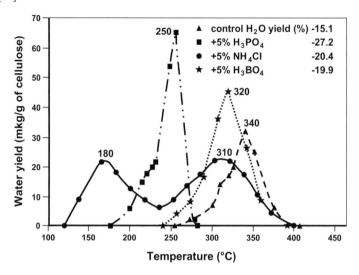

Figure 6.1 Variations in the water yield in the "Taircell" pulp thermodestruction process in the presence of dehydrating additives (the PyGLC method)

At present, studies on the production of LG-none from various types of cellulose-containing raw material are under way at the Latvian State Institute of Wood Chemistry, jointly with the Institute for Wood Chemistry and Chemical Technology of Wood [13,14].

A series of studies on the catalytic pyrolysis of biomass to obtain liquid products has been carried out at the Institute of Chemistry of Natural Organic Materials, the Siberian Division of the Academy of Sciences of Russia [15-18].

It has been shown that the thermal processing of aspen wood using catalysts (sulphuric acid, sulphates of iron, cobalt and aluminium, 3% of the dry raw material) in a flow of superheated steam results in an increase in the conversion degree of aspen wood (Table 6.2). The greatest effect of the catalytic treatment is found when heating to a temperature range of 200-300°C.

The highest levoglucosenone yield, i.e 8.7%, was obtained from catalytic pyrolysis in the presence of cobalt sulphate at 350°C in a flow of superheated steam.

Table 6.2 Conversion degree (% from dry substances) of aspen wood upon thermal treatment in the presence of catalysts

Temperature, C	Without catalyst	Catalysts			
		H_2SO_4	$Fe_2(SO_4)_3$	$CoSO_4$	$Al_2(SO_4)_3$
200	5.8	49.8	16.1	22.8	51.1
250	23.6	56.8	48.5	60.1	58.8
300	49.1	66.2	61.2	67.5	65.0
350	74.3	75.0	73.9	73.0	73.7

The possibility of obtaining levoglucosenone from the treatment of glucose in the temperature range 50-120°C in the presence of sulphuric acid in a medium of aprotic solvents (dimethyl, formamide, pyridine, dimethyl sulphoxide) has also been shown. The highest levoglucosenone yield, i.e. 44.3%, was obtained at a glucose to sulphuric acid ratio 1:2; the reaction time was 0.5 h [19].

6.1 References

1. Z.Rodionova, V.Savinykh and Z.Lebedeva. Directed Pyrolysis of Wood. Wood Chemistry and Boxing. Moscow, VNIPIEILesprom, 1982, No. 3, p. 36 (in Russian).
2. V.Savinykh, A.Kislitsin, Z.Rodionova and G.Domburg. On the effect of lignin and hemicelluloses on the thermal decomposition of cellulose upon wood pyrolysis. Processing of wood pyrolysis products. Proceedings of CNILKHI, Moscow, Lesnaja promyshlennost', 1976, pp. 43-47.
3. G.Domburg, I.Berzina, J.Kirsbaums and M.Gavars. On the problem of levoglucosenone structure. Khimiya drevesiny (Wood Chemistry), 1978, No. 6, pp.105-107 (in Russian).
4. G.Domburg, I.Berzina, E.Kupce and J.Kirsbaums. Investigation of a molecule conformation. Khimiya drevesiny (Wood Chemistry), 1980, No. 3, pp.99-102 (in Russian).
5. A.s.USSR 595319.
6. G.Dobele, T.Dizhbite, G.Rossinskaja and G.Telysheva. Thermocatalytic destruction of cellulose. In: J.F. Kennedy, G.O. Fillips, P.A.Williams (Eds), Cellulose and Cellulose Derivatives: Physico-Chemical Aspects and Industrial Application, Woodhead, Cambridge, 1995, p.125-129..
7. G.Dobele, G.Domburg. Thermocatalytic transformations of compounds modelling the carbohydrate bond. All-Union Conference on Thermal Analysis, Kiev-Uzhgorod, 1985, pp. 235-237 (in Russian).
8. G.Domburg, G.Dobele, G.Rossinskaya, V.Yurk'yan and B.Rone. Thermocatalytic transformation of cellulose and lignin in the presence of phosphorous acid. 4. Thermal degradation of cellobiose. Khimiya drevesiny (Wood Chemistry), 1986, No. 1, pp. 77-81 (in Russian).
9. G.Domburg, G.Dobele, G.Rossinskaya and B.Rone. Dehydration of glucose. Khimiya drevesiny (Wood Chemistry), 1988, No. 1, pp. 41-47 (in Russian).
10. G.Domburg, G.Dobele, G.Rossinskaya, V.Yurk'yan and B.Rone. (1988) Dehydration of cellulose under catalytic condition. Khimiya drevesiny (Wood Chemistry), 1988, No. 3, pp. 97-102 (in Russian).
11. G.Dobele, G.Rossinskaya and B.Rone. Thermodestruction of cellulose and obtaining of levoglucosenone. In: J.Kennedy and G.Williams, (Eds.). The Chemistry and Processing of Wood and Plant Fibrous Materials, Woodhead Publ., Cambridge, 1996, pp. 345-350.
12. G.Rossinskaya, G.Dobele and G.Domburg. Thermocatalytic transformation of cellulose and lignin in the presence of phosphorous acid. 3. Characteristics of the higher-boiling resin fraction of carbohydrates pyrolysis. Khimiya drevesiny (Wood Chemistry), 1986, No. 1, pp. 73-76 (in Russian).
13. G.Dobele, G.Rossinskaya, G.Telysheva, D.Meier and O.Faix. Cellulose dehydration and depolymerization reaction during pyrolysis in the presence of phosphoric acid. J. Anal. Appl. Pyrolysis, 1999, No. 49, pp. 307-317.
14. G.Dobele, G.Rossinskaya, T.Dizhbite, G.Telysheva, S.Radtke, D.Meier and O.Faix. Cellulose as a raw material for levoglucosenone production by catalytic pyrolysis. Cellucon'99, Joint Meeting with ITIT International Symposium, Recent Advances in Environmentally Compatible Polymers. March, 1999, pp. 24--26 Japan (in press).
15. B.Kuznetsov, A.Efremov, G.Slashchinin E.Korniets and L.Balakireva. Catalytic conversion of aspen wood in superheated steam in the presence of sulphuric acid and cobalt, ferric, and aluminium sulphate. Khimiya drevesiny (Wood Chemistry), 1990, No. 5, pp.51-56 (in Russian).

16. A.Efremov, G.Slashchinin, E.Korniets and V.Kuznetsov. Catalytic thermolysis of aspen wood in the presence of sulphuric acid and ferric, cobalt and aluminium sulphates applying high pressures. Khimija drevesiny (Wood Chemistry), 1990, No 5, pp. 57-60 (in Russian). .

17. A.Efremov, G.Slashchinin, E.Korniets, V.Sokolenko and B.Kuznetsov. Conversions of levoglucosenone in acidic media. Sibirskij khim. zhurnal (Siberian Chemical Journal), 1992, 6 Issue, pp. 34-39 (in Russian).

18. A.Efremov, V.Kuznetsov, A.Konstantinov, S.Kuznetsova and V.Krotova. New thermocatalytic methods of chemicals producing from lignocellulose materials in the presence of acid-type catalysts. Proceedings of 8th International Symposium on Wood and Pulping Chemistry, Helsinki, Finland, 6-9 June, 1995, vol. 1, pp. 689-696.

19. G.Slashchinin and A.Efremov. Preparation of levoglucosenone from glucose. Khimiya drevesiny (Wood Chemistry), 1991, No. 6, pp. 72-73 (in Russian).

7. CONCLUSIONS

The thermal conversion of biomass as an alternative solution to the problem of the use of renewable raw materials as a fuel is currently the subject of wide discussion.

The realisation of pyrolysis processes requires large capital investment. At existing petroleum prices, the obtaining of liquid and solid fuel from biomass may be justified for regions with cheap sources of raw material and energy, and a strong market for the products of pyrolysis.

Studies carried out on pilot plants in the former USSR have shown the possibility of developing complex technologies for the thermal processing of biomass, with waste-free production using all constituents of the raw material and yielding individual chemical compounds. These include the technologies considered in the present review for furfural and levoglucosan production. Wide scientific experience in the use of levoglucosan for synthesis of polymers and preparation of composite materials has shown the attraction of the commercialisation of processes for its production.

A dehydrated 1,6-anhydrosaccharide, levoglucosenone, detected in the 1970's, may only be obtained at a high yield by the catalytic pyrolysis of the cellulose-containing raw material. The uniqueness of the levoglucosenone structure offers strong possibilities for chemical synthesis of valuable biologically active compounds, including those identical to natural ones. The technology for producing levoglucosenone through fast pyrolysis can also become a method for the complex use of both biomass and secondary cellulose-containing raw materials.

The present review includes the classification of the components of pyrolytic oils, pathways for their use and technologies for obtaining individual compounds. However, work reviewed does not cover many avenues of investigation which have been pursued at a laboratory scale but not demonstrated at a larger scale. Most importantly this applies to non-traditional catalytic processes, whose implementation and optimisation require additional profound theoretical studies as well as the development of non-conventional equipment.

8. ACKNOWLEDGEMENTS

The authors would like to express very sincere thanks to Dr. D. Meier and colleagues (Institute of Wood Chemistry and Chemical Technology of Wood in Hamburg) for their help in design of figures to this review.

STUDY OF LEVOGLUCOSAN PRODUCTION - A REVIEW

Desmond Radlein, Ph.D.

President, RTI Ltd., 110 Baffin Place, Unit 5, Waterloo, ON, Canada, N2V 1Z7

1,6-Anhydro-β-D-Glucopyranose

EXECUTIVE SUMMARY

It is well known that the highest yield possible for a single chemical from pyrolysis of cellulosic biomass is levoglucosan. With the increasing interest in a sustainable development, it is therefore natural to seek its possible applications as a chemical or fuel feedstock. Interest in this possibility is shown by a number of recent conferences dealing with applications of anhydrosugars, e.g. the recent ACS Symposium "Levoglucosenone and Levoglucosans. Chemistry and Applications" (1).

Most of its chemical applications require the use of relatively pure crystalline levoglucosan while for use as an intermediate in the production of sugars for fermentation, e.g. to fuel ethanol, relatively impure solutions suffice.

The following are some salient facts concerning levoglucosan (1,6-Anhydro-β-**D**-Glucopyranose) and its present and potential status as an industrial material or material of commerce:

- Several processes have been patented for the pyrolytic generation, recovery and purification of levoglucosan from cellulose and other lignocellulosic materials. In particular CANMET has an interest in the patented University of Waterloo process for levoglucosan production and recovery from lignocellulose (2).
- By suitable pretreatment of lignocellulosic feedstocks the yields of levoglucosan as well as several other anhydrosugars can be enhanced. Five carbon anhydrosugars can be recovered from the hemicelluloses in the feed.
- The anhydrosugars are convertible to free sugars by hydrolysis producing fermentable solutions. This process, thermolytic saccharification, therefore represents an alternative to acid and enzymatic hydrolysis for the production of ethanol from lignocellulose. Feasibility studies suggest that it may well be economically competitive with these alternative methods.
- Levoglucosan has a number of chemical attributes that may make it desirable as a chemical feedstock. Examples are the use as a synthon in the manufacture of pharmaceuticals and related materials and biodegradable surfactants formed by combination of levoglucosan with naturally occurring fatty acids.
- Levoglucosan as well as certain of its derivatives can be polymerized to give a variety of polymers of potential interest in various kinds of applications.

CONCLUSIONS

- The existing market for levoglucosan is very small and very high priced. The situation is at least partly a reflection of the low yields and lack of scalability of the traditional methods of production by vacuum pyrolysis of starch and the difficulty of purifying levoglucosan at low cost. This has inhibited the development of large-scale uses.

- If a market could be established at the level of 10,000 to 50,000 tonnes per year, then the price range would probably have to be in the range of roughly $5 - $15/kg. It seems unlikely that high value pharmaceutical type applications could provide a basis for large commodity-scale production of levoglucosan.

- On the production side, most existing pyrolysis technologies can be adapted for levoglucosan production from lignocellulose. Fluidized beds are the preferred reactors for high heat transfer rate pyrolysis, but there are feeding problems associated with the low melting tendencies of pretreated lignocellulose, which cause bed agglomeration. Research is needed in the ways to deal with this, perhaps by modifying the pyrolysis regime (temperature and residence time).

- Nevertheless, it is considered that the principal obstacle towards large-scale production of crystalline levoglucosan is a cheap, reliable and efficient recovery procedure. In spite of many efforts over the years, it appears much remains to be accomplished in this aspect. Alternatively, uses for crude product mixtures should be sought.

- Successful development of many of the derivative chemical synthetic products must bring together the available expertise in the areas of production, chemistry and marketing. Furthermore, the timescale of development may be long so a significant investment over a period of time is required. This might be difficult to achieve, as no single organization appears to have all these necessary ingredients at present.

- However, in some cases, e.g. the glucoside surfactants or dextrins, markets already exist and it would be sufficient to demonstrate some technical benefits in order to compete with the alternative technologies. But, even here some improvements in the available pyrolysis technology will be required.

- In principle, the simplest products among those discussed are the glucoside surfactants, and the dextrin polymers. These are probably also the chemical products with the largest potential market volumes.

- Thermolytic saccharification for ethanol production appears to have the possibility to be a feasible alternative to enzymatic or acid hydrolysis technologies. However many technical and process uncertainties remain; which makes it very difficult to make a reliable assessment of the technology.

- Some of the issues to be resolved include:
  o Optimal pretreatment strategy.
  o Feeding of wet feedstock to a pyrolyzer.
  o Use of fluidized beds in regimes of low temperature and long residence times.
  o Optimal strategy for hexose/pentose fermentation
  o Improved hydrolysis of anhydrosugars to free sugars (use of pressurized water).
  o Development of adapted strains of microorganism for fermentation of pyrolytic liquors.
  o Direct fermentation of anhydrosugars

- If large-scale applications with higher than fuel value could be found for pyrolytic lignin, the economic feasibility of thermolytic saccharification would be greatly enhanced. One such prospect is its use in resin formulations, though it would have to be verified that the lignin from de-mineralized or hydrolysed wood was suitable; it may well differ in molecular weight from the normal pyrolytic lignin from untreated wood. In such a case, a "bio-refinery" could be envisaged in which ethanol and/or sugar solutions and lignin are the main products.

- To further the development, a sensitivity analysis of the various process options and possibilities should be carried out with a view to prioritization of the required developmental work. It should also be pointed that although only anhydrosugar solutions need be produced, their availability might prove to be a stimulus to other uses of levoglucosan.

1   BACKGROUND

Cellulose comprises the major part of all plant biomass. The source of all cellulose is the structural tissue of plants where it occurs in close association with hemicellulose and lignin, which together comprise the major components of plant fiber cells. It consists of long chains of β-glucosidic residues linked through the 1,4 positions. These linkages cause the cellulose to have a high crystallinity and thus low accessibility to enzymes or acid catalysts, rendering hydrolysis difficult. Hemicellulose is an amorphous hetero-polymer that is easily hydrolyzed. Lignin, an aromatic three-dimensional polymer, is interspersed among the cellulose and hemicellulose within the plant fiber cell.

In principle, it should be possible to produce ethanol by hydrolysis of the holocellulosic portion of biomass and fermentation of the resulting sugars. There are however considerable differences of opinion on the current commercial viability of available technologies. A discussion of the commercial issues may be found at: http://www.ceassist.com/economic.htm. There are four process technologies currently under development by industry for converting biomass to ethanol. They are:

1.   Dilute Acid Hydrolysis
2.   Concentrated Acid Hydrolysis
3.   Enzymatic Hydrolysis
4.   Syngas Fermentation

1.1   Dilute Acid Cellulose Conversion

This is by far the oldest technology for converting biomass to ethanol. The hydrolysis occurs in two stages to accommodate the differences between hemicellulose and cellulose. The first stage can be operated under milder conditions to maximize yield from the more readily hydrolyzed hemicellulose. The second stage is optimized to hydrolyze the more resistant cellulose fraction. The liquid hydrolyzates are recovered from each stage and fermented to ethanol. Residual cellulose and lignin left over in the solids from the hydrolysis reactors may be used as boiler fuel to produce steam or electricity. BC International (BCI), a start-up company based in Dedham, Mass., recently became the latest firm to promise commercial production of fuel-grade ethanol based lignocellulose. On Oct. 20, 1999, the company held a groundbreaking ceremony for a planned $90 million facility in Jennings, La., that is designed to convert bagasse, the fibrous residue of sugarcane refining, into 20 million gal of ethanol per year. BCI plans to use dilute acid hydrolysis during the startup phase of its facility in Jennings, Louisiana, which will use sugarcane bagasse as a feedstock. Further commercial and technological details may be found in the news article (3), along with an overall economic assessment (4).

The dilute acid process generally involves the use of 0.5% to 15% sulfuric acid to hydrolyze the cellulosic material. In addition, temperatures ranging from 90 °C - 300 °C, and pressures up to 800 psi are necessary to effect the hydrolysis. At high temperatures, the sugars degrade to form furfural and other undesirable by-products. The resulting glucose yields are generally low, less than 50%. Accordingly, the dilute acid process has not been successful in obtaining sugars from cellulosic material in high yields at low cost.

1.2 Concentrated Acid Cellulose Conversion

The concentrated acid processes have been somewhat more successful, producing higher yields of sugar. This technology dissolves and hydrolyzes cellulose into glucose sugar using concentrated sulfuric acid followed by dilution with water. The hydrolysis that occurs in the dilution step gives near theoretical sugar yields. The concentrated acid disrupts the bonds between the cellulose chains, making them susceptible to hydrolysis, which is accomplished by simply diluting the acid-cellulose mixture with hot water.

The process generally consists of the following stages:

1. prehydrolysis to hydrolyze the hemicellulose portion,
2. main hydrolysis to hydrolyze the cellulose, and
3. post hydrolysis to form glucose from oligosaccharides formed in step 2

The first step involves the addition of sulfuric acid to the biomass, which is then heated to at least 100 °C. to break down the hemicellulose. The result of this prehydrolysis step is a solution containing not only virtually all of the C_5 sugars, but also C_6 sugars. These C_6 sugars are thus not recovered if the C_5 sugar stream is not utilized, resulting in lower sugar yields. After the sugar stream produced by the prehydrolysis step is removed, concentrated acid is added to disrupt the crystalline lattice of the cellulose and form glucose. The sugars produced are then fermented to alcohols.

Typically 60% to 90% sulfuric acid is used to effect hydrolysis. These processes, although successful at producing sugar yields above 90%, have not been implemented commercially in the past due to the cost of concentrated sulfuric acid and its subsequent recovery, the difficulties encountered in handling concentrated sulfuric acid, and the need for equipment resistant to the acid at high temperatures. In addition, the higher the acid concentration used, the more energy required to concentrate the acid, resulting in the processes becoming more economically disadvantageous.

The process is rather complicated and commercialization requires that the steps be simplified, the energy consumption reduced, and the difficulties encountered in recycling spent acids eliminated. An additional problem faced in the commercialization of known acid hydrolysis processes is the production of large amounts of gypsum when the spent or used acid is neutralized. Besides, the low sugar concentrations resulting from the processes require the need for concentration before fermentation can proceed. Additionally, when hydrolysis is carried out at temperatures above 150 °C, compounds such as furfural that inhibit fermentation are produced from the degradation of pentoses.

Arkenol and Masada Resource Group in conjunction with the DOE have proposed commercial plants based on this technology. Arkenol, whose plant will be located in Sacramento County, holds a series of patents on the use of concentrated acid to produce ethanol. The economics of this opportunity are driven by the availability of a cheap feedstock, rice straw that poses a disposal problem. Masada will locate a municipal solid waste (MSW)-to-ethanol plant in Orange County, New York. The plant will process the lignocellulosic fraction of MSW into ethanol using technology based on a concentrated sulfuric acid process.

Arkenol gives the following inputs and outputs for a commercial stand-alone fuel-ethanol plant configuration assuming an available feedstock supply on a 330 days per year, twenty-four hours per day basis which has an average (holo)cellulosic content of 75%, having the following inputs and outputs (5):

Inputs		
Feedstock	454 dry tonnes per day	500 dry tons per day
Sulfuric Acid	21.45 tonnes per day	23.6 tons per day
Lime	8.25 tonnes per day	9.1 tons per day
Electricity	5,000 kW	5,000 kW
Steam	61,700 kg. per hour	136,000 lbs. per hour
Outputs		
Ethanol, 200 proof	227,000 liters per day	60,000 gallons per day
Carbon Dioxide	172.5 tonnes per day	190 tons per day
Lignin (50% moisture)	136.2 tonnes per day	150 tons per day
Gypsum (40% moisture)	27.2 tonnes per day	30 tons per day
Yeast (80% moisture)	45.2 tonnes per day	49.8 tons per day

1.3   Enzymatic Cellulose Hydrolysis

This technology involves the biochemical hydrolysis of cellulose with the cellulase enzyme. Actually this is a family with three major classes recognized: *endoglucanases*, which act randomly on soluble and insoluble glucose chains; *exoglucanases*, which include glucanhydrolases that preferentially liberate glucose monomers from the end of the cellulose chain and cellobiohydrolases that preferentially liberate cellobiose (glucose dimers) from the end of the cellulose chain and *β-glucosidases*, which liberate D-glucose from cellobiose dimers and soluble cellodextrins.

Although specificity and conversion are very high, rates are low and for practical purposes, substantial pretreatment of the substrate is required. Besides, since cellulase is not commercially available, the enzyme must be produced within the plant leading to considerable complication.

Further improvements in cellulase production technology and a significant increase in enzyme specific activity, so reducing cellulase cost, are thus probably required to make commercialization feasible. Increasing specific activity is the most effective way to reduce cellulase cost. It is also desirable to increase the optimal temperature for fermentation. Currently, fermentation organisms perform best at 30° - 35°C. If the temperature is increased to 55°C, advantage could be taken of the improved cellulase performance by reducing residence time. Furthermore, the higher temperature would help reduce the risk of contamination.

The first application of cellulases to wood hydrolysis in an ethanol process was to simply replace the acid hydrolysis step with an cellulase hydrolysis step. This is called separate hydrolysis and fermentation. The most important process improvement made for the enzymatic hydrolysis of biomass was the introduction of simultaneous saccharification and fermentation (SSF), which has recently been improved to include the co-fermentation of multiple sugar substrates. In the SSF process, cellulase and fermenting microbes are combined. As sugars are produced, the fermentative organisms convert them to ethanol. Enzymatic hydrolysis will be used in Iogen/Petro Canada's Ottawa, Canada project and is being explored for BCI's Gridley project (3).

1.4   Syngas Fermentation

Biomass can be converted directly or indirectly to synthesis gas (which consists primarily of carbon monoxide, carbon dioxide, and hydrogen) via a gasification process. After gasification, anaerobic bacteria are used to convert the gases to ethanol (6). Bioresource Engineering Inc. (7), has developed syngas fermentation technology that can be used to produce ethanol from cellulosic wastes with high yields and rates. The technology is based on the strict anaerobe *Clostridium ljungdahlii*, a bacterium discovered in 1987, capable of conversion of CO, H_2 and

CO_2 to a mixture of acetate and ethanol. It is also capable of growing on fructose as well as the C_5 sugars xylose, arabinose.

This is a very fascinating concept, as it would offer the possibility to utilize all the carbon present in the biomass to ethanol, and not just the portion present in the cellulose and hemicellulose. The feasibility of the technology has been said to have been demonstrated, and plans are under way to pilot the technology as a first step toward commercialization. However, the status of this project is unknown.

1.5   Thermolytic Saccharification

1.5.1   Introduction

Thermochemical saccharification involves the pyrolysis of alkali free lignocellulosic biomass to give high yields of anhydrosugars, which are subsequently converted by a relatively mild hydrolysis to free fermentable sugars. The concept has been re-investigated recently from the viewpoint of fast pyrolysis technologies, but it is by no means new. It was suggested as early as 1920 in Pictet's (8) Canadian patent. Later, a thorough scientific investigation led to the proposal for a potential industrial process by Shafizadeh et al. (9) in 1979 on the basis of the observation that prehydrolysis of hard- or softwoods leads to a very significant increase in the anhydrosugar yields upon pyrolysis.

As described by Shafizadeh, upon pyrolysis, cellulose undergoes intramolecular transglycosylation to provide "anhydrosugars, which are subsequently inter-converted partially re-polymerized by intermolecular transglycosylation to provide a tarry mixture of levoglucosan, its furanose isomer (1,6-anhydro-β-D-glucofuranose), and randomly linked oligo- and polysaccharides. The tar fraction also contains a variety of minor products resulting from competing dehydration, elimination and disproportionation reactions."

These studies were carried out using vacuum pyrolysis. Later work by Radlein et al. (10) established that, at least under fast pyrolysis conditions, in fact the predominant dimeric anhydrosugar present in cellulose-derived pyrolyzates was cellobiosan, which was a primary product. Boon et al subsequently confirmed this (11).

Cellobiosan

Shafizadeh (12) made a detailed study of the effect of the acid pretreatment of a lignocellulosic feedstock, Douglas-fir heartwood, on anhydrosugar yields. In general softwood hemicelluloses consist mainly of galacto-glucomannan with much smaller amounts of arabino-glucuronoxylan; i.e. they contain predominantly hexoses with only small amounts of pentose sugars like xylose and arabinose. Typical softwood pentosan (xylose + arabinose) contents range from 5 – 14 wt% of the moisture-free wood. This may be compared with hardwoods where the range of pentosan content is typically 18 – 23%, comprised mainly of glucuronoxylan with smaller amounts of glucomannan.

The Douglas fir wood was found to contain 1% arabinan, 6% xylan, 16% mannan, 48% glucan and galactan for an overall content of 7% pentosans and 64% hexosans. The remaining material is mainly lignin (27%) and the intrinsic ash content is small (0.1%). The wood was first extracted with chloroform then subjected to prehydrolysis to hydrolyze the hemicellulose sugars, and then residual lignocellulose (65% of the wood) was vacuum-pyrolyzed to yield char and an anhydrosugar rich tar containing a concentration of 29 wt% levoglucosan. The tar was hydrolyzed and the resulting hydrolysate combined with the prehydrolysate to give an overall hexose recovery of 38% of the wood. (This corresponds to 59% of the total hexose in the wood.) The results are summarized in the scheme below.

Other data by Shafizadeh (13) also indicate that for a variety of feedstocks, the average concentration of D-Glucose after hydrolysis of the pyrolyzates was approximately 1.5 ± 0.2 times the measured levoglucosan concentration. This suggests that the various anhydrohexosan monomers and oligomers are ultimately converted to glucose on hydrolysis.

As noted by Shafizadeh (13), *"The yields of levoglucosan are reduced in the above laboratory scale process, reflecting the problems of heat and mass transfer, which could be solved through the application of engineering methods, namely fluidized bed technology."* Apparently the first publication of the use of fluidized beds for pretreated lignocellulosics under fast pyrolysis conditions was by Radlein et al. (14).

Blazej and Kosik (15) have made a qualitative comparison of these various technologies for saccharification of lignocellulosics and gave the following assessment:

Parameter	Enzymatic Hydrolysis	Dilute Acid	Strong Acid	Thermolytic Saccharification
Mass Conversion	+	++	+++	++
Level of Technology	++	+	+	++
Driving Force	+	++	++	+++
Product Quality	++	+	+++	+
Energy Consumption	+	+	++	+++
Environmental Impact	+++	+	+	++
Economic Feasibility	+	+	+	++
Total Score	**11**	**9**	**13**	**15**

It was not clear what some of these terms like "Driving Force" meant. Nevertheless, while this is no doubt rather optimistic, their conclusion is that thermolytic saccharification deserves serious consideration as an approach to the production of fermentable sugars from biomass.

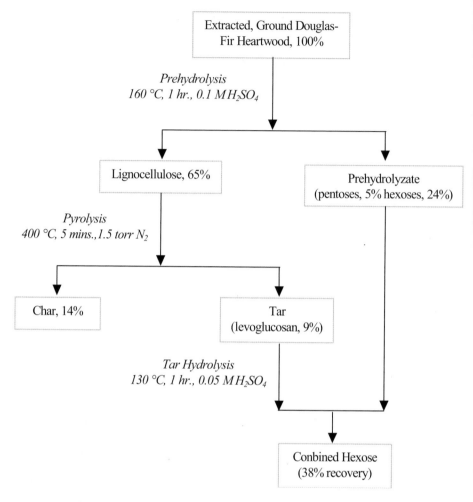

Shafizadeh Scheme for Thermo-Saccharification of a Softwood
(Percentages are wt% of dry wood feed)

1.5.2 Fast pyrolysis data

1.5.2.1 Hardwood

Workers at the University of Waterloo (14, 16, 17) later developed and extended Shafizadeh's proposal in a series of studies based on fast pyrolysis technology. Besides the differences in technology, the concept of pyrolysis of demineralized samples was introduced and provides an alternative pretreatment to instead of prehydrolysis. The presence of anhydro-oligosaccharides as primary pyrolysis products was also recognized for the first time. It has also been demonstrated that anhydrosugars could also be obtained from the hemicellulose portion of the biomass with adequate pretreatment and careful control of pyrolysis conditions. This has subsequently been confirmed. Thus, Lomax et al. (18) have studied various polysaccharides using in-source mass-spectrometric techniques. They confirm that under rapid pyrolysis oligosaccharides retaining the configuration of the glycosidic linkages are produced. They also conclude that the mass spectra are dominated by the sugars making up the backbone. The pendant side chains tend to be unstable and to undergo ring cleavage. In all cases, series of anhydro-oligosaccharides were observed but their relative efficiency of production was very variable depending on the precise conditions and the substrate structures. These observations remained true for pentosan containing substrates, except that a xylopyrannan produced many dehydration products as well. (This last result is in some contrast to our data, which suggests that relatively high yields of anhydroxylose were obtained by pyrolysis of demineralized biomass, both hard- and softwoods.)

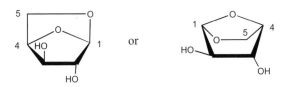

1,4-Anhydro-alpha-D-Xylopyranose

Radlein et al have made a comparison (19) of some of the better results for a representative hardwood, poplar-aspen. Like most hardwoods the hemicellulose content of aspen is dominated by glucuronoxylan giving a typical pentose content around 15%. Data are shown for raw, prehydrolysed and demineralized samples. Prehydrolysis (5% H_2SO_4 at 90 °C for 2 hrs.) removed essentially all the hemicellulose thereby enhancing the relative cellulose content of the residual lignocellulose. Demineralization was carried out by percolating cold 0.5 wt% H_2SO_4 through a column of ground wood, after which the acid was washed out with distilled water. The resulting pyrolysis data are summarized in Table 1.

Since hemicellulose is removed on prehydrolysis, it is useful to determine C_6 sugar yields on a total hexosan, instead of a wood, basis. In a typical hardwood glucuronoxylan, the xylose: glucose ratio is about 10:7. Assuming very roughly that the hemicellulose is *all* glucuronoxylan with the above C_5:C_6 ratio, and that the hemicellulose constitutes about 25 % of the wood, the total hexosan content (all glucan) would be the sum of that portions in the hemicellulose and cellulose respectively; about 59% for both the raw and demineralized wood samples. For the prehydrolysed sample the hexosan content derives solely from cellulose as the hemicellulose has been removed. Sugars are assumed to be the hydrolysis products of levoglucosan, cellobiosan and other oligosaccharides and their yield is multiplied by a factor of $180/162 = 1.11$ to correct for the increase in molecular weight in converting anhydrosugars to free sugars. This is the basis for computation of *"% Hexosan Conversion to Sugars"* in Table 1.

There is obviously a dramatic increase in sugar yield on going from raw wood to demineralized wood and further to prehydrolysed wood. Since this is also the order of increased preparation and process costs, then apart from other complicating issues relating to fermentability, the choice of feedstock will be a compromise between cost and yield.

Table 1 Comparison of Data for Thermochemical Saccharification of a

Table 1 Comparison of Data for Thermochemical Saccharification of a
Hardwood – Poplar-Aspen
(Yields in wt% of mf feed)

PreTreatment	None	Prehydrolysis	Demineralization
Additive	None	None	0.1 % $(NH_4)_2SO_4$
Pyrolysis Temperature (°C)	497	501	530
Cellulose Content (wt%)	49.1	62.8	~ 49
Hexosan Content (wt%)	~ 59	62.8	~ 59
Moisture in Feed	3.3	16.5	3.0
Yields			
Gas	10.8	6.4	7.4
Coke/Char	7.7	6.7	4.4
Organics (in liquid)	65.8	79.6	66.1
Water	12.2	0.9	16.2
Cellobiosan	1.3	5.7	2.5
Oligosaccharides	0.7	1.2	-
C_6 hexoses	1.7	5.8	2.5
Levoglucosan	3.0	30.4	18.7
1,6 Anhydroglucofuranose	2.4	4.5	2.8
Glycolaldehyde	10.0	0.4	3.4
Acetic Acid	5.4	0.2	1.4
Formic Acid	3.1	1.4	-
Other C_1-C_3 Oxo Compounds	4.3	1.2	> 3
Pyrolytic Lignin	16.2	19.0	21
% Hexosan Conversion to Sugars[†]	17	83.4	50

[†]See previous discussion

A most important observation is that demineralized and prehydrolysed feeds exhibit a tendency to liquefy at much lower temperatures than the untreated material. The "melt" is a high molecular weight material (20) that is probably a mixture or solution of molten lignin and carbohydrate polymer. It only relatively slowly decomposes to volatile matter comprising a mixture of aerosols and vapours. In a fluidized bed pyrolyzer, this is manifested as a tendency to coat the bed particles with melt, leading to clumping of particles and subsequent failure of fluidization. In the table, "Coke" refers to secondary solids that do not have the morphology of the feed particles as in pyrolysis of untreated feeds. Instead they appear to arise by carbonization of a liquid phase. This phenomenon has led to difficult and often severe operational problems with pretreated feeds. Thus feed rates had to be kept low to avoid tar overloading of the bed.

1.5.2.2   Softwood

Scott (21) has reported results for sugar yields via fast pyrolysis from a common Canadian softwood (Western Hemlock) as well.

Assume for simplicity that, as is typical of softwoods, the pentosan content of is only 8% and that the remaining hemicellulose sugars are in glucomannans with a mannose: glucose ratio of 4:1. Also assume that cellulose is 45% (vs. about 50% in hardwood) and total glucomannan about 17%; then the total hexosan content would be about 62% with a glucose : mannose ratio of around 3.6 : 1. (Experimentally, a total of 64% holocellulose was found.)

The same problems of gluing and bed failure were evident here as well. Piskorz et al. 22 have proposed a pyrolysis process variant based on the discovery that when pyrolysis of feedstocks that give high anhydrosugar yields is carried out in an oxidizing atmosphere, both low molecular weight oxygenates and, especially, lignin can be selectively oxidized with very little loss of anhydrosugar yield leading to much smoother operation and a substantial reduction in melt-related problems.

Table 2 shows results for a raw and a demineralized sample in an inert atmosphere as well as a demineralized sample in an oxidative atmosphere created by blending air with the fluidizing gas to bring its oxygen content to 6 vol%.

Table 2 Comparison of Data for Thermochemical Saccharification of a Hardwood – Western Hemlock.
Reaction Temperatures 450 - 490 °C, (Yields in wt% of mf feed)

Run no.	78	99	74	97
Pretreatment	None	None	Demineralization	Demineralization
Additive	None	None	0.1 % $(NH_4)_2SO_4$	0.1 % $(NH_4)_2SO_4$
Atmosphere	Inert	12 Vol % O_2	Inert	12 Vol % O_2
Hexosan Content (wt%)	~ 62	~ 62	~ 62	~ 62
Yields				
Gas	12.0	68.0	5.1	40.5
Coke/Char	18.2	12.7	17.3	13.3
Organics (in liquid)	56.5	32.3	64.7	54.8
Water	9.4	27.5	9.5	18.2
Cellobiosan	0.3	0	1.9	2.5
Levoglucosan	1.6	0.8	12.5	13.7
Anhydromannose	0.0	0	3.9	7.7
Glycolaldehyde	7.0	0	1.6	1.5
Acetic Acid	0.9	0.6	1.2	0
Other C_1-C_3 Oxo Compounds	4.9	2.9	~0	0
Pyrolytic Lignin	13.1	2.0	16.0	1.1
% Hexosan Conversion to Sugars	3	1	33	43

The anhydrosugar yield for Run 74 appears substantially lower than those above for poplar. However the analysis was incomplete as oligomers and 1,6-anhydroglucofuranose were not determined so that potential glucose will be underestimated. In fact, if use is made of Shafizadeh's correlation mentioned above and effective available glucose estimated by multiplying the levoglucosan concentration by 1.5, then much closer correspondence would be obtained.

It is also notable that the glucosan : mannosan ration in the pyrolysate Run 74 is 3.2 : 1, in good agreement with the intrinsic ratio in the wood. This suggests the simplification that hexosans are produced from hemicellulose at the same rate as from cellulose.

Evidently lignin is very selectively oxidized by oxygen thus reducing the amount of melt formed in the reactor resulting in improved 'runability". These results are the basis of a U.S. patent[2] assigned to Energy, Mines and Resources – Canada. However it should be stated that this could be a hazardous procedure with real potential for explosions to occur when operated on an industrial scale.

In the most recent study, Brown et al. (23) have studied the pretreatment and pyrolysis of herbaceous feedstocks like switchgrass and corn stover. For corn stover, the combined yield of levoglucosan and cellobiosan went from 1.6 wt% for the raw feed to 16.0 wt% in the demineralised feed. At the same time "lignin" decreased from 33.4 wt% to 17.7 wt%. An interesting feature of these tests was the use of very weak nitric acid (0.25wt% HNO_3) to effect demineralization. The acid was not washed out of at the end of the procedure, (which would be necessary in the case of H_sSO_4 to avoid excessive charring during pyrolysis). In fact Julien et al. (24) have found that residual Cl⁻ and SO_4^{2-} ions in cellulose enhanced char formation while residual HNO_3 did not. All and any such process simplifications lead to reduced costs and improved economic viability. The results for switch grass were similar.

1.5.3   Hydrolysis of Anhydrosugars

Acid hydrolysis of levoglucosan to free sugar generates not only glucose but also by-product disaccharides like isomaltose, cellobiose and gentiobiose.   These will usually also be fermentable so that aspect is not problematic. However, it does require elevated temperatures (~ 90 – 100 °C), 1 N sulfuric acid and extended times of the order of 0.5-1 hr (25). Ménard et al. (26) have studied the kinetics of this hydrolysis. Using ~ 3 N H_2SO_4, hydrolysis of pure levoglucosan was essentially complete in less than 30 mins. at 100 °C. At the same temperature, complete conversion required ~ 3 hrs. using 0.3 N acid.

On the other hand, Scott (27) found that boiling the pyrolyzate (80 – 100 °C) with 1 N H_2SO_4 for 2-4 hrs resulted in only about 50% recovery of sugars. This result might have been due to exposure to air. We found that a corn stover pyrolyzate yielded 1.4 g glucose per gram of levoglucosan consumed when heated in a sealed vessel with 0.2 N H_2SO_4 at 135 °C for 1.5 hrs. This is very close to the ratio reported by Shafizadeh cited above. On the other hand, conversion of levoglucosan was incomplete (only 60%) under these conditions, which agrees with the predictions of the kinetic data of Ménard (26).

It is clear that further work needs to be done to clarify the best conditions. Indeed the necessity to use acid of such strength and severity tends to defeat the purpose of having an acid-free technology. A more efficient and cheaper method is a highly desirable objective.

Recently, Katritzky et al. (28) have described just such a potential method. They reported that a 0.30 Molar aqueous solution of LG was quantitatively converted to predominantly glucose and traces of another glucose isomer at 205 °C. This result is probably because liquid water intrinsically becomes both more acidic and more basic at elevated temperatures so besides the normal thermal acceleration there is an enhanced catalytic effect. At 205 °C, the vapour pressure of pure water is 1906 kPa (276 psi) so the required conditions are relatively mild and thus, such a process step would be expected to be rather inexpensive. Unfortunately, kinetic data, which is required for equipment sizing, were not given so a precise cost estimate cannot be made. Nevertheless, this is a very promising approach requiring mild conditions and no added chemicals and certainly requires further investigation.

1.5.4   Energy requirements

It is generally found that while the pyrolysis reactions of untreated biomass are nearly thermoneutral, or even slightly exothermic, those of pure cellulose are somewhat endothermic when levoglucosan formation becomes the dominant decomposition mode. Suuberg et al. (29) have recently measured a heat of reaction of 0.54 kJ/g of volatiles evolved for pure cellulose. On the other hand they observed that pyrolysis could be driven in the exothermic direction by competing char-forming processes. Char formation was estimated to be exothermic by about -2 kJ/g of char formed.

Unlike pure cellulose, which gives only a negligible amount of char under fast pyrolysis conditions leading to maximal levoglucosan yield, pre-hydrolyzed and de-mineralized feeds do produce some char/coke, together with a concomitant amount of water. The net effect is that for anhydrosugar production also, the overall heat of pyrolysis should be very small so that the overall heat demand of pyrolysis is just the sensible heat requirement, typically ~ 1 MJ/kg (30).

It is curious and not at all clear why char/coke and water yields should be so high for de-mineralized feeds since, as indicated above, they are negligible from pure cellulose. A likely explanation is as follows: It is reasonable to presume that there would be a higher probability for condensation reactions, leading ultimately to char and water, to occur in the "melt-phase" compared to the solid phase. If this conjecture is true, there is the implication that, in principle, anhydrosugar yields can be even higher and char/coke/water yields even lower than those obtained so far. This point requires further investigation and methods of avoiding the limitation should be sought.

1.5.5   Process economics

There has been only very little work done of the feasibility of thermochemical saccharification as a potential industrial process and its comparison with the competing technologies (dilute and concentrated acid hydrolysis, enzymatic hydrolysis and gasification/fermentation). Indeed, only a couple of very preliminary studies appear to have been published. With the increasing activity in fast pyrolysis commercialization by companies like Dynamotive Technologies Corporation (Canada), Ensyn Engineering (Canada), Pyrovac (Canada) and BTG (Holland) among others, the time seems ripe to take the next step and carry out a proper comparison of these technologies with a view towards a pilot process if deemed favourable.

The simplest basis for comparison would be to isolate those parts of the feed preparation and saccharification steps unique to the particular process. All methods require some degree of comminution of the feed followed by some type of pre-treatment before saccharification. Thus ground and dried feed should be considered as input raw materials and the fermentable sugar solutions the main products. Effective market prices must then be estimated for these solutions, within the context of ethanol production.

For thermochemical saccharification, this approach leads to the following overall scheme:

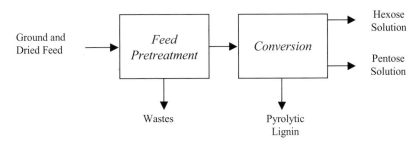

Waste streams may consist for example, of gypsum formed on neutralization of sulfuric acid solutions by lime together with waste water with non-zero BOD. "Pyrolytic lignin" is mainly a rather degraded form of lignin but may also contain some of the more hydrophobic carbohydrate-derived materials. It has several unique properties relative to typical pulping lignins and so many different applications have been proposed for it. However, as of this

217

moment there is no premium market and it seems safer, at best, to assign it a valuation based on its fuel value as determined by its heating value.

(In fact, pyrolytic lignin usually separates, on dilution of bio-oils with water, as a viscous semi-solid material when cold that is not easily handled in a plant. It may therefore be of practical interest to find methods to improve its handling properties. Teo and Watkinson (31) of the University of British Columbia describe a stable liquid lignin containing slurry that is readily pumpable and remains pumpable even when stored for a significant periods of time. It comprises lignin, water, fuel oil and a dispersing agent, which is a non-ionic surfactant, in amounts by weight of 35 - 60% lignin, 35 - 60% water and 0.5 - 30% oil and containing 50 - 50,000 ppm surfactant.)

Piskorz et al. (32) have made a preliminary study of the process economics for two different scenarios, demineralization and prehydrolysis. A 100 tpd plant was considered, the capital cost of which was estimated at $4.4M and $5.1M for the respective cases, including a working capital of 15%. Two cases of lignin valuation were considered, its fuel value at $120/tonne and a higher value of $300/tonne. The rate of return sensitivity was examined for cases when the product sugar solutions were valued at $150, $250 and $350/tonne of fermentable sugar solids. Since corn syrup sells at up to $500/tonne of contained glucose, dextrins and other sugars, these projected process appear reasonable. The resulting sensitivities of the Discounted Cash Flow pre-tax rates of return, based on a ten year life with no salvage value and no capital charges (i.e. 100% equity basis), to wood cost are shown in Figures 1 and 2 for the two scenarios; all prices are in 1997 Canadian dollars. The calculations are based on the results reported above for poplar-aspen, which is perhaps not likely to be a feedstock of immediate interest since much more softwood is available.

Both cases give reasonable to high rates of return even with lignin commanding a price based on its fuel value. So and Brown (33) have also recently reached a similar conclusion. They estimated a production cost of US$1.57/gal ethanol US$42/ton for wood and a rate of 25 million gal per year. This was said to be comparable to the cost for SSF (Simultaneous Saccharification and Fermentation) and for Dilute Acid Hydrolysis and Fermentation.

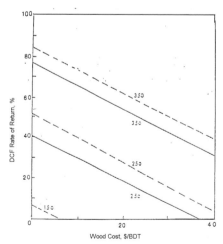

Figure 1   Rates of Return Sensitivity Chart for Demineralized Poplar for Fermentable Sugar Values of $150, $250 and $350/Tonne Solids.
(− − − − − − Lignin @ $300/tonne, ─────── Lignin @ $120/tonne.)

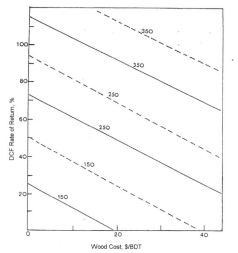

**Figure 2 Rates of Return Sensitivity Chart for Prehydrolysed Poplar for
Fermentable Sugar Values of $150, $250 and $350/Tonne Solids
(------ Lignin @ $300/tonne, ———— Lignin @ $120/tonne.)**

These studies are tentative at best. Many assumptions have been made, as there is a great deal of uncertainty about the technological details. Furthermore, the pricing of the sugar products appears optimistic. Ladisch et al (34) suggested in a 1991 study of the economics of enzymatic hydrolysis, that fermentable sugars must be produced at US$0.088–$0 11/kg in order for that technology to be competitive with sugars from corn. This translates to about C$140-180/tonne at present (1999 C$) values. Clearly, a more definitive answer is required for the pricing of the sugar solutions, as this is a crucial element in the economic analysis. More studies need to be carried out to eliminate these uncertainties. However, it does appear at least that thermolytic saccharification has a chance to be competitive with the alternate technologies of acid or enzymatic hydrolysis.

The basic fact is that if 50 % conversion of hexosans to sugars (demineralization case), and a 60 % hexosan content in the feed are assumed, then 1 tonne of wood would generate 50 US gals ethanol. I.e. the conversion rate would be about 20 kg wood /US gal ethanol. (This may be compared with corn starch where 1 tonne corn, containing 71% starch, typically yields 121 US gals. of ethanol.) Thus at $50/tonne feedstock cost, the feedstock cost will be about $1 per gal ethanol (C$0.26/L). This is significant and the operating cost would have to be held to an absolute minimum to make this feasible. Feedstock price and availability is likely to be a major consideration in any project. On the other hand, we have not accounted for conversion of pentosans or other anhydrohexoses, which would increase the effective glucose yield, in this calculation. In particular, if the Shafizadeh indicator factor of 1.4 kg glucose yield per kg levoglucosan is applicable to the prediction of glucose yields, the feedstock cost would fall to C$0.70 per US gal of ethanol. Furthermore, as the previous data indicates, much higher yields can be had with the prehydrolysis process variant, though at higher process cost. Finally, when considering the overall process to ethanol, credits for use of fermentation residues as animal feed may become available. In any case, it would be desirable to be able to reduce the net processing costs to no more than about C$0.50/US gal.

Several options are available for pretreatment and the process economics will be closely tied to choice made. Simple demineralization and prehydrolysis are two options discussed in some detail above. However, there are other possible alternatives that may be worth exploration. For example, Richards (35) has found that ion exchange of certain metal ions, particularly transition metal ions like iron and copper, lead to levoglucosan yields up to 16% (equivalent to 32% on a cellulose basis) on vacuum pyrolysis of cottonwood (*populus trichocarpa*). Even simple sorption of salts without prior demineralization led to significant, if rather smaller increases over raw untreated wood. However this is only a preliminary result and whether it is a viable option remains to be discovered.

A convenience when producing fermentation substrates instead of bio-oil fuels is that the feed does not have to be completely dried between the pretreatment and pyrolysis stages, since the water content of the product is then no longer a major issue and might even be advantageous for process effectiveness. The practicality of feeding wet raw material and the optimal moisture content needs to be evaluated for the preferred pyrolysis technology, especially bearing in mind that the reactor heat demand will increase and the condensation/quenching system may need to be modified.

1.5.6   Fermentation of Pyrolysates and their Hydrolysates

There is also much uncertainty surrounding the best conditions for fermentation of pyrolyzates or their free sugar hydrolysates. There are at least four cases that need to be considered.:

- Fermentation of the combined hexose/pentose sugars
- Separate (consecutive) fermentation of hexoses then pentoses (mainly xylose)
- Fermentation of hexoses only (mainly glucose, mannose and galactose
- Direct fermentation of anhydrosugars

Only very preliminary information is available for making a rational choice between these options.

The first option would obviously be desirable as it leads to a considerable simplification through replacement of one processing stage. There has been recent progress in the development of strains of microorganism that can simultaneously utilize both C_5 and C_6 sugars. For example, Zhang et al. (36), in a recent patent assigned to Midwest Research Institute, describe a new genetically modified strain of *Zymomonas mobilis* (Yeast Strain 1400 (LNH-ST)), which can ferment both glucose as well as C_5, sugars, (especially xylose and arabinose) stably and efficiently to ethanol (R&D Magazine, Sept. 1998, p134). The latter are the predominant products of the hydrolysis of hemicelluloses. Apparently there are pilot studies planned or underway for the utilization of this technology.

The second and third options have the merit that they mostly involve the use of existing technologies so development cost and time is reduced.
Of course specific questions arise as to the fermentability of pyrolysis liquors and the possible presence of toxic or inhibitory impurities. Some preliminary answers are available. Work carried out at the University of Waterloo indicates several conclusions: The prehydrolysates of aspen poplar, which are rich in xylose, could be fermented successfully by *Pichia stipitis* R after activated carbon treatment. Similarly, while fermentation was inhibited in the raw aqueous extract of poplar pyrolysates and prehydrolysates, upon hydrolysis, activated carbon treatment, and pH adjustment with lime, fermentation by bakers' yeast was rapid and quantitative in 1.5 hrs. Thus, either the yeasts could not attack the anhydrosugars or inhibitory compounds were present, or both. In a further test, after steam stripping of the raw aqueous extract and pH adjustment, *P. stipitis* R was able to ferment it, albeit slowly. This raises the possibility that anhydrosugars may have been fermentable directly, making this organism a candidate for adaptation to these feedstocks.

Thus, the fourth option seeks to remove the need of the anhydrosugar hydrolysis stage altogether, though the insufficient information is available and it is still very far from practical application.

Indeed, if a cheap and efficient hydrolysis method was found it would be questionable whether the effort to develop direct anhydrosugar fermentation technology would be worthwhile at this time. The status of direct anhydrosugar fermentation, including ethanol among other possible products, is discussed in more detail in the next section.

2   ANHYDROSUGAR FERMENTATION

Kitamura et al. (37) have studied the metabolism of levoglucosan in microorganisms. They refer to prior work indicating that many yeasts and fungi can use it as a carbon source and suggest that for eucaryotic microorganisms (i.e. yeasts and fungi) capable of utilizing levoglucosan, the first step involves phosphorylation to glucose-6-phosphate.   The responsible enzyme, termed "levoglucosan kinase" was isolated.   In earlier work they also showed that *Aspergillus terreus* converts levoglucosan to itaconic acid with the same yield and rate as in the conversion of glucose.   Subsequently (38) a new organism, *Arthrobacter sp.*, isolated from soil was found to hydrolyze levoglucosan to glucose by a novel enzyme, levoglucosan dehydrogenase).   Thus, somewhat surprisingly perhaps, levoglucosan specific enzymes do occur in nature; a fact that should no doubt have a strong bearing on the feasibility of genetic modification of organisms for levoglucosan specific activity.

Indeed an old German patent (39) suggests that levoglucosan or wood carbonization liquors containing it can be used as a nutrient in the production of yeast, e.g. *torula utilis*. Blazej and Kosik 15 reported that *aureobasidium pollulans* CCI-2771-14 showed good growth rates using anhydrosugars as a single carbon source on which basis they suggested that direct conversion of anhydrosugars to fodder yeast has great potential.

Prosen et al. (40) of the University of Waterloo studied the response of seven yeasts and four fungi to sugars and anhydrosugars in the pyrolysates of poplar. None of the fungi grew in the aqueous extract (AE) of the raw pyrolysis liquid but all grew equally well in both the aqueous extract after treatment with charcoal (AC) as well as in the acid hydrolysate (H) of this treated extract. Two yeast species did not grow in any of the samples but five grew well on both AE and AC. The hydrolysate was optimal in terms of biomass yield and ethanol production. Ethanol yields on the hydrolysate (H) were comparable to or better than those on glucose, but yields on AE and AC while significant, were considerably lower. The higher yields from H relative to glucose may be due to utilization of carbon species other than sugars.

These results establish that some yeasts and fungi are capable of utilizing levoglucosan in wood pyrolysis liquors though there are components that may be toxic or inhibitory to growth. A simple activated carbon cleanup appears to easily remove these.

3   LEVOGLUCOSAN AS A CHEMICAL FEEDSTOCK

Many possible industrial of levoglucosan have been identified. Some areas of applications include:

- Pharmaceutical and pesticides,
- Herbicides/plant growth regulators,
- Glycoconjugates, oligosaccharides,

- Glycoside surfactants,
- Dextrins,
- Stereoregular polymers,
- Biocompatible polymers e.g. methacrylate and ethylene oxide co-polymers,
- Liquid crystals,
- Non-hydrolysable glucose polymers and
- Epoxies,

all of which have been demonstrated in the chemical or patent literature. Some of these will be discussed in more detail below. The chemistry of levoglucosan has been extensively reviewed, e.g. by Cerny and Stanek (41). The following compilation of synthetic pathways is due to Cerny.

Reactions of levoglucosan

3.1 Chiral Synthon

Levoglucosan possesses several features that make it an attractive chiral raw material for the synthesis of a wide variety of materials. Like the ordinary sugar, it is a chiral compound but the internal acetal ring renders the molecule structurally rigid with the pyranose ring locked in the $^1C_4$ conformation. This has the following consequences:

- The three axial hydroxyl groups are now easily differentiated and a high degree of regiochemical and stereochemical control is therefore possible.
- The stereocentres are opposite to those found in the $^4C_1$ conformation of the ordinary sugar.
- The internal acetal reduces the number of protecting groups required in various transformations.
- The reduced number of hydroxyl groups enhances the solubility in organic solvents.
- Thermal stability is considerably increased allowing reactions under more severe conditions.

Applications in which these chirality properties have been exploited in the preparation of biologically active molecules have been studied and reported by Longley and McKinley (42), BC Research which may be consulted for further details. A further non-exhaustive list of examples may be found in the brochure published by RTI Ltd., Waterloo.

3.2 Biodegradable Surfactants

The utility of surfactants lies in the presence of a polar, water-soluble head group and a hydrophobic tail within the same molecule. There have been many patents recently concerning the production of biodegradable surfactants; especially for cosmetic and related applications where the higher costs associated with biologically derived raw materials are tolerable. Many of these are alkyl glycosides and are prepared from natural products: sugars, to provide the hydrophilic part and natural long-chain fatty acids or their derivatives to provide the hydrophobic part.

The focus of many of these patents is the purification of the product. Generally, the sugar must be reacted with the long chain acid, ester or alcohol using acid catalysis. The reactions are condensations that produce by-product water. The aqueous, acidic environment is deleterious to sugars and they readily degrade under these conditions generating many undesirable impurities. Thus, the surfactants must be purified and decolourized at substantial additional expense. This is difficult as they are high molecular weight, relatively non-volatile compounds, so are not readily separated by convenient techniques like distillation.

In contrast, theoretically, the combination of levoglucosan with fatty acids or alcohols is not a condensation reaction but rather an *addition* reaction with no water byproduct. It would be desirable to exploit this if possible, for making high purity alkyl glucosides. However, the anhydro ring in levoglucosan is very stable and difficult to open and a strong acid catalyst is required. Czerny and Stanek[41] include a review of the chemistry of ring opening of the 1,6-anhydride bond in hexosans (anhydro-hexoses) under various conditions along a variety of pathways. Some of these are illustrate on the following page.

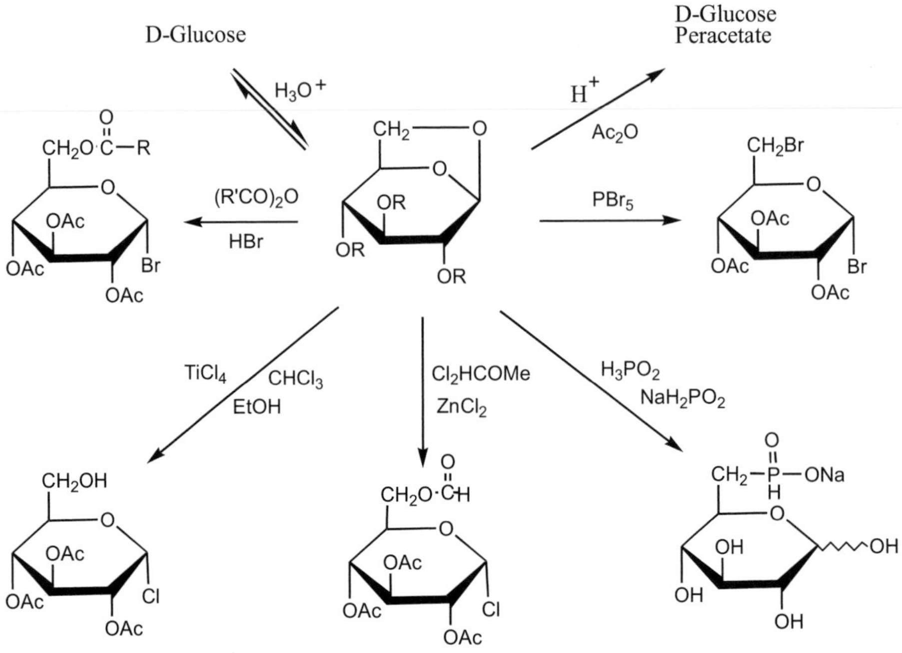

D-Glucose

D-Glucose
Peracetate

H$_3$O$^+$

H$^+$

Ac$_2$O

CH$_2$O·C—R
O
OAc
OAc   Br
OAc

(R'CO)$_2$O
HBr

CH$_2$——O
O
OR
OR
OR

PBr$_5$

CH$_2$Br
O
OAc
OAc   Br
OAc

TiCl$_4$ CHCl$_3$
EtOH

Cl$_2$HCOMe
ZnCl$_2$

H$_3$PO$_2$
NaH$_2$PO$_2$

CH$_2$OH
O
OAc
OAc   Cl
OAc

CH$_2$O·CH
O
O
OAc
OAc   Cl
OAc

CH$_2$—P—ONa
O
H
OH   OH
OH
OH

We have some preliminary data suggesting that the reaction between levoglucosan and a fatty alcohol (octanol) can be promoted by strong solid superacids, generating very small amounts of by-products. If the reaction proceeds quantitatively then the product is likely to be chiral, as it would proceed by a concerted S$_N$2 mechanism. The solid acid catalyst should be easily removable by filtration. This possibility needs to be explored further, since the processing advantages could offset even a higher price for levoglucosan than commercial sugars. If the product was also chiral then its value would be further enhanced, especially for biological applications. More work is required to confirm this observation and to explore optimal reaction conditions.

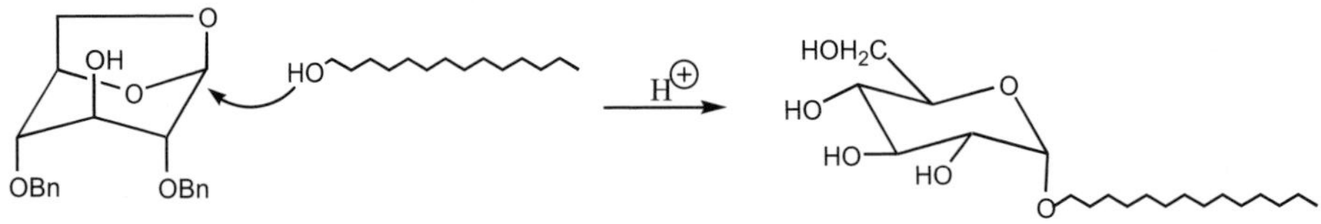

OH
O
OBn   OBn

HO~~~~~~~~

H$^{(+)}$

HOH$_2$C
HO
HO   O
HO
O~~~~~~~~

Stereoselective Alkyl Glucoside Synthesis

Another interesting possibility to use crude levoglucosan rich bio-oils for the production of low cost surfactants for use, for example, in the stabilization of bio-oil in diesel emulsion fuels.

It is also possible to generate a unique class of non-ionic surfactants in which the anhydrosugar ring remains intact by esterification of the free hydroxyl groups with long chain alkanoyl chlorides, as described by Ward and Shafizadeh (43).

224

3.3 Tosylate Intermediate

In many of its synthetic applications, the first step in the utilization of levoglucosan consists of generation of its 3,4-epoxide tosylate (1,6:3,4-dianhydro-2-O-tosyl-β-D-galactopyranose) because the opening of the epoxide ring is stereoselective and therefore allows facile differentiation of the three hydroxyl groups. It is worth considering whether the availability of this intermediate in large quantities might be a strong stimulus towards its greater utilization. Szeja describes a simplified synthesis (44).

1,6:3,4-dianhydro-2-O-tosyl-D-galactopyranose

4   POLYMERIC PRODUCTS

4.1   Dextrins

A variety of polymeric products incorporating levoglucosan have been described in the scientific and patent literature. Such applications represent a likely large-scale use for its use as a chemical feedstock. They include products in which LG is utilized as a monomer for synthesis of various types of polymers as well as classes of materials in which LG is reacted with other ingredients.

As a first example, in a patent assigned to Kraft Foods for a method of manufacture of a low-fat cheese, Jackson et al. (45) describe the use of what are referred to as "levoglucosans" as a low viscosity-bulking agent in low-fat processed cheese. The nature of these "levoglucosans" is rather poorly specified but they are oligomers which are said to contain a non-reducing end and have various α-1→4, α-1→6 and β-1→6, β1→3 and β-1→6 glycosidic bonds. They are also said to be artificially prepared or derived from starch. They provide the advantage that a 30 wt% solution will have a desirable viscosity in the range 10 to 20 cps. These solutions are added at about the 5-10 wt% level in the processed cheese product. The patent also indicates that such a solution is commercially available from Matsutani Chemical Industry Co., Ltd. Under the trade name FIBERSOL2™.

The important feature of this material is the non-reducing end group, which is an anhydroglucose unit. Thus, it may be regarded as an oligomer of levoglucosan and in principle could be synthesized there from.

In fact, the polymerization of anhydrosugars, and levoglucosan in particular, was extensively studied in the 1960's and 1970's. Much of that earlier work has been summarized in a review by Schuerch (46).

Levoglucosan undergoes uncatalyzed thermal polymerization in the melt at temperatures around 200 °C although it is necessary to rigorously exclude air as even trace amounts of oxygen catalyze its decomposition. Thus, Houminer and Patai (47) reported that while the melt turned brown in the presence of air, the reaction mixture only became yellowish after 2-3 hrs when the polymerization was carried out under argon. They also found the reaction to be autocatalytic which can be explained by the higher reactivity of the primary hydroxyl group, which is absent in the monomer, but one of which is present in the dimer (at C_6). This hydroxyl group can then attack the C_1 of a levoglucosan molecule generating a $(1 \rightarrow 6)$ glucosidic link. Indeed, it is known that levoglucosan polymers contain about 50% of C_6 links. In this scenario, dimerization itself would proceed via a nucleophilic attack of a hydroxyl of one levoglucosan molecule on the C_1 of another.

Structure of Levoglucosans (U.S. Patent 5,374,443, 1994)

This mechanism also suggest that there should be an inversion of configuration at C_1 of the attacked molecule (see mechanism below) and indeed, α-D-glucosidic linkages do predominate in the polymer; however, both α- and β-D-glucosidic do form. Houminer and Patai postulate that the latter arise from intermolecular transglycosylation reactions.

Dimerization of Levoglucosan (Houminer and Patai)

Furthermore, bearing in mind that all hydroxyl groups can react, albeit at different rates, it is not surprising that a variety of linkages may be found in the polymers and they tend to be multi-branched with a random distribution of glycosyl linkages. Indeed, they appear to be structurally similar to the condensation polymers of glucose, though possessing an anhydroglucose instead of a reducing end group. In general, the viscosity of levoglucosan polymer solutions is very small which is consistent with the properties mentioned earlier for the product FIBERSOL2™. It also suggests that the polymers are highly branched spherical molecules (48). *They may be regarded as a special class of dextrins and their applications are likely to be analogous.* They may be

226

expected to be especially advantageous when the stability bestowed by the non-reducing anhydrosugar ring end-group is of benefit.

The acid catalyzed polymerization of unsubstituted levoglucosan takes place at much lower temperatures, e.g. 115 - 120 °C, well below its melting point. It was first studied long ago (1918) by Pictet (49) but was explored in detail many years later. Schuerch et al used chloroacetic acid as catalyst in mole ratios to levoglucosan of between 1:10 to 1:20. The resulting polymers appear to be of similar chemical character to those from the purely thermal polymerization but higher molecular weights can be obtained. Schuerch obtained product polymers with molecular weights in the range 4000 – 300,000, depending of the precise temperature, catalyst ratio and duration of reaction. Furthermore, their solutions in water were straw coloured even without prior decolourization, indicating that by-products were much reduced.

Propagation in Thermal Polymerization of Levoglucosan

It may also be of practical interest that the aqueous solutions of polymer could be readily separated by alcohol addition into various molecular weight fractions. For instance a sample with average molecular weight 49,000 contained fractions of MW ranging from ~86,000 to ~ 22,000.

The methods of preparation referred to above of levoglucosan dextrins suggest that they could be produced at quite low cost from pure crystalline levoglucosan. Economic feasibility will therefore depend strongly on the cost of producing levoglucosan. In most of their applications however, the levoglucosan dextrins will be in competition with existing raw materials so there must be a clear and demonstrable advantage to use of levoglucosan, either in terms of raw material price or polymer product properties.

4.2   Stereoregular Polysaccharides

4.2.1   Dextrans

In general, polymerization reactions proceed by either a step growth or a chain growth mechanism. In the former, monomers combine to form dimers, dimers to tetramers and all oligomers of intermediate size combine at random to form larger molecules. In this case, products of high molecular weight are obtained only at very high conversion. In chain growth, a monomer is activated in an initiation step and reacts rapidly with other monomers in sequence so that high molecular weight products are obtained even at low conversion. Termination or chain transfer reactions give complete inactive products. The great synthetic advantage of anhydrosugars as monomers for polysaccharide synthesis, in comparison to ordinary sugars, is that they polymerize by a chain-growth mechanism.

Cationic polymerization mechanisms are generally characteristic of oxygen heterocycles and this is also the case with anhydrosugars. Thus, catalysts used have included not only protonic acids but also, preferably, Lewis acid like BF_3, $SnCl_4$, $ZnCl_2$, $FeCl_3$ and $TiCl_4$ among many others.

It has been found that very high molecular weight stereoregular polymers can be obtained by polymerization of *substituted* anhydrosugars. This offers far greater control over regioselectivity than the unsubstituted anhydrosugars. Such precision offers the possibility of synthesis of highly specialized polysaccharides with uses for instance, in biomedical applications[46]. A fairly recent review of the progress in the synthesis of artificial polysaccharides by ring-opening polymerization of substituted anhydrosugars, with special reference to their biological activities, is that of Uryu (50). The following examples give an idea of the kind of results that have been achieved:

- Levoglucosan tri-O-benzyl ether (i.e. 1,6-anhydro-2,3,4-tri-O-benzyl-β-D-glucopyranose) in which the three hydroxyl groups of levoglucosan have been substituted by benzyl ether groups, was polymerized with PF_5 as catalyst. A linear synthetic dextran, (1→6)-α-D-glucopyranan with the same structure as a natural dextran was obtained upon debenzylation of the product, according to the following scheme.

**Artificial Dextran Synthesis by Ring-Opening Polymerization
of a Substituted Levoglucosan**

228

- Comb-shaped polysaccharides having a glucose branch at the C4 position of each backbone glucose residue have been synthesized by the ring-opening polymerization of 1,6-anhydro disaccharides like cellobiosan among others. (Cellobiosan is usually a by-product of cellulose pyrolysis.)

Dextrans are stereoregular polysaccharides produced by certain bacteria grown on sucrose, containing a backbone of glucose units linked predominantly α-D-(1→6). They are obtained from the culture solution by precipitation with methanol. All dextrans are composed exclusively of α-D-glucopyranosyl units, differing only in degree of branching and chain length. Among others applications they are used as blood plasma extenders, as partial barley malt substitutes, in lacquers and in confectionary.

Stereoregular levoglucosan dextrans will clearly have a significant production cost - well above that of the dextrins – since polymerization require very high purity raw materials and solvents, protection and de-protection of hydroxyl groups, as well as carefully controlled polymerization conditions, including low temperatures (-75 °C), exclusion of air, etc. In their biomedical applications, they may command a high price but the volumes are likely to be relatively small. However, in most applications they will be in competition with existing raw materials so there must be a clear and demonstrable advantage to use of levoglucosan, either in terms of raw material price or polymer product properties.

4.2.2 Cyclodextrins

An intriguing speculative idea, only a possibility at this time, is to attempt the combination of the methods used for cationic polymerization with templated reaction conditions. The latter is in analogy to the synthesis by Dale (51) of cyclic oligoethers from ethylene oxide in the presence of a suitable inert salt (e.g. KPF_6) that functions as a template to induce ring closure. The resulting cyclic $\alpha(1,6)$-linked oligoethers would represent a levoglucosan based synthetic variant of the highly desirable, but expensive, cyclodextrins. For example, the following ring containing 4 glucosyl units is stereochemically possible.

Potential products such as this, which exploit unique features of anhydrosugars versus ordinary sugars, are the kind that is most likely to justify a levoglucosan chemical industry.

4.2.3 Glycolipids

Another interesting class of stereoregular polymers that have been synthesized from anhydrosugars in general and levoglucosan in particular are the glycolipids, which are mono-, oligo-, or polysaccharides having long hydrocarbon chains attached. They are widely distributed in nature, for instance they play an important role in specific recognition in cells. Exocellular glycolipids (emulsan) have become significant industrial products in recent years. Analogous synthetic materials have the prospect of specialty applications as carbohydrate liquid crystals, synthetic bio-membranes, polysaccharide coated liposomes, detergents for solubilization of membrane proteins and highly ordered helical superstructures (52). Several such materials have been synthesized from levoglucosan; for example, 1,6-anhydro-2,4-di-O-benzyl-3-O-octadecyl-β-D-glucopyranose was homo-polymerized followed by debenzylation to give the polysaccharide shown below.

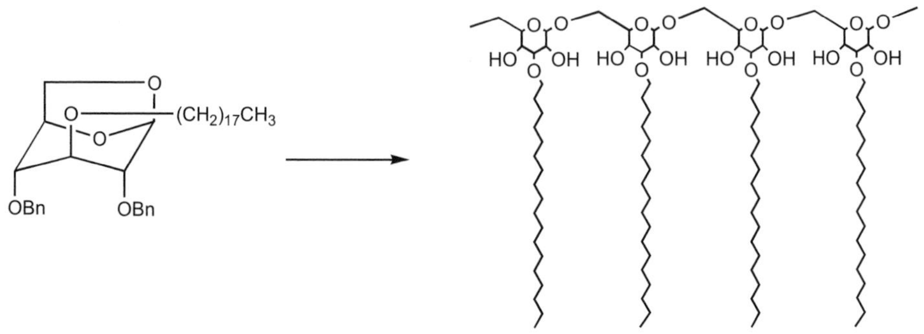

Stereoregular Glycolipid Synthesis

Again, these materials have a significant cost of production involving, as they do, multiple transformations. Their use is therefore likely to be small scale even though they may attract a good price.

4.2.4   Non-Hydrolysable Polymers

Yet another type of chiral, stereoregular polymer that has been obtained from levoglucosan is a class of non-hydrolysable polysaccharides which have been described by Berman et al. (53' 54). The non-hydrolysable property arises because the monomers are linked by ether bonds at C_2 and C_3 rather than an acetal bond at C_1. For example, the monomers are 1,6:2,3-dianhydro-4-O-alkyl-D-mannopyranoses, which are epoxy levoglucosan derivatives that can be synthesized by known methods from levoglucosan in four steps.

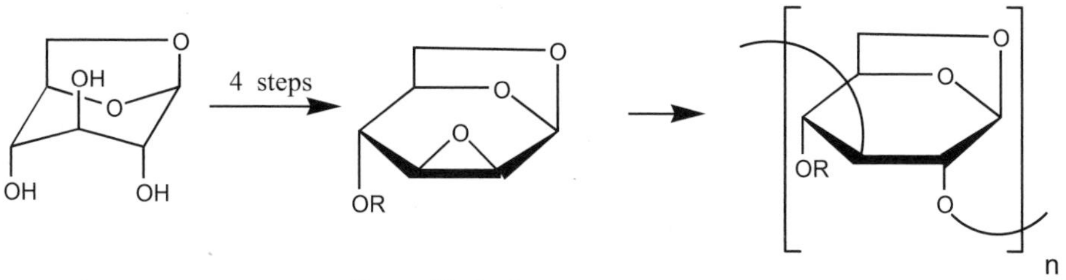

Levoglucosan   Epoxy Derivative   Polymer

Non-Hydrolyzable Polymers

The polymers may be subsequently transformed into interesting materials with potentially wide applications in the synthesis of elastomers, adhesives, coatings, liquid crystals, liposome formulations, etc. (55) For example, hydrolysis and reduction could lead to a family of chiral polysorbitols, materials not presently available. Alternatively, by suitable choice of the functionality R, the polymerization could be carried out regio- and stereoselectively to generate polymers in which the carbohydrate group is pendant to the main chain. These are illustrated below:

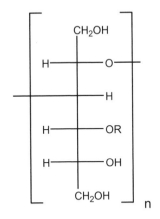

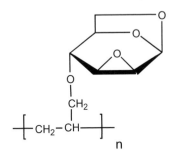

Chiral Polysorbitol   **Polymer with Pendant Sugar**

4.2.5   Copolymer Products

Copolymers of levoglucosan with other polymerizable monomers gives a range of products, which possess unique properties, often due to the rigid nature of the anhydrosugar ring. In some cases, this key difference leads to more desirable and in some cases to more unfavourable physico-chemical properties than free sugars. Some examples follow.

Shafizadeh (56) has described the properties of polyether polyols obtained by reacting the three hydroxyl groups of levoglucosan with a suitable alkylene oxide. Alternatively (57), an allyl halide is reacted with levoglucosan to give an allyl ether derivative that can then be epoxidized to give epoxy resin intermediates.

According to Pernikis at al. (62), esterification reactions make it possible to obtain simple and complex polyesters, polyurethanes and films, glues, plastic foams. Furthermore, the introduction of (levoglucosan) fragments unto polyesters improves heat stability and rigidity. Also, oligocarbonate methacrylates are characterized by high rates of polymerization under UV radiation, which allows their use in photo-polymerizing compositions. The polyfunctional polycarbonate methacrylates, in particular, have high optical rotatory powers. Their insignificant volume shrinkage and low index of birefringence makes them ideal for manufacture of optical discs. There are many other applications as well.

The figure overleaf, taken from Pernikis et al. (61), illustrates the variety of intermediates that have been studied.

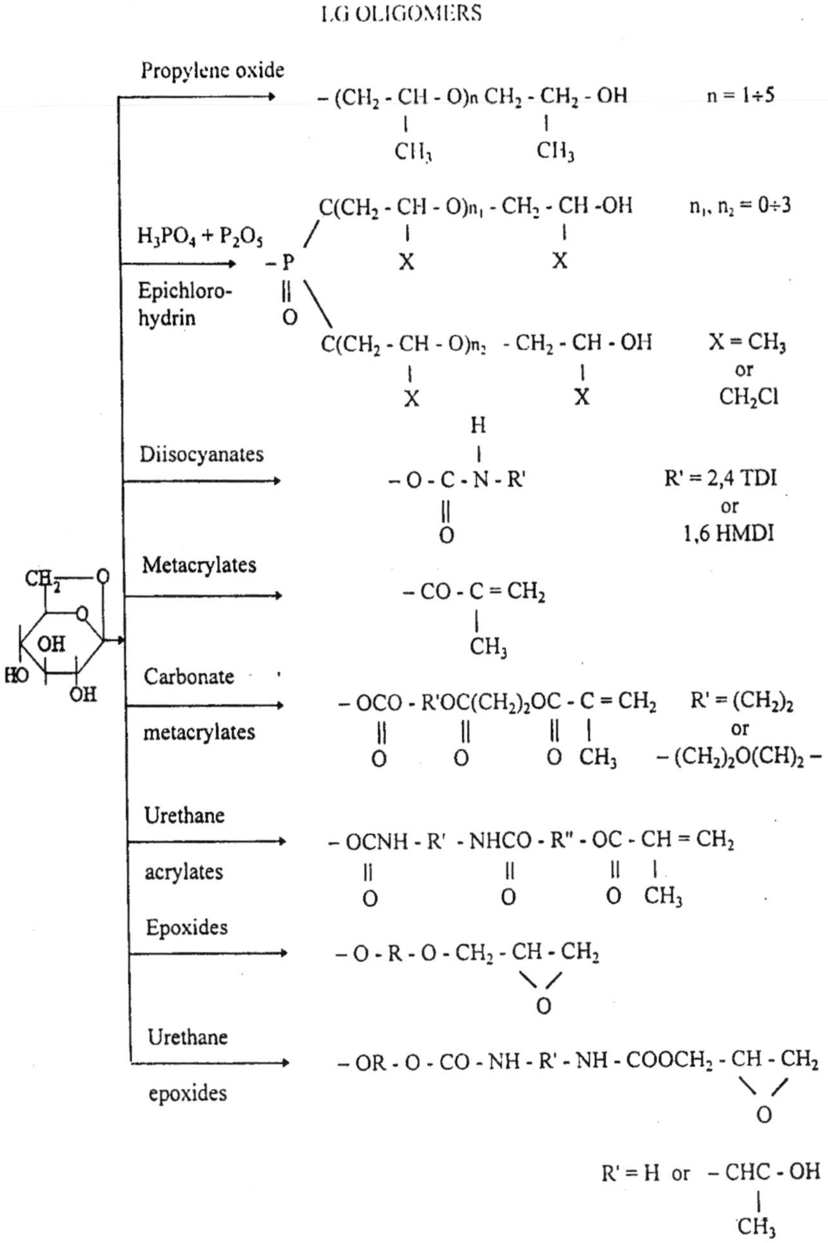

Propylene oxide

$$- (CH_2 - CH - O)n \; CH_2 - CH_2 - OH \qquad n = 1 \div 5$$
$$\qquad\qquad | \qquad\qquad\quad |$$
$$\qquad\qquad CH_3 \qquad\qquad CH_3$$

$H_3PO_4 + P_2O_5$

Epichloro-
hydrin

$$- P$$
$$\;\; \|$$
$$\;\; O$$

$$C(CH_2 - CH - O)n_1 - CH_2 - CH - OH \qquad n_1, n_2 = 0 \div 3$$
$$\qquad\quad | \qquad\qquad\qquad\quad |$$
$$\qquad\quad X \qquad\qquad\qquad\quad X$$

$$C(CH_2 - CH - O)n_2 \; - CH_2 - CH - OH \qquad X = CH_3$$
$$\qquad\quad | \qquad\qquad\qquad\quad | \qquad\qquad\quad or$$
$$\qquad\quad X \qquad\qquad\qquad\quad X \qquad\qquad CH_2Cl$$

Diisocyanates

$$\qquad\quad H$$
$$\qquad\quad |$$
$$- O - C - N - R' \qquad R' = 2,4 \; TDI$$
$$\qquad \| \qquad\qquad\qquad or$$
$$\qquad O \qquad\qquad\qquad 1,6 \; HMDI$$

Metacrylates

$$- CO - C = CH_2$$
$$\qquad\quad |$$
$$\qquad\quad CH_3$$

Carbonate

metacrylates

$$- OCO - R'OC(CH_2)_2OC - C = CH_2 \qquad R' = (CH_2)_2$$
$$\quad \| \qquad\quad \| \qquad\quad \| \;\; | \qquad\qquad or$$
$$\quad O \qquad\quad O \qquad\quad O \;\; CH_3 \qquad - (CH_2)_2O(CH)_2 -$$

Urethane

acrylates

$$- OCNH - R' - NHCO - R'' - OC - CH = CH_2$$
$$\quad \| \qquad\qquad\quad \| \qquad\quad \| \;\; |$$
$$\quad O \qquad\qquad\quad O \qquad\quad O \;\; CH_3$$

Epoxides

$$- O - R - O - CH_2 - CH - CH_2$$
$$\qquad\qquad\qquad\qquad \backslash \; /$$
$$\qquad\qquad\qquad\qquad O$$

Urethane

epoxides

$$- OR - O - CO - NH - R' - NH - COOCH_2 - CH - CH_2$$
$$\qquad\qquad\qquad\qquad\qquad\qquad\qquad\qquad \backslash \; /$$
$$\qquad\qquad\qquad\qquad\qquad\qquad\qquad\qquad O$$

$$R' = H \quad or \quad - CHC - OH$$
$$\qquad\qquad\qquad\qquad |$$
$$\qquad\qquad\qquad\qquad CH_3$$

5   LEVOGLUCOSAN PRODUCTION TECHNOLOGIES

A variety of pyrolysis technologies are applicable to the production of levoglucosan. The classical laboratory method uses batch vacuum pyrolysis of starch. It might seem unsuitable for large scale however as the starch melts and foams and eventually leaves a carbon residue which is difficult to remove. Nevertheless, only recently just such a process (Südzucker, Germany)

(58) has been patented though it should be pointed out that the focus of the patent was the purification of levoglucosan, an aspect discussed below. The process is said to have been practiced on a pilot plant scale (see figure below).

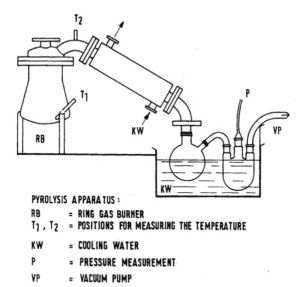

PYROLYSIS APPARATUS :
RB   = RING GAS BURNER
T_1, T_2 = POSITIONS FOR MEASURING THE TEMPERATURE
KW   = COOLING WATER
P   = PRESSURE MEASUREMENT
VP   = VACUUM PUMP

Presumably, more recent *continuous* vacuum pyrolysis technologies could also be adapted for the purpose 24.

Metzger et al. (59) have described the pyrolysis of cellulose and chitin in supercritical acetone at 47 bar and 509 K, giving yields comparable to those reported for fast pyrolysis (38 % levoglucosan and about 4% of each of its two principal isomers, 1,6-anhydro-β-D-glucofuranose and 1,4:3,6-dianhydro-α-D-glucopyranose from cellulose). The substituted levoglucosan, 2-acetamido-1,6-anhydro-2-deoxy-β-D-glucopyranose, was obtained from chitin in much smaller yield, ~6%. Interestingly, when the pyrolysis was carried out in a high boiling inert solvent, a glyme, anhydro-oligosaccharides like cellobiosan, cellotriosan etc., were formed from cellulose and the corresponding acetamido substituted compounds from chitin.

Chitin

2-acetamido-1,6-anhydro-
2-deoxy-glucopyranose

233

The scalability of this technique and its application to lignocellulose is not known, which, on the grounds of cost, would be the preferred feedstock for large-scale production.

An early patent by Esterer (60), assigned to the Weyerhauser Company, describes the entrained flow pyrolysis of cellulosic materials for levoglucosan production. Various entraining gases could be used, though steam was particularly recommended, at a ratio of 20-150 std. cu. ft. of gas per pound of lignocellulose. Yields from *untreated* sawdust were reported at ~ 10 – 25%, but this is almost certainly a gross overestimate based on a faulty analytical method. As Shafizadeh has pointed out 9, early analyses based on measurement of reducing sugar after hydrolysis are likely to be very much in error.

Workers at the Latvian State Institute of Wood Chemistry have reported an interesting approach (61). They pyrolyzed cellulose and lignocelluloses in a "cyclone reactor" (actually a long tubular coil) by transporting the feed through the reactor with superheated steam. Pyrolysis temperatures were in the range 370 – 410 °C and residence times were long, ~ 10 secs. The optimal mass ratio of steam to feed was found to be 4 – 5 kg/kg. The technology has been scaled up to a nominal capacity of 30 kg/hr (62). Levoglucosan yields were up to 50% on a cellulose basis with the best results from corncob lignocellulose. Pine wood, demineralized by washing with 5% sulfuric acid at 20 °C followed by washing to pH 2.0, gave 30% yields on a cellulose basis. These results are comparable to those obtained by fast pyrolysis except that, in contrast to the fast pyrolysis results, the total product tar from lignocellulose is completely water soluble, i.e. contains no pyrolytic lignin. Total tar yields were correspondingly lowered but the levoglucosan concentration in it could reach as high as > 70%. This is a quite remarkable and desirable feature in that lignin derived tar components were found to greatly increase the difficulty of separation and purification of levoglucosan. Indeed, as we discuss in a subsequent section, crystallization and purification, and not pyrolysis, are likely to be the major costs in the production of pure levoglucosan.

They attribute the results partly to the lower pyrolysis temperatures that, it is claimed, inhibit the volatilization of lignin, while the reduced kinetic rate of levoglucosan formation at the lower temperatures is compensated by longer residence times. It was also claimed that the "high thermal capacity" of steam was necessary to ensure a sufficiently high heating rate. However, this later claim seems weak since the heat capacity of steam on a mass basis is only one and a half times greater than that of nitrogen, say. It is interesting to speculate that steam may actually play a catalytic chemical role in cellulose depolymerization for, the heat transfer rates would be certainly far higher in a fluidized bed. This point requires further investigation.

Indeed fast pyrolysis for biomass conversion has tended to be focused on high temperature, and short residence time regimes since those conditions tend to maximize total liquid yield. It could therefore be interesting to modify existing fast pyrolysis processes to optimize for operation in the long residence time, low temperature regime for the specific case of levoglucosan production. Besides the inherent advantages for the levoglucosan crystallization, it could contribute to solving the problem of rapid melting and "gluing" of the sand, encountered in a fluidized bed under normal fast pyrolysis conditions for pre-hydrolysed lignocellulose.

Several different fast pyrolysis technologies are now available, and it is reasonable to suppose that they can all be adapted, with greater or lesser difficulty, for the high yield production of levoglucosan. Fluidized beds appear to be the most preferred of these, on account of their relative simplicity and good heat transfer characteristics. Since the various methods have been described recently in many different forums we refer to the reader to a recent review (63).

6   LEVOGLUCOSAN PURIFICATION

It is the author's belief that the main obstacle to levoglucosan production lies less in the pyrolysis technology and much more in the difficulty in its crystallization and purification from pyrolytic liquors.   This difficulty is illustrated by the abundance of published laboratory procedures and patents. A wide variety of solvents in different combinations, temperatures and concentrations have been mentioned as suitable for this purpose. While the difficulties are most acute for lignocellulose-derived tars, they do persist even from pyrolyzates from pure cellulose pyrolysis. If a simple, cheap, reliable and efficient method could be found it would represent an important breakthrough.

In Pictet's early patent 8, the vacuum pyrolyzate of a carbohydrate material is fractionally condensed, dehydrated and extracted into boiling acetone from which levoglucosan is crystallized. This is the classical "acetone method"; it suffers from the costs associated with use of a volatile and flammable solvent.

Esterer (64) in a patent assigned to Weyerhauser, has described the separation of levoglucosan from aqueous mixtures by first dewatering by azeotropic distillation then extraction into and crystallization from methyl isobutyl ketone. In another Weyerhauser patent by Peniston 65, on the other hand, impurities are precipitated out by adding a soluble salt of alkaline earths, aluminum, lead or zinc to the aqueous solution, the filtrate concentrated and levoglucosan crystallized from the water solution.

In the patent of Scott et al (2), the aqueous phase from water induced phase separation of the pyrolysate from the partial oxidative pyrolysis of lignocellulosics is first dewatered, then extracted into hot ethanol, treated with charcoal, and finally crystallized from ethanol.

Pernikis et al. (62) have described the crystallization of a "technical levoglucosan" (purity ~ 91-92%) from lignocellulosics tars using the acetone method whereby 2 parts of tar are combined with 1 part of acetone and allowed to crystallize. It was found that many of the impurities are insoluble in ethanol, and so it was found that the final product could be readily re-crystallized from this solvent. However, the overall recovery of levoglucosan was only ~ 55% for the technical grade and 35% overall for the purified product.

In the Südzucker patent referred to above 58, after neutralization with alkali, chromatographic methods using ion exchange resins were used to crystallize and purify levoglucosan. It was claimed that the method was suitable for "large scale production of levoglucosan in high purity". However, currently only fructose and glucose are separated (from corn starch hydrolysates) by large scale chromatography and the practicality of the proposed process is not known.

In Canadian patent application by Howard et al. (66) (BC Research) the pH of lignocellulosic pyrolysates is adjusted with base, filtered, neutralized, dehydrated then extracted into and crystallized from ethanol or acetone.

Most recently, Moens (67), in patents assigned to Midwest Research Institute describes a method by which lignocellulose is washed with hot acid and pyrolyzed at low temperature (350 – 375 °C), the pyrolysate of which is extracted with methyl isobutyl ketone, neutralized with base, freeze dried and finally extracted into and crystallized from ethyl acetate. This procedure appears highly complicated and is not likely to be inexpensive.

This wide variety and continuing search for improved methods suggests that a preferred method of separation and purification is probably very sensitive to the precise feedstock and pyrolysis process/conditions employed. Thus a general, truly satisfactory procedure in terms of both yield and cost remains elusive and is a matter of continuing research.

7 SCALE OF PRODUCTION AND PRICES FOR LOW MOLECULAR WEIGHT CARBOHYDRATES

In order to gain some perspective on the potential of levoglucosan as a chemical feedstock it is useful to compare typical prices of other low molecular weight carbohydrate as a function of their scale of production. The data in the tables below is taken from Lichtenthaler (68).

Scale (1000 tonnes)	Price Range (C$)
> 1000	0.6 – 1.6
>100	1 – 2.5
>50	1.3 – 5.9
>10	3.4 – 10.1

Prices of Various Low Molecular Weight Carbohydrates

Carbohydrate	World Production (tonne/yr)	Price (note a) (C$/kg)	Price (note b) (C$/kg)
Sugars			
Sucrose	123,000,000	0.63	1.16 (refined)
D-Glucose	5,000,000	0.97	1.64 (anhydr.)
D-Glucose			0.84 (hydr.)
Lactose	295,000	1.01	1.77
D-Fructose	60,000	1.69	1.26
Isomaltulose	35,000	3.38	
Maltose	3,000	4.23	
D-Xylose	16,000	10.10	
L-Sorbose	25,000	12.68	
D-Galactose	?	71.83	
Lactulose	10,000	76.05	
Sugar Alcohols			
D-Sorbitol	650,000	1.69	2.54
D-Mannitol	20,000	5.07	10.69
D-Xylitol	15,000	10.14	
Sugar Derived Acids			
L-Lactic Acid	100,000	3.40	2.54
D-Gluconic Acid	60,000	5.92	3.41
L-Tartaric Acid	?	8.45	7.25
L-Ascorbic Acid	60,000	10.14	16.10
Misc.			
Dextrin			1.03
NOTES			

NOTES
a Converted from prices in DM as given by Lichtenthaler. Prices are world market, bulk delivery in the ton range, 1997. 1DM = C$0.845.
b Converted from prices in $US taken from Chemical Marketing Reporter, Sept. 8, 1997. 1US$ = C$1.46.

These data are only a rough guide as there are evidently large regional variations in prices. They also indicate the extent to which the market will pay a higher price for desirable properties, in the absence of alternatives. The prices are further roughly summarized in the table below in which the production scale is related to a price range. Exceptions are sorbose, lactulose and ascorbic acid (vitamin C), high-value substances produced ultimately from sorbitol or lactose and used in therapeutic applications.

With respect to levoglucosan, there is a potential market as a synthon in pharmaceutical manufacture and this market would be willing to pay a very high price. However, although this could be a profitable activity on a small scale, the total market is likely to be a very small – perhaps of the order of a few tens of tonnes per year. It therefore seems unlikely that this market could be a good basis for the establishment of a levoglucosan industry.

On the other hand, if a market could be established at the level of 10,000 to 50,000 tonnes per year, then the price range would probably have to be in the range of roughly $5 - $15/kg since it would most likely then have to compete with other carbohydrates. Indeed there seems no reason at present to believe that there is likely to be any single "killer" application. Rather, there is likely to be a range of applications where in which it competes on either price or performance.

8   CONCLUSIONS AND RECOMMENDATIONS

- The existing market for levoglucosan is very small and very high priced. The situation is at least partly a reflection of the low yields and lack of scalability of the traditional methods of production by vacuum pyrolysis of starch and the difficulty of purifying levoglucosan at low cost. This has inhibited the development of large-scale uses.
- The lack of a significant current market is exemplified by the result of a market survey carried out in 1993 by the Waterloo Innovation Centre. In a recent quote from Glycon Biochemicals in Germany, levoglucosan was priced at DM 2668/kg (i.e. ~ US$1446/kg). Most recently, RTI received a request to quote for 1 tonne of pure levoglucosan, presumably intended for pharmaceutical manufacture, but this magnitude is currently outside our capability. Nevertheless, it seems unlikely that that market could provide a basis for large commodity-scale production.
- If a market could be established at the level of 10,000 to 50,000 tonnes per year, then the price range would probably have to be in the range of roughly $5- 15/kg.
- RTI has attempted to promote usage of levoglucosan through making small quantities (several kg) available for sale. The ethanol method was used for crystallization from cellulose pyrolysates. A high purity product was obtained but the recovery of pure levoglucosan was only approximately 50%. Simultaneously brochures were published illustrating some of the features of levoglucosan chemistry.
- Successful development of many of the derivative chemical synthetic products must bring together the available expertise in the areas of production, chemistry and marketing. Furthermore, the timescale of development may be long so a significant investment over a period of time is required. This might be difficult to achieve, as no single organization appears to have all these necessary ingredients at present.
- However, in some cases, e.g. the glucoside surfactants or dextrins, markets already exist and it would be sufficient to demonstrate some technical benefits in order to compete with the alternative technologies. But, even here some improvements in the available pyrolysis technology will be required.
- However, from the point of view of production for chemical feedstocks, there is a need for improved methods of crystallization, or, alternatively to find uses for crude product mixtures.

- In principle, the simplest products among those discussed are the glucoside surfactants, and the dextrin polymers. These are probably also the chemical products with the largest potential market volumes.
- Thermolytic saccharification for ethanol production appears to have the possibility to be a feasible alternative to enzymatic or acid hydrolysis technologies. However many technical and process uncertainties remain; which makes it very difficult to make a reliable assessment of the technology.
- Some of the issues to be resolved include:
  - Optimal pretreatment strategy.
  - Feeding of wet feedstock to a pyrolyzer.
  - Use of fluidized beds in regimes of low temperature and long residence times.
  - Optimal strategy for hexose/pentose fermentation
  - Improved hydrolysis of anhydrosugars to free sugars (use of pressurized water).
  - Development of adapted strains of microorganism for fermentation of pyrolytic liquors.
  - Direct fermentation of anhydrosugars
- If large-scale applications with higher than fuel value could be found for pyrolytic lignin, the economic feasibility of thermolytic saccharification would be greatly enhanced. One such prospect is its use in resin formulations, though it would have to be verified that the lignin from de-mineralized or hydrolysed wood was suitable; it may well differ in molecular weight from the normal pyrolytic lignin from untreated wood. In such a case, a "bio-refinery" could be envisaged in which ethanol and/or sugar solutions and lignin are the main products.
- To further the development, a sensitivity analysis of the various process options and possibilities should be carried out with a view to prioritization of the required developmental work. It should also be pointed out that although only anhydrosugar solutions need be produced, their availability might be prove to be a stimulus to other uses of levoglucosan.

9 ACKNOWLEDGEMENT

This contribution was submitted as a Final Report to PWGSC Contract No.23348-8-3247/001/SQ for Natural Resources Canada on March 24, 2000. Permission to include it in this Final Report as a contribution from Canada within the IEA Bioenergy Agreement is gratefully acknowledged.

10 REFERENCES

1   "Proc. Symp. On Levoglucosenone and Levoglucosans. ACS, August, 1992, Z.J. Witczak (ed.), ATL Press Mount Prospect, 1994.

2   (a) D.S. Scott, J. Piskorz, D. Radlein and P. Majerski, Process for the production of anhydrosugars from lignin and cellulose containing biomass by pyrolysis, U.S. patent 5,395,455 (1995); Assignee: Energy, Mines and Resources, Ottawa, Canada;
      (b) D. Scott and J., Piskorz, Process for the production of fermentable sugars from biomass, U.S. Patent 4,880,473 (1989); Assignee: Canadian Patents & Development Ltd., Ottawa, Canada,
      (c) D.S. Scott, J. Piskorz, D. Radlein and P. Majerski, "Improved Process for the Production of Anhydrosugars and Fermentable Sugars from Fast Pyrolysis Liquids", Canadian Patent 2,091,373, 1997.

3   M. McCoy, "Biomass Ethanol Inches Forward":
      http://www.ott.doe.gov/biofuels/news/12-7-98.html

4   "Economics": http://www.ceassist.com/economic.htm
5   "Our Technology: Concentrated Acid Hydrolysis": http://www.arkenol.com/tech01.html
6   K.T. Klasson, C.M.D. Ackerson, E.C. Clausen and J.L. Gaddy, "Biological Conversion of Synthesis Gas into Fuels", Int. J. Hydrogen Energy, 17, 281-288, 1992.
7   J.L. Gaddy, Bioresource Engineering Inc., 1650 Emmaus Road, Fayetteville AR 72701, USA. Phone: 501-521-2745, Fax: 501-521-2749.
8   A.Pictet, "Levoglucosane manufacture", Canadian Patent 195897, 1920.
9   F. Shafizadeh, R.H. Furneaux, T.G. Cochran, J.P. Scholl and Y. Sakai, "Production of Levoglucosan and Glucose from Pyrolysis of Cellulosic materials", J. Appl. Polymer Sci., 23, 3525-3539, 1979.
10   D. Radlein, A. Grinshpun, J. Piskorz and D.S. Scott, "On the presence of anhydro-oligosaccharides in the syrups from the fast pyrolysis of cellulose", J. Anal. Appl. Pyrol., 12, 39-49, 1987.
11   J.A. Lomax, J.M. Commandeur, P.W. Arisz and J.J. Boon, "Characterization of oligomers and sugar ring-cleavage products in the pyrolysates of cellulose", J. Anal. Appl. Pyrol., 19, 65-79, 1991.
12   F. Shafizadeh and T.T. Stevenson, "Saccharification of Douglas-fir Wood by a Combination of Prehydrolysis and Pyrolysis", J. Appl. Polymer Sci., 27, 4577-4585, 1982.
13   F. Shafizadeh, Proc. "A Comprehensive Pyrolytic Process for Conversion of Wood to Sugar Derivatives and Fuel", Proc. 1980 Annual Meeting, International Solar Energy Soc., p.122-125, 1980.
14   D.St.A.G. Radlein, J. Piskorz, A. Grinshpun and D.S. Scott, "Fast Pyrolysis of Pre-Treated Wood and Cellulose", Preprints, ACS, Div. Fuel Chem., Vol. 32, No. 2, pp29-35, 1987.
15   A. Blazej and M. Kosik, "Environmentally acceptable conversion technology for the biochemical utilization of lignocellulosics", in Cellulosics: Pulp, Fibre and Environmental Aspects, J.F. Kennedy, G.O. Phillips and P.A. Williams (eds.), Ellis Horwood, NY, 1993, pp355-364.
16   J. Piskorz, D. Radlein, D.S. Scott and S. Czernik, "Pretreatment of wood and cellulose for production of sugars by fast pyrolysis", J. Anal. Appl. Pyrol., 16, 127-142, 1989.
17   J. Piskorz, D. Radlein and D.S. Scott, Thermal Conversion of Cellulose and Hemicellulose to Sugars, Proc. Inter. Symp.on Advances in Thermochemical Biomass Conversion, Interlaken, 1992, A.V. Bridgwater (ed.), 1423-1440, Blackie Academic, London, 1994.
18   J.L. Lomax, J.J. Boon and R.A. Hoffmann, "Characterisation of polysaccharides by in-source pyrolysis positive- and negative-ion direct chemical ionization-mass spectrometry", Carbohyd. Res., 221, 219-233, 1991.
19   D. Radlein, J. Piskorz and D.S. Scott, "Fast pyrolysis of natural polysaccharides as a potential industrial process", J. Anal. Appl. Pyrol., 19, 41-63, 1991.
20   J. Piskorz, D. Scott and D. Radlein, Mechanisms of the Fast Pyrolysis of Biomass: Comments on Some Sources of Confusion, Minutes of the 2nd PYRA Meeting, La, Coruna, Nov. 1995, Appendix 19.
21   D.S. Scott, "Improved Process for the Conversion of Cellulose to Sugars by Fast Pyrolysis", Final Report, Waterloo Research Institute, 1995.
22   J. Piskorz, D. Radlein, P. Majerski and D.S. Scott, "Fast Pyrolysis of Pretreated Wood", Proc. Second Biomass Conference of the Americas, Portland, August, 1995 (NREL/CP-200-8098).
23   R.C. Brown, D. Radlein and J. Piskorz, submitted, Feb., 2000.
24   S. Julien, E. Chornet and R. Overend, "Influence of acid pretreatment (H2SO4, HCl, HNO3) on reaction selectivity in the vacuum pyrolysis of cellulose", J. Anal. Appl. Pyrol., 27, 43-052, 1993.

25   L. Reichel and H. Schiweck, Die Naturwissenschaften, **48**, 696, 1961.

26   H. Ménard, M. Grisé, A. Martel, C. Roy and D. Bélanger, "Saccharification de la biomasse par pyrolyse à pression réduite suivie d'une hydrolyse", 5'th Canadian R&D Seminar, S. Hasnain (ed.), Elsevier App. Sci. Publishers, London, 1984.

27   D.S. Scott, unpublished results.

28   *(a)* A.R Katritzky and S.M. Allin and M. Siskin, "Aquathermolysis: Reactions of Organic Compounds with Superheated Water", Acc. Chem. Rews., **29**, 399-406, 1996. *(b)* B. Kuhlmann, E.A. Arnett and M. Siskin, "Classical Organic Reactions in Pure Superheated water", J. Org. Chem., **59**, 3098-3101, 1994.

29   I. Milosavljevic, V. Oja and E.M. Suuberg, "Thermal Effects in Cellulose Pyrolysis: Relationship to Char Formation Processes", Ind. Eng. Chem. Res., **35**, 653-662, 1996.

30   D. Radlein, "Heat Requirement of Fast Pyrolysis", IEA Bioenergy: T13: Technoeconomics: 1998: 01, D. Beckmann, Final Report., Vol III/1-2, May 12, 1998, Appendix 4.

31   K.C.Teo and A.P. Watkinson, "Pumpable lignin fuel", United States Patent 5,478,366, 1995.

32   J. Piskorz, P. Majerski , D. Radlein, D.S. Scott, Y.P. Landriault, R.P. Notarfonzo and D.K. Vijh, "Economics of Fermentable Sugars from Biomass by Fast Pyrolysis" in *Making a Business from Biomass*, R.P. Overend and E. Chornet (eds.), Elsevier Science, N.Y., (1997), p. 823-833.

33   K. So and R.C. Brown, "Economic analysis of selected lignocellulose-to-ethanol conversion technologies", Appl. Biochem., and Biotechnol., **77-79**, 633-640, 1999.

34   M.R. Ladisch and J.A. Svarczkopf, "Ethanol production and the cost of fermentable sugars from biomass", Bioresources Technology, 36, 83-95, 1991.

35   G.N. Richards and G. Zheng, J. Anal. Appl. Pyrol., **21**, 133-146, 1991.

36   M. Zhang, Y. Chou, S.K. Picataggio and M. Finkelstein; "Single zymomonas mobilis strain for xylose and arabinose fermentation", United States Patent 5,843,760, 1998.

37   Y. Kitamura, Y. Abe and T. Yasui, "Metabolism of levoglucosan (1,6-Anhydro-β-D-glucopyranose) in Microorganisms", Agric. Biol. Chem., **55**, 515-521, 1991.

38   K. Nakahara, Y. Kitamura, Y. Yamagishi and H. Shoun, "Levoglucosan Dehydrogenase Involved in the Assimilation of Levoglucosan in *Arthrobacter sp.* I-552", Biosci. Biotech. Biochem., **58**, 2193-2196, 1994.

39   W. Schuchardt, Ger. 738,962, July 29, 1943.

40   E.M. Prosen, D. Radlein, J. Piskorz, D.S. Scott and R.L. Legge, "Microbial Utilization of Levoglucosan in Wood Pyrolysate as a Carbon and Energy Source", Biotechnol. and Bioengineering, **42**, 538-541, 1993.

41   M. Cerny and J. Stanek, Jr., "1,6-Anhydro Derivatives of Aldohexoses", Adv. Carbohydr. Chem. Biochem., **34**, 23-177, 1977.

42   C.J. Longley and J. McKinley, "Position Paper: levoglucosan ligands for use is asymmetric synthesis of biologically active molecules", DSS Contract No. 23440-0-9668/01SZ, 1994.

43   D.D. Ward and F. Shafizadeh, "Some esters of levoglucosan", Carbohydr. Res., **108**, 71-79, 1982.

44   W. Szeja, "A convenient synthesis of carbohydrate oxiranes via sulfonates", Carbohydr. Res., **183**, 135-139, 1988.

45   L.K. Jackson et al, "Method of Manufacture of a Low-Fat Cheese, U.S. Patent 5,374,443, 1994.

46   C. Schuerch, "Synthesis and Polymerization of Anhydro Sugars", Advances in Carbohydrate Chemistry and Biochemistry, Vol. **39**, 157-212, 1981.

47   Y. Houminer and S. Patai, "Thermal Polymerization of Levoglucosan", J. Polymer Sci., Part A-1, **7**, 3005-3014 1969.

48   J. daS. Carvalho, W. Prins and C. Schuerch, "Addition Polymerization of Anhydrosugar Derivatives", J. Am. Chem. Soc. **81**, 4045, 1959.

49   A. Pictet, Helv. Chim. Acta, **1**, 226-230, 1918.

50   T. Uryu, "Artificial Polysaccharides and their Biological Activities", Prog. Polymer Sci., **18**, 717-761, 1993.

51   J. Dale and K. Daasvatin, "Selective preparation of cation-complexing cyclic oligoethers from ethylene oxide by a template effect", J. Chem. Soc., Chem. Commun., 295, 1976.

52   K. Kobayashi, H. Ichikawa and H. Sumitomo, Macromolecules, **23**, 3708-3710, 1990.

53   E.L. Berman, "Regio- and stereospecific synthesis of a polyglucose with a new type of bond", Soviet Journal of Bioinorganic Chemistry (Engl. Transl.), 11, 608-612, 1986.

54   E.L. Berman, "Synthesis, characterization and chemical transformation of novel glucose polymers", Polymer Preprints, **31**, 28-29, 1990.

55   E.L. Berman, private communication, 1991.

56   F. Shafizadeh, "Polyethers of levoglucosan", U.S. Patent 3,305,542, 1967.

57   F. Shafizadeh, "Polyfunctional levoglucosan ethers", U.S. Patent 3,414,560, 1968.

58   M. Gander, K.M. Rapp and H. Schiweck, "Process for preparing 1,6-β-D-anhydroglucopyranose (levoglucosan) in high purity", U.S. Patent 5,023,330, 1991.

59   P. Köll, G. Borchers and J.O. Metzger, "Thermal degradation of chitin and cellulose", J. Anal. Appl. Pyrol., **19**, 119-129, 1991.

60*(a)* C.C. Heritage and A.K. Esterer, "Levoglucosan production by pyrolysis of cellulosic material", U.S. Patent 3,309,355, 1967;   see also *(b)* A.K. Esterer, "Pyrolysis of cellulosic material in concurrent gaseous flow", U.S. patent 3,298,928 (1967).

61   R. Pernikis, J. Zandersons and B. Lazdina, "Levoglucosan, its production by fast thermolysis of lignocellulose and use", Proc. 8'th Internat. Symp. On Wood and Pulping Chemistry, Vol. II, 1995.

62   R. Pernikis, J. Zandersons and B. Lazdina, "Obtaining of levoglucosan by fast pyrolysis of lignocellulose. Pathways of levoglucosan use" in *Developments in Thermochemical Biomass Conversion*, Vol. 1, A.V. Bridgwater and D.G.B. Boocock (eds.), Blackie, London, 1997.

63   J.P. Diebold and A.V. Bridgwater, "Overview of Fast Pyrolysis Technology for the Production of Liquid Fuels", in *Fast Pyrolysis of Biomass: A Handbook*, A. Bridgwater, S. Czernik, J. Diebold, D. Meier, A. Oasmaa, C. Peacocke, J. Piskorz and D. Radlein (eds.), CPL Press, Newbury, 1999, p.14-32.

64   A.K. Esterer, "Separating levoglucosan and carbohydrate acids from aqueous mixtures containing the same – by solvent extraction", Canadian Patent 818497, 1969.

65   Q.P. Peniston, "Separating levoglucosan and carbohydrate derived acids from aqueous mixtures containing the same – by treatment with metal compounds", Canadian Patent 804006, 1969.

66   J. Howard, C. Longley, A. Morrison and D. Fung, "Process for Isolating Levoglucosan from Pyrolysis Liquids", Canadian Patent Appl. 2084906, 1993.

67   L. Moens, *(a)* "Isolation of levoglucosan from lignocellulosic pyrolysis oil derived from wood or waste newsprint", U.S. Patent 5,432,276 1995; *(b)* "Isolation of levoglucosan from pyrolysis oil derived from cellulose", U.S. Patent 5,371,212, 1994.

68   F.W. Lichtenthaler, "Towards improving the utility of ketoses as organic raw materials", Carbohydrate Res., **313**, 69-89 (1998).

A REVIEW OF THE CHEMICAL AND PHYSICAL MECHANISMS OF THE STORAGE STABILITY OF FAST PYROLYSIS BIO-OILS

James P. Diebold, P.E.
Thermalchemie, Inc., 57 N. Yank Way, Lakewood, CO 80228 USA

ABSTRACT

It is necessary to understand the fundamental chemical and physical aging mechanisms to learn how to produce a bio-oil that is more stable during shipping and storage. This review provides a basis for this understanding and identifies possible future research paths to produce bio-oils with better storage stability. Included are 108 references.

The literature contains insights into the chemical and physical mechanisms that affect the relative storage stability of bio-oil. Many chemical reactions that normally are thought to require catalysis, proceed quite nicely without the addition of catalysts (or with catalysts indigenous to the bio-oil) during the long reaction times available in storage. The literature was searched for information describing the equilibrium constants and reaction rates of selected aging mechanisms, to determine if they are applicable to the relatively long times involved with storage. The chemical reactions reported to occur in pyrolytic liquids made from biomass are presented. As the bio-oil changes composition during aging, the mutual solubility of the components changes to make phase separation more likely to occur. With these insights into the aging mechanisms, the use of additives to improve storage stability is examined. Comparisons are then made to the storage stability of petroleum fuels. The review is summarized, conclusions are drawn, and recommendations are made for future research to improve the storage stability of bio-oils.

INTRODUCTION

Storage Stability Problem

Pyrolysis of biomass under conditions of rapid heating and short reactor residence times can produce a low viscosity, single-phase pyrolysis liquid (bio-oil) in yields reported to be over 70%. Most projected uses of bio-oil require that it retain these initial physical properties during storage, shipment, and use. Unfortunately, some bio-oils rapidly increase in viscosity during storage. Figure 1 shows this increase in viscosity for three different bio-oils made from three different hardwoods using different pyrolysis conditions, after aging three months at 35°C to 37°C (95°F to 99°F). These three bio-oils exhibit very different initial viscosities and also different rates of viscosity increase.

Figure 2 shows the effect of temperature on viscosity for three samples of a bio-oil made from poplar that had been aged at 90°C for 0, 8 and 20.5 hours. It is seen that aging effectively shifts the viscosity curve to right on the temperature axis, resulting in higher viscosities. The effect of aging on viscosity is greater at the lower temperatures of measurement. (Diebold and Czernik 1997) In this example, the change in viscosity appears to be about twice as high, if measured at 40°C rather than 50°C. At the higher temperature of measurement of 70°C, the effects of aging amount to an increase of only a few centipoise (mPas). The temperature of measurement is usually chosen by the researcher to be able to compare to petroleum fuel oil specifications in their country (e.g., 40°C in the U.S., while 40°C and 50°C have been used in Finland).

The aging effects are much faster at higher temperatures. Figure 3 shows that the rate of increase in viscosity of the hardwood bio-oils shown in Figure 1 and for a softwood bio-oil varied over four orders of magnitude from 0.009 cP/day at –20°C to over 300 cP/day at 90°C.

This is approximately a doubling of the viscosity-increase rate for each 7.3°C increase in storage temperature. The aging rate of softwood bio-oil is about the same as for hardwood bio-oils at 20°C, with some possible differences at lower storage temperatures. (However, the viscosity change during aging is very small below 20°C, making low-temperature aging rates subject to measurement errors.)

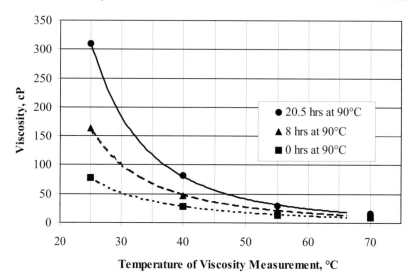

Figure 1   Effect of Temperature of Measurement on Apparent Aging of Poplar Hot-Gas Filtered Bio-Oil (Diebold and Czernik 1997)

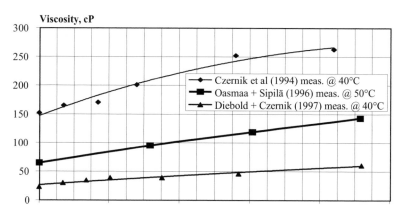

Figure 2   Aging of Bio-oils at 35°C to 37°C (cP = mPas)

Because the rates of viscosity change may be represented as Arrhenius exponential functions of the inverse of absolute temperature, chemical reactions appear to be involved. Figure 3 shows that it is very important that the bio-oils be quickly cooled after their production and then stored at low temperatures to maintain their initially low viscosity. The pyrolysis oils referred to in Figure 3 initially contained between 10 and 21 wt% water.

Although a loss of volatiles will increase the viscosity of bio-oil, the bio-oils shown in Figures 1, 2, 3, and 4 were carefully aged in sealed containers to prevent such losses. Using gel permeation

244

chromatography with uv detection of the aromatic compounds, the weight-average molecular weights of the aromatic compounds in aged bio-oils made from oak were determined (Czernik *et al* 1994). Figure 4 shows that molecular weight correlated very well with viscosity during aging, in this case with a linear-regression R^2 value of 0.96, for all of the aging data at 37°C, 60°C and 90°C treated as one data set. (The regression R^2 values are slightly improved if the data set is divided into three sets, one for each aging temperature.)

Figure 4 strongly implies that if a pyrolysis process more thoroughly cracks the bio-oil to lower molecular weights, then the initial viscosity would be desirably lower. Thus, it is important that partially pyrolyzed particles and droplets not be entrained prematurely from the reactor system, because if they are soluble in the bio-oil, they will cause an increase in the molecular weight and the viscosity.

It is apparent that chemical reactions take place in bio-oil during aging that increase the average molecular weight. Based on the good correlation for the aging data treated as one data set, it appears that relatively similar chemical reactions are occurring over this temperature range. This is the basis for conducting accelerated aging research at elevated temperatures and then applying those results to predict storage of bio-oils at lower temperatures. The advantage of accelerated aging tests is the short time required to demonstrate the aging properties of a particular bio-oil.

Bio-oil is not a product of thermodynamic equilibrium during pyrolysis, but is produced with short reactor times and rapid cooling or quenching from the pyrolysis temperatures. This produces a condensate that is also not at thermodynamic equilibrium at storage temperatures. Bio-oil contains a large number of oxygenated organic compounds with a wide range of molecular weights, typically in small percentages. During storage, the chemical composition of the bio-oil changes toward thermodynamic equilibrium at the storage conditions, resulting in changes in the viscosity, molecular weight, and co-solubility of the many compounds in bio-oil.

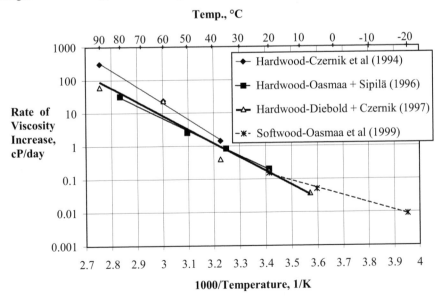

Figure 3 Rate of Viscosity Increase with Temperature During Storage of Bio-Oils

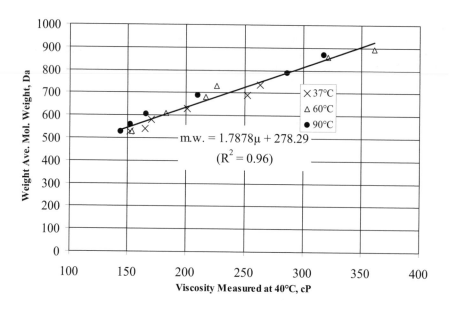

**Figure 4   Viscosity and Molecular Weight After Aging of a Bio-Oil
Made from Oak (data from Czernik et al 1994)**
(Molecular weight by GPC with UV detector)

In addition to simple viscosity increases, the single-phase bio-oil can separate into various tarry, sludgy, waxy, and thin aqueous phases during aging. Tarry sludges and waxes still in suspension have caused rapid plugging of fuel filters. These tarry sludges and waxes can form during storage in previously filtered bio-oils and in aqueous phases. The instability of bio-oils during storage appears to be more severe than that encountered with petroleum derived fuel oils, although there appear to be many similarities in their mechanisms.

Combustion Problems Due To Aging Or Excessive Heat

It is common practice to pre-heat fuel oils prior to combustion to lower their viscosity for better atomization. With diesel engines, the fuel is pumped through a pre-heater to the injector, where only a small fraction of the fuel is actually injected into the engine. The remainder of the hot fuel is normally recirculated back to the pump. This was found to be problematical with preheating bio-oils prior to their being injected into a diesel engine, with particulates growing in size from less than 10 µm to over 40 µm in this recirculation loop. Although filters were used to remove these particulates from the recycled fuel, reliable operation was only achieved after the removal of the recirculation loop and dumping of the excess hot bio-oil to a waste container. (Shihadeh 1998) This particle growth is thought to be due to polymerization reactions occurring in the heated bio-oil, although physical agglomeration of micelles would also explain this phenomena.

After preheating bio-oil to 90°C prior to atomization into a furnace, the 0.8-mm diameter holes in the fuel injector were plugged with deposits (Rossi 1994). Deposits were formed in the hot injectors, if they were not rinsed out with alcohol prior to shut down of the furnace (Gust 1997a)(Huffman and Freel 1997) or the diesel engine (Casanova 1994). Sludge deposits in the bottom of the fuel tank and in the fuel lines were flushed out with methanol (Leech and Webster 1999). The need for this alcohol rinse was cited as an impediment to the use of bio-oil in small combustion systems (Gust 1997a).

Although the solubility of bio-oil in diesel fuel is relatively slight, operating the combustion system for a short time with diesel fuel after shutting off the bio-oil flow precluded the formation of tarry deposits in the injector nozzles (Andrews et al 1997), but was not always effective (Gust 1997b). Although these deposits were often blamed on a loss of volatile components from the bio-oil during cool down, it is probable that there was also polymerization occurring as well.

COMPOSITION OF BIO-OILS

The composition of bio-oils is a complex inter-relationship of:
1. the biomass species used as feedstock (both organic and inorganic composition, including dirt and moisture)
2. organic nitrogen or protein content of the feedstock
3. the rate of heat transfer and final char temperature during pyrolysis
4. the extent of vapor dilution in the reactor
5. the time and temperature history of the vapors in the reactor
6. the temperature and time history of the vapors in the heated transfer lines from the pyrolysis reactor through the char removal equipment to the quench zone
7. whether the vapors pass through accumulated char (i.e., in hot-gas char filtration between back-flushing operations)
8. the efficiency of the char recovery system to separate the char from the bio-oil vapors prior to condensation
9. the efficiency of the condensation equipment to recover the volatile components from the non-condensable gas stream, e.g., water and low molecular weight esters, ethers, acetals, alcohols, aldehydes, etc.
10. whether the condensates have been filtered to remove suspended char fines
11. the water content of the bio-oil
12. the extent of contamination of the bio-oil during storage by corrosion or leaching of the containers
13. exposure to air during storage
14. the length of time of storage, and
15. the temperature of storage.

A thorough discussion of the effect of the reactor variables is outside of the scope of this review and these have already been treated in considerable detail (e.g., Diebold and Bridgwater 1997). The other variables are discussed in this review.

Organics In Bio-Oil

Because fast pyrolysis involves only the partial decomposition of biomass, the chemical composition of the resulting bio-oil is feedstock dependent. Biomass feedstocks rich in proteins would be expected to have high organic nitrogen contents. For example, bark, alfalfa, or grass cut for hay would be expected to produce bio-oils with higher nitrogen contents, compared to bio-oils produced from materials having low protein contents, e.g., straw or debarked wood. The presence of nitrogen compounds in bio-oils will adversely affect the NO_x content of the combustion products and the aging properties, as will be discussed.

Other influences of the feedstock species are found in the lignin. The phenolics in bio-oils are primarily derived from the lignin in the feedstock. A wood distillate made from hardwoods contained 55% phenolics with methoxy groups attached at both the number 2 and 6 positions of the phenolic molecule, i.e., syringols, and only 16% guaiacols with a methoxy at the number 2 position (Carraza et al 1994). Lignins from softwood tend to have one or no methoxy group attached to the number 2 position of the phenolic molecule (Lewis and Lansky 1989). Bark tends to have highly reactive tannins in it, as well as a high protein content.

As a consequence of the many variables in the pyrolysis of biomass and the storage of bio-oils, the reported compositions of bio-oil vary considerably. Over 400 organic compounds have been reported to be in pyrolysis liquids or wood smoke. The wood-smoke literature has a lot of detail on the presence of minor components that can impact on the flavors perceived (Maga 1987). The literature on the composition of fast pyrolysis oils was summarized by Milne, et al (1997). Table 1a and Table 1b summarize these compilations, which show the similarities in the qualitative composition of these pyrolysis-derived condensates (Diebold 1997). This implies that similar chemical reactions may occur in wood distillates, wood smoke, and bio-oil.

Of particular interest is the wide range reported for the composition of each of the organic components of bio-oil. For many of the compounds, this range exceeds a factor of ten to one. It is quite apparent that bio-oil is a poorly defined mixture of acids, alcohols, aldehydes, esters, ketones, sugars, phenols, guaiacols, syringols, furans, and multi-functional compounds, such as hydroxyacetic acid, hydroxyacetaldehyde, hyroxyacetone, 3-hydroxy-3-methoxy benzaldehyde, etc. The organic acids present cause bio-oil to be acidic, with a pH between about 2.3 to 3.0.

Inorganics In Bio-Oil

The inorganic or mineral content of biomass is found in many forms: in aqueous solution in association with various counter ions; bound to organic acids; as deposits; or related to various enzymatic compounds. The counter ions in solution include carbonates, oxalates, phosphates, silicates, chlorides, and sulfates. (French and Milne 1994) The char and inorganic contents of bio-oil are important to its aging characteristics, as they appear to catalyze polymerization reactions during storage, leading to viscosity increases and growth in the apparent diameter of the suspended char (Agblevor et al 1994 and 1995).

In the discussion to follow, it will be shown that some of these minerals are potential catalysts for reactions that are important in aging, e.g., chlorides of calcium, lithium, iron, magnesium, manganese, and zinc that catalyze acetal forming reactions.

The inorganic content of biomass forms ash during combustion, which can have a negative impact on its combustion applications (Miles et al 1996), but this report is primarily concerned with the inorganic constituents present in the bio-oil itself.

Fortunately, the minerals found in biomass remain in the condensed phase during fast pyrolysis, which concentrates the minerals in the char. During the production of bio-oil, char is entrained with the organic vapors and is separated from the vapors using equipment having varying efficiencies. Cyclonic separation is the easiest method to remove char, but even the best cyclones begin to lose their efficiency with char particles having diameters below about 10 μm. Hot-gas filtration is more efficient in removing smaller char particles, but a small amount of char fines passes through these filters as well. Filtration after the bio-oil has condensed can also remove char fines, but results in a high-ash sludge as a byproduct and does not remove nano-sized char particles nor minerals already solubilized by the acidic solution of bio-oil.

Table 2 shows representative inorganic elemental analyses made of wood and grass, char filtered from condensed bio-oil, and bio-oils made with char separated by cyclones or by hot-gas filtration. A comparison of the inorganic content of the bio-oil and the char recovered by filtration from the condensed bio-oil verifies that the inorganics were concentrated in the char. However, filtering the bio-oil (after dilution with methanol through a 2.5 μm filter after condensation) removed only 20 to 50% of the total inorganic content of the oil. This implies that the remaining inorganic elements are in suspended char particles less than 2.5 μm in diameter, or they are in solution. Adding water to separate the bio-oil into aqueous and tar phases, resulted in a disproportion of the inorganics, but not a clean separation. In fact, the potassium content of the aqueous phase was higher than for the tar phase. (Elliott 1994)

Table 1a Compounds Identified in Bio-Oils (Milne *et al* 1997) and Similar Pyrolysis Products (Diebold 1997)

Compound	Wood Distillate	Smoke Flavors	Bio-Oils wt%
Acids			
Formic (methanoic)	f,g,h,k,s,v	*Gl*,m,t	0.3-9.1
Acetic (ethanoic)	f,g,h,k,s,v	*Gl*,m,t	0.5-12
Propanoic	f,g,h,k,s,v	*Gl*,m	0.1-1.8
Hydroxyacetic	f, , , ,s,v	m	0.1-0.9
2-Butenic(crotonic)	f,g,h,k,s,v	m	---
Butanoic	f,g,h,k,s,v	t	0.1-0.5
Pentanoic (valeric)	f, ,h,k,s,v	*Gl*,m,t	0.1-0.8
2-Me butenoic	f, ,h,k,s	m	---
4-Oxypentanioc	, , ,s,v	*Gl*,m	0.1-0.4
4-Hydroxypentanoic	, , , , ,v		---
Hexanoic (caproic)	f, ,h, ,s,	m	0.1-0.3
Benzoic		m,t	0.2-0.3
Heptanoic	---	m	0.3
Esters			
Methyl formate	f, ,h, ,s,	m,t	0.1-0.9
Methyl acetate	f, ,h,k,s,v	m,t	---
Methyl propionate	f, ,h, , ,v	m	---
Butyrolactone	.g, , ,s,v	m,t	0.1-0.9
Methyl Crotonate	---	m	---
Methyl n-butyrate	f, ,h, , ,	m,t	---
Valerolactone	f, ,h,k,s,v	---	0.2
Angelicalactone	---	m	0.1-1.2
Methyl valerate	f, ,h, , ,	*Gl*,m,t	---

Compound	Wood Distillate	Smoke Flavors	Bio-Oils wt%
Alcohols			
Methanol	f,g,h,k,s,v	m,t	0.4-2.4
Ethanol	.g, , ,v	m,t	0.6-1.4
2-Propene-1-ol	f, ,h,k,s,	m,t	---
Isobutanol	f, ,h,k, ,v	m	---
3-Methyl-1-butanol	, , ,s,	---	---
Ethylene glycol	---	m	0.7-2.0
Ketones			
Acetone	f, ,h,k,s,v	m,t	2.8
2-Butenone		d	---
2-Butanone (MEK)	f,g,h,k,s,v	m	0.3-0.9
2,3 Butandione	, , , ,v	---	---
Cyclo pentanone	f, ,h, ,s,v	m,t	---
2-Pentanone	f, , , ,v	m	---
3-Pentanone	,h, , ,	m	---
2-Cyclopentenone	, , , ,v	*Gl*,m	---
2,3 Pentenedione	, , , ,v	m,t	0.2-0.4
3Me2cyclopenten2oll one	f, , , ,	m	0.1-0.6
Me-cyclopentanone	f, ,h, ,s,v	m	---
2-Hexanone	---	m	---
Cyclo hexanone	---	m	trace
Methylcyclohexanone	f, , ,	---	---
2-Et-cyclopentanone	---	t	0.2 -0.3
Dimethylcyclopentanone	---	m	0.3
Trimethylcyclopentenone	---	*Gl*	0.1-0.5
Trimethylcyclopentanone	---	m	0.2-0.4

Key to references is at the bottom of Table 1c.

Table 1b Compounds Identified in Bio-Oils (Milne *et al* 1997) and Similar Pyrolysis Products (Diebold 1997)

Compound	Wood Distillate	Smoke Flavors	Bio-Oils wt%
Guaiacols			
2-Methoxy phenol	f,g,h, ,s,v	G,*Gl*,m,t	0.1-1.1
4-Methyl guaiacol	, , , , ,v	G,*Gl*,m,t	0.1-1.9
Ethyl guaiacol	f,g,h, , ,v	G,*Gl*,m,t	0.1-0.6
Eugenol	, , , , ,v	*Gl*,m,t	0.1-2.3
Isoeugenol		*Gl*,m,t	0.1-7.2
4-Propylguaiacol	f, ,h, , ,v	G,*Gl*,m,t	0.1-0.4
Acetoguiacone	---	---	0.8
Propioguiacone	---	---	0.8
Syringols			
2,6-DiOMe phenol	f,g,h, , ,v	G,*Gl*,m,t	0.7-4.8
Methyl syringol	f, ,h, , ,v	G,*Gl*,m,t	0.1-0.3
4-Ethyl syringol	, , , , ,v	G,*Gl*,m,t	0.2
Propyl syringol	f, ,h, , ,v	G,*Gl*,m,t	0.1-1.5
Syringaldehyde	, , , , ,v	G,*Gl*,m,t	0.1-1.5
4-Propenylsyringol	, , , , ,v	G,*Gl*,m,t	0.1-0.3
4-OH-3,5-diOMe phenyl ethanone		G, *Gl*	0.1-0.3
Sugars			
Levoglucosan	, , , ,s,	---	0.4-1.4
Glucose	---	---	0.4-1.3
Fructose	---	---	0.7-2.9
D-xylose	---	---	0.1-1.4
D-Arabinose	---	---	0.1
Cellubiosan	---	---	0.6-3.2
1,6 Anhydroglucofuranose		---	3.1
Phenols			
Phenol	f, ,h, ,s,v	G,*Gl*,m,t	0.1-3.8
2-Methyl phenol	f, ,h, ,s,v	G,*Gl*,m,t	0.1-0.6
3-Methyl phenol	f, ,h, ,s,v	G,*Gl*,m,t	0.1-0.4
4-Methyl phenol	f, ,h, , ,v	G,*Gl*,m,t	0.1-0.5
2,3 Dimethyl phenol	, , , ,s,v	G,*Gl*,m,t	0.1-0.5
2,4 Dimethyl phenol	f, ,h, ,s,v	G,*Gl*,m,t	0.1-0.3
2,5 Dimethyl phenol	, , , ,s,v	G,*Gl*,m,t	0.2-0.4
2,6 Dimethyl phenol	, , , ,s,v	G,*Gl*,m,t	0.1-0.4
3,5 Dimethyl phenol	f, ,h, ,s,v	m,t	---
2-Ethylphenol	f, , , , ,	G,*Gl*,m,t	0.1-1.3
2,4,6 TriMe phenol	---	m	0.3
1,2 DiOH benzene	f, ,h,k,s,v	G,*Gl*,m,t	0.1-0.7
1,3 DiOH benzene	---	m	0.1-0.3
1,4 DiOH benzene	---	---	0.1-1.9
4-Methoxy catechol	---	G,*Gl*,m,t	0.6
1,2,3 Tri-OH-benzene	, ,h, ,s,	t	0.6
Aldehydes			
Formaldehyde	, ,h, , ,	m,t	0.1-3.3
Acetaldehyde	f,g,h,k,s,v	*Gl*,m,t	0.1-8.5
2-Propenal (acrolein)	---	m	0.6-0.9
2-Butenal	, , , , ,v	m	trace
2-Methyl-2-butenal	---	*Gl*,m	0.1-0.5
Pentanal	f, ,h, , ,v	m	0.5
Ethanedial	---	---	0.9-4.6

Key to references is at the bottom of Table 1c.

Table 1c Compounds Identified in Bio-Oils (Milne *et al* 1997) and Similar Products (Diebold 1997)

Compound	Wood Distillate	Smoke Flavors	Bio-Oils wt%
Furans			
Furan	, , , ,v	m,t	0.1-0.3
2-Methyl furan	f, ,h, , ,	m,t	0.1-0.2
2-Furanone		*Gl*	0.1-1.1
Furfural	f, ,h,k,s,v	G,*Gl*,m,t	0.1-1.1
3-Methyl-2(3h)furanone	---	G,*Gl*	0.1
Furfural alcohol	, , ,s,v	*Gl*,m,t	0.1-5.2
Furoic acid	,h,k,s,	m	0.4
Methyl furoate	,g, , ,v	m,t	---
5-Methylfurfural	f, ,h,k,s,v	G,m,t	0.1-0.6
5-OH-methyl-2-furfural	, , , ,s,	m	0.3-2.2
Dimethyl furan	f, ,h, , , ,	m	---
Misc. Oxygenates			
Hydroxyacetaldehyde	---	m,t	0.9-13
Acetol (hydroxyacetone)	, , ,s,v	m,t	0.7-7.4
Methylal	f, , ,k,s,	m	---
Dimethyl acetal	, ,h,k,s,	m	---
Acetal	---	---	0.1-0.2
Acetyloxy-2-propanone	---	m	0.8
2-OH-3-Me-2-cyclopentene-1-one	---	m	0.1-0.5
Methyl cyclopentenolone	---	m	0.1-1.9
1-Acetyloxy-2-propanone	---	G,*Gl*	0.1
2-Methyl-3-hydroxy-2-pyrone	---	---	0.2-0.4
2-Methoxy-4-ethylanisole	---	---	0.1-0.4
4-OH-3-methoxybenzaldehyde	---	G,*Gl*,m	0.1-1.1
Maltol	,g, , ,s,	G,*Gl*,m,t	---

Compound	Wood Distillate	Smoke Flavors	Bio-Oils wt%
Alkenes			
2-Methyl propene	---	---	---
Dimethylcyclopentene	---	---	0.7
Alpha-pinene	, , ,s,	---	---
Dipentene	, , , ,s,	---	---
Aromatics			
Benzene	, ,k, ,v	m	---
Toluene	f, ,h,k,s,v	G,m	---
Xylenes	f, ,h,k,s	---	---
Naphthalene	, , ,s,v	t	---
Phenanthrene	---	t	---
Fluoranthene	---	t	---
Chrysene	, ,k, ,	t	---
Nitrogen Compounds			
Ammonia	,h,k,s,	---	---
Methyl amine	f, ,h,k,s,	m	---
Pyridine	, ,h, ,s,	m	---
Methyl pyridine	, ,h, ,s,	---	---

Key to References

f---Fraps (1901)	k---Klar (1925)
g---Goos and Reiter (1946)	m---Maga (1987+1988)
G---Guillén *et al* (1995)	s---Stamm + Harris (1953)
Gl---Guillén and Ibaragoitia (1996)	t---Tóth + Potthast (1984)
h---Hawley (1923)	v---Vergnet + Villeneuve (1988)

Filtering liquid bio-oil derived from switchgrass through a series of five progressively finer filters showed that most of the calcium and about half of the potassium were present in suspended char or sludge particles above 10μm. However, the remaining potassium was not removed with 0.7μm filters. (Agblevor *et al* 1994) Similarly, the calcium level was reduced only from 540 ppm down to 311 ppm using 0.1 μm filters to filter an oak bio-oil. The potassium level was reduced even less, from 440 ppm to 402 ppm. (Oasmaa *et al* 1997). A significant amount of the inorganic material was either associated with particles less than 0.1 μm in size, or was dissolved by the acidic bio-oil and in solution.

Table 2 also shows the much lower inorganic content of bio-oil produced by filtering the hot vapors prior to condensation, compared to the removal of char with cyclones. The best job of hot-gas filtering to date at NREL resulted in less than 2 ppm alkali and 2 ppm alkaline earth metals in the bio-oil (Scahill *et al* 1997). It appears that to obtain very low levels of inorganics in bio-oil, it is necessary to remove the char particles prior to condensation of the pyrolysis vapors.

At these low sodium levels, significant sodium contamination can be leached from common laboratory glassware by the acidic bio-oil, as the sodium ion appears to be mobile within the sodium borosilicate glass. There was a steady increase in sodium content from 8 ppm to 17 ppm as the bio-oil from switchgrass was progressively filtered through finer filters, while the potassium and calcium contents decreased to constant values below 10μm (Agblevor *et al* 1994). This anomalous increase in sodium content with increased processing could be explained by the leaching of sodium from laboratory glassware. In addition, the alkali contamination from dust in the air and residual alkali contents of purified water can be significant as these low levels. These considerations require special reagents and handling of the samples in clean, inert containers.

Other inorganic contaminants found in acidic bio-oil appear to be from processing or storage equipment made of iron or galvanized steel, or from the attrition of heat-transfer sands. For example, extraneous contamination levels of 2270 ppm iron, 1950 ppm zinc, and 80 ppm lead were reported by Elliott (1986) to be in an early SERI (NREL) bio-oil, which had been produced using a galvanized scrubbing tower for condensate recovery. (This galvanized scrubber was taken out of service shortly after the analyses were made.) The level of silicon in a bio-oil made in an entrained sand reactor was 330 ppm (Oasmaa *et al* 1997), compared to 112 ppm for a bio-oil made in an entrained flow reactor without sand (Elliott 1994).

Very few analyses have been performed on pyrolysis bio-oil and char for chlorine. Table 2 shows that the chlorine content of the hot-gas-filtered hardwood bio-oils varied considerably, from 0.3 to 2 equivalents of chlorine per equivalent of potassium, sodium, and calcium (Diebold *et al* 1996)(Scahill *et al* 1997). The chlorine content of switchgrass oils did not change significantly during filtration of the bio-oil and was in the range of 1200 ppm to 1600 ppm, while the metal ion content decreased. (Agblevor *et al* 1994). The equivalents of chlorine per equivalent of alkali or alkaline earth metals increased from 3.4 up to 8.5 during this filtration. This suggests that the residual inorganics may be present as chlorides and that much of the chlorine may be in solution. Chloride ions in solution will have an adverse effect on corrosion of many metals, including stainless steels like SS304.

Compared to coal and many crude oils, biomass has a low sulfur content. An inspection of Table 2 reveals that the small amount of sulfur in the biomass feed becomes concentrated in the char, rather than in the bio-oil.

Table 2 Inorganics in the Chars and Bio-Oils Made from Various Biomass at NREL with Char Removal by Cyclones or by Filtration

Reference	Elliott 1994				Agblevor et al 1994			Diebold et al 1996		Seahill et al 1997
Feedstock	Oak		Southern Pine		Switchgrass			Hybrid Poplar		
Material	Bio-Oil	+2μm Char in Bio-Oil	Bio-Oil	+2μm Char in Bio-Oil	Feed	+10μm Char in Bio-Oil	Bio-Oil -0.7μm	Feed	Bio-Oil (Run 175)	Bio-Oil (Run M2-10)
Char Removal Method	Cyclones	Cyclones + Oil Filtr.	Cyclones	Cyclones+ Oil Filtr.	---	Cyclones + Oil Filtr.	Cyclones + Oil Filtr.	---	Hot-Gas Filter	Hot-Gas Filter
Char, %	0.74	---	0.13	---	---	---	---	---	---	---
Ash, %	0.09	---	0.03	---	4.92	15.3	<0.05	0.77	0.01	0.007
Calcium, ppm	160	4580	160	8100	2400	7100	2.2	1550	2.2	1
Silicon	112	---	93	3452	---	---	1.4	---	---	---
Potassium	55	1300	10	667	8500	8500	175	1200	2.7	1
Iron	86	---	47	1772	---	---	---	---	---	---
Aluminum	55	---	41	---	---	---	---	---	2.6	0.3
Sodium	2	60	<0.1	372	31	690	17	27	7.2	0.9
Sulfur	<60	---	<50	349	---	---	---	---	---	---
Phosphorus	<50	---	<50	550	---	3600	0.6	---	---	---
Magnesium	<55	---	<45	903	---	---	---	---	---	0.7
Nickel	<22	---	<20	288	---	---	---	---	---	---
Chromium	<17	---	<17	524	---	---	---	---	---	---
Zinc	28	---	14	258	---	---	---	---	---	---
Lithium	25	---	7	110	---	---	---	---	---	---
Titanium	17	---	5	130	---	---	---	---	<0.2	---
Manganese	15	---	6	353	---	---	---	---	0.063	0.04
Copper	---	---	---	39	---	---	---	---	---	---
Barium	<3	---	<2	170	---	---	---	---	---	---
Vanadium	---	---	---	---	---	---	---	---	0.002	<0.01
Chlorine	---	---	---	---	6800	10600	1600	---	7.9	11
Equiv. Cl per equiv. K+Na+Ca	---	---	---	---	0.57	0.50	8.5	---	0.3	2

PROBABLE CHEMICAL MECHANISMS OF STORAGE INSTABILITY

It is beyond the scope of this review to attempt to discuss all of the possible reactions taking place in a mixture of up to 400 organic compounds. However, it is instructive to review several of the generic chemical reactions thought to play an important part in the aging reactions of bio-oil. Many of these reactions are said by organic chemistry textbooks to require catalysis, but in the long times available in storage, additional catalysis may not be needed or the catalysts may be already present in the bio-oil. Referring to the original literature is often necessary to find information on reactions that take place too slowly to generate commercial interest. It is interesting to note that industry has been already controlling some of these chemical reactions during the commercial shipping and storage of chemicals, which could have relevance to the aging of bio-oil.

It is thought that the majority of the important reactions that occur within bio-oil involve:
1) organic acids with alcohols to form esters and water
2) organic acids with olefins to form esters
3) aldehydes and water to form hydrates
4) aldehydes and alcohols to form hemiacetals, or acetals and water
5) aldehydes to form oligomers and resins
6) aldehydes and phenolics to form resins and water
7) aldehydes and proteins to form oligomers
8) organic sulfur to form oligomersunsaturated compounds to form polyolefins, and
9) air oxidation to form more acids and reactive peroxides that catalyze the polymerization of unsaturated compounds.

Reactions 1 through 5 can form products in thermodynamic equilibrium with the reactants, which means that a change in temperature, or relative amounts of water and other reactive compounds will upset the equilibrium and initiate compositional changes. Reactions 4 through 10 can form resins or polyolefins and may be irreversible at likely bio-oil storage conditions.

Reactions of Organic Acids

Esterification

The reaction of alcohols with organic acids forms esters and water:

$$ROH + R'C\overset{O}{\underset{OH}{\diagup}} \rightleftharpoons R'C\overset{O}{\underset{OR}{\diagup}} + H_2O$$

where R and R' are alkyl groups. Esters are well known for their aromas. In the case of wine and liquors, these ester forming reactions can take place over the course of several years. These reversible reactions are catalyzed by the presence of acids, which is of interest in bio-oil with its naturally low pH of between 2.3 and 3.0.

Table 3 shows the volatility expressed as the normal boiling point for esters and solvents likely to be present in bio-oils. Esters are relatively volatile compounds. For comparison, the boiling point of acetone at 56.5°C lies between that of ethyl formate and methyl acetate. The boiling point of methyl formate at 32°C is similar to that of diethyl ether. The vapor pressure at 0°C of methyl formate is 27 kPa (200 mm Hg) and that of methyl acetate is 8.8 kPa (66 mm Hg). If methyl formate were present in the pyrolysis products, it would not be recovered from the pyrolysis gases in most condensation trains because of this volatility.

The formation of esters from organic acids and alcohols is thermodynamically favored (i.e., an equilibrium constant greater than 1). Table 3 also shows that the equilibrium constants reported for the liquid phase vary from 4 to 5.2 for the lower molecular weight esters from primary

254

alcohols and about half that for those from secondary alcohols. The heat of esterification is relatively small, so the equilibrium constants are nearly independent of temperature. (Simons 1983) With equal molar concentrations initially of the alcohol and the acid, an equilibrium constant of two corresponds to a conversion of 59 mol % of the reactants, while an equilibrium constant of 4 corresponds to a conversion of 67 mol %.

The vapor-phase equilibrium constants are quite different from the liquid-phase equilibrium constants. Using values for the standard free energy at 25°C from the literature (DIPPR 1998) for the reactants and the products in the gas phase for the esterification reactions, the free energy for esterification in the gas phase was calculated. The thermodynamic equilibrium constants were then calculated from:

$$\ln K_e = -\Delta F / RT$$

Values for these calculated vapor-phase equilibrium constants are given in Table 3. The equilibrium constant of 367 for ethyl acetate in the vapor state corresponds to a conversion of 95% of the reactants.

Table 3   Normal Boiling Points of Probable Alcohols, Acids, and Esters in Bio-Oils, Liquid Phase Equilibrium Constants (Simons 1983), and Vapor Phase Equilibrium Constants for Ester Formation from Alcohol and Organic Acids at 25°C (calculated from Gibbs free energy in DIPPR (1998))

Compound	nbp, °C	Thermodynamic Equilibrium Constant Liquid	Vapor
Methyl Formate	32.0	---	60
Ethyl Formate	54.3	---	174
Methyl Acetate	57.8	5.2	243
Methanol	64.7	---	---
Ethyl Acetate	77.1	4	367
Ethyl Alcohol	78.4	---	---
Methyl Propionate	79.8	---	68
Iso-propanol	82.5	---	---
Iso-propyl acetate	88.4	2.4	400
Propanol	97.8	---	---
Ethyl Propionate	99.1	---	213
Formic Acid	100.6	---	---
Propyl Acetate	101.6	4.1	200
Acetic Acid	118.1	---	---
Propyl Propionate	122.5	---	31
Propionic Acid	141.1	---	---

Note that the calculated equilibrium constant for the formation of ethyl acetate from ethanol and acetic acid in the vapor phase is nearly two orders of magnitude higher than for the same reaction in the liquid phase.

Early esterification studies by Berthelot and Péan de Saint-Gilles in the 1800's used sealed tubes with various relative vapor headspaces. They found that the equilibrium was shifted to higher yields of ester with more headspace. In the 1920's, vapor-phase esterification reactions were re-examined. Three different equilibrium mixtures of acetic acid, ethanol, water, and ethyl acetate (in the absence of known catalysts) were refluxed for about one hour. Reactions occurred in the vapor phase at temperatures as low as 53°C to form about the same composition of vapors from the three initially very different equilibrium compositions. The composition of the vapors had shifted to more heavily favor the formation of the ester. Equilibrium constants of over 400 were calculated, taking into account the dimerization of acetic acid in the vapor phase. (Edgar and Schuyler 1924)

The temperature and liquid residence time used by Edgar and Schuyler (1924) are commonly encountered during the scrubbing of hot fast-pyrolysis vapors and gases downstream of the pyrolysis reactor with re-circulated condensates. Under these conditions, this suggests the probable formation and subsequent loss of the volatile esters in the non-condensable gases. The losses of these volatile esters could be a major cause of problems with a lack of mass-closure in fast pyrolysis systems using this type of scrubbing, because the off-gases are not usually analyzed for esters nor other oxygenated volatiles. These volatile losses would be increased with the use of large amounts of carrier gases for fluidization or entrainment in the pyrolysis reactor. Larger amounts of carrier gases reduce the partial pressure of the volatiles, requiring a lower temperature for condensation of the volatiles.

With no added catalysts, the reaction of acetic acid in an excess of methanol in the **liquid** phase appears to be a second order reaction with respect to acetic acid. At 70°C, a mixture of 1 N acetic acid in purified methanol had achieved 50% completion after about 27 hours. The presence of sodium acetate or ammonium acetate buffers reduced the reaction rate; after 27 hours the reaction had only achieved 10% completion at 70°C. With methanol and acetic acid at 0.125 N to 1.0 N in anisol as the solvent, the esterification rate appeared to be proportional to the concentration of methanol times the concentration of acetic acid raised to the 1.5 power. (Rolfe and Hinshelwood 1934)

At 25°C with no added catalysts, the bimolecular reaction rates of acetic acid and methanol were extrapolated to be: twice that of acetic acid and benzyl alcohol; 88 times faster than for methyl alcohol and benzoic acid; and 173 times faster than for benzyl alcohol and benzoic acid. This reflects the slower reaction rate of the larger molecules. (Hinshelwood and Legard 1935).

With 0.005N hydrochloric acid (HCl) (182 ppm or a pH of 2.3) as the catalyst at 30°C, the esterification of a stoichiometric mixture of methyl alcohol with formic acid was 50% completed in a little over about three minutes. The reaction rate of the mixture then slowed considerably and approached equilibrium after about 40 minutes. The catalyzed reaction rate for methanol with formic acid was: 14 times higher than with acetic acid; 18 times higher than with propionic acid, and 28 times higher than with butyric acid and with higher acids. (Smith 1939)

Experimentally the rate of hydrolysis of esters is a function of the pH. Figure 5 shows that the rate of hydrolysis of ethyl acetate varied continuously over 5 orders of magnitude, when the pH was changed from –0.9 to 3.7 (Euranto 1969). The rate of the esterification (forward reaction) can be calculated from the rate of hydrolysis (reverse reaction) rate multiplied by the thermodynamic equilibrium constant.

Organic acids react with olefins to form iso-esters, but without the formation of water as a byproduct. This reaction is normally catalyzed with a strong acid, e.g., sulfuric or zeolites (Sato 1984), in order to achieve reaction rates of interest commercially. However, without the presence of added catalysts, olefins (except ethylene) react slowly with dilute acetic acid at elevated temperatures to form iso-esters. Thus, the reaction of propene with acetic acid yields isopropyl acetate. Hydrolysis of the iso-ester results in the formation of the iso-alcohol. (Suida 1931)

Relatively anhydrous formic acid under reflux conditions without added catalysts reacted with unsaturated, long-chain fatty acids to form the expected iso-ester. These iso-esters were then hydrolyzed to form the iso-alcohol. (Knight *et al* 1954)

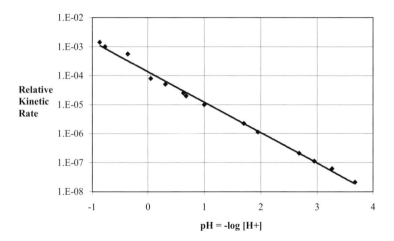

**Figure 5   Rate of Hydrolysis of Ethyl Acetate and pH
(Euranto 1969)**

Transesterification

The transesterification reaction is the exchange of alcohol and acid groups in a mixture of two or more esters:

$$RC\overset{O}{\underset{OR'}{}} + R''C\overset{O}{\underset{OR'''}{}} \rightleftharpoons RC\overset{O}{\underset{OR'''}{}} + R''C\overset{O}{\underset{OR'}{}}$$

This reaction is catalyzed similarly to esterification. In a mixture of esters such as exist in bio-oil, transesterification is thought to occur. Because esterification is reversible, it is also possible to react acids and alcohols with esters to form new esters, acids, and alcohols.

Reactions of aldehydes

Homopolymerization

Aldehydes can react with each other to form polyacetal oligomers and polymers:

$$nRC\overset{O}{\underset{H}{}} + H_2O \rightleftharpoons H(-C\underset{R}{\overset{H}{|}}O-)_nOH$$

The poly(oxymethylene) polymer has limited solubility in water. The presence of methanol in aqueous formaldehyde solution decreases the value of n, which is used advantageously to stabilize formaldehyde commercially. The methanol content of commercial formaldehyde solutions normally varies between 6% and 15%, although lower methanol contents of 1% or less are available. These solutions develop measurable amounts of methylal during prolonged aging. Other alcohols used to stabilize formaldehyde solutions are ethanol, propanol, isopropanol, glycols, and glycerol. (Walker 1953)

Additives to stabilize formaldehyde solutions include hydroxypropyl methyl cellulose, methyl and ethyl cellulose, and isophthalobisguanamine at levels up to 100 ppm. Conversely, if the formaldehyde polymer is stabilized with the proper end-groups (e.g., esterification of the hydroxyl end group with acetic anhydride), it can be a useful polymer such as DuPont's Delrin. Without the use of special polymerization techniques, the value of n is limited to about 100. (Gerberich *et al* 1980).

The addition reaction of aldehydes to form hydroxy aldehyde dimers is known as the aldol reaction, which proceeds nicely under basic conditions and very slowly in acidic conditions (Cason 1956). Formaldehyde condenses to form hydroxyacetaldehyde, glyceraldehyde, dihydroxy acetone, ketotetrose, and various aldotetroses in the presence of a small amount of organic acid and the oxides, hydroxides, carbonates and organic salts of lead, magnesium, zinc, and the alkaline earth metals. Formaldehyde was dissolved in an essentially anhydrous solvent and acidified with the equivalent of 0.08 to 0.19 wt% of acetic acid (after neutralization of the catalyst). Organic acids useful to catalyze this reaction were said to be formic, acetic, propionic, glycolic, benzoic, oxalic, and etc. Solvents used included methanol, ethanol, dioxan, n-butanol, and ethylene glycol. Yields of syrups or precipitates were 55% to 93% of the formaldehyde charged, after heating to 100°C to 138°C for 20 to 100 minutes. No examples were given for an aqueous reaction medium, nor for an uncatalyzed reaction. Also, no quantitative analyses were given for the products. (Lorand 1942)

Furfural forms a resinous tar under acidic conditions at rates proportional to the concentration of the acid times that of furfural. Temperatures studied were 50°C to 300°C. Acids used included hydrochloric acid at 1800 ppm. (Williams and Dunlop 1948).

Hydration

Aldehydes or ketones mixed with water react to form hydrates, also referred to as glycols:

$$RCR' + H_2O \rightleftharpoons R\underset{OH}{\overset{OH}{C}}R'$$

where R and R' are hydrogen or alkyl groups. The rate of formation of aldehyde hydrates was said to be quite rapid (Adkins and Broderick 1928a). The reaction of formaldehyde in an excess of water had a half-life of only 70 ms at 22°C. The ratio of the hydrate to formaldehyde was 2200 at 22°C. (Sutton and Downes 1972).

The equilibrium constants (Molar basis) for water and aldehyde reacting to form the hydrate are 41 for formaldehyde, 0.018 for acetaldehyde, and 0.000025 for acetone (Carey 1996). For a bio-oil containing 25 wt% water and 3 wt% formaldehyde in equilibrium, 99.9 mol% of the formaldehyde would be in the hydrate form. For a similar bio-oil with 3 wt% acetaldehyde, 24 wt% of the acetaldehyde would be in the hydrate form. Only a trace of the hydrate of acetone (0.04% of the acetone weight) would be present as the hydrate, in a bio-oil with 3 wt% acetone and 25 wt% water.

Hemiacetal Formation

When an alcohol is mixed with an aldehyde, a significant amount of heat is liberated and the formation of a hemiacetal takes place by the following reaction:

$$ROH + R'\underset{H}{\overset{O}{C}} \rightleftharpoons R'\underset{OH}{\overset{OR}{C}}-H$$

This reaction takes place relatively quickly and without the need for catalysis. Hemiacetals were studied that were made from: ethyl alcohol and acetaldehyde; isopropyl alcohol and acetaldehyde; ethyl alcohol and benzaldehyde; and ethyl alcohol and methoxy benzaldehyde. Refractive index and density measurements were used to confirm the rapid formation of hemiacetals. In those cases where the refractive index of the hemiacetal varied considerably from idealized mixtures of the reactants, it was deduced that equilibrium compositions favored the hemiacetals. (Adkins and Broderick 1928a and 1928b).

In the absence of catalyst, phenol reacts with formaldehyde to form the phenyl hemiformal, although to a much lesser degree than the methyl hemiformal. Formaldehyde reacts in the absence of catalysts with the hydroxyl groups of sugars, starches, and cellulose to form loosely bound hemiformals. (Walker 1953)

Commercially available formaldehyde is mixed with water and methanol to chemically stabilize it. In aqueous solutions, the amount of free, monomeric formaldehyde is less than 100 ppm. It is known that this forms "...a complex equilibrium mixture of methylene glycol ($CH_2(OH)_2$), poly(oxymethylene glycols), and hemiformals of these glycols". The water and methanol serve to stabilize the formaldehyde and prevent the formation of polyformaldehyde resins. (Gerberich *et al* 1980)

Acetalization

Aldehydes and alcohols react to form acetals as shown by the following reaction:

$$2ROH + R'\overset{O}{\underset{H}{C}} \rightleftharpoons R'\overset{OR}{\underset{OR}{C}}H + H_2O$$

A modification to this acetal formation is with a diol such as ethylene glycol ($HOCH_2CH_2OH$), in which one mole of glycol rather than two moles react with the aldehyde to form a cyclic diether (Street and Adkins 1928).

$$HOCH_2CH_2OH + R'\overset{O}{\underset{H}{C}} \rightleftharpoons R'\overset{O-CH_2}{\underset{O-CH_2}{C}}H | + H_2O$$

In the presence of acid catalysts, stable formals are formed by the reaction of aldehydes and the hydroxyl groups of sugars, starches, and cellulose. Cellulose reacted with formaldeyde and an acid catalyst creates a crosslinked polymer that resists swelling in water and dyes. Cotton cloth treated with formaldehyde and an acid catalyst will hold creases better and is less apt to shrink. (Walker 1953)

In the vapor phase at 25°C, the equilibrium constant for the formation of 1,1 dimethoxy methane (methylal) is 3402 and for 1,1 diethoxy ethane (acetal) is 36, based on aldehyde and alcohol reactants and the Gibbs free energy of formation (DIPPR 1998). Without catalysts in the vapor phase at high temperatures, the reaction of methanol with saturated aldehydes up through heptaldehyde and the reaction of formaldehyde with saturated primary or secondary alcohols are very rapid. These reactions were explored in a plug flow reactor at stoichiometric ratios of alcohols to aldehydes, temperature histories of 1.3 to 6 seconds at 350°C or 1.4 to 2.5 seconds at 460°C. The reported yields of acetals were between 15 wt% and 69 wt%, based on the weight of starting materials. There was apparently very little, if any, cracking of the materials to non-condensable gases under these reaction conditions. The mass balances of just the liquids closed to better than 98%, except when nitrogen carrier gas was used resulting in a 93% mass balance. (Frevel and Hedelund 1954) The actual average temperature of the reactants may not have been as high as stated, as no details were given on how and where the temperature was measured.

These conditions for the vapor phase formation of acetals are very close to those used to produce bio-oil. This implies that freshly made bio-oil should contain a large amount of acetals, especially with a long residence time at just below the pyrolysis temperature, e.g., in cyclonic separators, hot-gas filters, or in the free-board volume above a fluidized bed. Because the normal boiling temperature of methylal is only 42°C and its vapor pressure at 0°C is 17.2 kPa (129 mm Hg), most of the volatile methylal formed will probably not be collected in a typical condensation system.

Catalysts for acetal formation at lower temperatures in the condensed phase have been shown to include many of the salts potentially present in biomass as ash, e.g. chloride salts of aluminum, calcium, iron (ferric), lithium, magnesium, manganese, and zinc; copper sulfate, sodium bisulfite; and monosodium phosphate. It was observed that successful acetalization catalysts also form alcoholates. Non-catalytic salts for this reaction include: carbonates and chlorides of sodium and potassium; calcium and sodium sulfates; and calcium and sodium acetates. (Adams and Adkins 1925, Adkins and Nissen 1922) Catalysts for formal formation include formates of iron, zinc, and aluminum (Walker 1953).

Calcium chloride was shown to be an effective acetal catalyst at a concentration as low as 0.2 wt%, although higher concentrations were more effective. Part of the function of calcium chloride was to remove water from the reaction mixture to achieve a higher conversion, but the presence of other dehydration salts, e.g., sodium chloride or zinc chloride, served to reduce the catalytic effect of calcium chloride. (Adams and Adkins 1925)

Trace amounts of strong mineral acids also catalyze the acetalization reaction, e.g., hydrochloric acid (HCl), phosphoric acid, and nitric acid. In the presence of 5 mg HCl per mole of aldehyde (114 ppm), the times required to achieve equilibrium at 25°C with alcohols were: 1 to 2 hours with furfural and unsaturated aldehydes; and two days for saturated aldehydes (Minné and Adkins 1933). The rate of formation of acetal increased with the concentration of acid; the reactions were complete within an hour or so with hydrochloric acid at a concentration as low as 900 ppm or within five to ten minutes with 9000 ppm HCl (Adams and Adkins 1925).

Later studies showed that with only 13 ppm HCl as catalyst, that the acetal formation with acetaldehyde had gone to 50% of completion after 50 minutes with methanol, 60 minutes for ethanol and *iso*-propanol, and 120 to 180 minutes for butanol (Adkins and Broderick 1928b). With only 3 ppm HCl, a mixture of ethanol and furfural achieved about 75% of the equilibrium amount of acetal within one day. Non-equilibrium mixtures of furfural acetal, water, and ethanol with only 0.5 ppm HCl formed essentially equilibrium compositions, after an unspecified time (Adkins *et al* 1931). Assuming 100% ionization of dilute HCl, 3 ppm HCl and 0.5 ppm HCl correspond to pH's of 4.0 and 4.9 respectively. Clearly, only trace amounts of HCl are necessary to catalyze the formation of acetals.

Strong organic acids such as oxalic acid (pK_a = 1.3) and tartaric acid (pK_a = 3.0) also catalyze acetalization, but acetic acid (pK_a = 4.7) does not catalyze this reaction. (Adams and Adkins 1925) Although, oxalic and tartaric acids are not among the acids listed by Maga (1987) nor by Milne *et al* (1997), 2-hydroxy benzoic acid was listed by Maga (1987), which has a pK_a equal to that of tartaric acid. So, it is possible that some of the stronger organic acids in bio-oil catalyze acetal reactions, along with chloride ions in solution and some chloride salts.

Table 4 shows that the structure of the alcohol and of the aldehyde affect the equilibrium constant (K_e) of acetal formation. In general, the acetals from primary alcohols and saturated primary aldehydes are favored. Primary alcohols have K_e's about 10 times higher than secondary alcohols. Secondary alcohols have K_e's about 10 times higher than tertiary alcohols. Likewise, if the aldehyde group was attached to a primary carbon (*n*-butanal), the K_e was about 100 times higher, than if the aldehyde group was attached to a secondary carbon (*iso*-butanal). Using the reaction of ethanol with aldehydes containing four carbon atoms as an example, saturated aldehydes had a K_e about 100 times higher than unsaturated aldehydes and 30 times higher than for cyclic unsaturated aldehydes (furfural). (Minné and Adkins 1933) With a mixture of various alcohols and aldehydes, the number of possible acetals is quite high, as different alcohols could react with the same aldehyde.

The rate of acetal formation has no relationship to the relative value of the equilibrium constant. In fact, methanol and butanol, which have high equilibrium constants with acetaldehyde, react relatively slowly to form the acetals. In the presence of 13 ppm HCl, the reaction rates of secondary alcohols were about twice as fast as with primary alcohols. Tertiary alcohols were about twice as fast as secondary alcohols. Acetaldehyde reacted up to twice as fast as butyraldehyde. The reaction rates of furfural, benzaldehyde, cinnamic aldehyde, and heptaldehyde with alcohols were extremely fast and could not be accurately measured. (Adkins and Adams 1925).

Transacetalization

Because the formation of acetals is reversible, it is possible to react alcohols with acetals or different acetals with each other to generate new acetals. If a low molecular weight alcohol or aldehyde were added to acetals formed from high molecular weight alcohols and acids, the average molecular weight of the original mixture would be expected to decrease.

Phenol / aldehyde reactions and resins

In the absence of catalysts, phenol reacts as an alcohol with formaldehyde to form the hemiformal, although not as favored as the methyl formal from methanol. The phenyl hemiformal is an intermediate in the formation of *ortho*-methylol phenol:

In the presence of acid catalysts, the ortho-methylol phenol is very reactive with other phenols to attach to the ortho or para position with the elimination of water. In the presence of added acid catalysts and methanol, methyl phenyl formal can be made from the phenyl hemiformal. (Walker 1953)

In the presence of acid catalysts, phenols and substituted phenols react with aldehyde hydrates to form novolak resins and water.

Use of strong acids such as oxalic or mineral acids results in thermoplastic novolak resins with molecular weights between 500 and 5000, with the methylene linkages being a mixture of ortho and para. The reaction rate is inversely proportional to the water content and proportional to the concentration of the reactants and catalyst. Other aldehydes react in a manner similar to formaldehyde, although typically at much lower rates. (Kopf 1996)

In contrast to reactions under alkaline conditions, the reaction rate of various phenolic compounds with formaldehyde are relatively the same under acidic conditions. The condensation reaction of the methylol phenol in the presence of acid catalysts is much faster than the rate of formation of the methylol phenol itself. Because the rate of reaction of formaldehyde with the polymer is much slower than with monomeric phenolics, novolaks have very little three-dimensional crosslinking between the chains. The relatively non-crosslinked novolak chains maintain their solubility in solvents, e.g., acetone. (Granger 1937)

Table 4   Equilibrium Constants for Liquid Acetal Formation (at 25°C)
(Minné and Adkins 1933) and Normal Boiling Point of Resulting Acetals
(K$_e$ based on concentrations expressed on a mole fraction)

1 ALCOHOL	2 Aldehyde	K$_e$	nbp of Acetal, °C
Methanol	Formaldehyde	---	42
"	Ethanal (Acetaldehyde)	1.95	64
"	Propanal	---	83
"	Propenal (Acrolein)	---	89
"	n-Butanal	4.22	---
"	Furfural	0.100	---
Ethanol	Formaldehyde	---	87
"	Ethanal (Acetaldehyde)	1.21	102.2
"	Hydroxyacetaldehyde (Glycolic)	0.28	167
"	Propanal	1.21	123
"	Propenal (Acrolein)	0.176	125
"	2-Butenal (Crotonaldehyde)	0.011	---
"	n-Butanal	1.21	---
"	Furfural	0.041	---
"	Benzaldehyde	0.083	---
"	3-Phenyl Propenal (Cinnamaldehyde)	0.013	---
n-Propanol	Ethanal (Acetaldehyde)	1.06	---
"	n-Butanal	1.95	---
2-Propen-1-ol	Ethanal (Acetaldehyde)	0.84	---
iso-Propanol	Ethanal (Acetaldehyde)	0.107	---
"	Propanal	0.057	---
"	n-Butanal	0.069	---
"	iso-Butanal	0.018	---
"	Furfural	0.009	---
	Benzaldehyde	0.005	---
	3-Phenyl Propenal (Cinnamaldehyde)	0.005	---
n-Butanol	Ethanal (Acetaldehyde)	3.08	---
2-Butanol	"	0.128	---
t-Butanol	"	0.019	---
Benzyl Alcohol	"	0.94	---
β-Methoxyethanol	"	0.94	---
β-Ethoxyethanol	"	0.94	---

At higher pH's between 4 and 7, phenol and formaldehyde react in the presence of catalysts to form "high ortho" novolaks. Catalysts for this reaction are some aluminum salts and divalent metals such as calcium, zinc, magnesium, manganese, lead, copper, and nickel. These metals are solubilized as the acetate salts. Under these conditions, about 10% of the product was dibenzyl ether. (Kopf 1996)

With no catalysts present, the dimethylol cresols were shown to slowly react to form resins, achieving 50% of theoretical reaction within 24 hours at 100°C. These resins readily precipitated from aqueous solutions and appeared to be novolaks in nature. (Granger 1937)

The active sites for the reaction of formaldehyde with phenol are at the 2, 4, and 6 positions, i.e., the ortho and para positions. For catechol (1,2 dihydroxy benzene), all remaining sites are available to react with formaldehyde (Kelley et al 1997). The 2,6-dimethoxy phenols (syringols) present in bio-oils derived from hardwoods have only the para site remaining for reaction with aldehydes, which will result in chain-termination of the resultant phenol/aldehyde oligomer. If the phenol/formaldehyde reaction is important during the storage of bio-oils, then bio-oils derived from hardwoods should store better than soft-wood bio-oils, due to this chain-termination ability of the syringols.

Furfural has the reactivity of its aldehyde functional group to form polymers with phenol. The initial reaction products are analogous to those from phenol and formaldehyde (Dunlop and Peters 1953).

Polymerization of furan derivatives

Fufural alcohol undergoes condensation reactions with the evolution of water, analogous to the condensation of *ortho*-methylol phenol, i.e., with linkages at the 2 and 5 ring positions (adjacent to the oxygen of the furan ring). This reaction is catalyzed by lactic acid, hydroxy acetic acid, formic acid, calcium chloride, and strong acids. The polymerization takes place in 1 to 1½ hours at 100°C to produce an acetone soluble resin. This resin can be crosslinked with added formaldehyde using the same catalysts to form an insoluble resin. (Harvey 1944a) Furfuryl alcohol, in the presence of water and an acid catalyst also opens the furan ring, to form levulinic acid. (Dunlop and Peters 1953)

Thermosetting resins can also be made from the reaction of furfural alcohol and formaldehyde in a one-step process to produce the intermediate acetone soluble resin, which crosslinks upon further heating. The reaction takes place at a pH between 1.5 and 3.5, using the same acid catalysts as for the two step process. (Harvey 1944b)

Dimerization of organic nitrogen compounds

The chemical reaction of aldehydes with proteins is used advantageously in liquid smoke applications to give the meat a desirable brown color. Aldehydes that are very reactive for this purpose include hydroxyacetaldehyde (glycoaldehyde), ethanedial (glyoxal), and methyl glyoxal (Riha and Wendorff 1993). These reactions can lead to cross-linking of the proteins (Acharya and Manning 1983):

$$HOCH_2\overset{O}{\underset{H}{C}} + 2H_2N\text{-Protein} \longrightarrow \underset{H}{\overset{CH_2\text{-NH-Protein}}{C}}=N\text{-Protein} + 2H_2O$$

Pyrolysis operators who have had bio-oil on their skin are well aware of the brown stains rapidly left by this reaction. In bio-oil, these reactions lead to an effective dimerization of the proteins to increase the average molecular weight.

Sulfur containing compounds

Although bio-oil normally contains very low levels of sulfur, the presence of organic sulfur counteracts the formation of peroxides. Especially effective for this are nonyl sulfide and thiophenol (phenyl mercaptan). In the reaction of mercaptans (RSH) with peroxides, the hydrogen is abstracted and the disulfide is formed which nearly doubles the molecular weight of the sulfur compounds. Alternatively, the RS* free-radical can attach to unsaturated bonds to also form larger molecules. The benefit is the termination of the free-radical to prevent further chain reactions. (Mushrush and Speight 1995)

Unsaturated organic reactions

Alcohol addition

Unsaturated aldehydes such as acrolein are very reactive due to the conjugation of the vinyl group with the carbonyl group. In addition to forming acetals with alcohols, acrolein also can react with alcohol in an addition reaction that is catalyzed by acids or bases to form the alkoxy aldehyde, unsaturated acetal, or alkoxy acetal (Etzkorn *et al* 1991):

$$H_2C=CHC\underset{H}{\overset{O}{\diagdown}} + 3ROH \longrightarrow ROCH_2CH_2C\underset{OR}{\overset{OR}{\diagup}} + H_2O$$

where R is an alkyl group.

Olefinic condensation

Unsaturated compounds can react with each other to form polyolefins. Examples of unsaturated compounds in bio-oil listed in Table 1 include acrolein, 2-Butenone, propenyl substituted phenolics, and 2-propene-1-ol. Acrolein homo-polymerizes in the presence of anionic, cationic, or free-radical agents to form an insoluble, highly cross-linked resin, in the absence of inhibitors. Acrolein is commercially stabilized with acetic acid and 0.10 wt% to 0.25 wt% hydroquinone. (Etzkorn *et al* 1991)

In strongly acidic (3N HCl) solutions overnight, 2-methoxy-4-propenylphenol turned into a solid precipitate, 2-methoxy-4-allylphenol formed a brown viscous gel, and 4-allyl anisole and 4-propenyl anisole formed brown oils (Polk and Phingbodhippakkiya 1981).

Carboxylic acids catalyze olefinic condensation reactions, but hydroperoxides that form free-radicals are the most deleterious oxygenated species in petroleum liquids (Mushrush and Speight 1995). The free-radicals catalyze olefin condensation reactions. Sources of free-radicals in the bio-oil include organic hydroperoxides, organic peroxides and nitrogen compounds. Nitrogen compounds in the bio-oil would be derived from proteins in the original feedstock.

Oxidation

The exposure of bio-oil to air can oxidize the alcohols and aldehydes to carboxylic acids. An example of this reaction is the auto-oxidation of ethanol in wine to vinegar (acetic acid) after exposure to air, which is perceived as a sour taste after exposure overnight at room temperature or after storage with a defective or cracked cork.

However, a more important reaction impacting the storage of pyrolysis oils is that of forming hydroperoxides and alkylperoxides by autoxidation with air. For example, the formation of α-hydroperoxy ether from diethyl ether is a hazard after the ether container has been opened. These peroxides are not very stable and spontaneously decompose to form free-radicals. In a concentrated form, many of the organic peroxides are explosive in nature.

Organic peroxides or hydroperoxides can be formed by the reaction of air with olefins, ethers, ketones, nitrogen compounds, aldehydes, and organic acids. Peroxides are sources of free-radicals, which can catalyze olefinic polymerization. The ease of formation of hydroperoxides is related to the ease in breaking of the C-H bonds involved. Thus, for hydrocarbons the relative ease of forming hydroperoxides increases in the following order *n*-alkanes < branched alkanes < aralkanes $\cong$ alkenes < alkynes (Hiatt 1971). The "drying" of linseed oil upon exposure to air is an example of air oxidation that catalyzes the unsaturated oil to polymerize it to a solid varnish.

The importance of the formation of free-radicals from peroxides is their ability to catalyze the polymerization of olefins. Thus, exposure to air would be expected to increase formation of polyolefins during the storage of bio-oil.

Gas Forming Reactions

Carbon dioxide

The reactions considered thus far have been those that increased the molecular weight of the organic material, with the possible co-production of water. Most carboxylic acids are relatively stable relative to decarboxylation, but there are exceptions. Some di-carboxylic acids are relatively unstable and decarboxylate to form the mono-acid and carbon dioxide at moderate temperatures:

$$\underset{\underset{HO}{} \quad \underset{R'}{} \quad \underset{OH}{}}{C-C-C} \longrightarrow \underset{\underset{R'}{} \quad \underset{OH}{}}{H\,C-C} + CO_2$$

Where R and R' are H or alkyl groups. If R and R' are both hydrogen (malonic acid), the decarboxylation proceeds readily at 150°C. (Carey 1996) Although it is widely believed that this easy decarboxylation occurs only when there are two carboxyl groups attached to the same carbon atom (Cason 1956), a brief examination of the physical properties of organic compounds reveals that many di-carboxylic compounds decompose at low temperatures, e.g., 1,2 di-carboxyl ethylene (maleic acid) decomposes readily at 135°C, 1,2 di-carboxyl-2-hydroxy ethane (Malic acid) decomposes at 140°C, and 2-hydroxy acetic acid (glycolic acid) decomposes before boiling (Perry *et al* 1997).

Similarly, β-keto acids also decompose at low temperatures to form the ketone and carbon dioxide:

$$\underset{\underset{OH}{}}{RCCH_2C} \longrightarrow RCCH_3 + CO_2$$

where R is H or an alkyl group. If R is $-CH_3$, the decarboxylation proceeds readily at 25°C. (Carey 1996)

An alternate route to forming carbon dioxide in bio-oil is from ferulic acid, which was said to be a primary pyrolysis product of lignin. With thermogravimetric analysis (TGA) at 6°C/min, re-crystallized ferulic acid started to decompose noticeably at 200°C to 4-vinylguiacol and carbon dioxide:

$$\underset{HO}{H_3CO} \underset{}{\bigcirc} CH=CHC\overset{O}{\underset{OH}{}} \longrightarrow \underset{HO}{H_3CO} \underset{}{\bigcirc} CH=CH_2 + CO_2$$

The activation energy was much lower than for normal carbon-carbon bond breakage and the decomposition rate was faster in the presence of oxygen, causing the authors to postulate that free-radical reactions were involved. (Fiddler *et al* 1967) It is possible that free-radicals in aqueous solutions of bio-oil catalyze this reaction at much lower temperatures during the long times available in aging.

These three possible routes for the formation of carbon dioxide during storage involve relatively unstable organic compounds. The low temperature instability of these compounds would seem to preclude their survival through the high temperatures of pyrolysis. This suggests that they would have to be formed during the aging process, but their presence in bio-oil has not been reported in reviews of the composition of bio-oils (Milne et al 1997) nor of liquid smoke flavors (Maga 1987).

Hydrogen

If there are metals such as iron, zinc, aluminum, etc. present with the bio-oils, it can be assumed that they would react to form the organic acid salt and free hydrogen. This hydrogen might not react with unsaturated organic compounds in the absence of hydrogenation catalysts and it could escape as a gas to the headspace of the storage container.

Insights To Be Gained From The Chemical Mechanisms Of Aging

In bio-oil, the various equilibrium chemical reactions discussed above are constantly adjusting to changes in temperature, water content, volatile content, etc. Using the thermodynamic equilibrium constants for the liquid phase of 0.018 for acetaldehyde hydrate formation, 5.2 from Table 3 for methyl acetate and 1.95 from Table 4 for 1,1 dimethoxyethane, the equilibrium composition was calculated for a pseudo bio-oil. The base-case, pseudo bio-oil initially contained 25 wt% water, 3% acetaldehyde, 2% methanol, and 8% acetic acid. It was assumed that the volumes of the water and bio-oil were additive and that the density change during aging of the bio-oil solutions was negligible. The linear mass balances for the four initial compounds and the three non-linear equilibrium expressions involving the products were simultaneously solved by a trial and error method, using a simple computer spreadsheet.

Figure 6 shows the effect of changing the initial water content on the final equilibrium composition, by water removal or addition to the original base-case pseudo bio-oil. As predicted by equilibrium concepts, increasing the water initially in the pseudo bio-oil from 5 wt% to 30 wt% decreased the methyl acetate content (from 2.8 wt% to 0.98 wt% of the base-case initial bio-oil, respectively). As the equilibrium amount of methyl acetate decreased with more initial water, the acetic acid content increased from 5.7 wt% to 7.2 wt%, respectively. The content of acetaldehyde hydrate increased from 0.3 wt% to 1.21 wt% with this same increase in initial water content. As the initial water content increased, the net yield of water decreased from 0.6 wt% to -0.09%.

In a complex, real bio-oil, the number of possible chemical reactions is very high, especially when one considers the large number of complex oligomers present. However, the trends calculated to occur with representative, low-molecular-weight compounds are instructive for speculation on reactions with the complex oligomers. It is the reactions between the oligomers that are responsible for nearly all of the viscosity increases observed during aging of bio-oil.

For example, an increase in formation of aldehyde hydrates with higher water contents, leaves less free aldehydes available to react with hydroxy containing oligomers to form acetal linked polymers. Increased water contents encourages the hydrolysis of ester-linked polymers to form lower molecular-weight acidic and hydroxy compounds. Recalling the relationship of viscosity and molecular weight shown in Figure 4, reducing polymerization reactions will reduce the viscosity increase during aging.

Adding methanol to bio-oil shifts the equilibrium toward more esters, acetals, and water. Figure 7 shows the effect of adding methanol to the same base-case, psuedo bio-oil used for Figure 6. With the addition of 20 g methanol per 100 g of the base-case pseudo bio-oil, 57% of the acetic acid present at thermodynamic equilibrium was calculated to be converted to methyl acetate, and 68 % of the acetaldehyde was converted to 1,1 dimethoxyethane. The yield of methyl acetate was calculated to be 6.2 wt% of the pseudo bio-oil and that of 1,1 dimethoxyethane was 4.3 wt%.

It should be noted that the increase in molecular weights of these methyl esters and acetals are relatively very low, compared to the possible doubling or tripling of molecular weights that occurs in a real bio-oil with oligomeric hydroxy compounds reacting with oligomeric acids and aldehydes. This suggests the addition of low molecular weight reactants to react with oligomeric bio-oil components to form medium molecular weight oligomers, rather than high molecular weight polymers.

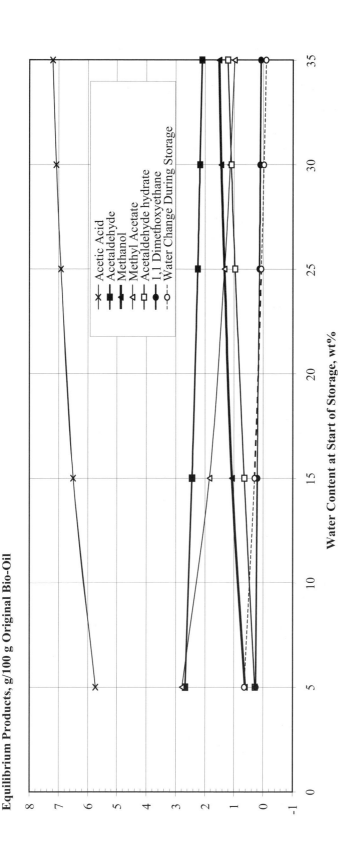

Equilibrium Products, g/100 g Original Bio-Oil

Legend:
- ─✕─ Acetic Acid
- ─■─ Acetaldehyde
- ─▬─ Methanol
- ─△─ Methyl Acetate
- ─□─ Acetaldehyde hydrate
- ─●─ 1,1 Dimethoxyethane
- ─○─ Water Change During Storage

Water Content at Start of Storage, wt%

Figure 6. Calculated Equilibrium Composition of Pseudo Bio-Oil with Initial Water Content
(25% Water, 3% Acetaldehyde, 2% Methanol, and 8% Acetic Acid in Original Pseudo Bio-Oil)

OBSERVED CHEMICAL REACTIONS IN WOOD DISTILLATES, WOOD SMOKE, OR BIO-OIL

Wood Distillates

Making esters from pyrolysis condensates goes back to the days of wood distillation, using slow pyrolysis in a large updraft retort. The condensates were allowed to settle into aqueous and tar phases. The tar phase was distilled to recover the volatiles, including acids, prior to higher temperature distillation of the oils to recover creosote oil and pitch. The tar-distilled volatiles were returned to the aqueous phase for recovery of acids as bottoms and crude methanol spirits as overheads in another distillation step. Methanol, methyl acetate, and 2-butanone (methyl ethyl ketone) were recovered from the crude methanol spirits. The acidic bottoms were mixed with either lime to make a brown calcium-acetate precipitate or with ethanol (with sulfuric acid as catalyst) to make ethyl acetate and ethyl formate esters on a commercial basis (Nelson 1930).

Using Georgia Tech's updraft gasifier with a feedstock of pine, a condenser oil and a draft-fan oil were recovered. The draft fan was located downstream of the condenser. Some very curious results were reported for the aging of these two oils. It was reported that the viscosity of the condenser oil at 25°C decreased after aging for eight months at ambient temperature, but increased viscosity after storage at 0°C. With the draft fan oil, the viscosity at 25°C was reported to be lower after eight months at 0°C than initially, but a higher viscosity was observed after storage at ambient temperatures. These conflicting results on the effect of temperature are not the trends others have experienced with the aging of bio-oils at moderately low temperatures, which suggests that perhaps phase separations were involved. Data were presented for the accelerated aging of the condenser oil at 110°C, from which aging rates (measured at 40°C) of 100 to 235 cP/day were derived, which is in agreement with the magnitude of extrapolations of the aging rates shown in Figure 3. (Knight *et al* 1977).

Also using Georgia Tech's updraft gasifier with a feedstock of pine, a condenser oil and a draft-fan oil were recovered. The condenser oil was vacuum distilled at 22°C/6.6mmHg to 128°C/15mmHg. Gas chromatography (GC) of the fresh distillate and again after 5½ months of aging at room temperature, showed a drop in the content of acids, alcohols, aldehydes, and phenolics with unsaturated-alkyl groups. There was an increase in water content with time. It was concluded that esters, acetals, and polyolefinic phenolic polymers were formed during the aging process. (Polk and Phingbodhippakkiya 1981)

Also using an updraft gasifier, the condensates were gravity separated into oil and aqueous phases. The aqueous phase was distilled to drive off much of the water (and volatiles), resulting in a water-insoluble oil phase forming in the bottom of the still. The two oil phases were then mixed together and esterified with methanol. Heating and the option of catalysis were recommended to increase the reaction rates. The reaction mixture was optionally distilled to reduce the water content of the esterified oils (this would also drive the esterification reaction to completion). Alternatively, the methanol was just added to the oil phase to reduce the viscosity, with esterification thought to occur during storage. (Capener and Low 1982) It appears that during distillation of the aqueous phase, that the formic acid and much of the acetic acid would have been distilled over with the water phase. The remaining organic acids with low volatility in the oil phase would then have formed esters with the added methanol. No examples were given in this patent, so the efficiency of the esterification cannot be judged.

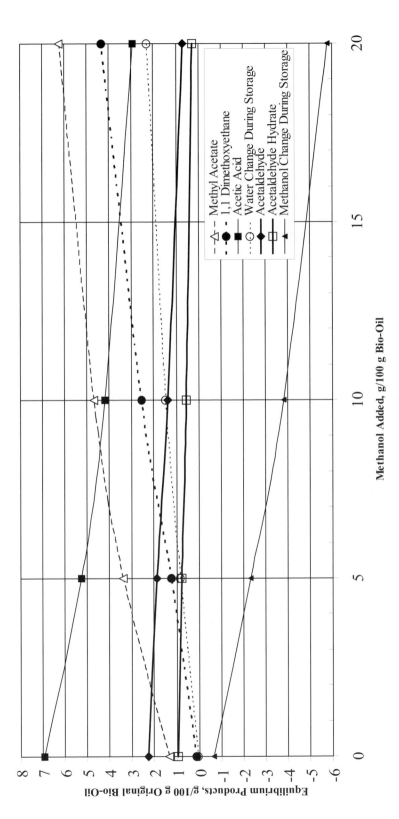

Figure 7   Calculated Equilibrium Composition of Psuedo Bio-Oil with Added Methanol
(25% Water, 3% Acetaldehyde, 2% Methanol, and 8% Acetic Acid in Original Bio-Oil)

Wood Smoke

The formation of esters and transesterification in whole smoke condensates (made from hickory sawdust) during aging has been confirmed to occur without the addition of catalysts. Figure 8 shows that methyl acetate formation occurred over a period of 25 days, whereas methyl formate was at equilibrium after 3½ days. It was also observed that 2-butenone disappeared after 72 hours, but with no conjecture made as to the products. (Doerr *et al* 1966) A close examination of Figure 8 shows the effect of transesterification with the equilibrium amount of the first-formed methyl formate decreasing slightly, while the amount of methyl acetate increased. This reflects the decrease in the amount of methanol present after the formation of significant amounts of methyl acetate and the resultant shift in the equilibrium quantity of methyl formate.

Concentration, molarity

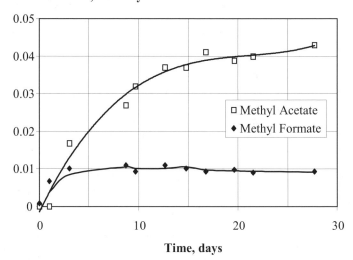

Time, days

Figure 8   Non-Catalytic Esterification in Whole Smoke Condensate at 25°C (Doerr *et al* 1966)

Bio-oils

Aging

Using a bio-oil made from oak at the National Renewable Energy Laboratory (NREL), the viscosity and weight average molecular weight both increased with time. Gel permeation chromatography (detecting only the aromatic compounds) showed that aging decreased the amount of lower molecular weight material (mw between 100 and 300) and increased the higher molecular weight material (mw between 300 and 10,000). The weight-average molecular weight of the aromatic fraction increased from 530 to 990. Figure 4 showed that the viscosity of the bio-oil could be used to predict the weight average molecular weight during aging, after calibration for a specific bio-oil.

Analysis for water content using Karl Fischer titration with Hydra-Point Comp 5 showed an increase in water content of 1.6%, from 16.1 % to 17.7% during aging at 60°C and 90°C. Analysis by FTIR of the bio-oils revealed an increase in water, an increase in carbonyls, and ethers and/or esters after aging. The carbonyls may have been the product of oxidation (or possibly from the hydrolysis of acetals to form aldehydes and alcohols). It appeared that esterification or etherification were a probable mechanism of condensation during storage. (Czernik *et al* 1994)(Czernik 1994) The formation of additional aldehyde hydrates or an equilibrium shift from acetals to hemiacetals may have kept the water content from increasing more.

Analysis by gas chromatography with mass-spectrometric detection (GC/MS) of bio-oil made from poplar at NREL revealed that the levels of hydroxyacetaldehyde, propenal (acrolein), 2-methoxy-4-propenylphenol, and 5-methylfurfural quickly decreased during aging in both neat bio-oil samples and methanol/oil samples. Adding methanol to bio-oil prior to aging showed a decrease in acetic acid and an increase in methyl acetate after aging at 90°C, verifying that the esterification reaction is active in bio-oils. Analysis by nuclear magnetic resonance (NMR) showed the increase in ether and ester bonds during aging. (Diebold and Czernik 1997).

Room temperature aging data for an oil made from mixed hardwoods for a 12 month period were presented, but with no speculation on chemical mechanism. The viscosity reached an unusual asymptote after ten months of aging. (Bakhshi and Adjaye 1994) The leveling off of the viscosity may have been due achieving thermodynamic equilibrium or to the formation of micelles and phase separation.

Esterification and acetalization

The addition of alcohol and a mineral acid catalyst to pyrolysis tars and bio-oils resulted in the formation of esters, hemiacetals, and acetals, after reaction at room temperature for two to five hours. Alcohols used were methanol, ethanol, and propanol. By using molecular sieves to selectively dehydrate the reaction mixture, the thermodynamic equilibrium was shifted to heavily favor the esters and acetals. Because the water was removed from the reaction mixture, the heating value of the remaining oil product was increased. With an aged sample of hardwood bio-oil, 0.16 kg ethanol was chemically bound per kg of organics in the bio-oil. (Radlein *et al* 1995) No viscosity information was given and no comparisons were made of the reaction rate with mineral acid catalyst to the non-catalyzed rate.

Some bio-oil researchers have reported very carefully made mass balances in which the condensing equipment was washed with methanol or acetone to recover all of the bio-oil present. The bio-oils were then heated under vacuum to distill off the solvent, resulting in very high recovered bio-oil yields. It was assumed that the solvent did not react with the bio-oil. (Palm *et al* 1993)(Scott and Piskorz 1984) Based on the above discussion of bio-oil composition and chemistry, these solvents may have been reacting with the bio-oil to form ketals or acetals and esters, with a probable loss of the volatile products and an increase in yield of the less volatile products.

Hydrogenation

A low temperature, mild catalytic hydrogenation was performed on bio-oil from Georgia Tech's entrained flow pyrolysis reactor. The bio-oil originally contained 29 wt% moisture, 38% oxygen (dry basis), and had a viscosity of 10 cP at 60°C. The intent was to merely stabilize the oil, prior to more extensive hydrogenation. The hydrogenation was carried out at 274°C and 14 Mpa (2025 psia) over a CoMo catalyst. A total of 60 L of hydrogen was consumed per liter of bio-oil, resulting in a 69% by volume conversion to a tar having a viscosity of 14,200 cP measured at 60°C. Although a small amount of hydrogen was consumed, there was a significant amount of carbon dioxide formed (a 7% conversion of the carbon to gas) and 34% by volume yield of an aqueous phase containing 8.2% of the carbon. The oxygen content of the bio-oil had been reduced from 38% to 15% in the hydrotreated product, on a dry basis, with a 31% loss in volume and an increase in viscosity by a factor of 1420. (Baker and Elliott 1988)

Later, more extensive experimentation was performed with mild hydrogenation at 150°C to 450°C in a two-stage reactor with sulfided NiMo and CoMo catalysts at pressures less than 21

Mpa (3000 psi). This hydrogenation removed about 50% of the oxygen in bio-oil, but also resulted in a viscosity increase of about three orders of magnitude to make a tar-like product. In addition, there was an aqueous phase after hydrogenation that contained significant amounts of organic compounds. Only after more extensive hydrogenation in the second stage reactor, that reduced the oxygen content to less than 10%, was the viscosity of the product oil lower than the original bio-oil (Elliott and Neuenschwander 1997). It appears that much of the initial oxygen reduction in the tar-like product was due to the extraction of the polar oxygenated organic compounds by the aqueous phase from the tar phase, removing with their viscosity reducing (solvent) properties from the organic phase.

A bio-oil having 40.7% oxygen (dry basis) was hydrogenated at 15 Mpa (2200 psi pressure), 275°C, with a sulfided NiMo catalyst. Aging the hydrogenated bio-oil, having 23% residual oxygen and a nearly neutral pH of 6.5, showed good viscosity stability after 24 hours at 100°C. However, the viscosity of the hydrogenated bio-oil was about 1000 times higher than the aged, non-hydrogenated bio-oil (Conti *et al* 1996 and 1997).

These high-pressure studies were not promising that a very mild hydrogenation of bio-oil to just saturate reactive unsaturated olefinic molecules can be easily done, without drastically increasing the viscosity.

Hydrotreating a vacuum-pyrolysis oil under mild conditions, the molecular weight of the oil increased by thermal polymerization and decreased by hydrogenolysis by varying degrees, depending upon the reaction conditions. The thermal polymerization was speculated to be due to reaction of aldehydes with carbohydrates, catalyzed by the carboxylic acids present. The best conditions for stabilization of the bio-oil for subsequent hydrodeoxygenation were found to be at the lowest temperature and pressure investigated, 80°C and 600 psig for 120 minutes with a ruthenium catalyst. Under these conditions the average molecular weight decreased from 1775 Da to 1040 Da, with very little loss of oxygen and a 10% increase in hydrogen content. It was thought that reactive aldehyde groups in sugars were being reduced to their corresponding alcohols. (Gagnon and Kaliaguine 1988) No speculation was made concerning possible hydrogenation of olefinic compounds. Viscosity and aging characteristics were not reported.

Milder hydrogenation conditions have also been investigated at pressures from a few mbar to 100 bar, temperatures between 20°C and 200°C, and with various catalysts: ruthenium, copper, copper-chromite, palladium, platinum, cobalt-molybdenum, nickel, and nickel oxide. The best conditions for hydrotreating was with fresh bio-oils, dry Raney nickel catalyst, 200 to 300 kPa (2-3 bar) hydrogen pressure, and 80°C. Most of the ketones and aldehydes were reduced to alcohols, but acetic acid was unchanged under these mild conditions. No mention was made of the fate of the unsaturates in the bio-oil, e.g., propenyl phenolics. (Meier et al 1997) The viscosity of these mildly hydrogenated bio-oils was said to be very high (Meier 1999).

Mild hydrogenation to saturate the olefins and to convert acids and esters to the corresponding alcohols is being developed. A catalyst based on CoMo on a carbon support gave some conversion of ethyl decanoate to ethanol and decanol. (Maggi *et al* 1996) No results were given for the hydrogenation of bio-oils.

Polymerization with formaldehyde

Starting with bio-oils made from redwood or preferably pine, a series of extractions recovered a fraction very high in phenolics and "neutrals" (P/N). The bio-oil was first extracted with ethyl acetate, and the extract treated with an aqueous solution of sodium bicarbonate to extract the organic acids into the aqueous phase. The solvent remaining in the organic solution was evaporated to recover the P/N material. The P/N material was used as an inexpensive replacement for phenol in acid-catalyzed phenol/formaldehyde novolak resins. With a mixture of 50% phenol and 50% P/N, the amount of formaldehyde required to make a good novolak resin was reduced by a third, reflecting the oligomeric nature of P/N and its aldehyde content. In the presence of about 2% HCl, the novolak reaction was completed in about one hour at 90 to 95°C. (Chum and Kriebich 1991)

The use of up to 20% of a pine bio-oil in a urea-formaldehyde resin has been reported for use in making particle board, with no loss in strength. (Meier et al 1998) No mention was made of any partitioning of the bio-oil into different fractions prior to being used in the formulation.

Air oxidation

The effect of the small amount of air in the headspace above a hardwood bio-oil while aging has been investigated. Eight glass containers were 90% filled with a bio-oil, leaving a headspace equal to 10% of the volume. If all of the headspace were filled with air, the maximum initial amount of oxygen present in the headspace would have been 0.025 wt% (250 ppm) of the bio-oil. This represents the upper limit of air, because some of this headspace would be filled by volatiles from the bio-oil. In half of the containers, no attempt was made to remove the air above the bio-oil. In the other half of the containers, the headspace was purged with nitrogen. The tightly closed samples were then exposed to accelerated aging at 80°C for 25 hours. No significant differences were observed in the viscosity increase nor in the water content for the two sets of bio-oils. (Oasmaa et al 1997) Further testing with and without stirring of closed containers showed no significant differences (Oasmaa 1999). From these data it appears that it is not necessary to purge the air from the headspace of a nearly full, closed container during storage.

Effect of entrained char

Many of the reaction mechanisms in the proceeding discussion were catalyzed by the inorganic content of the char, e.g. calcium, magnesium, aluminum, zinc, etc. The presence of char added to bio-oil greatly increased its rate of viscosity increase (Agblevor et al 1994 and 1995). It has also been reported that previously filtered bio-oil that had been aged, needed to be filtered again after aging due to the apparent growth or agglomeration of tar particles (Shihadeh 1998), possibly using char particles as catalytic condensation sites.

Off-gassing during storage

Aging pyrolysis oils results in the evolution of permanent gases. During severe accelerated aging tests of bio-oil at 90°C, gas pressure built up in the closed containers. The gases saturated the bio-oil at the autogenous pressure. After removing the bio-oil samples from the storage oven, cooling, and weighing, the sample vials were opened very slowly to prevent the foaming over of the bio-oil. This was analogous to opening a bottle of a warm carbonated beverage, after being shaken. The gases were not analyzed. (Diebold and Czernik 1997)

The headspace of a barrel of pyrolysis oil was sampled after six months of closed storage at ambient conditions that existed in shipping and while waiting in customs. A positive pressure had built up in the barrel. Analysis by gas chromatography (GC) revealed the gas contained 29 vol% CO_2, 1% CO, 1% CH_4, 61% N_2, and 6% O_2 for a lack of closure of 2%. (Peacocke et al 1994) (The 2% missing material puts an upper limit on the possible amount of the unreported organic vapors present.) Assuming that all of the nitrogen was originally present as air in the headspace, 10 percentage points of O_2 was missing from the headspace gas. This missing oxygen could have been responsible for only a fraction (10 percentage points) of the 29 percentage points of CO_2 present. This small weight of oxygen actually missing could easily have been absorbed by the relatively large weight of bio-oil present in the barrel.

During the repeated operation of a viscometer, which utilized a timed, falling gold-plated ball, anomalies were observed after an induction period with a pine bio-oil at 40°C and higher temperatures. These anomalies were shown to have been caused by gas or vapor bubbles adhering to the ball, which gave the ball an increased buoyancy and longer falling times. By periodically degassing the sample, the minimum time for the ball to fall was ascertained and assumed to relate to the true viscosity of the bio-oil. The bubbles appeared to form even after 2½ hours, which was interpreted as being indicative of chemical reactions occurring. No analyses were made of the bubbles, but after six weeks of storage of one of the bio-oils at ambient temperatures, the gases in the headspace were found to have equimolar amounts of carbon dioxide and carbon monoxide. (Radovanovic *et al* 1997) The normal boiling point of methyl formate is 32°C and methylal is 42°C, so they could have been a significant part of the vapors in the bubbles forming at 40°C and higher.

Gas or vapor bubbles were also observed while using capillary tubes to determine viscosity. However, these bubbles were easily removed by preheating the fluid at temperatures above ambient. Viscosity measurements were successfully made with hardwood bio-oil at temperatures up to 80°C. (Oasmaa *et al* 1997)

Chemical reactions that would have produced carbon monoxide or methane in the head space are not listed in the model reactions considered in the previous section of this review. The chemical reactions identified that could produce carbon dioxide result in a products having lower molecular weights than the reactants, i.e., they are not polymerization reactions. This consideration appears to de-couple gas forming reactions from those identified as molecule building or polymerization reactions. However, since the decarboxylation of ferulic acid appears to be involve free radicals, olefinic polymerization maybe a byproduct.

However, if bio-oil does produce gases during storage, as indicated in the aging tests, then venting and proper disposal of these gases will be necessary. These gases (Peacocke *et al* 1994) have a molecular weight of over 33, making them 13% denser than air. This means that they would not disperse readily after venting and would tend to collect in low lying areas. The gases could cause asphyxiation and contain sufficient carbon monoxide to be deadly even after some dilution, but their heating value is too low for them to be considered a fuel. One disposal method would be to oxidize the off-gases using automotive catalytic muffler technology. Depending upon the amount of organic vapors in the off-gases and relative heat losses, supplemental fuel might have to be added for catalytic oxidation, because the adiabatic combustion temperature of these permanent off-gases would be only about 320°C.

METHODS TO SLOW AGING IN BIO-OILS

Solvent Addition To Reduce Viscosity And Aging Rates

It was recommended some time ago to add low viscosity water, methanol, or acetone to combat increases in viscosity after aging of updraft gasifier oils (Polk and Phingbodhippakkiya 1981). This was apparently to dilute the polymers being formed. No mention of chemical reactions with the solvents taking place was made, nor was any data presented to demonstrate the effectiveness of these solvent additions. Other early work with updraft gasifier oils used methanol to reduce the viscosity, with esterification thought to be involved during storage (Capener and Low 1982). Adding 10% furfural to a very thick pyrolysis oil, reduced its viscosity from 7000 cSt to 767 cSt (Salvi and Salvi, Jr. 1991).

The addition of water and methanol to a hardwood bio-oil was systematically investigated for their impact on viscosity. Water contents up to 30 wt% and methanol contents up to 10 wt%

were investigated and both were found to reduce the viscosity of the mixtures, with methanol slightly more effective. The viscosity at 25°C was lowered from 76 cP with 20 wt% water and down to 12 cP with 30% water and 10% methanol. (Diebold *et al* 1996)

Similarly, the viscosity at 25°C of a hardwood bio-oil was reduced from 1127 cP to 199 cP as the moisture content increased from 17.4 wt% to 30 wt%. More importantly, the rate of viscosity increase after aging four months at room temperature decreased from 3.3 cP/day with 20 wt% water, to 0.9 cP/day with 25% water, to about 0.05 cP/day at 30% water. Thus, it appeared that the presence of water at 25 wt% to 30 wt% was stabilizing the bio-oil from aging effects. (Tiplady *et al* 1996)

The addition of 2 wt%, 5%, 10%, and 20% of ethanol to a hardwood bio-oil was investigated. The initial viscosity of the bio-oil at 50°C was decreased from 50 cSt with no ethanol to 10 cSt with 20% ethanol (cSt = the viscosity in cP divided by the density in g/cc). After aging at 20°C for up to 4 months, Table 5 shows that the aging rate decreased an order of magnitude from 0.12 cSt/day with no ethanol to a barely perceptible value of about 0.01 cSt/day with 20% ethanol. After aging at 50°C for up to 7 days, the viscosity increase of the neat bio-oil was 3½ cSt/day compared to 0.4 cSt/day for a mixture of 5 wt% ethanol in bio-oil, for a reduction in the aging rate by a factor of 0.12. (Oasmaa *et al* 1997)

The stabilization of a hardwood bio-oil by the addition of up to 10 wt% of methanol, ethanol, acetone, ethyl acetate, a 1/1 mixture of acetone and methanol, or a 1/1 mixture of methanol and methyl *iso*-butyl ketone was investigated. The initial viscosities at 40°C were decreased from 30 cP for the neat bio-oil to about 15 cP for the bio-oil/solvent mixtures. These experiments used accelerated aging at 90°C to demonstrate the ability to preheat the stabilized bio-oil prior to combustion and to reduce the amount of time required for aging studies. The viscosities for the aged neat bio-oil and the bio-oil/solvent mixtures were found to vary quite linearly with time. Table 5 shows the dramatic impact these solvents had on slowing of the rate of aging or viscosity increase. Methanol was found to be the cheapest and the most effective organic solvent on a weight basis to reduce the rate of viscosity increase at 90°C; the aging rate was reduced by a factor of 0.057 with 10% methanol, compared to the neat bio-oil aged at the same temperature. (Diebold and Czernik 1997)

Table 5 shows that the beneficial effect (on aging) of adding solvents to bio-oil to decrease the rate of aging is progressively more effective at higher storage temperatures. Bio-oils from both hardwoods and softwoods benefit in the reduction of the aging rates with added solvents.

With 0% and 5 wt% methanol in the bio-oil, the increase in viscosity with time was decreased slightly by increasing the water content from 20% to 30%. With 10% methanol, the aging rate was independent of water content over this same range. It was concluded that the effect of the solvents was more than would be expected from a physical dilution. (Diebold and Czernik 1997)

Adding methanol shortly after the production of a hardwood bio-oil was significantly better than adding it after aging at 90°C for 20.5 hours. It appears that not all of the aging reactions are in thermodynamic equilibrium and some reactions cannot be reversed by adding low molecular weight reactants. However, dramatic decreases in viscosity were attained by the addition of methanol to previously aged oil. (Diebold and Czernik 1997)

Similarly, adding 5% methanol to a fresh softwood bio-oil, which was then aged at room temperature for three months had a viscosity increase of only 15%. Adding the same amount of methanol after the same aging resulted in a viscosity increase of 28%. (Oasmaa 1999)

From a molar perspective, ethyl acetate and acetone were twice as effective as methanol in lowering the aging rate of a hardwood bio-oil. It was thought that the solvents were probably reacting to form low molecular weight products, at the expense of polymerization reactions. For example, ethyl acetate could hydrolyze to form ethyl alcohol and acetic acid; acetone can react under acidic conditions to form cyclic ketals with two adjacent hydroxyl groups on a polyhydroxyl organic molecule. It was shown that the cost of adding methanol to the bio-oil is at least partially compensated by the increased energy content of the mixture. (Diebold and Czernik 1997)

The phase stability of bio-oils during storage is increased by the addition of the proper solvents. Between 2 and 10% methanol was added to bio-oils that had previously phase separated to result in homogenous liquids. (Oasmaa 1999)

Table 5   Effect of Adding Solvents on Aging Rates

Reference	Bio-Oil Sample	Aging Temp °C	Aging Rate	Aging Rate Relative to Neat Bio-Oil
Oasmaa *et al* 1997	VTT Batch 10/95 (Oak+Maple)	20	0.12 †	1.0
"	Batch 10/95 + 5% Ethanol	20	0.04†	0.3
"	Batch 10/95 + 20% Ethanol	20	0.008†	0.07
Oasmaa 1999	AFBP 10/2 (Hot Gas Filtered Pine)	20	0.15†	1.0
"	AFBP 10/2 + 10% MeOH	20	0.029†	0.19
"	PDU8 (Pine)	20	0.19*	1.0
"	PDU8 + 2% MeOH	20	0.065*	0.34
"	PDU8 + 5% MeOH	20	0.062*	0.33
"	PDU8 + 10% MeOH	20	0.016*	0.083
Oasmaa *et al* 1997	VTT batch 10/95	50	3.6†	1.0
"	Batch 10/95 + 5% Ethanol	50	0.43†	0.12
Diebold+ Czernik 1997	NREL Run 175 (Poplar)	90	60*	1.0
	Run 175+ 5% MeOH	90	12*	0.20
"	Run 175+10% Ethyl Acetate	90	8.6*	0.14
"	Run 175+5%MiBK+5% MeOH	90	6.0*	0.10
"	Run 175+10% Ethanol	90	5.3*	0.091
"	Run 175+5% MeOH+5% Acetone	90	4.8*	0.080
"	Run 175+10% Acetone	90	4.6*	0.077
"	Run 175+10% MeOH	90	3.4*	0.057

MeOH is methanol; MiBK is methyl *iso*-butyl ketone
† cSt/day measured at 50°C
* cP/day measured at 40°C

The efficient collection of volatile components during the production of the bio-oil, will result in a bio-oil with more low molecular weight components having lower viscosity, better solvency properties, and possibly better storage properties. Because the volatiles will not condense until their vapor pressure equals their partial pressure, these volatiles are collected more efficiently with less carrier gas present to dilute them.

Hemiacetal, acetal, and ester linkages could be important in the aging process of bio-oils. Because they react toward thermodynamic equilibrium, the addition of a low molecular weight alcohol to a mixture of high molecular weight hemiacetals, acetals, and esters would shift the equilibrium composition to a mixture having a lower molecular weight and viscosity. This could be important in the stabilization of bio-oil by four mechanisms: 1) the reduction in the concentration of reactive aldehydes present by converting more of them to less reactive, relatively low to moderate molecular-weight hemiacetals and acetals; 2) the transacetalization of large hemiacetals and acetals to form lower molecular weight hemiacetals and acetals; 3) conversion of organic acids to low molecular weight esters, and 4) transesterification of large esters to form lower molecular weight esters. The second and fourth mechanisms could be considered depolymerization of the original high molecular-weight hemiacetals, acetals, and esters. An example of the fourth mechanism is the reaction of canola oil (a high molecular weight triglyceride) with methanol to form lower molecular weight mono methyl esters (bio-diesel) and glycerol.

Mild Hydrogenation

From a cost perspective, it would be ideal to mildly hydrogenate the bio-oil to only saturate the reactive aliphatic unsaturated compounds, e.g., the propenyl group of 2-methoxy-4-propenyl phenol (eugenol). However, it is well known that the hydrogenation of unsaturated vegetable oils creates a viscous greasy semi-solid material used for imitation lard and butter. Thus, saturating the olefins present in bio-oil might be expected to increase the viscosity of the product oil.

Wood distillates from an updraft gasifier were hydrogenated with a catalyst of Pd on carbon at hydrogen pressures less than 405 kPa (four atmospheres). This mild hydrogenation eliminated 4-hydroxy-3-methoxy styrene, and reduced the content of 2-methoxy-4-allyl phenol (isoeugenol) and 2-methoxy-4-propenyl phenol (eugenol). The actual viscosity of the hydrotreated oil was not given. The hydrogenated oil was apparently quite viscous, as it was diluted with m-cresol, and then aged at room temperature. After 8 months of aging, a dilute solution of 20% of the as-produced oil in m-cresol increased in viscosity by 72%, whereas the viscosity of 20% hydrogenated oil in m-cresol increased by a factor of only 21%. Thus, the viscosity increase was reduced by a factor of 0.29 by the hydrogenation, strongly implying that olefin condensation was one of several mechanisms involved with aging. (Polk and Phingbodhippakkiya 1981)

At this time, it is speculated that a very mild hydrogenation of bio-oil could eliminate olefinic condensation as one of the several aging mechanisms. However, if this results in a large increase in viscosity, as it apparently has thus far, this processing step would not be advantageous. Future hydrogenation experimentation in this area needs to be directed toward minimizing this viscosity increase during very mild hydrogenation.

Limiting Access to Air and Antioxidants

Although the small amount of air in the head-space of a sample did not appear to catalyze aging reactions, the result could have been different with more air available to the bio-oil. For example, vigorous mixing of the bio-oil in an open container prior to sampling could entrain air into the bio-oil to result in the saturation of that bio-oil with air, even if allowed to sit until the entrained bubbles rise to the surface. This increased level of oxygen in the bio-oil could produce enough peroxides to catalyze the polymerization of the olefins. It would be prudent to limit the access of air to the bio-oil to minimize the possibility of forming peroxides that catalyze olefinic polymerization.

Many phenolics are known to be good free-radical traps and are used to protect olefins from the polymerizing effect of free-radicals. However, based on the reported disappearance of olefinic compounds in bio-oils and wood smoke, it appears that bio-oils do not contain enough free-radical trap molecules for stabilization.

The phenolics deliberately used as free-radical traps have the ability to form free-radicals that are stabilized by resonance, thus stopping the chain reaction of polymerization. Commercially, hydroquinone is used as a free-radical trap to stabilize acrolein during storage and shipping. Very small quantities of hydroquinone at 0.10 to 0.25 wt% are used for this purpose in the presence of acetic acid at a pH of 6. (Etzkorn *et al* 1991) It may be more cost effective to add a small quantity of an anti-oxidant, e.g., hydroquinone, to stabilize the olefins in bio-oil, rather than to use mild hydrogenation to saturate the olefins.

PHYSICAL MECHANISMS OF PHASE INSTABILITY
Co-Solvency of Bio-Oil Components

The mutual solubility of a mixture of compounds is related to their chemical structures and their relative amounts. In general, compounds having lower molecular weights are more soluble than higher molecular weights. Compounds with similar polarities or similar molecular structures tend to be soluble in each other. Examples of highly polar materials are water, alcohols, and organic acids. Less polar materials include esters, ethers, and phenolics. Non-polar materials include hexane and other hydrocarbons. Compounds that favor mutual solubility between compounds with different polarities have both a polar and a non-polar aspect, e.g. a long hydrocarbon chain (low polarity) with an alcohol group (high polarity) at the end, such as *n*-butanol. In general, bio-oils containing more oxygen on a dry basis can keep more water in solution than bio-oils with lower oxygen contents. Some of that water appears to be in the form of aldehyde hydrates, but much of it is probably just hydrogen bonded to polar organic functionalities, e.g., carboxylic acids and alcohols.

Because a higher amount of hydrogen bonding of a compound increases its heat of vaporization, this property is used to define the solubility parameter, δ:

$$\delta = [(\Delta H_v - RT) / V]^{0.5}$$

where ΔH_v is the molar heat of vaporization in J/kmol, R is the gas constant (8314 J/kmol/K), T is the temperature in Kelvin, and V the molar volume (m^3/kmol). This solubility parameter is also called the Hildebrand parameter and was developed for use with non-polar materials. Materials with similar solubility parameters tend to be soluble in each other. (Barton 1983)

For multi-functional component mixtures, it is useful to expand the solubility parameter to include a better description of the organic compounds involved. It is assumed that the solubility is a combination of contributions from the non-polar or dispersion interactions, the polar interactions, and the hydrogen bonding interactions of the molecules:

$$\delta_t^2 = \delta_d^2 + \delta_p^2 + \delta_h^2$$

where δ_t is the total solubility parameter, δ_d is the dispersive solubility parameter, δ_p is the polar solubility parameter, and δ_h is the hydrogen bonding solubility parameter. These are called Hansen solubility parameters. The Hansen total solubility parameter is qualitatively similar to the Hildebrand parameter. (Barton 1983)

The Hansen parameters are used to predict the solubility of solutes (j) and solvents (i). If the Hansen parameters for the solvent fall within the "spherical volume" of the solute, then solubility is predicted. This is satisfied, if the distance from the center of the solvent sphere to the center of the solute sphere (R_{ij}) is smaller than the radius of the solute sphere (R_j):

$$R_{ij} = [4(\delta_{di} - \delta_{dj})^2 + (\delta_{pi} - \delta_{pj})^2 + (\delta_{hi} - \delta_{hj})^2]^{0.5}$$

$$R_{ij} < R_j$$

Tables 6 and 7 show the Hansen solubility parameters for 37 solvents and 11 polymeric materials (Barton 1983) commonly reported to be in bio-oils or thought to be similar to some compounds in bio-oils. Water (when mixed with organic compounds) has the highest δ_p at 17.8 MPa$^{0.5}$, which contributes along with a high δ_h of 17.6 MPa$^{0.5}$ and a high δ_d of 19.5 MPa$^{0.5}$ to the highest δ_t shown on the table of 31.7 MPa$^{0.5}$. In contrast, consider n-hexane with a δ_p and a δ_h of zero, which results in a low δ_t of 14.9 MPa. In between these two extremes are the organic compounds reported to be found in bio-oil. The polymer solubility radius, R_j, is based the observed solubility of 0.5 g of polymer in 5 cm^3 of solvent. (Barton 1983)

To predict the solubility of a polymer in a mixture of various solvents, each solubility parameter (δ_d, δ_p, and δ_h) for each solvent is multiplied times its volume fraction (vol.fr.$_i$) and added, e.g.,

$$\delta_{dm} = \Sigma_i (\delta_{di} * \text{vol.fr.}_i)$$

where δ_{dm} is the dispersive solubility parameter for the solvent mixture, and δ_{di} is the dispersive solubility parameter for the ith solvent. (Barton 1983)

Table 6 Hansen Solubility Parameters (Barton 1983) for Solvents in Bio-Oil or of Potential Interest to Bio-Oil Producers

COMPOUND	δ_{di}, MPa$^{0.5}$	δ_{pi}, MPa$^{0.5}$	δ_{hi}, MPa$^{0.5}$	δ_{ti}, MPa$^{0.5}$
ALCOHOLS Methanol	15.1	12.3	22.3	29.6
Ethanol	15.8	8.8	19.4	26.5
1,2–ethanediol (ethylene glycol)	17.0	11.0	26.0	32.9
2-Propen-1-ol (allyl alcohol)	16.2	10.8	16.8	25.7
1-Propanol	16.0	6.8	17.4	24.5
2-Propanol	15.8	6.1	16.4	23.5
Propanediol	16.8	9.4	23.3	30.2
1,2,3 Propanetriol (glycerol)	17.4	12.1	29.3	36.1
1-Butanol	16.0	5.7	15.8	23.1
Furfuryl alcohol	17.4	7.6	15.1	24.3

Table 6   continued

COMPOUND	δ_{di}, MPa$^{0.5}$	δ_{pi}, MPa$^{0.5}$	δ_{hi}, MPa$^{0.5}$	δ_{ti}, MPa$^{0.5}$
ALDEHYDES Acetaldehyde	14.7	8.0	11.3	20.3
Butanal	14.7	5.3	7.0	17.1
Furfural	18.6	14.9	5.12	24.4
Benzaldehyde	19.4	7.4	5.3	21.5
ACIDS Formic	14.3	11.9	16.6	24.9
Acetic	14.5	8.0	13.5	21.4
1-Butanoic	14.9	4.1	10.6	18.8
Benzoic	18.2	7.0	9.8	21.8
ESTERS Methyl acetate	15.5	7.2	7.6	18.7
Ethyl formate	15.5	7.2	7.6	18.7
Ethyl acetate	15.8	5.3	7.2	18.1
Ethyl lactate	16.0	7.6	12.5	21.6
n-Butyl acetate	15.8	3.7	6.3	17.4
γ-Butyrolactone	19.0	16.6	7.4	26.3
Ethyl cinnamate	18.4	8.2	4.1	20.6
ETHERS				
Methylal	15.1	1.8	8.6	17.5
Furan	17.8	1.8	5.3	18.6
Methoxy benzene (anisol)	17.8	4.1	6.8	19.5
Dibenzel ether	17.4	3.7	7.4	19.3
KETONES				
Acetone	15.5	10.4	7.0	20.0
Methyl ethyl ketone	16.0	9.0	5.1	19.0
Methyl isobutyl ketone	15.3	6.1	4.1	17.0
PHENOLICS				
Phenol	18.0	5.9	14.9	24.1
1,3 Benzendiol (resorcinol)	18.0	8.4	21.1	29.0
3-methyl phenol (m-cresol)	18.0	5.1	12.9	22.7
o-Methoxy phenol	18.0	8.2	13.3	23.8
HYDROCARBONS				
n-Hexane	14.9	0.0	0.0	14.9
Benzene	18.4	0.0	2.0	18.6
Toluene	18.0	1.4	2.0	18.2
Styrene	18.6	1.0	4.1	19.0
Turpentine	16.4	1.4	0.4	16.5
Naphthalene	19.2	2.0	5.9	20.3
WATER WITH ORG. LIQUIDS	19.5	17.8	17.6	31.7

Table 7. Hansen Parameters for Polymers (Barton 1983) Possibly Relevant to bio-oils

Polymer	δ_{dj} MPa$^{-0.5}$	δ_{pj} MPa$^{-0.5}$	δ_{hj} MPa$^{-0.5}$	R_j MPa$^{-0.5}$
Cellulose Acetate (Cellidore® A. Bayer)	18.6	12.7	11.0	7.6
Coumarone-indene resin (Parlon® P-10, Hercules)	19.4	5.5	5.8	9.6
Epoxy (Epikote® 1001, Shell)	20.4	12.0	11.5	12.7
Ester gum (Ester gum BL, Hercules)	19.6	4.7	7.8	10.6
Furfuryl alcohol resin (Durez® 14383, Hooker Chem.)	21.2	13.6	12.8	13.7
Isoprene elasomer (Ceriflex® IR305, Shell)	16.6	1.4	-0.8	9.6
Phenolic resin, resole (Phenodur® 373U, Chemische Werke Albert)	19.7	11.6	14.6	12.7
Phenolic resin, pure (Super Beckacite® 1001, Reichhold)	23.3	6.6	8.3	19.8
Poly(ethyl methacrylate) (Lucite® 2042, Du Pont)	17.6	9.7	4.0	10.6
Polystyrene (Polystyrene LG, BASF)	21.3	5.8	4.3	12.7
Saturated polyester (Desmophen® 850, Bayer)	21.5	14.9	12.3	16.8
Terpene resin (Piccolyte® S-1000, Penn. Ind. Chem.)	16.5	0.4	2.8	8.6
Water (with organic liquids)	19.5	17.8	17.6	14.7

Changes In Mutual Solubility With Aging

During aging, chemical reactions change the polarity of the components of bio-oil. For example, esterification converts highly polar organic acid and alcohol molecules into esters having relatively lower polarity and extremely polar water. The formation of acetals shifts the composition away from acetaldehyde hydrates, releasing the water of hydration **and** the water formed with the acetal. Acetals are in the relatively non-polar family of ethers. Thus, the polarity of the organic material is decreased at the same time that the water content is increased. This increasing difference in polarity among the compounds in the aged bio-oil increases the tendency for phase separation. Although ethyl acetate was used to extract the phenolics and other aromatics from bio-oil (Chum and Black 1990)(Chum and Kreibich 1991), adding only 10% ethyl acetate to bio-oil allowed more severe aging to take place before phase separation than similar amounts of alcohols or ketones (Diebold and Czernik 1997).

With the formation of larger molecules during aging, their mutual solubility decreases. All of these things work to increase the probability that phase separation will occur into a light, highly polar aqueous phase and a less polar heavier organic phase. Additional lighter waxy phases and heavier sludge phases have also been reported to form during storage, as a consequence of the decrease in mutual solubility during aging (Diebold and Czernik 1997).

Micelles, Suspensions, And Emulsions

As the aging bio-oil becomes saturated with respect to large molecules having low polarity, it was postulated that there is a segregation of polar molecules from less polar molecules, with multi-functional molecules between to form micelles, analogous to petroleum fuel oils. (Radlein 1999) The low-polarity molecules can be attracted to the surface of suspended char particles, which act as condensation nuclei.

With micelle formation, the initial liquid phase separation is in the form of very small droplets having a very large surface area. The initial droplets aggregate and then coalesce into larger droplets having smaller surface area, until there are sufficient multi-polarity molecules to form a stable coating on the surface of the droplets. Depending upon the nature and quantity of the multi-polarity molecules present, the stabilization of the two-phase liquid mixture can occur before or after the droplets become large enough to plug filters or to phase separate. The action of the multi-polarity molecules is to stabilize the emulsion. Molecules that are particularly good at stabilizing emulsions are called emulsifiers. Small solid particles at the interface of the droplets can aid in stabilization of the emulsion. (Friberg and Jones 1994)

Emulsifiers are commonly ionic or non-ionic. The ionic emulsifiers would be exemplified by metal soaps or sulfur-containing detergents, neither of which would form desirable combustion products. Consequently, it is the non-ionic emulsifier that is of interest for bio-oil. Non-ionic emulsifying materials consist of linear block co-polymers, with each block having a different solubility characteristic. Each block of the co-polymer is long enough to form a loop or a tail into one of the phases, or to form a "train" at the interface. An example of such a co-polymer would be formed by alternating blocks of poly (ethylene glycol) and poly alkyl aryl ether. (Friberg and Jones 1994)

Only a very small fraction of bio-oil is soluble in typical diesel fuels. Mixtures of bio-oil in #2 diesel oil have been made that were stable emulsions for over 90 days. The emulsifiers were of the non-ionic type, with hydrophilic to lipophilic block (HLB) ratios between 4 and 18. The emulsifiers were derived from block co-polymers of fatty acids and polyoxyethylene glycol, or fatty acids, sorbitol, and polyoxyethylene glycol or polyethoxyethylene glycol with long aliphatic chains. Mixtures of up to 40% bio-oil were emulsified with diesel fuel, using a combination of two emulsifiers each at the 1% level. The emulsions were a clear phase, implying micelles that were too small to refract light (sub-micron). (Ikura *et al* 1998)

It appears feasible to extend the useful life of bio-oil, as an apparently single-phase liquid during aging, through the use of appropriate emulsifying agents. The use of emulsifiers, e.g., lipophilic-hydrophilic block co-polymers, may be advantageous to stabilize bio-oil as a stable emulsion during aging. If the resulting emulsion is aqueous-phase continuous, it would be expected to have a low viscosity similar to the aqueous phase.

Off-Gassing During Aging

Depending upon the production procedure, the headspace gases that evolve during aging could have been pyrolysis gases that were selectively absorbed by the bio-oil in the condensation train during production. For example, cold methanol is commercially used to absorb carbon dioxide from gas streams in the Rectisol process. At 10°C and atmospheric pressure, 5 volumes of carbon dioxide can be absorbed per volume of methanol (Kohl and Riesenfeld 1974), or a little over 1 wt% carbon dioxide. The pyrolysis gases would be desorbed with a rise in temperature or with a chemical change in the bio-oil that reduced their solubility. For example, the reaction of methanol to form esters and acetals could reduce the solubility of carbon dioxide in bio-oil. The desorbing carbon dioxide would tend to strip the volatile esters and acetals out of the bio-oil, as well as other absorbed gases.

COMPARISONS OF THE STORAGE INSTABILITY MECHANISMS OF BIO-OILS AND OF PETROLEUM OILS

Petroleum derived fuel oils consist principally of hydrocarbons. The hydrocarbons present include saturated aliphatics (paraffins), unsaturated aliphatics (olefins), saturated ring compounds (naphthenes), and unsaturated ring compounds (aromatics). These different hydrocarbon families have decreased solubility in each other, as their molecular weight increases. Many compounds in petroleum fuel oils have several of the families represented in a single large molecule. For example, a long aliphatic chain attached to an aromatic, e.g. heptyl benzene. This multi-functionality increases the mutual solubility of the hydrocarbons present. With the exception of the olefins, hydrocarbons are relatively stable at storage conditions. (Mushrush and Speight 1995)

Crude petroleum typically has very low levels of olefins. Refinery operations to lower the molecular weight and viscosity of petroleum include visbreaking, thermal cracking, coking, and catalytic cracking. These cracking reactions result in the formation of unsaturated or olefinic fragments of the original crude oil. (Meyers 1996)(Gary and Handwerk 1994)

Crude petroleum has varying amounts of elements other than carbon and hydrogen, depending upon the source. These other elements are principally oxygen, sulfur, nitrogen, vanadium, iron, and nickel. The metals are typically not volatile and are concentrated in the tar fractions used for residual fuel oils, asphalt, and petroleum coke. The other heteroatoms can be removed from the distillates by hydrogenation, a process that also saturates the olefins. Highly refined fuels such as gasoline and diesel fuel contain very small amounts of heteroatoms, e.g. oxygen, sulfur, and nitrogen. With heavier, less refined fuel oils, the relative amounts of heteroatoms increase. (Mushrush and Speight 1995)

The storage stability of hydrocarbon fuels is a function of their heteroatom and olefin contents. Many organic sulfur compounds have a tendency to dimerize. Oxygen and nitrogen compounds act as catalysts for the polymerization of olefins. This has been particularly troublesome in hydrocarbon fuels derived from oil shale and coal. The oxygen may have been in the original crude oil, or it may have been originally absorbed from the atmosphere. Once in the oil, dissolved oxygen can react with hydrocarbons to form carboxylic acids, or with olefins to form hydroperoxides and peroxides. It is the formation of the hydroperoxides that is the key to many instability problems in petroleum fuels. (Mushrush and Speight 1995)

The olefinic polymers formed during aging have a lower volatility and do not evaporate in carbureted gasoline fuel systems, leaving behind deposits called gums. Heteroatoms make up a disproportionately large amount of such deposits. In heavier fuels that are atomized rather than evaporated, the olefinic polymers increase the viscosity and difficulty of atomization. In extreme cases of polymerization during storage, the fuel becomes saturated with the polymers and sludge forms on the bottom of the storage tank. (Mushrush and Speight 1995)

As a fuel oil becomes saturated with various polymers having different solubilities, they form heterogeneous molecular aggregations or micelles. The nucleous of micelles contains the insoluble hydrocarbon family, typically an aromatic asphaltene. If there are sufficient multifunctional compounds (typically oxygenated aromatics, called resins) present to coat the nucleus of the micelle, a stable suspension of the colloidal micelles can form. In a complex mixture like fuel oils, there can be several layers of compounds having different mutual solubilities present in a micelle. If the micelles in a stable suspension are small enough to pass through filters, then their presence does not create operational problems to the fuel-oil user. (Friberg and Jones 1994)

Changes in temperature can destabilize the emulsion. Incompatibility, when two different fuel oils are mixed, can also destabilize the emulsions present. An unstable emulsion passes sequentially through micelle flocculation or aggregation, agglomeration into larger droplets, and gravitational separation, i.e., sludge formation. Incompatibility can also occur if two different fuel oils are mixed that result in the phase separation of one of the components. (Mushrush and Speight 1995)

Additives to increase the stability of petroleum fuel oils include emulsifying agents to maintain stable emulsions. Naturally present phenols in low concentrations act as anti-oxidants, but at higher levels they participate in oxidative coupling and sludge formation. Anti-oxidant additives like zinc dialkyldithiophosphate, aromatic amines, and alkylated phenols are used to stabilize lubricating oils. Sulfur compounds act as free-radical traps, but also participate in cross linking reactions. Excluding air from storage tanks is suggested to reduce oxidative aging effects. (Mushrush and Speight 1995)

SUMMARY

Bio-oil is an ill-defined mixture of water, char, and oxygenated organic compounds that include the following functional groups: organic acids; aldehydes; esters, acetals; hemiacetals; alcohols; olefins; aromatics; phenolics, proteins, and sulfur compounds. The actual composition of a bio-oil is a complex function of the feedstock, pyrolysis technique, char removal system, condensation system, and storage conditions employed.

These various organic compounds have the ability to react during storage to produce oligomers and polymers that have an increased viscosity and reduced solubility in the bio-oil. In particular, there is evidence to suggest that during storage of bio-oil:
a)   acids and alcohols react to form esters and water
b)   aldehydes react with water to form hydrates
c)   aldehydes react with alcohols to form hemiacetals, acetals (ethers) and water
d)   aldehydes react to form oligomers and resins
e)   aldehydes react with phenolics in the acidic bio-oil to form novolak resins and water
f)   aldehydes react with proteins to form dimers
g)   mercaptans react to form dimers
h)   olefins polymerize to form oligomers and polymers, and
i)   atmospheric oxygen reacts with many of the organics present to form peroxides which catalyze the polymerization of olefins and the addition of mercaptans to olefins.
The organic acids and the elements commonly found in the char can act as catalysts for many of these reactions. Bio-oils with lower char contents tend to have lower aging rates.

Methods found to stabilize the potential reactions within bio-oil have been the addition of solvents and hydrogenation. The addition of inexpensive methanol or ethanol has been found to stabilize bio-oil made from both softwoods and hardwoods. Compared to the neat bio-oil, the rate of viscosity increase was decreased by a factor of 0.057 with 10% methanol, making the mixtures relatively stable even at elevated temperatures. The small residual aging effects are speculated to be due to primarily to olefinic-addition reactions. These residual aging effects will be important only for long term storage conditions and would be minimized with the proper anti-oxidant or stabilizer.

Unfortunately, complete hydrogenation to remove oxygen is relatively expensive. Mild hydrogenation to saturate the olefins and to convert aldehydes to alcohols is expected to be less expensive, but has been reported to cause phase separation and to drastically increase the viscosity of the organic phase. The use of anti-oxidants to prevent the olefinic and sulfur-based polymerization reactions may be more cost effective than hydrogenation to stabilize bio-oil.

The small amount of air in the headspace of storage containers 90% full does not appear to contain enough oxygen to affect the aging rate of bio-oils. However, if mixing is done in containers open to the atmosphere, bio-oil may absorb enough oxygen to form organic peroxides in sufficient quantity that could affect aging rates. Preventing the exposure of bio-oil to oxygen in the atmosphere is a prudent action to take. Adding a small amount of anti-oxidant may be a good precautionary measure to take to prevent the effects of oxygen, e.g., oxygen that may infiltrate the storage tank as it "breathes" air during its daily cycle of expansion and contraction due to heating and cooling.

Gas evolution during aging has been reported by several research groups. It is not clear at this time how much of this gas was pyrolysis gases that had been adsorbed by the bio-oil and how much of it was actually produced by aging reactions. Changes in gas solubility in the bio-oil can be due to temperature changes and to changes in the composition of bio-oil. It is improbable that the chemical reactions that produce carbon dioxide are responsible for the increase in molecular weight and viscosity observed during aging.

The concepts used to describe mutual solubility, phase separation, and co-solvents for petroleum derived fuel oils appear to be qualitatively useful to describe bio-oil. In particular, the concepts of micelles and emulsions are useful to understand phase separation in bio-oils. It is probable that with some research, non-ionic block polymers can be identified that can be used as additives to make stable emulsions of bio-oil, as it ages and phase separates. Ideally, these stable emulsions would have the low viscosity of the aqueous phase with the viscous tars in suspension (aqueous continuous).

CONCLUSIONS AND RECOMMENDATIONS

It has been demonstrated that the addition of low-molecular-weight solvents at the 10% level has a remarkably stabilizing effect on bio-oils made from both hardwoods and softwoods. Adding reactive solvents shifts the stoichiometry away from polymers to oligomers capped with the reacted solvent group. Because pyrolysis is known to produce many of these volatile solvents in small quantities, it would appear to be critical to the storage stability of the bio-oil that the volatiles be recovered in the condensation train and not lost in the pyrolysis gases used for process fuel. The formation of esters in the vapor space of hot spray towers should be investigated to determine if volatile ester solvents are being formed and lost in the pyrolysis gases.

In addition, there may be pyrolysis conditions that can be optimized to produce more of the beneficial solvents, e.g., acetals. The goal of future research in pyrolysis should be to produce a bio-oil having a low viscosity, a high energy conversion (based on the "lower heating value"), and a low aging rate. Trade-off studies could be made concerning the purchase of solvent and any losses in energy conversion required to produce a naturally stable bio-oil having a low viscosity. However, if the volatile content of the bio-oil is increased significantly, parameters like the flash point of the bio-oil should be checked to see if any safety issues have been compromised.

With 10% methanol in bio-oil, the viscosity increased only from 20 to 22 cP over a 4-month period when stored at 20°C. This would extrapolate to a viscosity of 30 cP after storage for 12 months. Ethanol at 20% had a similar stabilizing effect. (Oasmaa et al 1997)(Oasmaa 1999) With 10% methanol added to bio-oil, the viscosity at 40°C rose from about 13 cP to an interpolated 15 cP after preheating for 12 hours at 90°C, e.g., to reduce the viscosity for ease of atomization. (Diebold and Czernik 1997). This demonstrated stability needs improvement only for extreme storage situations of elevated temperature, extreme preheating environments, or extended periods of time. Because biomass harvesting is an annual or more often event, in many cases it may not be necessary to have more stability in storage than that obtained by the simple addition of about 10% to 20% of an alcohol.

Because the addition of solvents is so effective in slowing the apparent aging reactions, it is important to learn more about the mechanisms involved. There may be synergistic effects with combinations of solvents to maximize the co-solubility aspects of the bio-oil solution. Some modeling with the Hansen parameters could be well worthwhile, prior to actual aging experiments.

To attain more stability in storage and in severe preheating situations, it may be desirable to use hydrogenation. However, even mild hydrogenation can cause phase separation and an increase in the viscosity of the organic phase. The low pressure, very mild hydrogenation of bio-oils to only saturate the olefins present should continue to be investigated. Optimizing the viscosity and storage stability of the resultant bio-oils would be the goal of this mild hydrogenation research, instead of maximizing the amount of oxygen removed. In addition, effort should be expended into the use of anti-oxidants to preclude the need for hydrogenation for stability.

Emulsifying agents need to be developed to prevent the settling out of tars from bio-oils during extreme aging. These emulsifiers probably will be non-ionic in nature and consist of block polymers, e.g., that alternate water-soluble (polar) blocks with phenolic-soluble (less polar) blocks.

As is the case with petroleum products, it will be advantageous to develop an "additives package" to address the many sources of bio-oil instability during storage. This is envisioned to include solvents to reduce polymerization by esterification, acetalization, and phenol/formaldehyde reactions, anti-oxidants to reduce olefin polymerization reactions, and emulsifiers to prevent phase separation problems. This additives package would most likely be a proprietary blend, with the ingredients being a trade secret. However, the effectiveness of the additives package would make a nice paper, with comparisons to neat bio-oil and to bio-oil with methanol added.

Because it appears that aldehyde hydrates are present in bio-oil, it would be instructive to know whether the Karl Fischer method analyses report the water present as hydrates or only the free water.

Research needs to determine the composition and relative quantity of gases and vapors produced during the storage of bio-oil made from both hardwoods and softwoods, including volatile oxygenates that have low flash point temperatures. Inadequate venting of a storage tank could lead to a build up of pressure that could rupture the tank catastrophically. Any gases vented will need to be deodorized and made non-toxic. The catalytic air-oxidation of these off-gases appears promising and needs to be investigated.

3 ACKNOWLEDGMENTS

The financial support to perform this review was provided equally by PyNe (managed by Prof. A.V. Bridgwater, Director of the Bio-Energy Research Group, Department of Chemical Engineering and Applied Chemistry, Aston University, Birmingham, UK) and by the Biomass Power Program (managed by Mr. Kevin Craig) at the National Renewable Energy Laboratory, Golden, CO of the U.S. Department of Energy with purchase order 165134 of June 3, 1999. This support is gratefully acknowledged.

The encouragement of Dr. Stefan Czernik of NREL, Mr. Jan Piskorz of Resource Transforms International, Dr. Dietrich Meier of the Institute of Wood Chemistry, and Ms. Anja Oasmaa of the Technical Research Centre (VTT) of Finland is also gratefully acknowledged.

REFERENCES

Acharya, A.S. and Manning, J.M. (1983) "Reaction of Glycoaldehyde with Proteins: Latent Crosslinking Potential of α-Hydroxyaldehydes," *Proc. Natl. Acad. Sci. USA*, 80, pp. 3590-3594.

Adams, E.W. and Adkins, H (1925) "Catalysis in Acetal Formation," *J. Am. Chem. Soc.*, 47, pp. 1358-1367.

Adkins, H. and Nissen, B.H. (1922) "A Study of Catalysis in the Preparation of Acetal," *J. Am. Chem. Soc.*, 44, pp. 2749-2755.

Adkins, H. and Adams, E.W. (1925) "The Relation of Structure, Affinity, and Reactivity in Acetal Formation," *J. Am. Chem. Soc.*, 47, pp. 1368-1381.

Adkins, H. and Broderick, A.E. (1928a) "Hemiacetal Formation and the Refractive Indices and Densities of Mixtures of Certain Alcohols and Aldehydes," *J. Am. Chem. Soc.*, 50, pp. 499-503.

Adkins, H, and Broderick, A.E. (1928b) "The Rate of Synthesis and Hydrolysis of Certain Acetals," *J. Am. Chem. Soc.*, 50, pp. 178-185.

Adkins, H.; Semb, J.; and Bolander, L.M. (1931) "Some Relationships of the Ratio of Reactants to the Extent of Conversion of Benzaldehyde and Furfuraldehyde to Their Acetals," *J. Am. Chem. Soc.*, 53, 1853.

Agblevor, F.A.; Besler, S.; and Evans, R.J. (1994) "Inorganic Compounds in Biomass Feedstocks: Their Role in Char Formation and Effect on the Quality of Fast Pyrolysis Oils," in *Proceedings of Biomass Pyrolysis Oil Properties and Combustion Meeting,* September 26-28, Estes Park, CO, T.A. Milne, ed. National Renewable Energy Laboratory, Golden, CO, NREL-CP-430-7215, pp. 77-89.

Agblevor, F.A.; Besler, S.; and Evans, R.J. (1995) "Influence of Inorganic Compounds on Char Formation and Quality of Fast Pyrolysis Oils," abstracts of the ACS 209th National Meeting, Anaheim, CA, April 2-5.

Andrews, R.G.; Zukowski, S.; and Patniak, P.C. (1997) "Feasibility of Firing and Industrial Gas Turbine Using a Bio-Mass Derived Fuel," in *Developments in Thermal Biomass Conversion*, A.V. Bridgwater and D.G.B. Boocock, eds., Blackie Academic and Professional, London, pp. 495-506.

Bahkshi, N.N. and Adjaye, J.D. (1994) "Properties and Characterization of Ensyn Bio-Oil," *Proceedings of Biomass Pyrolysis Oil. Properties and Combustion Meeting*, September 26-28, Estes Park, CO, T.A. Milne, ed., National Renewable Energy Laboratory, Golden, CO, NREL-CP-430-7215, pp. 54-66.

Baker, E.G. and Elliott, D.C. (1988) "Catalytic Hydrotreating of Biomass-Derived Oils," in *Pyrolysis Oils from Biomass. Producing, Analyzing and Upgrading*, ACS Symposium Series 376, American Chemical Society, Washington, D.C., pp. 228-240.

Barton, A.F.M. (1983) *CRC Handbook of Solubility Parameters and Other Cohesion Parameters*, CRC Press, Boca Raton, FL, pp.139-165.

Capener, E.L. and Low, J.M. (1982) "Method and Apparatus for Converting Solid Organic Material to Fuel Oil and Gas," U.S. Patent 4,344,770.

Carey, F.A. (1996) *Organic Chemistry, 3rd edition*, McGraw-Hill, NY.

Carraza, F.; Rezende, M.E.A.; Pasa, V.M.D.; and Lessa, A. (1994) "Fractionation of Wood Tar," in *Advances in Thermochemical Biomass Conversion*, A.V. Bridgwater, ed., Blackie Academic and Professional, London, pp.1465-1474.

Casanova Kindelan, J. (1994) "Comparative Study of Various Physical and Chemical Aspects of Pyrolysis Bio-Oils Versus Conventional Fuels, Regarding their Use in Engines," in *Proceedings of Biomass Pyrolysis Oil Properties and Combustion Meeting*, September 26-28, Estes Park, CO, T.A. Milne, ed., National Renewable Energy Laboratory, Golden, CO, NREL-CP-430-7215, pp. 343-354.

Cason, J. (1956) *Essential Principles of Organic Chemistry*, Prentice-Hall, Inc., Englewood Cliffs, NJ, p. 369.

Chum, H.L. and Black, S.K. (1990) "Process for Fractionating Fast-Pyrolysis Oils, and Products Derived Therefrom," U.S. Patent 4,942,269.

Chum, H.L. and Kriebich, R.E. (1991) "Novolak Resin from Fractionated Fast-Pyrolysis Oils," International Patent Application WO 91/09892.

Conti, L.; Scano, G.; Boufala, J.; and Mascia, S. (1997) "Bio-Crude Oil Hydrotreating in a Continuous Bench-Scale Plant," in *Developments in Thermal Biomass Conversion*, A.V. Bridgwater and D.G.B. Boocock, eds., Blackie Academic and Professional, London, pp. 622-632.

Conti, L.; Scano, G.; Boufala, J.; and Mascia, S. (1996) "Experiments of Bio-Oil Hydrotreating in a Continuous Bench Scale Plant," in *Bio-Oil. Production and Utilization*, A.V. Bridgwater and E.N. Hogan, CPL Press, Newbury, U.K., pp. 198-205.

Czernik, S. (1994) "Storage of Biomass Pyrolysis Oils," in *Proceedings of Biomass Pyrolysis Oil Properties and Combustion Meeting*, September 26-28, Estes Park, CO, T.A. Milne, ed., National Renewable Energy Laboratory, Golden, CO, NREL-CP-430-7215, pp. 67-76.

Czernik, S.; Johnson, D.K.; and Black, S. (1994) "Stability of Wood Fast Pyrolysis Oil," *Biomass and Bioenergy*, 7, No. 1-6, pp. 187-192.

Diebold, J.P.; Scahill, J.W.; Czernik, S.; Phillips, S.D.; and Feik, C.J. (1996) "Progress in the Production of Hot-Gas Filtered Biocrude Oil at NREL," in *Bio-Oil Production and Utilization*, A.V. Bridgwater and E.N. Hogan, eds., CPL Press, Newbury, UK, pp. 66-81.

Diebold, J.P. (1997) *A Review of the Toxicity of Biomass Pyrolysis Liquids Formed at Low Temperatures*, National Renewable Energy Laboratory, Golden, CO, NREL/TP-430-22739, 35 pp. (81 references) Also appears in *The IEA Bioenergy Handbook on Biomass Pyrolysis*, (1998), T. Bridgwater, S. Czernik, J. Leech, D. Meier, A. Oasmaa, and J. Piskorz, eds., *IEA Bioenergy Task XIII Pyrolysis Activity Final Report*, Aston University, Birmingham, U.K. Also in *The Fast Pyrolysis of Biomass: A Handbook*, A. Bridgwater, S. Czernik, J. Diebold, D. Meier, A. Oasmaa, C. Peacocke, J. Piskorz, and D. Radlein, eds., CPL Press, Newbury, UK, pp. 135-163.

Diebold, J.P. and Bridgwater, A.V. (1997) "Overview of Fast Pyrolysis of Biomass for the Production of Liquid Fuels," in *Developments in Thermal Biomass Conversion*, A.V. Bridgwater and D.G.B. Boocock, eds., Blackie Academic and Professional, London, pp. 5-26. (93 references) Also appears in *The IEA Bioenergy Handbook on Biomass Pyrolysis*, (1998), T. Bridgwater, S. Czernik, J. Leech, D. Meier, A. Oasmaa, and J. Piskorz, eds., *IEA Bioenergy Task XIII Pyrolysis Activity Final Report*, Aston University, Birmingham, U.K. Also in *The Fast Pyrolysis of Biomass: A Handbook*, A. Bridgwater, S. Czernik, J. Diebold, D. Meier, A. Oasmaa, C. Peacocke, J. Piskorz, and D. Radlein, eds., CPL Press, Newbury, UK, pp. 14-32.

Diebold, J.P. and Czernik, S. (1997) "Additives to Lower and Stabilize the Viscosity of Pyrolysis Oils During Storage," *Energy and Fuels*, 11, pp. 1081-1091.

Doerr, R.C.; Wasserman, A.E.; and Fiddler, W. (1966) "Composition of Hickory Sawdust Smoke, Low Boiling Constituents," *J. Agr. Food Chem.*, 14, No. 6, pp.662-665.

DIPPR (1998) *Physical and Thermodynamic Properties of Pure Chemicals: Data Compilation*, T.E. Daubert, Danner, R.P., H.M. Sibul, and C.C. Stebbins, eds., AIChE's Design Institute for Physical Property Data, Taylor and Francis, Bristol, PA, 3548 pp.

Dunlop, A.P. and Peters, F.N. (1953) *The Furans*, ACS Monographs Series, Reinhold Publishing Corp., NY, 400+ pp.

Edgar, G. and Schuyler, W.H. (1924) "Esterification Equilibria in the Gaseous Phase," *J. Am. Chem. Soc.*, 46, pp.64-75.

Elliott, D.C. (1986) *Analysis and Comparison of Biomass Pyrolysis/Gasification Condensates-Final Report*, Pacific Northwest Laboratory, Richland, WA, PNL-5943, UC-61D, p. 37 and p. 41.

Elliott, D.C. (1994) "Water, Alkali, and Char in Flash Pyrolysis Oils," *Biomass and Bioenergy*, 7, No. 1-6, pp. 179-186.

Elliott, D.C. and Neuenschwader, G.G. (1997) "Liquid Fuels by Low-Severity Hydrotreating of Biocrude," in *Developments in Thermal Biomass Conversion*, A.V. Bridgwater and D.G.B. Boocock, eds., Blackie Academic and Professional, London, pp. 611-621.

Etzkorn, W.G.; Kurland, J.J.; and Neilsen, W.D. (1991) "Acrolein and Derivatives," in *Kirk and Othmer's Encyclopedia of Chemical Technology, Vol. 1, 4th ed.*, John Wiley and Sons, NY, pp.232-251.

Euranto, E.K. (1969) "Esterification and Ester Hydrolysis," in *The Chemistry of Carboxylic Acids and Esters*, S. Patai, ed., Interscience Publishers, London, p. 507.

Fiddler, W.; Parker W.E.; Wasserman, A.E.; and Doerr, R.C. (1967) "Thermal Decomposition of Ferulic Acid," *J. Agric. Food Chem.*, 15, No. 5, pp. 757-761.

Fraps, G.S. (1901) "The Composition of Wood Oil," *American Chemical Journal*, 25, pp. 26-53.

French, R.J. and Milne, T.A. (1994) "Vapor Phase Release of Alkali Species in the Combustion of Biomass Pyrolysis Oils," *Biomass and Bioenergy*, 7, No. 1-6, pp. 315-325.

Frevel, L.K. and Hedelund, J.W. (1954) "Acetals and Process for Making the Same," U.S. Patent 2,691,684.

Friberg, S.E. and Jones, S. (1994) "Emulsions," in *Kirk and Othmer Encyclopedia of Chemical Technology. 4th ed., Vol. 9,* John Wiley and Sons, NY, pp. 393-413.

Gagnon, J. and Kaliaguine, S. (1988) "Catalytic Hydrotreatment of Vacuum Pyrolysis Oils from Wood," *Ind. Eng. Chem. Res.*, 27, pp. 1783-1788.

Gary, J.H. and Handwerk, G.E. (1994) *Petroleum Refining Technology and Economics, 3rd ed.*, Marcel Dekker, Inc., NY, 460 pp.

Gerberich, H.R.; Stautzenberger, A.L.; and Hopkins, W.C. (1980) "Formaldehyde," in *Kirk and Othmer's Encyclopedia of Chemical Technology, Vol.11, 3rd ed.,* John Wiley and Sons, NY, pp.231-250.

Goos, A.W. and Reiter, A.A. (1946) "New Products from Wood Carbonization," *Ind. Eng. Chem.*, 38, 2, pp.132-135.

Granger, F.S. (1937) "Condensation of Phenols with Formaldehyde. Resinification of Phenol Alcohols," *Ind. Eng. Chem.*, 29, pp. 860-866.

Guillén, M.D.; Manzanos, M.J.; and Zabala, L. (1995) "Study of a Commercial Liquid Smoke Flavoring by Means of Gas Chromatography/Mass Spectrometry and Fourier Transform Infrared Spectroscopy," *J. Agric. Food Chem.*, 43, pp. 463-468

Guillén, M.D. and Ibaragoitia, M.L. (1996) "Relationships Between the Maximum Temperature Reached in the Smoke Generation Processes from *Vitus vinifera L.* Shoot Sawdust and Composition of the Aqueous Smoke flavoring Preparations Obtained," *J. Agric. Food Chem.*, 44, pp. 1302-1307.

Gust, S. (1997a) "Combustion Experiences of Flash Pyrolysis Fuel in Intermediate Size Boilers," in *Developments in Thermal Biomass Conversion*, A.V. Bridgwater and D.G.B. Boocock, eds., Blackie Academic and Professional, London, pp. 481-488.

Gust, S. (1997 b) "Combustion of Pyrolysis Liquids," in *Biomass Gasification and Prolysis. State of the Art and Future Prospects*, M. Kaltschmitt and A.V. Bridgwater, eds., CPL Press, Newbury, UK, pp. 498-503.

Harvey, M.T. (1944a) "Acid Condensation Product of Formaldehyde and Acid Condensation Polymerization Product of Furfuryl Alcohol and Method for Preparing the Same," U.S. Patent 2,343,973.

Harvey, M.T. (1944b) "Novel Furfural Alcohol-Formaldehyde Acid Condensation Resinous Product and Method for Preparing the Same," U.S. Patent 2,343,972.

Hawley, L.F. (1923) *Wood Distillation*, The Chemical Catalog Company, Inc., N.Y., pp. 64-72.

Hiatt, R. (1971) "Hydroperoxides," in *Organic Peroxides*, D. Swern, ed., Wiley Interscience, NY, pp. 1-134.

Hinshelwood, C.N. and Legard, A.R. (1935) "The Factors Determining the Velocity of Reactions in Solution. Molecular Statistics of the Esterification of Carboxylic Acids," *J. Chem. Soc.,* p. 587-596.

Huffman, D.R. and Freel, B.A. (1997) "RTP™ Biocrude: A Combustion / Emissions Review," in *Developments in Thermal Biomass Conversion*, A.V. Bridgwater and D.G.B. Boocock, eds., Blackie Academic and Professional, London, pp. 489-494.

Ikura, M.; Mirmiran, S. Stanciulescu, M.; and Sawatzky, H. (1998) "Pyrolysis Liquid-in-Diesel Oil Microemulsions," U.S. Patent 5,820,640.

Kelley, S.S.; Wang, X-M, Myers, M.D.; Johnson, D.K.; and Scahill, J.W. (1997) "Use of Biomass Pyrolysis Oils for Preparation of Modified Phenol Formaldehyde Resins," in *Developments in Thermal Biomass Conversion*, A.V. Bridgwater and D.G.B. Boocock, eds., Blackie Academic and Professional, London, pp. 557-572.

Klar, M. (1925) *The Technology of Wood Distillation*, Chapman & Hall, London (translated to English by Alexander Rule), pp. 60-72.

Knight, H.B.; Koos, R.E.; and Swern, D. (1954) "New Method for Hydroxylating Long-Chain Unsaturated fatty Acids, Esters, Alcohols, and Hydrocarbons," *J. Am. Chem. Soc.*, 31, No. 1, pp. 1-5.

Knight, J.A.; Hurst, D.R.; and Elston, L.W. (1977) "Wood Oil from Pyrolysis of Pine Bark-Sawdust Mixture," in *Fuels and Energy from Renewable Resources*, D.A. Tillman, ed., Academic Press, pp. 169-194.

Kohl, A.L. and Riesenfeld, F.C. (1974) *Gas Purification. 2nd ed.*, Gulf Publishing Co., Houston, TX, p. 691-700.

Kopf, P.W. (1996) "Phenolic Resins," in Kirk and Othmer's *Encyclopedia of Chemical Technology. Vol. 18, 4th ed.*, John Wiley and Sons, NY, pp. 604-609.

Leech, J. and Webster, J.A. (1999) "Storage and Handling of Flash Pyrolysis Oil," in *Fast Pyrolysis of Biomass: A Handbook,* A. Bridgwater, S. Czernik, J. Diebold, D. Meier, A. Oasmaa, C. Peacocke, J. Piskorz, and D. Radlein, eds., CPL Press, Newbury, UK, pp.66-68.

Lewis, N.G. and Lantzy, T.R. (1989) "Lignin in Adhesives. Introduction and Historical Perspectives," in *Adhesives from Renewable Resources*, R.W. Hemingway, A.H. Conner, and S.J. Branham, eds., ACS Symposium Series 385, American Chemical Society, Washington, D.C., pp.13-26.

Lorand, E.J. (1942) "Preparations of Lower Sugars," U.S. Patent 2,272,378.

Maga, J.A. (1987) "The Flavor Chemistry of Wood Smoke," *Food Reviews International*, 3(1&2), pp.139-183.

Maga, J.A. (1988) *Smoke in Food Processing*, CRC Press, Boca Raton, FL, 160 pp.

Maggi, R.; Centeno, A.; and Delmon, B. (1996) "Stabilization by Mild Catalytic Hydrogenation of Bio-Oils Produced by Fast Pyrolysis," abstracts of the 9th European Bioenergy Conference, Copenhagen, Denmark, June 24-27, p. 80.

Meier, D. (1999) private communication to the author on May 25.

Meier, D.; Bridgwater, A.V.; Di Blasi, C.; and Prins, W. (1997) "Integrated Chemicals and Fuels Recovery from Pyrolysis Liquids Generated by Ablative Pyrolysis," in *Biomass Gasification and Pyrolysis State of the Art and Future Aspects*, A.V. Bridgwater and M.K. Kaltschidtt, eds., CPL Press.

Meier, D.; Crespo Rodriguez, M.I.; Bridgwater, A.V.; Nakos, P.; di Blasi, D.; Prins, W.; Slonso, J.; and Samolada, M. (1998) *Catalytic Pyrolysis of Biomass for Improved Liquid Fuel Quality. Publishable Final Report. January 1996- June 1998*, Contract JOR CT95-0081.

Meyers, R.A. (1996) *Handbook of Petroleum Refining Processes, 2nd Ed.*, McGraw-Hill, NY, 802 pp.

Miles, T.R.; Miles, T.R., Jr.; Baxter, L.L.; Bryers, R.W.; Jenkins, B.M.; and Oden, L.L. (1996) *Alkali Deposits Found in Biomass Power Plants. A Preliminary Investigation of their Extent and Nature*, National Renewable Energy Laboratory, Golden, CO, NREL/TP-433-8142 or Sandia National Laboratories, Livermore, CA, SAND 96-8225.

Milne, T.A.; Agblevor, F.; Davis, M.; Deutch, S.; and Johnson, D. (1997) "A Review of the Chemical Composition of Fast Pyrolysis Oils, in *Developments in Thermal Biomass Conversion*, A.V. Bridgwater and D.G.B. Boocock, eds., Blackie Academic and Professional, London, pp. 409-424.

Minné, N. and Adkins, H. (1933) "Structure of Reactants and the Extent of Acetal Formation," *J. Am. Chem. Soc.*, 55, pp. 299-309.

Mushrush, G.W and Speight, J.G. (1995) *Petroleum Products: Instability and Incompatibility*, Taylor and Frances, Washington, D.C., 390 pp.

Nelson, W.G. (1930) "Waste-Wood Utilization by the Badger-Stafford Process," *Ind. Eng. Chem.*, 22, No. 4, pp. 312-315.

Oasmaa, A. and Sipilä, K. (1996) "Pyrolysis Oil Properties: Use of Pyrolysis Oil as Fuel in Medium Speed Diesel Engines," *Bio-Oil Production and Utilization*, A.V. Bridgwater and E.N. Hogan, eds., CPL Press, Newbury, UK, pp. 175-185.

Oasmaa, A.; Leppämäki, E.; Koponen, P.; Levander, J.; and Tapola, E. (1997) *Physical Characterisation of Biomass-Based Pyrolysis Liquids*, Technical Research Centre of Finland, VTT Publication 306, 46 pp. plus appendices.

Oasmaa, A. (1999) Private communication to the author on June 23.

Palm, M.; Piskorz, J.; Peacocke, C.; Scott, D.S.; and Bridgwater, A.V. (1993) Fast Pyrolysis of Sweet Sorghum Bagasse in a Fluidized Bed," *Proceedings of the First Biomass Conference of the Americas. Energy, Environment, Agriculture, and Industry. Vol. II*, August 30- September 2, Burlington, Vt., National Renewable Energy Laboratory, NREL/CP-200-5768, DE93010050, pp.947-963.

Peacocke, G.V.C.; Russell, P.A.; Jenkins, J.D.; and Bridgwater, A.V. (1994) "Physical Properties of Flash Pyrolysis Liquids," *Biomass and Bioenergy*, 7, No. 1, pp.169-177.

Perry, R.H.; Green, D.W.; and Maloney, J.O., eds. (1997) *Perry's Chemical Engineers' Handbook, 7th ed.*, McGraw-Hill, N.Y., pp. 2-28 to 2-47.

Polk, M.B. and Phingbodhippakkiya, M. (1981) *Development of Methods for the Stabilization of Pyrolytic Oils*, Municipal Environmental Research Laboratory, U.S. Environmental Protection Agency, Cincinnati, OH, 45268, EPA-600/2-81-201, 74 pp.

Radlein, D.St.A.G.; Piskorz, J.K.; and Majerski, P.A. (1996) "Method of Upgrading Biomass Pyrolysis Liquids for Use as Fuels and as a Source of Chemicals by Reaction with Alcohols," *European Patent Application* EP 0 718 392 A1.

Radlein, D. (1999) "The Production of Chemicals from Fast Pyrolysis Bio-Oils," in *The Fast Pyrolysis of Biomass: A Handbook*, A. Bridgwater, S. Czernik, J. Diebold, D. Meier, A. Oasmaa, C. Peacocke, J. Piskorz, and D. Radlein, eds., CPL Press, Newbury, UK, pp. 164-188.

Radovanovic, M; Venderbosch, R.H.; Prins, W.; and van Swaaij, W.P.M. (1997) "Thermal Stability of Biomass Pyrolysis Liquids: a New Approach," in *Biomass Gasification and*

Pyrolysis. State of the Art and Future Prospects, M. Kaltschmitt and A.V. Bridgwater, eds., CPL Press, Newbury, UK, pp.460-470.

Riha, W.E. and Wendorff, W.L. (1993) "Browning Potential of Liquid Smoke Solutions: Comparison of Two Methods," *J. of Food Science*, 55, No. 9, pp. 671-674.

Rolfe, A.C. and Hinshelwood, C.N. (1934) "The Kinetics of Esterification. The Reaction Between Acetic Acid and Methyl Alcohol," *Trans. Faraday Soc.*, 30, pp. 935-944.

Rossi, C. (1994) "Bio-Oil Combustion Tests at ENEL," in *Proceedings of Biomass Pyrolysis Oil Properties and Combustion Meeting,* September 26-28, Estes Park, CO, T.A. Milne, ed., National Renewable Energy Laboratory, Golden, CO, NREL-CP-430-7215, pp. 321-328.

Salvi, G. and Salvi, Jr., G. (1991) *Pyrolytic Products Utilisation Assessment Study*, commission of the European Communities, Contract No. EN3B-0191-1 (CH).

Sato, H. (1984) "Process for the Production of Carboxylic Acid Esters," U.S. Patent 4,465,852.

Scahill, J.W.; Diebold, J.P.; and Feik, C. (1997) "Removal of Residual Char Fines From Pyrolysis Vapors by Hot Gas Filtration," in *Developments in Thermochemical Biomass Conversion*, A.V. Bridgwater and D.G.B. Boocock, eds, Blackie Academic and Professional, London, pp. 253-266.

Scott, D.S. and Piskorz, J. (1984) "The Continuous Flash Pyrolysis of Biomass," *Can. J. Chem. Eng.*, 62, pp. 404-412.

Shihadeh, A.L. (1998) *Rural Electrification from Local Resources: Biomass Pyrolysis Oil Combustion in a Direct Injection Diesel Engine*, PhD Thesis at the Massachusetts Institute of Technology, p. 34.

Simons, R.M. (1983) in *Encyclopedia of Chemical Processing and Design, Volume 19*, J.J. McKetta and W.A. Cunningham, eds., Marcel Dekker, N.Y., p. 381.

Smith, H.A. (1939) "Kinetics of the Catalyzed Esterification of Normal Aliphatic Acids in Methyl Alcohol," *J. Am. Chem. Soc.*, 61, pp. 254-260.

Stamm, A.J. and Harris, E.E. (1953) *Chemical Processing of Wood*, Chemical Publishing Co., N.Y., pp. 444-447.

Street and Adkins (1928) "The Effect of Certain Beta Substituents in the Alcohol Upon Affinity and Reactivity in Acetal Formation," *J. Am. Chem. Soc.*, 50, 162.

Suida, H. (1931) "Manufacture of Isopropyl- and Homologous Esters of Aliphatic Acids and Isopropyl Alcohol and Homologous Alcohols," U.S. Patent 1,836,135.

Sutton, H.C. and Downes, T.M. (1972) "Rate of Hydration of Formaldehyde in Aqueous Solution," *J. Chem. Soc. Chem. Comm.*, 1, pp. 1-2.

Tiplady, I.R.; Peacocke, G.V.C.; and Bridgwater, A.V. (1996) "Physical Properties of Fast Pyrolysis Liquids from the Union Fenosa Pilot Plant," in *Bio-Oil. Production and Utilization*, A.V. Bridgwater and E.N. Hogan, eds., CPL Press, Newbury, U.K., pp. 164-174.

Tóth, L. and Potthast, K. (1984) "Chemical Aspects of the Smoking of Meat and Meat Products," *Advances in Food Research*, 29, pp. 87-158.

Vergnet, A-M and Villeneuve, F. (1988) "Techniques Analytiques Applicables Aux Gaz et Jus de Pyrolyse de la Biomasse Tropicale," *Cahiers Scientifiques Bois et Forets, des Tropiques*, Centre Technique Forestier Tropical, Nogent-Sur-Marne, France, No. 9, pp. 3-66.

Walker, J.F. (1953) *Formaldehyde*, 2nd ed., ACS Monograph Series, Reinhold Publishing Corp., NY, 575 pp.

Williams, D.L. and Dunlop, A.P. (1948) "Kinetics of Furfural Destruction in Acidic Aqueous Media," *Ind. Eng. Chem.*, 40, No. 2, pp. 239-241.

TRANSPORT, HANDLING AND STORAGE OF
FAST PYROLYSIS LIQUIDS

Cordner Peacocke
Conversion And Resource Evaluation Ltd, 3 Glen Road, Holywood
Co. Down, N. Ireland BT18 0HB

SUMMARY

Pyrolysis liquids are being developed for fuel and chemical applications. As these developments proceed, liquids are increasingly being transported by air, water, rail and road. To this end, this report addresses the legislative requirements and regulations for the safe transport of pyrolysis liquids.

Pyrolysis liquids are not listed on the UN approved carriage list for dangerous or hazardous goods; therefore, the most appropriate not otherwise specified [N.O.S.] classification was determined. The classification is:

UN 1993 FLAMMABLE LIQUID, N.O.S. (Fast Pyrolysis Liquid), 3, 1°(a), 2°(a)

This classification should be used on all packages containing biomass derived fast pyrolysis liquids. The packaging group is "1" or "X", as is commonly used. Labelling and packaging are vital requirements of transportation of all quantities of pyrolysis liquids. Protocols for the labelling of packages and containers of all sizes are given with the aim of compliance with transport regulations in the EU, Canada and the USA. In conjunction with the requirements for packaging and labelling, guidance on the details to be enclosed on the transportation documents are given, with appropriate MSDS for the liquids. Guidance on the handling of pyrolysis liquids and storage are given and preliminary procedures for the treatment of spills.

Further work is required to determine the degradability of pyrolysis liquids, appropriate transportation codes also need to be derived for other pyrolysis liquids fractions, and those derived from non-biomass materials, e.g. plastics, wastes, etc. Further work is required to determine procedures for dealing with spills and how spill areas can be remediated.

1.   INTRODUCTION

Pyrolysis liquids are now being actively produced for research, testing and evaluation purposes, for use as a chemical feedstock, source of individual chemicals and as an alternative fuel for use in boilers, engines and turbines.

As pyrolysis technologies advance and utilisation of the liquids increases, there will be a greater demand for the transportation of the liquids, by all possible routes – air, road, rail and water, or a combination of routes. To ensure that the liquids are transported in a safe and environmentally secure manner, all due care and attention must be taken to ensure that the appropriate national and international regulations pertaining to the transport of the pyrolysis liquids are met. To this end, it is likely that pyrolysis liquids will be classed as "dangerous" or a "hazardous" substance for transportation purposes.

To this end, Conversion And Resource Evaluation Ltd. was commissioned by the Pyrolysis Network [PyNe] to assess the relevant regulation applicable to the transportation of pyrolysis liquids. In conjunction with this, the packaging requirement for the liquids for each mode of transport and for various quantities was to be considered.

The tasks to be carried out, were:
- Derive an appropriate transportation code for pyrolysis liquids,
- Derive procedures for the treatment of spills [small and large],
- Produce a guide for the preparation of samples for shipment from a few grams to tonnes, with protocols for labelling, packaging and shipment.

The structure of the report to fulfil these tasks is:
- International regulations on transport   Section 2
- Properties of pyrolysis liquids and UN classification   Section 3
- Packaging and labelling protocols   Section 4
- Handling and storage of pyrolysis liquids   Section 5
- Treatment of spills   Section 6

The Department of Transport, Environment and Regions [Transport of Dangerous Goods Division] and the UK Health and Safety Executive were consulted initially to discuss the classification of liquids for transportation purposes. (1). Based on discussions with these organisations, their opinion was that pyrolysis liquids would be classed as a "dangerous good" [or hazardous material], due to the chemical composition and its flammable properties.

The biomass pyrolysis community may feel that the classification of pyrolysis liquids as "dangerous" material does not reflect the true nature of the liquids. However, due the variability of liquids produced by different processes, using variable feedstocks, the chemical composition of the liquids is variable and physical properties are significantly different. This variability in properties has been taken into account in classifying the liquids. This report focuses on whole fast pyrolysis liquids and not derived fractions or products thereof. Treated liquids, e.g. hydrotreated, fractionated, etc. or liquids produced in other pyrolysis processes may require a separate classification, which is outside the scope of this work.

2. INTERNATIONAL REGULATIONS ON THE TRANSPORTATION OF DANGEROUS GOODS

The scope of this legislative review pertains to the transport of goods in the EU, USA and Canada for all modes. This report cannot cover all the national regulations in force, however, most national transportation regulation are based on, or use the UN Regulations, as described below [see Section 2.1].

2.1 UN Regulations

At the United Nations level, all work related to the transport of dangerous goods is co-ordinated by the Economic and Social Council [ECOSOC] Committee of Experts on the Transport of Dangerous Goods, which produces the "Recommendations on the Transport of Dangerous Goods", also called the "Orange Book" (2). These Recommendations and Regulations are addressed not only to all Governments for the development of their national requirements for the domestic transport of dangerous goods, but also to international organisations such as:

- The International Maritime Organisation [IMO];
- The International Civil Aviation Organisation [ICAO] and;
- Regional commissions such as the Economic Commission for Europe [ECE];

for regulations and international/regional agreements or conventions governing the international transport of dangerous goods by sea, air, road, rail and inland waterways. Although the ECOSOC Committee of Experts on the Transport of Dangerous Goods is not a UN/ECE body, but is a body with activities of world-wide scope, the secretariat of the UN/ECE is responsible for its service. The UN Recommendations on the Transport of Dangerous Goods therefore addresses the following main areas:

1. List of dangerous goods most commonly carried and their identification and classification,
2. Consignment procedures: labelling, marking, and transport documents,
3. Standards for packagings, test procedures, and certification,
4. Standards for multimodal tank-containers, test procedures and certification.

These recommendations contain all basic provisions for the safe carriage of dangerous goods, but they may have to be completed by additional requirements that are applicable at national level or for international transport depending on the mode or modes of transport envisaged. These recommendations are presented in the new form of Model regulations so that they can be more easily transposed into national or international legislation (2).

2.2 International Carriage of Dangerous Goods by Air

There is no EU Agreement concerning the international carriage of dangerous goods by air. The ICAO issues technical instructions for the transport of dangerous goods by air (3) and regulations are produced by the International Air Transport Association [IATA] and these regulations will typically be those followed by cargo carriers for international transport (4). The IATA Dangerous Goods Regulations manual is based on the ICAO Technical Instructions. For example, the Civil Aviation Authority [CAA] in the UK has produced a guide for the transport of dangerous goods based on the IATA regulations (5).

The core of the IATA Dangerous Goods Regulations is the alphabetical list of substances classified as dangerous goods. This list:
• Specifies whether the substance may be carried by air and, if so, on which type of aircraft;
• States the quantity limitations applicable;
• Identifies the kind of packaging required;
• Indicates which dangerous goods label must be used;
• Gives the dangerous goods class for each substance;
• Shows the official UN number of the substance described and details the containers that may be used for each substance. The construction specifications of each type of package are given together with details of the performance tests that are required before they can be certified for use.

Dangerous goods can be transported safely by air transport provided certain principles are strictly followed. Air transport incorporates additional operational requirements that provide a harmonised system for airlines to accept and transport dangerous goods safely and efficiently. Users of the IATA Dangerous Goods Regulations are assured that they are meeting all legal requirements for shipping dangerous goods by air internationally and its use is therefore strongly recommended.

The IATA Regulations include a detailed list of individual articles and substances specifying the United Nations classification of each article or substance and their acceptability for air transport as well as the conditions for their transport. Since no listing can be complete, the list also includes many generic or "not otherwise specified [N.O.S.]" entries to assist in the classification of those articles or substances not listed by name. Pyrolysis liquids, as noted in Section 3.3 would be classed at this stage as an" N.O.S." substance.

Packaging is the essential component in the safe transport of dangerous goods by air, similarly for all the other modes as discussed below. The IATA Dangerous Goods Regulations provide packing instructions for all dangerous goods acceptable for air transport with a wide range of options for inner, outer and single packagings. The packing instructions normally require the use of UN performance-tested specification packagings, however these are not required when dangerous goods are shipped in Limited Quantities under the provisions of Limited Quantity "Y" Packing Instructions. The quantity of dangerous goods permitted within these packaging is strictly limited so as to minimise the inherent risk presented by the dangerous goods should an incident occur.

2.3   International Carriage of Dangerous Goods by Water

As noted above the IMO is responsible for the provision of guidance on the safe transport of dangerous goods on water. The International Maritime Dangerous Goods [IMDG] Code is accepted as an international guide to the transport of dangerous goods by sea and is recommended to governments for adoption or for use as the basis for national regulations (6). It is intended for use not only by the mariner but also by all those involved in industries and services connected with shipping and contains advice on terminology, packaging, labelling, stowage, segregation, handling, and emergency response.

This edition incorporates Amendment 29-98, adopted by IMO's Maritime Safety Committee [MSC] in May 1998. The classification, identification, labelling and packaging of dangerous goods is standardised across all modes of transport. New provisions have been added to cover the carriage of goods in open-top containerships and many of the existing sections of the Code have been revised and updated. The Amendment includes changes to harmonise the IMDG Code with the tenth edition of the United Nations Recommendations on the Transport of Dangerous Goods, which has now been updated (2).

2.4   International Carriage of Dangerous Goods by Rail

The transport of dangerous goods by rail is serviced by the Intergovernmental Organisation for International Carriage by Rail [OTIF]; it is responsible for ensuring harmonisation between ADR [Regulations concerning the international carriage of dangerous goods by road - see Section 2.5.1 below], RID [Regulations concerning the international carriage of dangerous goods by rail – see Section 2.5.3] and ADN [Regulations concerning the international carriage of dangerous goods by water – see Section 2.5.2]. RID gives similar guidance to ADR with some additional provisions for transport. The reader is advised to consult the RID regulations for specific information and requirements.

2.5   EU Transport Regulations

ECE subsidiary bodies deal with the transport of dangerous goods. These bodies are subsidiary bodies of the Inland Transport Committee, and therefore they are concerned only with inland transport, i.e. road, rail and inland waterway. These bodies are The Working Party on the Transport of Dangerous Goods [known as WP.15], which is responsible for:

* The European Agreement concerning the International Carriage of Dangerous Goods by Road [ADR] and;
* The European Provisions concerning the International Carriage of Dangerous Goods by Inland Waterways [ADN] and;

- The Joint Meeting of the Working Party on the Transport of Dangerous Goods and the RID Safety Committee also called the RID/ADR/ADN Joint Meeting.

The RID/ADR/ADN Joint Meeting is serviced jointly by the ECE secretariat and the secretariat of the Intergovernmental Organisation for International Carriage by Rail [OTIF]; it is responsible for ensuring harmonisation between ADR, RID [Regulations concerning the international carriage of dangerous goods by rail) and ADN.

2.5.1 European Agreement concerning the International Carriage of Dangerous Goods by Road [ADR] (7)

ADR is based on the UN Recommendations on the Transport of Dangerous Goods as regards the listing and classification of dangerous goods, their marking and labelling and packaging standards, but it also contains much more detailed provisions as regards:

1. The types of packaging which may be used,
2. The consignment procedures,
3. Transport equipment [vehicle to be used, vehicle construction and equipment],
4. Transport operation [training of drivers, supervision, emergency procedures, loading and unloading, placarding of vehicles].

The ADR is intended primarily to increase the safety of international transport by road, but it is also an important trade facilitation instrument. Except for dangerous goods which are totally prohibited for carriage, and except when carriage is regulated or prohibited for reasons other than safety, the international carriage of dangerous goods by road is authorised by ADR on the territory of Contracting Parties provided that the conditions laid down in annexes A and B are complied with. There are at present 34 Contracting Parties to ADR, including all of the EU member states, USA and Canada.

2.5.2 European Agreement concerning the International Carriage of Dangerous Goods by Inland Waterways [ADN] (8)

The status of the European Provisions for the International Carriage of Dangerous Goods by Inland Waterways is different from that of ADR as ADN is only a recommendation directed to Governments for their national regulations and to river commissions for regulating the international carriage of dangerous goods on specific inland waterways under their responsibility. One well-known example of such regulations is the "Regulations for the Carriage of Dangerous Substances on the Rhine [ADNR]" developed by the Central Commission for the Navigation of the Rhine [CCNR]. A draft European agreement concerning the international carriage of dangerous goods by inland waterways ["ADN" agreement] has recently been completed under the joint auspices of the UN/ECE and the CCNR (8).

2.5.3 European Agreement concerning the International Carriage of Dangerous Goods by Rail [RID] (9)

The RID is also intended primarily to increase the safety of international transport by rail. Except for dangerous goods which are totally prohibited for carriage, and except when carriage is regulated or prohibited for reasons other than safety, the international carriage of dangerous goods by rail is authorised by RID on the territory of Contracting Parties provided that the conditions laid down in annexes A and B of the Agreement are complied with (9).

2.6 USA and Canada Regulations

2.6.1 USA Regulations on the Transport of Hazardous Materials

The United States Department of Transportation [USDOT] regulates the transportation of Hazardous Materials to, from and through the United States. Title 49, Code of Federal

Regulations, Subtitle B "Other Regulations Relating to Transportation" Parts 100 - 199 set forth the standards for Hazardous Materials transportation, commonly known as "49 CFR" and these are readily available (10). There are significant differences between 49 CFR and other international regulations such as ICAO, IATA and IMO, despite efforts to reduce these differences and the US Regulations change regularly. All shipments of Hazardous Materials to, from and through the United States must comply with all aspects of Parts 100 through 199 of these regulations and others.

These regulations are administered by the Research and Special Programs Administration of the Department of Transportation [RSPA] and enforced on a federal level by the Federal Aviation Administration [FAA] for air transport, the Federal Highway Administration [FHWA] for ground transportation and the United States Coast Guard [USCG] for water transportation. Enforcement on a local level is usually the State Police or Highway Patrol. Exporters, importers, shipping lines, airlines, forwarders & couriers have to comply with 49 CFR, and failure can lead to delays in transport or in the case of non-compliance to potentially massive fines.

Title 49 Part 171.11 allows the use of [ICAO] Technical Instructions to be used for air shipments as long as certain additional requirements set forth in Title 49 are also complied with. Title 49 Part 171.12 allows the use of the International Maritime Organization's International Maritime Dangerous Goods Code [IMDG] for import and export shipments by vessel as long as certain additional requirements set forth in Title 49 are also complied with.

2.6.2   Canadian Regulations on the Transport of Hazardous Materials

Transport Canada is responsible for the transportation of dangerous goods in Canada legislated by the Transportation of Dangerous Goods [TDG] Regulations (11). These regulations are based on the UN Regulations and are very similar in content. Under these regulations, a detailed manifest must accompany all waste shipments: those that originate in Canada, as well as those that enter Canada from another country. This manifest allows easy tracking of shipments until they arrive at their destination or exit the country. An emergency response plan must also be submitted to Transport Canada. The following information is also required for shipments within Canada:
1.   Summary of Emergency Response Plan;
2.   The reference number of the Emergency Response Plan filed with Transport Canada and;
3.   24 hour telephone number.

These must be included on the transportation documents with the consignment.

2.7   Note on Transport Regulations

It should be noted that the requirements of Annexes A and B of ADR have been annexed to the European Union Council Directive 94/55/EC on the approximation of the laws of the Member States with regard to the transport of dangerous goods, and therefore these requirements have become applicable not only to international transport of dangerous goods but also to domestic traffic in all countries of the European Union as from 1 January 1997 (12, 13). It is therefore incumbent on those transporting pyrolysis liquids to ensure compliance with the UN Regulations, ADR, RID and ADN. In North America, in particular the USA, there is still usage of older codes and frequently containers may have "NA" [North America] rather than "UN" on the packages.

Within the context of the contract, and time constraints due to the budget, not all of the transportations routes could be fully assessed, however, provided the requirements for packaging are met, then the supplier of pyrolysis liquids can assist the handler in compliance with the relevant codes of practice and international regulations. Consignors and transportation companies must comply with an extensive set of codes, guides, regulations that are beyond the scope of this work.

3. PROPERTIES OF FAST PYROLYSIS LIQUIDS

The nature of pyrolysis liquids means that there in no "generic" analyses to cover the wide spectrum of liquids producible from biomass. In the UN or EU regulations, there is no classification for pyrolysis liquids or its derivatives, fractions or by-products.

As noted in Section 1, based on discussions with the Department of the Environment, Transport and Regions – Dangerous Goods Branch, a submission to the EU to have pyrolysis liquids listed in ADR, RID and ADN could take 2-3 years, therefore a self assessment of the substance must be made, using the methods described in the UN Manual (2). It is also likely that the outcome of a submission would be that pyrolysis liquids would be classed as a "dangerous substance" or "hazardous material". The assessment of a substance as a dangerous good considers the chemical and physical properties of pyrolysis liquids are summarised below.

3.1 Physical Properties of Pyrolysis Liquids

Specific physical properties, if not known, can be determined to UN test methods (14). As pyrolysis liquids are not listed as a substance in their own right in ADR or the UN Regulations, the liquids can be categorised with a N.O.S. classification. The physical properties of the fast pyrolysis liquids used are given in Table 1, based upon typical values in the literature [see Appendix I for references used]. Detailed chemical analysis is required as discussed in Section 3.2 below.

Table 1: Applicable physical properties of fast pyrolysis liquids (15)

Physical property	Fast pyrolysis liquid
Moisture content	~20-40wt%
Ph	~2-3
Specific gravity	~1.2
Dynamic Viscosity [cp @ 40°C]	~50 cp
Kinematic viscosity [cSt]	20-1000 @ 25°C
	15-500 @ 40°C
Flash point [°C]	50-70
Pour point [°C]	-23

From these properties, pyrolysis liquids may be generally classed a Class 3 substance-Flammable Liquid. The exact specification is then related to its chemical composition to determine its level of hazard, as described in Section 3.2.

3.2 Chemical composition of biomass fast pyrolysis liquids

There are numerous references in the literature with chemical analyses of pyrolysis liquids from a variety of sources, including slow pyrolysis tars, fast pyrolysis liquids and fractions of the liquids. As fast pyrolysis liquids may be raw, treated, filtered, and derived from a variety of biomass feedstocks, a "worst case" must be taken for the liquids composition, i.e. high chemical variability. The classes of compounds which those found in fast pyrolysis liquids in the UN Manual are mainly in Class 3, but some are also Class 6.1 – Toxic substances [phenols, etc.] which are in concentrations > 0.1 wt%. The chemical compositions of some fast pyrolysis liquids are given in Table 2.

Table 2 GC-MS Analysis of three Fast Pyrolysis Liquids (16)

Chemical	Aston	BTG	IWC
(5H)-Furan-2-one	0.63	0.34	0.61
2,4- and 2,5-Dimethyl phenol	0.04	0.07	0.37
2.5-Dimethoxytetrahydrofuran (cis)	0.12	-	0.47
2-Furaldehyde	-	0.35	-
2-Furfuryl alcohol	-	0.01	-
2-Hydroxy-1-methyl-1-cyclopentene-3-one	0.22	0.13	0.15
4-Ally- and 4-Propyl syringol	-	-	0.19
4-Ethyl guaiacol	0.12	0.08	0.07
4-Methyl guaiacol	0.73	0.30	0.13
4-Methyl syringol	-	-	0.27
4-Vinyl guaiacol	0.05	0.03	0.05
5-Hydroxymethyl-2-furaldehyde	0.37	0.28	0.00
Acetic acid	2.65	2.56	4.23
Acetoguaiacone	0.16	0.17	0.07
Acetol	5.78	3.57	3.54
Eugenol	0.18	0.13	0.05
Guaiacol	0.44	0.20	0.14
Homovanillin	0.16	0.13	0.08
Hydroxyacetaldehyde	10.40	10.89	7.07
Isoeugenol (cis)	0.21	0.13	0.06
Isoeugenol (trans)	0.55	0.20	0.27
Levoglucosan	4.47	4.46	3.20
m-Cresol	0.26	0.08	-
o-Cresol	0.04	0.08	-
p-Cresol	0.02	0.08	-
Phenol	0.05	0.13	0.03
Syringol	-	-	0.29
Vanillin	0.21	0.29	0.09
Water	21.4	18.6	10.0
CHEMICALS	27.9	24.7	21.4
TOTAL	49.3	43.3	31.4

Aston – Aston University, UK [fluid bed reactor, softwood feedstock]
BTG – Biomass Technology Group, the Netherlands [rotating cone reactor, mixed wood
  feedstock]
IWC - Institute of Wood Chemistry, Germany [fluid bed reactor, mixed wood feedstock]

Due to the presence of Class 6.1 compounds in concentrations greater than 0.1 wt%, pyrolysis liquids are classed as 3(a) overall classification, using a cross classification to derive the most appropriate classification for complex mixtures. If the concentration of acetic acid is below 10wt%, there is no need to add additional labelling to highlight corrosiveness in the liquids. Using the chemical and physical data, further requirements relating to the hazard level posed by the liquids can be assessed and these are presented in Appendices II and III and are discussed later.

3.3   UN Code for Pyrolysis Liquids

To this end, the following generic classification is proposed based on the guidance in the UN Manual and the ADR for basic labelling purposes. The most appropriate UN not otherwise specified [N.O.S.] classification that can be used is:

UN 1993 FLAMMABLE LIQUID, N.O.S., 3, 1°(a), 2°(a), I

Only two chemical groupings need to be indicated for the components comprising the most significant risk. If needed, to ensure all possible chemical hazards are identified on a transportation document, conforming to UN 1993, the following code is recommended:

UN 1993 FLAMMABLE LIQUID, N.O.S., (Fast Pyrolysis Liquid) 3, 1°(a), 2°(a), (b), 3°(b), 5°(c), 31°(c), 1

This code may be interpreted as follows:

UN 1993	UN code for a flammable liquid [generic]
FLAMMABLE LIQUID	Flash point between 23-61°C for this liquid
N.O.S.	Not Otherwise Specified [i.e. not a listed substance in the UN Manual]
Fast pyrolysis liquid	Identify the name commonly found in the literature or reference texts. Trade names cannot be used.
3	Class number for a flammable liquid
1°(a), 2°(a), (b), 3°(b), 5°(c), 31°(c)	Chemicals or chemical groups in the liquid. Only the key two hazardous chemicals or chemical groups need to be identified
1	Packing Group [1 is for highly hazardous substances and usually an X is used on the UN approved package and on placards]

This classification may be used until an application is made to the European Union for the inclusion of pyrolysis liquids on the dangerous goods list. Pyrolysis liquids do contain other chemicals not included in the list noted above, however, the most hazardous chemicals are the ones that need to be identified. This classification is valid for liquids where the concentration of acetic acid is below 10wt% in the fast pyrolysis liquids. More details on the requirements for transportation of shipments are given in the following Section 4 and Appendices II and III.

This transportation code should be used on all labels and for all shipments, in particular on the transportation documents and the MSDS. For tank containers and for bulk shipments over 500 litres, placarding is also used and this is discussed in Section 4.

4.   PACKAGING OF PYROLYSIS LIQUIDS

The most crucial aspect of transport of dangerous goods is packaging. Appendix II gives the fuller details of packaging codes and weight restrictions for specific UN approved packaging types. Pyrolysis liquids are shipped in small samples of the order of a few mg to tonne quantities, in a variety of receptacles [single package or combination packaging] and for different purposes. This section highlights the following:

1.	Receptacle requirements to comply with UN regulations	Section 4.1
2.	Labelling of and packaging for all sizes of shipments	Section 4.2
3.	Empty packages and mixed packaging	Section 4.3

4.1 Receptacle Requirements to Comply with UN Regulations

Consignments of fast pyrolysis liquids are acceptable for international transport provided they meet the specifications of packaging described in the UN regulations (2). Packaging relates to samples of liquids of all sizes, however, depending on the classification of the substance as a hazardous material, there are limitations to the quantities, which may be shipped per package. Packages may also be single, e.g. drums, or combination packages, e.g. plastic bottles inside a cardboard box. In summary, the physical requirements for packaging are in Table 3.

Table 3   Minimum Package requirements for Pyrolysis Liquids

Packing group	I or "X" is used on UN approved packaging
Receptacle required minimum test pressure	250kPa g
Degree of filling of receptacle [at 15°C]:	90%
Hazard symbols	models 3, 6.2 and 11 [see Section 4.3.2]
Transportation document	see Appendix IV
MSDS	see Appendix V
Other comments	acid resistant material must be used

There are also particular volume and weight limits, depending on the package, materials of construction and type. The wide range of combinations is discussed in Appendix II, but a summary of the limits for inner packages is given in Table 4.

Table 4. Limitations on inner packages

Type of inner packaging	Maximum permissible capacity [l]
Glass, porcelain or stoneware packaging	5
Plastic packaging	30
Metal packaging	40
Other types of small packaging, e.g. tubes	1

Inner packages are the containers with the liquids inside; outer packages are for the containment of the sample and prevention of damage to the inner package.

4.2 Packaging Requirements

Pyrolysis liquids can be transported in varying quantities, from mg to tonnes. Due to the hazardous nature of pyrolysis liquids, it is possible that shipment quantities will be limited to 10,000 l per tank, although this would need to be clarified during classification. Some general guidance is given on the package specification for a range of shipment sizes.

4.2.1 Very small quantities [< 1 l]

For small samples, it is recommended that polypropylene [or Nalgene™] bottles be used with a cap insert inside the neck. This type of plastic packaging is extremely resilient to compression and damage. Glass bottles and sample vials should be avoided where possible, unless satisfactorily packed with adsorbent and a support material to reduce the potential for breakage. All very small quantities should be shipped as a combination package- an inner package in a cardboard box to UN standard with a suitable fabric adsorbent in the package.

4.2.2 Small quantities [1-10 l]

For larger quantities, individual 1 l containers should be used, either with a cap insert, or small plastic drums with non-removable heads. A stockist of UN approved packaging can provide a suitable receptacle and it is recommended that for quantities of less than 10 l, an outer package is used, e.g. a cardboard or a wooden box filled with adsorbent. If for example a 10 l plastic drum was used, it would be preferable to place it for shipment in an outer package, e.g. a steel drum or wooden box filled with adsorbent. Such small drums can also be shipped as a single package with the appropriate transport requirements.

4.2.3 Moderate quantities [10-450 l [max 400 kg]]

For moderate quantities, the UN limitations mean that maximum volumes are only 30 l for plastic packages [drums] or 40 l for metal drums [see Table 4]. Metal drums should be stainless steel, or a PTFE lined mild steel drum if used as an inner package. Mild steel drums will be attacked by the pyrolysis liquids and this could lead to drum failure. For single packages, e.g. drums, the maximum weight is 400 kg; therefore, standard drums/barrels can be used, provided they are stainless steel or polypropylene plastic barrels [or lined mild steel drums] and are appropriately filled. Drums with non-removable heads are recommended.

In addition to the packaging types noted above, the other type of packaging, which is occasionally used for pyrolysis liquids, is the Intermediate Bulk Container [IBC]. An IBC is a rigid, or flexible portable packaging, other than those specified in Appendix A.5 of the UN guide (2). According to the requirements of ADR, **IBCs are not to be used for Packing Group I liquids.**

4.2.4 Large quantities [> 400 kg]

Large samples need to be transported in larger containers or tanks. Containers are defined quite specifically in the ADR regulations [in decreasing volume] as:

Large container	internal volume more than 3 m^3
Small container	> 1 m^3, but less than 3 m^3.
Tank [alone]	means a tank-container, or a fixed tank or a demountable tank, or an element of a battery vehicle having a capacity more than 1 m^3.
Fixed tank	capacity more than 1 m^3, which is structurally attached to a vehicle, or is an integral part of the frame of such vehicle.
Tank-container	means an article of transport equipment [inc. tank swap bodies] conforming to the definition of the term container [marginal 10014] and built to contain liquids, gaseous, powdery or granular substances but having a capacity of more than 0.45m^3.
Demountable tank	tank, other than a fixed tank, a tank-container or an element of a battery vehicle, which has a capacity not more than 0.45 m^3, is not designed for carriage of goods without breakage of load, and normally can only be handled when empty.

The additional requirement for tank/containers is the use of a placard on road containers, displayed the appropriate UN code of 1993 on the bottom and 33X on the top [see Section 4.3.2]. Containers and tanks will be the preferred method with time for larger quantities for land transport. Again, acid resistant containers and tanks are required.

4.3   Labelling of Packages

4.3.1   Marking

Each package shall be clearly marked with the substance identification number of the goods to be entered in the transport document, preceded by the letters "UN". For pyrolysis liquids, this classification is 1993.

4.3.2   Danger labels

Packages containing substances or articles of this class shall bear a label conforming to model No. 3 [Class 3 – Flammable liquids] as shown in Figure 1 below:

Figure 1   Label model No. 3 [Class 3 – Flammable liquids] (black on red background)

Packages containing substance of 11° to 19°, 32° and 41° shall in addition bear a label conforming to model No. 6.1 [Class 6 – Toxic substances], as shown in Figure 2 overleaf:

Figure 2   Label model No. 6.1 [Class 6 – Toxic substances] (black on white background)

There is a requirement for label model no. 6.1, due to the presence of phenols in the liquids. Labels No. 3 and No. 6.1 shall be diamond shaped and measure at least 100 x 100 mm. They have a line of the same colour as the symbol appearing on the label 5 mm inside the edge and running parallel to it. If the size of the package so requires, the dimensions of the label may be reduced, if they remain clearly visible [seeTable 5]. Labels to be affixed to vehicles, to tanks of more than 3 m³ or to large containers shall measure not less than 250 x 250 mm.

Packages containing receptacles, the closures of which are not visible from the outside and packages containing vented receptacles or vented receptacles without outer packaging shall in addition bear on two opposite sides a label conforming to model 11, as shown in Figure 3. The label size should be 148mm x 210 mm high, or reduced sizes in this ratio depending on the package size..

Figure 3   Label model no. 11 – This way up

Any label required to be carried on a package shall be securely fixed to the package with its entire surface in contact with it and the label shall be clearly and indelibly printed. The colour and nature of the marking shall be such that the symbol [if any] and wording stand out from the background to be readily noticeable and the wording shall be of such a size and spacing as to be easily read. The package shall be so labelled that the particulars can be read horizontally when the package is set down normally. The dimensions of the labels required for packages are given in Table 5.

Table 5   Label model no. 11 Size Requirements for Packages

Capacity of Package	Dimensions of label
Not exceeding 3 litres if possible	at least 52 x 74 mm
Exceeding 3 litres but not exceeding 50 litres	at least 74 x 105 mm
Exceeding 50 litres but not 500 litres	at least 105 x 148 mm
Exceeding 500 litres	at least 148 x 210 mm

In addition, for tankers, or other large bulk transport, placards are typically used for road and rail transport. The placard dimensions are typically a minimum of 30 cm high by 40 cm wide, numerals to be a minimum of 10 cm high. The requisite codes for a placard are:

Substance Identification No. [Lower part]	Name of Substance	Hazard Identification No. [Upper part]
1993	Flammable Liquid [Fast Pyrolysis Liquid]	33X

A placard for large quantities [> 500 l] is shown in Figure 4 below, with the appropriate transportation codes.

305

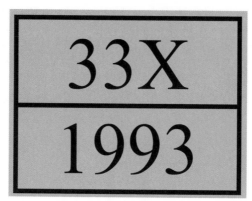

**Figure 4   Placard for transportation in containers
and bulk carriage [> 500 l] (black on orange)**

4.4   Mixed and Empty packaging

4.4.1   Mixed packaging

Fast pyrolysis liquids should not be shipped with other dangerous goods. Up to 5 litres [inner packaging] may be shipped with goods not subject to the provisions of ADR, provided they do not react dangerously with each other. Pyrolysis liquids shall not be packed together with substances and articles of classes 1 and 5.2 [explosives substances and articles and organic peroxides respectively, other than hardeners and compound systems] and class 7 material [radioactive material].

Only 0.5 l of fast pyrolysis liquids per inner packaging and 1 litre per package, which are classed under hazard group (a), may be shipped with up to 5 litres of Class 3 substance not exceeding 5 litres if they classified under (b) or (c). This is provided that mixed packaging is also permitted for substances or articles of these classes, and/or with goods, which are not subject to the provisions of ADR, provided they do not react dangerously with each other. Dangerous reactions are classed as:

1.   Combustion and/or giving off considerable heat,
2.   Emission of flammable and/or toxic gases,
3.   Formation of corrosive liquids,
4.   Formation of unstable compounds.

If wooden or fibreboard boxes are used, a package shall not weigh more than 100 kg [provided marginals 2002 (6) and 2002 (7) relating to the inclusion of adsorbent packaging and the provision of a leak proof outer package and that all individual packages are clearly singly labelled (2)].

4.4.2   Empty packaging

In the case of empty tank vehicles, empty demountable tanks and empty tank containers, uncleaned, this description shall be completed by adding the words, "Last load", together with the name and item number of the goods last loaded:

Last load 1993 Flammable liquid [Fast pyrolysis liquids], 1°(a)

This should help to reduce misuse of empty containers for other incompatible liquids.

5.   HANDLING AND STORAGE OF PYROLYSIS LIQUIDS

Handling of fast pyrolysis liquids is taken in this context to apply to the usage of the material and the precautions that should be adhered to minimise harm or risks from exposure to the liquids. The legal requirements for the bulk handling of materials is not discussed here, as this is again subject to national guidance and is summarised in Appendix IV. MSDS sheets, which are required to accompany the transportation document, are in Appendix V. Careful consideration must be given to site layout and the loading and unloading of the liquids to minimise spills and potentially harmful releases to the environment. This is discussed in Section 6.

6.   TREATMENT OF SPILLS

Spills can range in size and their prevention, the problems associated with spillages can be minimised by good handling, storage practices, and preventative measures.   Spillages of hazardous substances can arise as a result of poor storage systems, the use of unsuitable or defective containers, during refilling of tanks and other containers, or as a result of human error. Measures to deal with spillages will depend upon:

(a)   The hazard of the substance;
(b)   The quantities involved;
(c)   The location of the spillage; and
(d)   The potential for occupational ill health and/or pollution of ground, water and/or air arising from the spillage.

Procedures for dealing with both small and large spillages should be outlined in the supplier's safety data sheet for the substance concerned and, where the substance is classified as a "substance hazardous to health", in a health risk assessment. For pyrolysis liquids, there is extremely limited data or trials on the treatment of spillages of pyrolysis liquids.

In most cases, small spillages can be dealt with immediately by absorption in sand, sawdust or proprietary absorbent granules and subsequent disposal to a waste container. Large spillages will need a considerable degree of attention to ensure, in particular, that substances do not enter a drainage system or natural watercourse in concentrated form. This may entail containment of the spillage using drain cones, sand bags, polythene sheeting and a range of other materials, so that it can eventually be pumped into a disposal container. Never wash spillage of fast pyrolysis liquids to surface water drains. It should be noted that a material used to clean up a spillage of fast pyrolysis liquids must be disposed of as a special waste.

The potential for water and ground pollution because of spillages very much depends upon the handling systems operated on site. The following recommendations with regard to materials handling are made:

1. All loading and discharge points should be designated, marked and isolated from the surface water drainage system.
2. Routes of transfer for all materials should be identified and the complete route should be protected against spillages to the surface water system.
3. Avoid underground pipework where possible, as faults are more difficult to detect and can lead to groundwater contamination.
4. Avoid manual handling where possible to reduce the risk of human error and accidents.
5. Yard areas used for materials handling or materials handling processes must be isolated from the surface water drainage system by bunding. Roofing such an area is an advantage to prevent the accumulation of rainwater, fire regulations permitting.

6. Always use appropriate containers for different materials. They should be sturdy, in good condition, clearly labelled and not liable to leak.
7. Minimise the necessity for materials handling and transfer where possible to reduce the risk.

To handle spills, the following preliminary procedures are proposed:

Small Quantities[<1000ml]

1. Wear rubber gloves and suitable eye and face protection.
2. Cover contaminated area with sawdust, or other suitable inert adsorbent, e.g. vermiculite or montmorilonite
3. Take up sawdust and place in closed container.
4. Transport to approved landfill or incinerator for disposal

Large Quantities [>1000 ml]

1. Evacuate area.
2. Wear rubber boots, rubber gloves, suitable eye/face protection and NIOSH/MSHA approved respirator.
3. Cover contaminated area with sawdust, or other suitable inert adsorbent, e.g. vermiculite or montmorilonite
4. Take up sawdust or vermiculite and place in closed container. Transport to approved landfill or incinerator.
5. Where necessary, contain large spills with sandbags and other appropriate containment, e.g. adsorbent socks as found in emergency spill kits.

Where possible for large spills, a spill kit is advised to contain the spill and prevent its incursion to local watercourses. The recovery of pyrolysis liquids as an adsorbed waste may mean that they must be treated as "special waste" for disposal. Where possible, adsorbent should be recovered and stored in sealed containers for subsequent disposal.

7.   CONCLUSIONS

Fast pyrolysis liquids may be classed as a flammable liquid for the purposes of transport by any mode. The use of the UN manual for the assessment of pyrolysis liquids has allowed a N.O.S. classification to be used to ensure that liquids, when shipped in UN approved packaging, or packaging conforming to UN requirements is meeting the relevant international transportation regulations. The use of the UN code will allow samples to be shipped in a manner complying with the regulations currently in force in the EU, Canada and the USA.

Consignors must provide MSDS and a transportation document containing certain basic requirements, which have been summarised in this report. The use of a supplier of UN approved packaging will assist in the compliance of suppliers of liquids with packaging requirements. The requirements for such packages have been addressed in this report.

1. An appropriate UN code can be used for the shipment of pyrolysis liquids, conforming to international transportation regulations, primarily by road.
2. Labelling requirements have been summarised to ensure compliance with transportation regulations for all sizes of shipments
3. Volume and weight limitations on a range of shipment sizes have been specified.
4. MSDS and transportation documentation requirements have been specified
5. Preliminary procedures for the treatment of spills have been made.

8. RECOMMENDATIONS

The following recommendations are made:

1. Data on pyrolysis liquids is provided to the DETR – Dangerous Goods Branch to permit an application to the EU for the listing of pyrolysis liquids,
2. Data is required on the toxicity of pyrolysis liquids to ensure that MSDS can provide sufficient information for handlers and transporters of the liquids,
3. Assess relative quantities of adsorbents required to deal with spills and devise procedures for the treatment of spills under a variety of conditions.
4. Carry out work to remediate spill areas and assess environmental consequences.

9. DISCLAIMER

Conversion and Resource Evaluation Ltd accepts no responsibility for the subsequent use of the information contained herein, or for the result of loss or liability related to the use of this report. The authors accept no responsibility for opinions or conclusions based upon the report, which were not expressed by the report authors.

10. GLOSSARY

49 CFR	US Regulations relating to standards for Hazardous Materials transportation
ADN	The European Provisions concerning the International Carriage of Dangerous Goods by Inland Waterways
ADNR	Regulations for the Carriage of Dangerous Substances on the Rhine
ADR	EU Agreement on the Carriage of Dangerous Goods by Road
CAA	Civil Aviation Authority
CCNR	Central Commission for the Navigation of the Rhine
COSHH (UK)	Control of Substances Hazardous to Health Regulations 1994
DETR (UK)	Department of the Environment, Transport and Regions, UK
ECE	Economic Commission for Europe
ECOSOC	Economic and Social Council Committee of Experts
FAA (USA)	Federal Aviation Administration
FHWA (USA)	Federal Highway Administration
HSWA (UK)	Health and Safety at Work Act
IATA	International Air Transport Association
IBC	Intermediate Bulk Container
ICAO	International Civil Aviation Organisation
IMDG	International Maritime Dangerous Goods
IMO	International Maritime Organisation
MSDS	Material Safety Data Sheets
MSHA	Mine Safety and Health Administration, USA
NIOSH	National Institute of Occupational and Safety Hazards, USA
OTIF	Organisation for International Carriage by Rail
PTFE	Polytetrafluoroethene
RID	EU Agreement on the Carriage of Dangerous Goods by Rail
RID/ADR/ADN	The Joint Meeting of the Working Party on the Transport of Dangerous Goods and the RID Safety Committee
RSPA (USA)	Research and Special Programs Administration of the Department of Transportation, USA
TDG (Canada)	Transportation of Dangerous Goods Regulations, Canada
USCG	United States Coast Guard
USDOT	United States Department of Transportation

APPENDIX I: CHEMICAL AND PHYSICAL PROPERTIES OF PYROLYSIS LIQUIDS

I.1   Chemical Properties of Fast Pyrolysis Liquids

There are numerous references that include chemical analyses of fast pyrolysis liquids from a variety of sources, including slow pyrolysis tar, fast pyrolysis liquid and fractions thereof. As fast pyrolysis liquid may be raw, treated, filtered, catalytically derived, upgraded and dependent on the process parameters, a "worst case" must be taken for the liquid composition, i.e. high variability. As noted, fast pyrolysis liquid falls within Class 3, based on its physical properties, however, this is further complicated by the presence of other chemicals that are in Class 6.1 – Toxic substances [phenols, etc.] which are in concentrations above 0.1 wt%. The general chemical classes which the key components of pyrolysis liquids may be categorised in are given in Table 6.

Table 6   Chemical groups in the ADR and their Class identification [present in pyrolysis liquids]

Class 3 Groups	Chemical, chemicals or type and identification number
2301 A 1°(a)	1089 acetaldehyde, 3295 hydrocarbons, liquids, n.o.s., 2389 furan
2301 A 2°(a)	1993 flammable liquid, n.o.s.
2301 A 2°(b)	1224 ketones, n.o.s., 1989, aldehydes, flammable, n.o.s.
2301 A 3°(b)	1088 acetal
2301 B 17°	1230 methanol
2301 E 31°(c)	1989 aldehydes, flammable, n.o.s.
2301 E 31°(c)	2607 acrolein dimer, stabilised
2301 E 31°(c)	2245 cyclopentanone
2301 E 31°(c)	1224 ketones, n.o.s.
2301 E 32°(c)	Others, slightly toxic
Class 6.1 Groups	
2601 A 8°(a)	1092 acrolein, inhibited, 1143 crotoaldehyde, stabilised
2601 B 13°(b)	1199 furaldehydes
2601 B 14°(b)	2821 phenol solution
2601 B 14°(c)	2874 furfuryl alcohol

Due to the presence of Class 6.1 compounds, pyrolysis liquids are classed as highly hazardous, i.e. packing group I or X

Bibliography – Chemical Compounds

1.   Achladas, G.E., "Analysis of Biomass Pyrolysis Liquids: Separation and Characterisation of Phenols", J. Chrom., 1991, vol. 542, no. 2, pp. 263-275.
2.   Amen-Chen, C., Pakdel, H. and Roy, C., "Separation of phenols from eucalyptus wood tar", Biomass and Bioenergy, 1997, vol. 13, no. 1-2, pp. 25-37.
3.   Bighelli, A., Tomi, F., Casanova, J., "Computer Aided Analysis by Carbon-13 NMR Spectroscopy of Artificial Mixtures of Phenolic Derivatives Present in Biomass Pyrolysis Mixtures", Biomass for Energy and Industry: 7th EC Conference, Hall, D.O., Grassi, G., Scheer, H., (Eds.), Plenun Press, 1994, pp. 1008-1012.
4.   Elliot, D.C., "Analysis and Comparison of Biomass Pyrolysis/Gasification Condensates-Final Report", PNL-5943 UC-61D, US Department of Energy under contract De-AC06-76RLO 1830, 1986.

5. Elliot, D.C., "Chemical Analysis of Biomass Fast Pyrolysis Oil", Proceedings: Biomass Pyrolysis Oil Properties and Combustion Meeting, Milne, T.A. (Ed.), NREL, 1995, pp. 27-33.

6. Evans, R. J. and Milne, T.A., "Molecular Characterisation of the Products of Pyrolysis. 2. Applications", Energy & Fuels, vol. 1, no. 4, p. 311-319, ACS, 1987.

7. Evans, R. J., Milne, T. A., "Pyrolysis Oils from Biomass: Producing, Analyzing, and Upgrading", ACS Symposium Series No. 376, Soltes, E.J. and Milne, T.A., (Eds.), ACS Washington, DC, 1988, pp. 311-327.

8. Ghetti, P., Ricca, L. and Angelini, L., 'Thermal analysis of biomass and corresponding pyrolysis products", Fuel, 1996, vol. 75, no. 5, pp. 565-573.

9. Longley, C.J., Howard, J. and Fung, D., "Levoglucosan recovery from cellulose and wood pyrolysis liquids", Advances in Thermochemical Biomass Conversion, Bridgwater, A.V., (Ed.) , Blackie, 1994, pp. 1441-1451.

10. Material Safety Data Sheet prepared by G. Underwood for Red Arrow Products Inc., USA.

11. McKinley, J., "Final report biomass liquefaction: centralized analysis", B.C. Research, Vancouver, B.C., Canada, 1989.

12. Menard, H., Gaboury, A., Belanger, D. and Roy, C., "High-performance liquid chromatographic analysis of carboxylic acids in pyroligneous liquors", J. Anal. Appl. Pyr., Feb 1984, vol. 6, no. 1, pp. 45-57.

13. Menard, H., Roy, C., Gaboury, A., Belanger, D. and Chauvette, G., "Analyse totale du pyroligneux provenant de populus tremuloides par HPLC (Comprehensive Analysis of the Pyroligneous Acid Obtained from 'Populus Tremuloides' by HPLC)", Proceedings - Bioenergy R&D Seminar, NRC, 1982, pp. 331-336.

14. Milne, T.A., Evans, R.J. and Filley, J., "Molecular Beam Mass Spectrometric Studies of HZSM-5 Activity during Wood Pyrolysis Product Conversion", Research in Thermochemical Biomass Conversion, Bridgwater, A.V. and Kuester, J.L., (Eds.), Elsevier, 1988, pp. 910-926.

15. Piskorz J., Radlein D., Scott D.S. and Czernik S., "Liquid Products from the Fast Pyrolysis of Wood and Cellulose", Research in Thermochemical Biomass Conversion, A. V. and Kuester, J .L., (Eds.), Elsevier, 1988, pp. .557-571.

16. Piskorz, J., Scott, D.S., Radlein, D. and Czernik, S., "New applications of the Waterloo Fast Pyrolysis process", Biomass Thermal Processing, Hogan, E., Robert, J., Grassi, G. and Bridgwater A. V., (Eds.), CPL Press, UK, 1992, pp.64-73.

17. Scott, D.S., Piskorz, J. and Radlein, D., "The yields of chemicals from biomass based fast pyrolysis oils", Energy from Biomass and Wastes 16th, 1992.

18. Solantausta, Y, Diebold, J, Elliott, D.C, Bridgwater, A.V, Beckman, D, "Assessment of Liquefaction and Pyrolysis Systems", Technical Research Centre of Finland, Espoo, 1993.

19. Vanderhage, E.R.E., Boon, J.J., "Online curie-point pyrolysis-high-performance liquid-chromatographic mass-spectrometric analysis of lignin polymers", J. Chrom. A, 1996, vol. 736, no. 1-2, pp. 61-75.

20. Williams, P.T. and Horne, P.A., "Analysis of aromatic hydrocarbons in pyrolytic oil derived from biomass", J. Anal. Appl. Pyr., 1994, vol. 31, pp. 15-37.

Bibliography - Physical properties

1. Bakhshi, N. and Adjaye, J.D., "Properties and Characteristics of Ensyn bio-oil", Proceedings Biomass Pyrolysis Oil Properties and Combustion Meeting, Milne, T.A. (Ed.), US DoE, 1995, pp. 54-66.

2. Besler, S., Horne, P.A, Williams, P.T., "The fuel properties of biomass derived pyrolytic oils and catalytically upgraded products", Renewable Energy Technology and the Environment: Biomass Technology Wind Energy, Sayigh, A.A., (Ed.), Pergamon Press, vol. 3, pp. 1341-1345.

3.   D"Alessio, J., Lazzaro, M., Massoli, P. and Moccia, V., "Thermo-optical investigation of burning biomass pyrolysis oil droplets", Twenty-Seventh Symposium (International) on Combustion, Burgess, A.R. and Dryer, F.L., (Eds.) 1999, vol. 1 and 2, pp. 1915-1922.

4.   Diebold, J.P. Milne, T.A., Czernik, S., Oasmaa, A., Bridgwater, A.V., Cuevas, A., Gust, S., Huffman, D. and Piskorz, J., "Proposed specifications for various grades of pyrolysis oils", Developments in Thermochemical Biomass Conversion Bridgwater, A.V. and Boocock, D.G.B., (Eds.), Blackie Academic and Professional, 1996, vol. 1, pp. 1-26.

5.   Karaosmanoglu, F. and Tetik, E., "Fuel properties of pyrolytic oil of the straw and stalk of rape plant", Renewable Energy, 1999, vol. 16, no. 1-4, pp. 1090-1093.

6.   Krumdieck, S.P. and Daily, J.W., "Evaluating the feasibility of biomass pyrolysis oil for spray combustion applications", Combustion Science and Technology, 1998, vol. 134, no. 1-6, pp. 351-368.

7.   Oasmaa, A. and Sipilä, K., "Pyrolysis Oil Properties - Use of Pyrolysis Oil as Fuel in Medium-Speed Diesel Engines", Bio-oil Production and Utilisation. Proceedings of the 2nd EU-Canada Workshop on Thermal Biomass Processing, Eds. Bridgwater, A.V. and Hogan, E.N., 1995, pp. 175-185.

8.   Oasmaa. A., Leppämäki, E., Koponen, P., Levander, J. and Tapola, E., "Physical characterisation of biomass-based pyrolysis liquids - application of standard fuel oil analyses", VTT, Espoo, Finland, 1997.

9.   Peacocke, G.V.C., Russell, P.A., Jenkins, J.D. and Bridgwater, A.V., "Physical Properties of Flash Pyrolysis Liquids", Biomass and Bioenergy, Elsevier Science Ltd., 1994, vol. 7, no. 1-6, pp. 169-177.

10. Sipilä, K., Kuoppala, E., Fagernas, L. and Oasmaa, A., "Characterization of biomass-based flash pyrolysis oils", Biomass and Bioenergy, 1998, vol. 14, no. 2, pp. 103-113.

APPENDIX II: PACKAGING OF PYROLYSIS LIQUIDS FOR TRANSPORT

It is most likely that producers of pyrolysis liquids will purchase packaging for the storage of liquids for transport or pay a company to transport the liquids. It is important that the supplier of the sample complies with UN approved guidelines for the package and thus avoid the need for expensive testing of other packages.

By using, an approved supplier of UN packaging; there will usually be no need to test a package. It is strongly recommended that packaging be purchased from companies supplying UN approved packaging. Most companies, institutions and research groups do not have the resources to test other packaging, which would have to comply with the UN Regulations. As noted, pyrolysis liquids are a Class 3 substance – Flammable Liquid. All of the guidance provided below is tailored to fast pyrolysis liquids. This section therefore deals with the exact coding on a packaging suited to transport of fast pyrolysis liquids. The term "marginals" are referred to and these are found in the ADR requirements and are included for reference (2). The codes are given to assist suppliers of pyrolysis select the correct packaging, or combination of packaging to ensure compliance.

II.1   Coding of for packaging conforming to UN Regulations

Packages for compliance with the UN regulations must display a range of codes as described below. Guidance is given here to assist receivers of packages and those unfamiliar with the coding system fir UN approved packaging. There are 10 parts to the code. The code number consists of in the first two parts:
1.   An Arabic numeral indicating the kind of packaging, e.g. drum,
2.   A capital letter or letters indicating the nature of the material

Where necessary, an Arabic numeral indicating the category of packaging within the type which the packaging belongs.

For pyrolysis liquids, the following packaging are recommended with codes:
1.   Drum
4.   Box
5.   Bag
6.   Composite packaging

Type of material:
A.   Steel (all types and surface treatments)
H.   Plastics material, including expanded plastics material
P.   Glass, porcelain or stoneware

e.g. 1H is a plastic drum

Pyrolysis liquids are in Packing Group I. The code number of the packaging shall therefore be followed by an X, indicating the groups of substances for the design type is approved for.
Each packaging shall bear markings that are durable, legible and placed in a location and of such a size relative to the packaging as to be readily visible. For packages with a gross mass of more than 30 kg, the markings or a duplicate thereof shall appear on the top or on a side of the packaging. Letters and numerals and symbols shall be at least 12 mm high, except for packaging of 30 litres or 30 kg capacity or less, when they shall be at least 6 mm in height and for packaging of 5 litres or 5 kg or less when they shall be of an appropriate size. The markings for new packaging manufactured in conformity with the approved design type consists of 8 requirements [A-G]:

A   the symbol UN for packaging conforming to UN regulations [marginal 3510]:

For metal packaging on which the marking is stamped, the letters UN may be applied instead of the symbol. The symbol ADR [or RID/ADR for packaging approved for rail transport as well as road transport]

B   the packaging code number, e.g. 1H1 for a plastic drum with a non-removable head

C   a code in two parts:

    (i)   A letter denoting the packaging group- for pyrolysis liquids this is X [packing group I]

    (ii)   For single packages without inner packages, intended to contain liquids, which have successfully passed the hydraulic pressure test, the relative density, rounded off to the first decimal place of more than 1.2, for which the design type has been tested. This information may be omitted if the relative density is not more than 1.2. For pyrolysis liquids, the relative density is typically 1.2; therefore, this information must be displayed.

    (iii)   For packaging intended to contain inner packaging, and removable head packaging, intended to contain liquids having a viscosity at 23°C of more than 200 mm^2/s [200cSt], the maximum gross mass in kg. This may apply to some pyrolysis liquids, depending on the water content.

D   Either a letter "S" denoting that the packaging is intended to contain liquids having a viscosity at 23°C of more than 200 mm^2/s [200cSt], or inner packaging, or where a hydraulic pressure test has been successfully passed, the test pressure rounded down to the nearest 10 kPa.

E   The year of manufacture [last two digits], in addition for packaging of types 1H and 3H, the month of manufacture; this part of the marking may be affixed in a different place from the other particulars.

F   The mark of the State in which the approval was issued;

G   Either a registration number and the name of the mark of the manufacturer or some other packaging identification mark specified by the competent authorities.

For remanufactured metal drums, if there is no change in the packaging type and no replacement or removal of integral structural components, the required markings need not be permanent [e.g. embossed]. Every other remanufactured metal drum shall bear the markings in A-E in a permanent form [e.g. embossed] on the top head or side. Metal drums made from materials [e.g. stainless steel] designed for reuse repeatedly may bear the markings indicated in F-G in a permanent form [e.g. embossed]. Container providers are referred to the exact requirements of the UN protocols as given on p. 466-467 of Vol. 1 (1).

An example is a steel drum [1A1]:

1A1/X1.2/S/99   codes A (i), B-E inclusive for pyrolysis liquids, containing pyrolysis liquids [viscosity > 200 cSt at 23°C], drum made in 1999

NL/VL123   e.g. codes F and G for a company in the Netherlands

Codes suitable for pyrolysis liquids for inner or primary storage, suited to pyrolysis liquids are summarised below and are given to assist providers of pyrolysis liquids in the identification of shipments and ensure compliance with regulations for international shipment.

A. Packaging conforming to marginal 3510 (1) and marked "UN"				
Kind	Material	Category	Code	Marginal
1. Drums	A. Steel	Non-removable head	1A1	3520
		Removable head	1A2	3520 *1
	H. Plastics	Non-removable head	1H1	3526
		Removable head	1H2	3526 *1
3. Jerricans	A. Steel	Non-removable head	3A1	3522
		Removable head	3A2	3522 *1
	H. Plastics	Non-removable head	3H1	3526
		Removable head	3H2	3526 *1
4. Boxes	A. Steel	-	4A	3532 *1
		With liner	4A	
	B. Aluminium	-	4B	3532 *1
		With liner	4B	
	C. Natural wood	Ordinary	4C1	3527 *1
		Sift-proof walls	4C2	
	D. Plywood	-	4D	3528 *1
	F. Reconstituted Wood		4F	3529 *1
	G. Fibreboard	-	4G	3530 *1
	H. Plastics	Expanded	4H1	3531 *1
		Solid	4H2	
5. Bags	H. Plastics film	-	5H4	3535
6. Composite Packaging	H. Plastics Receptacles	In steel drum	6HA1	
		In steel crate *2 or box	6HA2	
		In aluminium drum	6HB1	
		In aluminium crate or box	6HB2	
		In wooden box	6HC	3537
		In plywood drum	6HD1	
		In plywood box	6HD2	
		In fibre drum	6HG1	
		In fibreboard box	6HG2	
		In plastics drum	6HH1	
		In solid plastics box	6HH2	
B. Packaging which may conform to marginal 3510 (1) or (2)				
6. Composite packaging	P. Glass, porcelain or stoneware receptacle	In steel drum	6PA1	
		In steel crate *2 or box	6PA2	
		In aluminium drum	6PB1	
		In aluminium crate or box	6PB2	
		In wooden box	6PC	3539
		In plywood drum	6PD1	
		In wickerwork hamper	6PD2	
		In fibre drum	6PG1	
		In fibreboard box	6PG2	
		In expanded plastics packaging	6PH1	
		In solid plastics packaging	6PH2	

C. Conforming only to marginal 3510 (2) and marked "ADR" [or RID/ADR]				
0. Light gauge metal packaging	A. Steel	Non-removable head	0A1	3540
		Removable head	0A2	

Note: *1   according to marginal 3538, these packaging can be used as outer packaging
           for combination packaging.

        *2   crate are packaging with incomplete surfaces.

Limits are imposed on the packages in terms of weight:

A. Packaging Conforming to marginal 3510 (1) and marked "UN"				
Kind	Material	Category	Maximum Capacity [l]	Maximum net mass [kg]
1. Drums	A. Steel	Non-removable head	450	400
		Removable head		
	H. Plastics	Non-removable head	450	400
		Removable head		
3. Jerricans	A. Steel	Non-removable head	60	120
		Removable head		
	H. Plastics	Non-removable head	450	400
		Removable head		
4. Boxes	A. Steel	-		
		With liner		
	B. Aluminium	-		
		With liner		
	C. Natural wood	Ordinary		400
		Sift-proof walls		400
	D. Plywood	-		400
	F. Reconstituted Wood			400
	G. Fibreboard	-		400
	H. Plastics	Expanded		60
		Solid		400
6. Composite Packagings	H. Plastics receptacles	In steel drum	250	400
		In steel crate *2 or box	60	75
		In aluminium drum	250	400
		In aluminium crate or box	60	75
		In wooden box	60	75
		In plywood drum	250	400
		In plywood box	60	75
		In fibre drum	250	400
		In fibreboard box	60	75
		In plastics drum	250	400
		In solid plastics box	60	75

For inner packaging, the following may be used:

1.   Glass, porcelain or stoneware packaging with a maximum permissible capacity of 5 litres
     for liquids
2.   Plastic packaging with a maximum permissible capacity of 30 l for liquids
3.   Metal packaging with a maximum permissible capacity of 40 l for liquids
4.   Other types of small packaging such as tubes maximum permissible capacity of 1 l for
     liquids

B. Packaging which may conform to marginal 3510 (1) or (2)				
Kind	Material	Category	Maximum Capacity [l]	Maximum net mass [kg]
6. Composite packagings	P. Glass, porcelain or stoneware receptacle	In steel drum	60	75
		In steel crate *2 or box		
		In aluminium drum		
		In aluminium crate or box		
		In wooden box		
		In plywood drum		
		In wickerwork hamper		
		In fibre drum		
		In fibreboard box		
		In expanded plastics packaging		
		In solid plastics packaging		
C. Conforming only to marginal 3510 (2) and marked "ADR" [or RID/ADR]				
0. light gauge metal packaging	A. Steel	Non-removable head	40	50
		Removable head		

II.2 Filling of Receptacles

It is not recommended to fit packages containing pyrolysis liquids with vents for overpressure. Liquids shall only be filled into packages that have appropriate resistance to the internal pressure that may be developed under normal conditions of carriage. Packaging marked with the hydraulic test pressure [as prescribed in UN Regulations, marginal 3512 {1}] shall be filled only with liquid having a vapour pressure:

1. Such that the total gauge pressure in the packaging (i.e. the vapour pressure of the pyrolysis liquids plus the partial pressure of air or other inert gases, less 100 kPa) at 55°C determined on the basis of the maximum degree of filling [90%] and at a filling temperature of 15°C, will not exceed 2/3 of the marked test pressure; or
2. At 50°C, less than 4/7 of the sum of the marked test pressure plus 100 kPa; or
3. At 55°C less than 2/3 of the sum of the marked test pressure plus 100 kPa

For packing group I, which pyrolysis liquids would be classed as, the required minimum test pressure is 250kPa. The degree of filling [as a percentage of the capacity of the packaging at 15°C] is 90%

APPENDIX III: HANDLING AND STORAGE – GENERAL GUIDANCE

III.1 Handling of Pyrolysis Liquids

Handling and storage activities are a common feature of the majority of workplaces. They are also one of the principal causes of death and injury due to this interface between people and the wide range of materials handled. This section sets out the measures necessary on the part of employers, employees, manufacturers, designers, importers and suppliers of pyrolysis liquids used at work to ensure safety and the absence of risks to health in connection with the use, handling, storage and transport of pyrolysis liquids. Handling and storage covers a very broad range of areas, including:

1. The use of fixed and mobile handling equipment;
2. Manual handling operations;
3. Design of the workplace;
4. The provision of a suitable working environment;
5. Specific requirements for the handling and storage of identified hazardous substances;
6. Controls on the use of hazardous materials;
7. Specific requirements for labelling of hazardous substances; and
8. The selection, provision and use of personal protective equipment.

The law on handling and storage is diverse, ranging from the more general requirements under the Health and Safety at Work Act [HSWA] to the specific requirements of the Control of Substances Hazardous to Health [COSHH] Regulations 1994 and the Highly Flammable Liquids and Liquefied Petroleum Gases Regulations 1972 in the UK for example. Each country has its own legislation and guidance for the handling of goods and therefore national information should be used where possible.

The handling and storage of materials has, in many cases, great potential for pollution incidents, particularly in the case of hazardous materials that may be discharged by natural seepage to water and land resulting in groundwater pollution in particular. Within the EU, this issue is tasked within the new Integrated Pollution Prevention and Control [IPPC] regulations, which came into force on 1st August 2000.

Most national regulations apply, in the case of the handling and storage of materials, with particular reference to:
(a) Risk assessments;
(b) Implementation of management systems for the effective planning, organising, controlling, monitoring and review of any preventive and protective measures arising from a risk assessment;
(c) Appointment of competent persons;
(d) Establishment of emergency procedures to be followed in the event of serious or imminent danger;
(e) Provision of comprehensible and relevant information;
(f) Consideration of human capability;
(g) Provision of health and safety training; and
(h) Consultation with safety representatives.

III.2 Storage of hazardous substances

Before storing and handling pyrolysis liquids, it is imperative to consult sources of hazard data, typically the MSDS or other available sources (17). The chemical compatibility of hazardous materials must be given particular consideration. Potentially reactive material must be stored separately [mixing may occur due to spillage, leakage or accident e.g. during a fire]. The following precautions are necessary to ensure the safe handling and storage of dangerous goods and/or chemical substances with pyrolysis liquids:

1. Meticulous standards of housekeeping should be maintained at all times;
2. Smoking and the consumption of food or drink should be prohibited in any area in which substances are used or stored e.g. laboratory, bulk chemical store;
3. Staff must be reminded regularly of the need for good personal hygiene, in particular washing of hands after handling chemical substances;
4. The minimum quantities only should be stored in the working area; extra bulk storage may be required separately and well away from the work area;
5. Containers and transfer containers should be clearly and accurately marked;
6. Chemical substances should always be handled with care and carriers used for Winchester and other large containers;
7. Fume cupboards should operate with a minimum face velocity of approximately 0.4 m/sec when measured with the sash opening set at 300 mm maximum, and performance should be checked frequently in accordance with the COSHH Regulations;
8. Staff should always wear personal protective clothing and equipment e.g. eye protection, face protection, aprons, gloves, wellington boots, whenever handling or using dangerous chemical substances;
9. Any injury should be treated promptly, particularly skin wounds and abrasions; and
10. Responsibility for safe working should be identified at senior management level, and written procedures published and used in the training of staff.

III.2.1 Bulk chemical storage [drums, barrels, tanks and similar containers]

In the design and use of bulk storage facilities, the following aspects need attention:

1. The range and quantities of substances to be stored;
2. Dependent upon (1) above, the degree of segregation by distance of:
   a. The store from any other building; and
   b. Certain chemical substances within the store from other chemical substances stored.

Purpose-built chemical stores should be of the detached single-storey brick built type or constructed in other suitable materials, such as concrete panels, with a sloping roof of weatherproof construction. The structure should have a notional period of fire resistance of at least one hour.

Other features include:

(a) Permanent ventilation by high and low level air bricks set in all elevations, except in those forming a boundary wall; low-level air bricks should be sited above door sill level;
(b) Access doors constructed from material with at least one hour notional period of fire resistance; doorways should be large enough to provide access for fork lift trucks, with ramps on each side of the door sill (also to contain any internal spillage); separate pedestrian access, which also serves as a secondary means of escape, should be provided;
(c) An impervious chemical-resistant finish to walls, floors and other surfaces;
(d) Artificial lighting by sealed bulkhead or fluorescent fittings, to provide an overall luminance level of 300 lux;
(e) Provision of adequate space, with physical separation and containment for incompatible substances, each area to be marked with the permitted contents, the hazards and the necessary precautions, and incorporating an area for the storage of empty containers;
(f) Fire separation of individual areas sufficient to prevent fire spreading;
(g) Provision of the following equipment in a protected area outside the store:
   1. Fire appliances (dry powder and/or foam extinguishers);
   2. Fixed hose reel appliance;

3.	Emergency shower and eyewash station with water heating facility to prevent freezing;
4.	Personal protective equipment i.e. safety helmet with visor, impervious gloves, disposable chemical-resistant overall, with storage facilities for same; and
5.	Respirator and breathing apparatus in a marked enclosure;

(h)	A total prohibition on the use of naked flames and smoking, appropriate warning signs should be displayed;

(i)	A prohibition on the use of the store for storage of other items or for any other purpose; and

(j)	Provision of racking or pallets to enable goods to be stored clear of the floor.

III.2.2   External drum storage

Drums, barrels, carbuoys and other similar containers of pyrolysis liquids should be stored in the external air on an impervious and durable surface, which is in excess of 4 m to any risk area, bund or open boundary. The area should be protected by a bund wall, dished or ramped to contain spillages, with the walls and floor impervious to the materials stored. The bunded area should contain no drains or valves. Vehicular access to such areas should be protected by a ramp or a channel ensuring, of course, that the ramp itself does not cause regular spillages.

Generally, no container should be stored within 2 m of any window, escape route or door. Much will depend on the nature of the substances stored and the design of the storage area. Where a storage area is constructed with fire resistant walls, these distances can be reduced.

Ensure that overflow pipes on all tanks discharge within the bunded area. Remember any tank situated on a roof may drain to the surface water system via the guttering therefore avoid roof storage wherever possible. Flammable liquids should be stored in a purpose-built external flammable materials store and not in a warehouse. Much will depend upon the quantities to be stored. Small quantities should be stored in a lockable metal cupboard, suitably marked.

Consider the storage of chemical drums too. These must be within a bunded area to contain any spillages. Ensure vehicular access to such areas is protected too, by a ramp or a channel, but ensure the use of the ramp does not itself cause regular spillages! Have automatic cut-offs on all delivery pipes to prevent spillage due to overfilling.

III.2.3   Underground storage tanks

A wide range of flammable liquids are stored in underground tanks, although at present there is no requirement for underground storage of pyrolysis liquids. The following general precautions are necessary:
1.	The tank should be subject to regular examination and test by a competent person;
2.	The tank should be located in an area free from vehicular traffic as far as possible; where this is not possible, the tank compartment will need extra reinforcement and protection;
3.	Permanent venting should be provided to allow for the release of waste gases and to prevent excessive pressure in the tank;
4.	The tank should be located away from buildings to prevent subsidence; the distance from buildings will depend on the nature of the substance stored;
5.	Operators should be trained in safe entry procedures, emergency rescue procedures and in the use of breathing apparatus;
6.	Operators should, where atmospheric testing identifies a risk, wear breathing apparatus, together with rescue harness with lifeline attached; and
7.	Someone should be stationed outside the tank to keep watch and communicate with people inside and, if necessary, take charge of rescue procedures.

Access to underground storage tanks and work in confined spaces has always been a high risk activity due to the possibility of dangerous gas or vapour concentrations, rust, which consumes oxygen, and oxygen-deficient atmospheres. For examples, the Confined Spaces Regulations 1997 [UK] require employers to:

(a)   Avoid entry to confined spaces, for example, by doing the work from outside;

(b)   Follow a safe system of work, e.g. a Permit to Work system, if entry to a confined space, such as an underground tank, is unavoidable; and

(c)   Put in place adequate emergency arrangements before work starts, which will also safeguard rescuers.

Underground tanks used for the storage of oils, solvents and effluents are a common source of groundwater pollution.   Such tanks require regular examination, maintenance and testing. Pressure testing of the tanks should be undertaken to identify leaks. Where possible, new underground tanks should be double skinned or be 'housed' in a concrete structure (to reduce corrosion and provide secondary containment). Particular attention should be paid to the location of underground connecting pipework. Where possible, underground pipework should be located within impervious ductwork.

III.3   Health risk assessments

Employers must, where employees may be exposed to substances hazardous to health, make a suitable and sufficient assessment of the risks (a health risk assessment) created by that work to employees and the steps that need to be taken. The assessment must be reviewed if it is no longer valid or there has been a significant change in the work to which the assessment relates, and any changes, as a result of the review, must be made.

III.3.1   Prevention or control of exposure to substances hazardous to health:

Exposure to pyrolysis liquids and vapours must either be prevented or, where this is not reasonably practicable, controlled. Except in the case of a carcinogen or biological agent, prevention or adequate control shall be by means other than the provision of personal protective equipment.

III.3.2   Use of control measures, etc:

Every employer shall take all reasonable steps to ensure that any control measure is properly used or applied. Every employee shall make full and proper use of any control measures, return PPE to any accommodation provided, and report defects in control measures to his employer.

III.3.3   Maintenance, examination and test of control measures, etc:

Employers must ensure that any control measure is maintained in efficient state, in efficient working order and in good repair and, in the case of PPE, in a clean condition.

III.3.4   Monitoring exposure at the workplace:

Where appropriate, the employer shall ensure that there is a suitable procedure for monitoring the exposure of employees, including the keeping of records of such monitoring procedures.

III.3.5   Health surveillance:

Where appropriate, the employer shall ensure that employees who are liable to be exposed are under suitable health surveillance.

III.3.6   Information, instruction and training etc:

Where employees are exposed to the risk of exposure, those employees must be provided with such information, instruction and training as is suitable and sufficient for them to know the risks of such exposure and the precautions that should be taken.

III.3.7 Prevention and control strategies

Every employer shall ensure that the exposure of his employees to substances hazardous to health is either prevented or, where this is not reasonably practicable, adequately controlled.

So far as is reasonably practicable, the prevention or adequate control of exposure of employees to a substance hazardous to health, except to a carcinogen or biological agent, shall be secured by measures other than the provision of personal protective equipment.

Control strategies include:
1. Enclosure/containment: can the materials be handled so that individuals never need come into contact with them? Total enclosure or containment of the process may be possible by the use of bulk tanks and pipework to deliver a liquid directly into a closed production vessel. Complete enclosure is practicable if the substances are in liquid form, used in large quantities, and if the range of substances is small.
2. Isolation/separation: can the process be put somewhere else? The isolation of a process may simply mean putting it into a small locked room, thereby separating the workforce from the risk, or could involve the construction of a chemical plant in a remote geographical area. The system of isolation is required to prevent access effectively, or certainly restrict access only to those who need to be there.
3. Ventilation systems: ventilation is an important control strategy [see also Section. III.2.1]. Here it is necessary to distinguish between natural ventilation and mechanical ventilation systems.

Handling of pyrolysis liquids can lead to the release of vapours and gases during use, either in production, transfer or monitoring, therefore consideration must be given to exposure to the gases, due to a variety of means. Local exhaust ventilation may therefore be required. Local exhaust ventilation [LEV] systems these take two principal forms, receptor systems and captor systems:
1. Receptor systems: in a receptor system, the contaminant enters the system without inducement. The fan in the system is used to provide airflow to transport the contaminant from the hood/enclosure through the ducting to a collection system. The hood may form a total enclosure around the source, for example, a laboratory fume cupboard. Receptor hoods receive contaminants as they flow from their origin under the influence of thermal currents.
2. Captor systems: with a captor system, the air which flows into the hood captures the contaminant at some point outside the hood and induces its flow into the system. The rate of flow of air into the hood must be sufficient to capture the contaminant at the furthermost point of origin, and the air velocity induced at this point must be high enough to overcome any tendency the contaminant may have to go in any direction other than into the hood.

Contaminants emitted with high energy (large particles with high velocities) will require high velocities in the capturing stream.

III.3.8 Dilution ventilation

In certain cases, it may not be possible to extract a contaminant close to the point of origin. If the quantity of contaminant is small, uniformly evolved and of low toxicity, it may be possible to dilute the contaminant by inducing large volumes of air to flow through the contaminated region. Dilution ventilation is most successfully used to control vapours, for example, organic vapours from pyrolysis liquids.

III.3.9   Personal protective equipment [PPE]

The use of various forms of PPE, including respiratory protective equipment [RPE] is never a perfect solution to preventing exposure to hazardous substances. As a control strategy, it relies heavily on the operator wearing the correct PPE/RPE all the time he is exposed to the risk and people simply will not do this. In the majority of cases, the provision and use of PPE should be seen as an extra form of protection where other forms of protection, as indicated above, are operating.

Recommended PPE for using pyrolysis liquids are:
1.   Safety boots or shoes with protective steel toecaps
2.   Acid and solvent resistant gloves
3.   Coveralls
4.   Safety glasses or goggles
5.   Breathing apparatus if dealing with spills [organic vapour filter mask for short term exposure]

APPENDIX IV: TRANSPORT DOCUMENT FOR PYROLYSIS LIQUIDS

IV.1 EU Transportation requirements

The transport document must contain the following information for any dangerous goods; however, the information relevant to pyrolysis liquids is included for future reference [assumed in this case initially for road transport]:

A description of the goods including the substance identification no. [where available]	UN 1993 Flammable Liquid, N.O.S., (Fast pyrolysis Liquid) Biomass derived liquid produced by fast pyrolysis. Flash point 50-70°C.
The class	3
The item number together with any letter	1°(a), 2°(a), 2°(b), 3°(b), 5°(c), 17°(a), 17°(b), 19°(a), 31°(c), 32°(c)
The initials ADR or RID	ADR, RID as appropriate
The number and description of the packages	As applicable
The total quantity of the dangerous goods [as a volume or gross mass or as a net mass]	As applicable
The name and address of the consignor	As applicable
The name and address of the consignee(s)	As applicable
A declaration as required by the terms of any special agreement	Not applicable
Instructions to be implemented in the case of an accident	See below

IV.2 Instructions in Case of Accidents or Spills

No specific procedures have been adequately developed for pyrolysis liquids, however, the unusual properties of the liquids allows some general comments to be made to assist in the minimising of any hazard to the environment, or personnel involved with dealing with an accident. Advice on dealing with spills is given in Section 6 and in the MSDS [Appendix V].

Instruction must be provided in writing for the driver of a vehicle, or other transporter by the consignor. As a precaution against any accident or emergency that may occur or arise during carriage, the transporter shall be given instructions in writing, specifying concisely for each dangerous substance or article carried for each group of dangerous goods presenting the same dangers to which the substance(s) or article(s) carried belong(s):

1. The name of the substance or article or group of goods, the class and the identification number or for a group of goods, the identification numbers of the goods for which these instructions are intended or applicable;
2. The nature of the danger inherent in these goods as well as the measures and personal protection to be applied by the driver;
3. The general actions, e.g. to warn road users and passers-by and call the fire brigade/police, etc.
4. The additional actions to deal with minor leakages or spillages to prevent their escalation, if this can be achieved without personal risk;
5. The special actions for special products, if applicable;
6. The necessary equipment for general and if applicable additional and/or special actions.

Instruction to be provided to the Transporter [Provisional]

I   Load
Pyrolysis liquids, class 3, UN 1993 [package identification numbers to be added here].
Hazard Identification number: 33X

II   Nature of Danger
- Harmful by inhalation, in contact with skin and if swallowed.
- Do not breathe vapours or fume from combusting or heated liquids.
- Irritating to eyes, respiratory system and skin.
- Corrosive [pH 1.5-3].
- Flash point of 50-70°C for the raw liquids.
- Possible mutagen, contains potentially carcinogenic compounds.
- Avoid continuous exposure.

III   Personal Protection
Wear gloves, full eye protection [goggles] and organic vapour filter mask if a significant spillage occurs. Wear coveralls if dealing with a spill.

IV   General Actions by transporter
1. Stop engine
2. No naked light, no smoking
3. Cordon off area if land release occurs and warn others nearby
4. Notify appropriate fire and police authorities as soon as possible

V   Additional And/or Special Actions by the transporter
To handle spills, the following equipment is recommended

Small Quantities [<1 l]
- Wear rubber gloves and suitable eye and face protection.
- Cover contaminated area with a suitable inert adsorbent preferably, e.g. vermiculite, or montmorilonite.
- Take up adsorbent and place in closed container.
- Transport to approved landfill or incinerator.

Large Quantities [>1 l]
1. Evacuate area.
2. Wear rubber boots, rubber gloves, suitable eye/face protection and NIOSH/MSHA approved respirator.
3. Cover contaminated area with a suitable inert adsorbent preferably, e.g. vermiculite, or montmorilonite. Take up adsorbent and place in closed container. Transport to approved landfill or incinerator.

VI   Special Action - Fire

Extinguishing Media: Water, Carbon Dioxide, Foam, Powder

Special Fire fighting Precautions:
1. Wear self-contained breathing apparatus and protective clothing to prevent contact with eyes and skin.
2. Do not inhale smoke from fire.
3. Use water spray to cool fire exposed containers.

First Aid

1.   In case of contact with eyes, flush with copious amounts of water for 15 minutes. Remove contaminated clothing and seek medical assistance.
2.   In case of contact with skin, flush with copious amounts of water.   Remove contaminated clothing.
3.   If inhaled remove to fresh air. If breathing is difficult, give oxygen. If not breathing, give artificial respiration and call for medical assistance.
4.   If swallowed, wash out mouth with water. Consume water to dilute. Call for medical assistance immediately.

Additional information
Carriage of a suitable adsorbent is recommended, i.e. dry sawdust or an inert adsorbent such as vermiculite. It is recommended to carry approx. 0.5-kg adsorbent/kg pyrolysis liquids.

Note:
These particulars to be entered in the document shall be drafted in an official language of the forwarding country, and also, if that language is not English, French or German, in English, French or German, unless international road transport tariffs, if any, or agreements concluded between the countries concerned in the transport operation, provide otherwise.

IV.3   North America Transportation Requirements

In North America, there are slightly different procedures for the transport of dangerous goods and the U.S. DOT has established a set of regulations that requires a series of warning signs, labels, package markings, and shipping documents, similar to the UN system. In Canada, an emergency response plan must be submitted to Transport Canada.

APPENDIX V: MATERIAL AND SAFETY DATA SHEETS FOR PYROLYSIS LIQUIDS

NOTE:

There is at present no official recognition of the information contained within this document.

The recommended transportation code for pyrolysis liquids is:

UN 1993 Flammable liquids, N.O.S., [Fast pyrolysis liquids], 3, 1°(a), 2°(a), I

The user of these notes is also advised to refer to the following UN Regulations:

United Nations Recommendations on the Transport of Dangerous Goods Model Regulations, 11th Revised Edition, ISBN 92-1-139067-2, January 2000.

Consult closely with a UN Approved supplier of packaging materials.

Material Safety Data Sheets

1.0   Chemical Identification
Pyrolysis liquid (also known as pyrolysis oil, bio-crude-oil, bio-oil bio-fuel-oil, pyroligneous tar, pyroligneous acid, wood liquids, wood oil, liquid smoke, wood distillates)

2.0   Composition/Information on Ingredients
Complex mixture of highly oxygenated hydrocarbons. A mixture of three to four hundred chemicals derived by the thermal decomposition of biomass. A typical pyrolysis oil may be composed as follows:

Composition	wt%
organic acids	5-10%
anhydrosugars	0-5%
ketones	0-5%
phenolics	up to 25%
hydrocarbons	0-5%
water	up to 35%

3.0   Synonyms
Bio-oil, pyrolysis liquid, pyroligneous oil, bio crude oil, pyroligneous liquid, liquid wood, bio-fuel oil, wood tar, wood oil.

4.0   Hazards Identification
Label Precautionary Statements

- Harmful by inhalation, in contact with skin and if swallowed.
- Irritating to eyes, respiratory system and skin.
- Corrosive (pH 1.5-3).
- Flash point of 50-70°C for the raw liquids.
- Possible mutagen, contains potentially carcinogenic compounds.
- Keep container tightly closed in a cool well ventilated place.
- Do not empty into drains.

- Wear suitable gloves and eye/face protection when handling.
- Avoid continuous exposure.

5.0 *First Aid Measures*
- In case of contact with eyes flush with copious amounts of water for 15 minutes. Remove contaminated clothing.
- In case of contact with skin flush with copious amounts of water. Remove contaminated clothing.
- If inhaled remove to fresh air. If breathing is difficult give oxygen. If not breathing give artificial respiration.
- If swallowed, wash out mouth with water. Consume water to dilute. Call doctor immediately.

6.0 *Fire Fighting Measures*
Extinguishing Media
Water, Carbon Dioxide, Foam, Powder

Special Fire fighting Precautions
- Wear self contained breathing apparatus and protective clothing to prevent contact with eyes and skin. Do not inhale smoke from fire.
- Use water spray to cool fire exposed containers.

7.0 *Accidental Release Measures*

7.1 *Small Quantities (<1 l)*
- Wear rubber gloves and suitable eye and face protection.
- Cover contaminated area with a suitable inert adsorbent preferably, e.g. vermiculite, or montmorilonite.
- Cover contaminated area with a suitable inert adsorbent preferably, e.g. vermiculite, or montmorilonite. Take up adsorbent and place in closed container.
- Transport to approved landfill or incinerator.

7.2 *Large Quantities (>1 l)*
- Evacuate area.
- Wear rubber boots, rubber gloves, suitable eye/face protection and NIOSH/MSHA approved respirator.
- Cover contaminated area with a suitable inert adsorbent preferably, e.g. vermiculite, or montmorilonite. Take up adsorbent and place in closed container.
- Transport to approved landfill or incinerator.

8.0 *Handling and Storage*
- Store in sealed container in darkness at room temperature.
- Immiscible with water above 50% weight concentration. Soluble only in solvents such as acetone or ethanol. Immiscible in hydrocarbon solvents.
- Reacts with mild steel and impure copper due to high acidity. Attacks buna rubber and in some cases causes rubber seals to swell.

9.0 *Exposure Controls/Personal Protection*
- Wear appropriate NIOSH/MSHA approved respirator, rubber gloves, rubber apron, and suitable eye/face protection.
- Use in a well ventilated area or fume cupboard.

- Safety shower and eye bath.
- Do not breath vapour.
- Wash thoroughly after handling.

10.0   Physical and Chemical Properties
Dark brown, viscous liquid with a smoky odour.

[All values are typical, not definitive, values]

Boiling Curve:	Start point 90-100°C
	NB. When heated above 100°C pyrolysis oil forms a solid char.
Setting Point:	-26°C
Specific Gravity:	1.2 @ 25°C
Flash Point:	50-70°C
Auto ignition Temperature:	110-120°C
Upper Explosion Level:	Unknown
Lower Explosion Level:	Unknown
Vapour Pressure:	similar to water
Vapour density:	Unknown

11.0   Stability and Reactivity
Incompatibilities

      Heat - 50% char formed upon continuous heating above 100°C

Hazardous Combustion or Decomposition Products.

Fumes of:

      Carbon dioxide

      Carbon monoxide

      Light organics-2-propenal, acetic acid, formaldehyde, methanol, acetaldehyde

12.0   Toxicological Information
- Liquids produced at reactor temperatures greater than 600°C contain condensed polyaromatic ring compounds which have mutagenic effects.
- Harmful if swallowed, inhaled, or absorbed through the skin.
- Vapour is irritating to the eyes, mucous membranes and upper respiratory tract.
- Can sensitise skin.
- Laboratory test have shown mutagenic properties.

13.0   Ecological Information
Very high BOD-no exact value yet. Do not discharge to ground or water sources

14.0   Disposal Considerations
Burn in a chemical incinerator observing appropriate environmental regulations.

15.0   Transport Information.
Transport in sealed, acid resistant, translucent container in cool conditions. Avoid prolonged exposure to strong sunlight. Fill receptacle to a maximum of 90 %

APPENDIX VI: COMPANIES WHO SUPPLY UN APPROVED PACKAGING IN THE EU, USA AND CANADA [AND SOME ASSOCIATED EU STATES]

The following organisations are given only as examples of suppliers of UN approved packaging and are not an endorsement, or recommended by Conversion And Resource Evaluation Ltd., but as a consultative guide to pyrolysis liquids producers and handlers.

Packaging Suppliers	Telephone/Fax	Drums	Boxes	Others
AUSTRIA				
Karl Pawel 1100 Vienna	0043/1-602 1322 0043/1-603 2528 office@pawel.at www.pawel.at		4D 4G	
BELGIUM				
No Nail Boxes (Europe) S.A. L-9001 Ettelbruck (G.D. Luxembourg)	352-819281 3520810517 no-nailboxes@vo.lu		4D	
CANADA				
Environmental Packaging Systems Ltd. Dartmouth, NS	902-461-1300 902-466-6889 eps@ep-systems.ns.ca			Div. 6.2 (PI 620/650)
Centre de Conformité Dorval, QC	514-636 8146 514-636 3522 www.thecompliancecenter.com	1A1 1A2 1H1	4C1 4D 4G 4GV	IP1 IP2 IP5 IP10 3H1 3H2 6HG2 PIH Div. 6.1
Nefab Inc. Peterborough, ON	705-748 4888 705-748 0034		4D 4DV	11D
Reliance Products L.P. Winnipeg, MB	204-633-4403 204-694-5132	1H1 1H2	4G	IP2 6HG2
Saf-T-Pak Inc. Edmonton, AB	780-486 0211 780-486 0235 www.saftpak.com		4G 4H2	Div.6.2 (PI 602/650)
The Compliance Centre, Dartmouth, NS	902-468-8885 902-468-8886 www.thecompliancecenter.com	1A1 1A2 1H1	4C1 4D 4G 4GV	IP1 IP2 IP5 IP10 3H1 3H2 6HG2 PIH Div. 6.1
The Compliance Centre, Mississauga, ON	905-890-7227 905-890-7070 www.thecompliancecenter.com	1A1 1A2 1H1	4C1 4D 4G 4GV	IP1 IP2 IP5 IP10 3H1 3H2 6HG2 PIH Div. 6.1
DENMARK				
Dangerous Goods Management Aps. Copenhagen	45-3 252 7690 45-3 252 7890 dgmcph@posts.tele.dk		4G	Explosafe+UN A4
Nefab Danmark A/S 2600 Glostrup	43-96 41 00 43-96 46 10		4D 4DV	
FINLAND				
Oy Nefab AB SF-00160 Helsinki	9-17 08 02 9-17 00 06		4D 4DV	11D

Packaging Suppliers	Telephone/Fax	Drums	Boxes	Others
FRANCE				
Air Pack 13700 Marignane	04-42 77 17 57 04-42 31 44 42		4G 4H1	IP1 IP2 IP3 IP3A IP5 Div. 6.2 (PI 602/650)
Air Sea France 84094 Avignong 9	490 86 54 46 490 16 91 20 isovation@wanadoo.fr	1A1 1A2 1H1 1H2 1G	4D 4DV 4G 4GV 4H 4H2	IP1 IP2 IP3 IP3A IP5 3H1 6HA1 Div. 6.2 (PI 602/650)
E3 Airembal 95 Roissy C.D.G.	01-48 62 49 99 01-48 62 55 43		4B1 4B2 4C1 4D 4G 4GV 4H1	Div.6.2 (PI 602/650)
E3 Cortex 77230 Thieux	01-60 26 86 00 01-60 26 84 62 info@e3cortex.fr		4B1 4B2 4C1 4D 4G 4GV 4H1	Div.6.2 (PI 602/650)
Emballage SFE Roissy CDG Airport	01-48 62 20 20 01-48 62 39 40		4D 4G	
Nefab S.A.R.L. 41300 Salbris	254 96 82 55 254 96 35 77		4D 4DV	11D
No Nail Boxes (Europe) S.A. L-9001 Ettelbruck (G.D. Luxembourg)	352-81 92 81 352-81 05 17 no-nailboxes@vo.lu		4D	
RCO 75792 Paris Cedex 16	01-44 34 75 58 01-44 34 75 48		4G 4Y	
RCO Specialités 60350 Attichy	03-44 42 73 99 03-44 42 73 98		4G 4Y	
SoFRIGAM 92000 Nanterre	14 66 98 500 14 72 59 844 sofrigam@magic.fr www.sofrigam.com		4G	
Sotralentz S.A. 67320 Drulingen	388 01 60 50 388 01 60 60	1H1 1H2		3H1
GERMANY				
Air Sea Deutschland Göttingen	551-504 410 551-504 4129	1A1 1A2 1H1 1H2 1G	4D 4DV 4G 4GV 4H 4H2	IP1 IP2 IP3 IP3A IP5 3H1 6HA1 Div. 6.2 (PI 602/650)
Alex Breuer GmbH 50677 Köln	0221-934749-0 0221-394748-9		4D 4G	Div 6.2 (PI 602/650)
Andresen & Jochimsen EXPORTPAK GmbH., 22525 Hamburg	040-547 2720 040-547 27282		4D 4G	
Dangerous Goods Management (Deutschland), Kelsterbach	6107 71723 6107 717244	1A2	4D	3H1 Explosafe+UN A4

Packaging Suppliers	Telephone/Fax	Drums	Boxes	Others
DVG 90552 Röthenbach/Pegnitz	0911-95 78 70 0911-95 78 730		4C1 4C2 4D	
Nefab GmbH 41836 Hückelhoven	2433-45 09 0 2433-45 09 90		4D 4DV	11D
Schütz Werke GmbH & Co. KG. 56242 Selters	2626 77-0 2626 77-365 sandra.schmitz@shuetz.net	1A1 1A2 1H1 1H2		3H1 31HA1
Siepe GmbH 50170 Kerpen	2273-56921 2273-56979	1A1 1A2 1H2		6HA1
Techno-pack GmbH 65451 Kelsterbach	6107-1657 6107-61884	1A1 1A2 1H1 1H2 1G	4D 4DV 4G 4GV 4H 4H2	IP1 IP2 IP3 IP3A IP5 3H1 6HA1 Div.6.2 (PI 602/650)
Wellpappe Ansbach, Duropack GmbH, 91522 Ansbach	981-18 80 981-18 81 10		4G 4GV	
Wellpappenfabrik GmbH 67269 Grünstadt-Sausenheim	06359-806 341 06359-806 441		4G	
GREECE				
Freight Plus S.A. 16777 Helliniko-Athens	1964 6271 1961 7677 fplus@otenet.gr www.freightplus.gr	1A1 1A2 1H1 1H2 1G	4D 4DV 4G 4GV 4H 4H2	IP1 IP2 IP3 IP3A IP5 3H1 6HA1 Div. 6.2 (PI 602/650)
Air Sea Forward Packaging Dublin	1-833 2281 1-833 1370	1A1 1A2 1H1 1H2 1G	4D 4DV 4G 4GV 4H 4H2	IP1 IP2 IP3 IP3A IP5 3H1 6HA1 Div.6.2 (PI 602/650)
Dgp Irl Ltd.. Cork	021-31 31 30 021-31 33 02 trident@indigo.ie	1A2T 1H1 1H2 1H2T	4D 4DV 4G 4GV 4H1 4H2	IP1 IP2 IP3 IP4 IP5 6HA1 Div.6.2 (PI 602/650)
ISRAEL				
PACHMAS Metal, Platic and Fibre Industries 38980 Ein-Hahoresh	6-625 0204 6-625 0209 shivuk_p@packmas.co.il www.pachmas.co.il	1A1 1A2 1H1 1H2 1G		3H1 6HA1 31H1
ITALY				
Air Sea Italia SRL Rome	0524 528418 0524 520159	1A1 1A2 1H1 1H2 1G	4D 4DV 4G 4GV 4H 4H2	IP1 IP2 IP3 IP3A IP5 3H1 6HA1 Div.6.2 (PI 602/650)
Dangerous Goods Management (Italia) Milan	02 9537 5677 02 9537 5148 dgmmil@iol.it	1A1 1G	4DV 4GV	3H1 Explosafe+UN A4
Fustiplast Spa. 24040 Bottanuco (BG)	035-90 74 84 035-90 62 48 fustiplat@uninet.com.it	1H1 1H2		3H1

Packaging Suppliers	Telephone/Fax	Drums	Boxes	Others
Momor SPA 20056 Grezzago (MI)	02-9093 651 02-9096 9772 info@mamor.it	1H1 1H2		31H1
Nefab S.r.l. 201 37 Padermo Dugnano, Milan	02-990 48520 02-990 48521		4D 4DV	11D
Pirola e Passerini Srl 20125 Milano	02-2885 1266 02-261 3858 vicini@pirolaepasserini.com		4D 4DV 4G 4GV	5M2 Div. 6.2 (PI 602/650)
Tanks International S.r.l. 24049 Verdellino Zingonia (BG)	035 884 587 035 482 0597	1A1 1A2		6HA1
NETHERLANDS				
CarePack Holland v.o.f. 1438BA Schilphol Skypark	020 354 0787 020 354 0650 carepack@globalxs.nl	1A1 1A2 1A2T 1G 1H1 1H2 1H2T 1H1	4B 4D 4G 4GV 4H 4H2 4HV	IP1 IP2 IP3 IP3A IP5 3A1 3H1 5H2 6HA1 13H3 Div.6.2 (PI 602/650)
Dangerous Goods Management b.v. Amsterdam	20 449 6565 20 449 6575 info@dgm.nl	1A1 1A2 1G 1H1 1H2 3H1	4D 4DV 4G 4GV 4H	3H1 3H2 Most IP 1-8 Class 6.2 Explosage+UN A4
Nefab Benelux BV 7411 HC Deventer	05706-10808 05706-10814		4D 4DV	11D
No Nail Boxes (Europe) S.A. L-9001 Ettelbruck (GD Luxembourg)	352 819281 352 810517 no-nailboxes@vo.lu		4D	
Van Leer Nederland BV 3633 AK Vreeland	0294-238911 0294-232441 HugoVanDenBerg@VanLeer.com	1A1 1A2 1G		6HA1 31HA1
NORWAY				
Dangerous Goods Management a/s Stavanger	51 71 55 40 51 71 55 41 arve@dgm.no	1A1 1A2 1H1 1H2	4D 4G	IP1 IP2 IP3
Nefab Emballasje A/S 1067 Oslo	022-90 56 90 022-90 56 99		4D 4DV	11D
Vestfold Skips & Industriservice a.s. Tonsberg/Oslo	33 31 88 88 33 31 88 89 vest.ships@vf.telia.no	1A1 1A2 1H1 1H2 1G	4D 4DV 4G 4GV 4H 4HV	IP1 IP2 IP3 IP3A IP5 3H1 6HA1 Div. 6.2 (PI 602/650)
PORTUGAL				
Air Sea D.G. Representocoes S.A. Lisbon	1-796 5530 1-793 9393	1A1 1A2 1H1 1H2 1G	4D 4DV 4G 4GV 4H 4H2	IP1 IP2 IP3 IP3A IP5 3H1 6HA1 Div.6.2 (PI 602/650)

Packaging Suppliers	Telephone/Fax	Drums	Boxes	Others
SPAIN				
Air Sea Espana 28001 Madrid	91-542 93 67 91-542 93 67	1A1 1A2 1H1 1H2 1G	4D 4DV 4G 4GV 4H 4H2	IP1 IP2 IP3 IP3A IP5 3H1 6HA1 Div. 6.2 (PI602/650)
Blagden Packaging Femba SA 08902 L'Hospitalet de Llobregrat	93 331 93 00 93 421 13 36 feba@femba.com	1A1 1A2		
Dangerous Goods Management 28850 Torrejon de Ardoz	91 676 2660 91 677 1072 dgm@lander.es	1A1 1A2 1H1 1H2	4D 4DV 4G 4GV	3H1 Class 6.2 Explosafe+UN A4
EmMerPe S.L. 28042 Madrid	91 329 1583 91 329 4782	1A1 1A2 1H1 1H2	4D 4DV 4G 4GV	3H1 Class 6.2 Explosafe+UN A4
Nefab, S.A. 28906 Getafe (Madrid)	91-696 69 11 91-696 74 42		4D 4DV	11D
Reyde, S.A. 08820 El Prat de Llobregat (Barcelona)	93 478 7600 93 478 7296 reyde@mundivia.es	1A1 1A2 1H1 1H2	4G	IP2 3H1 6HA1 31HA1
SWEDEN				
Nefab Emballage AB 822 92 Alfta	0271-59000 0271-59010		4D 4DV	11D
Nefab PlyPak AB 566 21 Habo	036 410 25 036 468 71		4D 4DV	11D
Nefab RePak AB 822 92 Alfta	0271-59000 0271-59040		4D 4DV	11D
Noax AB Stockholm	8-500 25185 8-500 29956		4G 4H2	Div.6.2 (PI 602/650)
UNITED KINGDOM				
Air Sea Containers Ltd. Birkenhead	0151-645 0636 0151-644 9268 www.air-sea.co.uk	1A1 1A2 1H1 1H2 1G	4D 4DV 4G 4GV 4H 4H2	IP1 IP2 IP3 IP3A IP5 3H1 6HA1 Div.6.2 (PI 602/650)
Bibby Sterilin Ltd. Staffordshire ST15 OSA	1785 812121 1785 815066 jwardell@bibby-sterilin.com		4G	3H1
Dangerous Goods Management Ltd. Aberdeen	1224 773 776 1224 773 622 info@dgm.co.uk		4D 4DV 4G 4GV	Explosafe+UN A4
Dangerous Goods Management Ltd. London	181 577 8566 181 577 8588 steve@dgm.co.uk		4D 4DV 4G 4GV	Explosafe+UN A4
DGP UK Ltd. Coventry	01203 602060 01203 602003 trident@indigo.ie	1A2T 1H1 1H2 1H2T	4D 4DV 4G 4GV 4H1 4H2	IP1 IP2 IP3 2IP4 IP5 6HA1 Div. 6.2 (PI 602/650)
Nefab (UK) Ltd. Northampton	01604-76 63 77 01604-76 59 21		4D 4DV	11D

Packaging Suppliers	Telephone/Fax	Drums	Boxes	Others
Norlab Instruments Ltd. Aberdeen	01224-724849 01224-723564	1A1 1A2 1H1	4G	6HA1
UNITED STATES				
Action Pack Inc. Bristol, PA	800 755 9764 215 788 1760	1A1 1A2 1H1 1H2 1G	4C1 4D 4DV 4G 4GV 4H	IP1 IP2 IP3 11G 6HA1 6PH1 Div.6.2 (PI 602/650)
Air Sea Containers Inc. Miami, FL	305-599 9123 305-599 1668 sales@airseacontainters.com	1A1 1A2 1H1 1H2 1G	4D 4DV 4G 4GU 4GV 4H	IP1 IP2 IP3 IP3A IP5 3A1 3H1 6HA1 Div.6.2 (PI 602/650)
Air Sea Atlanta Inc. Atlanta, GA	404-351 8600 404-351 4005	1A1 1A2 1H1 1H2 1G	4D 4DV 4G 4GV 4H 4H2	IP1 IP2 IP3 IP3A IP5 3H1 6HA1 Div.6.2 (PI 602/650)
All-Pak Inc. Bridgeville, PA	412-257 3000 412-257 3001	1A1 1A2 3A1	4G 4GV	Div.6.2 (PI 602/650)
Allflex Hazardous Material Packaging Inc. Ambler, PA	800 448 2467 215 643 3339 sales@allflex.com		4G 4GV	
CARGOpak Corp. Raleigh, NC	919 873 9440 919 873 9465 www.cargopak.com	1A1 1A2 1H1 1H2	4DV 4G 4GV	IP1 IP2 IP3
Cin-Made Packaging Group, Inc. Cincinnati. OH	513 681 3600 513 541 5945		4G 4GX	IP1 IP2 IP3A Div.6.2 (PI 602/650)
Dangerous Goods Consultant Houston, TX	281-821 0859 281-821 6558 hazmat@webtv.net www.hazmatman.com	1A1 1A2 1H1 1H2	4C1 4C2 4D 4G 4GV	3H1 3H2
Dangerous Goods Management, Inc. Houston, Tx	281 442 8434 281 442 6055 jeanp@dgm-usa.com	1A1 1H1 1H2	4D 4DV 4G 4GV	IP1 IP2 IP3 Explosafe+UN A4
DG Supplies Inc. Cranbury, NJ	800 347 7879 609-860 0285 sales@dgsupplies.com	1A 1G 1H	4G 4GV	IP1 IP2 IP3 3A1 3H1 Div. 6.2 (PI 602/650)
EXAKT Technologies Inc. Oklahoma City, OK	800 866 7172 405 848 7701		4G	Div.6.2 (PI 602/650)
Federal Industries Corp. Minneapolis, MN	800-523 9033 612-476 8155 chemtran@aol.com	1A1 1A2 1H1 1H2	4C1 4C2 4G 4GV 4GX 4GU	IP1 IP2 IP3 1P3A IP5 IP6 3A1 3H1 3H2 5H4 6HA1 6HG2 PIH Div 6.1, Div.6.2 (PI 602/650)
Freund Can Company Chicago, IL	773-224 4230 773-224 8812	1A1 1A2 1H1 1H2	4G 4GV	3H1 6HA1

Packaging Suppliers	Telephone/Fax	Drums	Boxes	Others
General Container Corp. Somerset, NJ	732 435 0020 732 435 0040 gencon@eclipse.net www.generalcontainer.com	1A1 1A2 1B1 1H1 1H2 1G	4G 4GV 11G	3H1 31A1 31H1 31HA1 6HA1
HAZMATPAC, Inc. Houston, TX	800-923 9123 713-923 1111 hazmatpac@hazmatpac.com www.hazmatpac.com	1A1 1A2 1H1 1H2	4C1 4C2 4G 4GU 4GV 11G	IP1 IP2 IP3 IP4 IP5 IP6 3H1 3H2 PIH Div. 6.1, Div 6.2 (PI 602/650)
Industrial Crating & Packing, Inc. Seattle, WA	425 226 9200 425-226 9205 indcrate@earthlink.net		4G	
Inmark, Inc. Atlanta, GA	800-646 6275 404-267 2021	1A1 1A2 1H1 1H2	4C1 4C2 4G	IP1 IP2 IP3 3H1 3H2 6HG2 Div. 6.2 (PI 602/650)
Labelmaster Chicago, IL	800-621 5808 800-723 4327 sales@labelmaster.com	1A1 1A2 1H1 1H2	4D 4DV 4G 4GU 4GV	3A1
LPS Industries Moonachie, NJ	800 242 7628 201 438 1326	1A1 1A2	4G 4GV	IP1 IP2 IP4 IP5 6HA1
Nefab Inc. Schaumburg, IL	847 985 1600 847 985 3200		4D 4DV	11D
O. Berk International Union, NJ	908-687 7720 908-687 5157 obj@oberk.com		4G	IP1 IP2 IP3A Div. 6.2 (PI 602/650)
Polyfoam Packers Corp. Wheeling, IL	847-398 0110 847-398 0653 info@polyfoam.com www.polyfoam.com		4G	Div.6.2 (PI 602/650)
ProPack, Inc. Essington, PA	610-521 4050 610-521 8737 dg.propack@erols.com	1A1 1A2 1G 1H1	4G	3H1
Russel-Stanley Borp. Bridgewater, NJ	908 203 9500 908 203 1944 info@russell-stanley.com	1A1 1A2 1H1 1H2		3H1 3H2 6HA1
Skolnik Industries, Inc. Chicago, IL	773 745 0700 773 735 7257 info@skolnik.com	1A1 1A2		6HA1
The Compliance Centre Niagara Falls, NY	716 283 0002 716 283 0119 www.thecompliancecenter.com	1A1 1A2 1H1	4C1 4D 4G 4GV	IP1 IP2 IP3 IP10 3H1 3H2 6HG2 PIH Div. 6.1
The Compliance Centre Houston, TX	713 722 0023 713 722 0026 www.thecompliancecenter.com	1A1 1A2 1H1	4C1 4D 4G 4GV	IP1 IP2 IP3 IP10 3H1 3H2 6HG2 PIH Div. 6.1
Tri State Steel Drum Corp. SO. Kearny, NJ	973 344 2625 973 344 7151	1A1 1A2 1G 1H1 1H2		6HA1

10. REFERENCES

1. Approved Supply List [4th Edition] - Information Approved for the Classification and Labelling of Substances and Preparations Dangerous for Supply, Stationery Office, 1998, ISBN 0-7176 1641X.

2. United Nations Recommendations on the Transport of Dangerous Goods Model Regulations, 11th Revised Edition, ISBN 92-1-139067-2, January 2000.

3. Technical Instructions for the Safe Transport of Dangerous Goods by Air, 1999-2000 Edition. Doc 9284-AN/905, ICAO.

4. Dangerous Goods Regulations CD-ROM [ref#9515-41] IATA Dangerous Goods Regulations manual, available from IATA, Montreal, Quebec, Canada.

5. Technical Instructions for the Safe Transport of Dangerous Goods by Air [Doc 9284-AN/905 and supplement] from Civil Aviation Authority, Printing and Publications Service, Greville House, 37 Gratton Road, Cheltenham, Gloucestershire GL50 2BN.

6. International Maritime Dangerous Goods Code (IMDG Code) - 1994 [including Amendment 29-98], IMO-213E, 1999, IMO.

7. European Agreement concerning the international carriage of dangerous goods by road [ADR] and protocol of signature, United Nations, New York and Geneva, ECE/TRANS/130 [Vols 1 and 2], 1998, ISBN 92-1-139062-1.

8. European Provisions concerning the International Carriage of Dangerous Goods by Inland Waterway [ADN], ECE/TRANS/WP.15/148, 2000 ISBN 92-1-139059-1.

9. Regulations concerning the international carriage of dangerous goods by rail [RID] 1999 edition [Annex 1 to Appendix B to the Convention concerning international carriage by rail [COTIF]], HM Stationery Office 1998, ISBN 0 11 552032 5.

10. See Internet site http://hazmat.dot.gov/ for useful information and North America legislation.

11. The Export and Import of Hazardous Wastes Regulations and the Transportation of Dangerous Goods [TDG] Regulations, 1992, Transport Canada, Ottawa, 1992.

12. Council Directive 96/49/EC on the approximation of the laws of member states with regard to the transport of dangerous goods by road, Official Journal of the European Communities, 12 December 1994, 37 L319/7-13 [ADR Framework Directive].

13. Council Directive 96/55/EC on the approximation of the laws of member states with regard to the transport of dangerous goods by rail, Official Journal of the European Communities, 17 September 1996, 39 L235/25-30 [RID Framework Directive].

14. Recommendations on the Transport of Dangerous Goods - Manual of Tests and Criteria - Third revised edition, ISBN 92-1-139068-0, January 2000, Vol. 2.

15. Bridgwater, A.V., "Production of high-grade fuels and chemicals from catalytic pyrolysis of biomass", Catalysis Today, 1996, vol. 29, no. 1-4, pp. 285-295.

16. Meier D., Oasmaa A. and Peacocke G.V.C., "Properties of Fast Pyrolysis Liquids: Status of Test Methods", Developments in Thermochemical Biomass Conversion, A.V. Bridgwater and D.G.B. Boocock, (Eds.), Blackie, vol. 1, pp 391-408.

17. Diebold, J.P., "A Review of the toxicity of biomass pyrolysis liquids formed at low temperatures", Fast pyrolysis of biomass: A handbook, Bridgwater, A.V., Czernik, S., Diebold, J., Meier, D., Oasmaa, Peacocke, C., Piskorz, J. and Radlein, D., (Eds.), CPL Press, 1999, pp. 135-163.

Evaluation of Bio-Energy Projects

P Thornley and E Wright
PB Power, Amber Court, William Armstrong Drive
Newcastle upon Tyne, NE4 7YQ, UK

1. EVALUATION OF BIO-ENERGY PROJECTS WITH SPECIAL REFERENCE TO FAST PYROLYSIS AND LIQUIDS

1.1 Executive Summary

PB Power has prepared this report as a guide to the assessment of pyrolysis projects by investors when demonstration scale plants are progressed to commercial scale. The report describes how the financing of small scale renewable energy projects is usually undertaken under a limited recourse project finance scenario. It also identifies some of the key technical areas where researchers should focus their efforts in order to maximize the chances of obtaining successful finance and in order to minimize the perception of risk.

The key to enabling successful financing of any novel process plant project is ensuring that sufficient records have been kept of both process and plant details for the demonstration projects,. As well as illustrating plant configuration, it is of fundamental importance that accurate heat and mass balances can be tabled and proven for the proposed plant. In terms of project feasibility it is also important to ensure that projects demonstrate sufficiently long periods of continuous operation.

Environmental issues frequently play a major role in determining the feasibility or otherwise of small scale projects. It must be ensured that the pyrolysis project being considered is capable of meeting existing *and forthcoming* legislation. This may be difficult to demonstrate for a technology which is in the early stages of development, since it is not until the first plants are operated that regulators will monitor achievable emission levels and possibly seek to enforce modifications in certain areas. It is *essential* that developers have carried out environmental emissions analyses and evaluated the potential for pollution from the proposed plant. In the long term, pyrolysis will only continue to be authorized by regulators if it can be demonstrated to have an environmental impact as minimal as competing technologies utilizing the same fuel: eg combustion of wood with flue gas treatment.

The actual operation of a pyrolysis plant is also likely to give rise to some initial problems. An appropriate operator skills base may not be readily available and careful consideration should be given to how this will be overcome. Technology developers should remember that it is unlikely that future operators will have the same level of in-depth pyrolysis knowledge as the process developers.

The characterization of any pyrolysis oils is important, particularly with respect to toxicity. Long term trials of the oil in the relevant prime mover should also be demonstrated.

Secure fuel supply agreements should be put in place for the plant. It is important that developers focus on what output/performance can be anticipated from the plant when supplied with fuel which is out of specification.

Wood pyrolysis and power generation from the resultant pyrolysis oils has many novel technological aspects which may be perceived as risky. It is possible, however, to mitigate these risks via an appropriate development and demonstration programme.

2.   INTRODUCTION

PB Power (PBP) has undertaken an assessment of the possible routes for financing fast pyrolysis processes on behalf of Pyrolysis Network (PyNe). A guide to the techniques and procedures typically employed in assessing bio-energy projects was required so that PyNe might better understand the criteria a potential investor might impose in deciding whether or not to support a particular pyrolysis project. This work is being undertaken as part of the implementation focus of the PyNe group in which they seek to identify and, where possible, address barriers to the implementation of pyrolysis projects.

The report considers the technical, economic, market, financial and commercial aspects of the evaluation which would typically be carried out by a lender's technical advisor and identifies which aspects of fast pyrolysis could give particular cause for concern. Possible routes to minimizing these concerns are indicated, where appropriate.

This report aims to provide a guide to the overall assessment of bio-energy projects and is intended to assist those involved in research, development and demonstration projects to understand the criteria applied by investors making implementation decisions. In addition, by suggesting possible mitigants for the various risks identified, developers should be able to identify key areas which may be focused upon to maximize the prospects of successful project funding and implementation.

This report has considered fast pyrolysis of biomass in a general sense and has not focused on any particular technology or process configuration. It has, however, been necessary to make some basic assumptions regarding the "project" being evaluated in this report. It has been assumed that the project consists of a plant which receives virgin or residual biomass (eg not contaminated wood) and converts this via pyrolysis to a combustible pyrolysis oil, plus other by-products. The pyrolysis oil is then processed and combusted in a prime mover, such as a gas turbine or reciprocating engine. The by-products may be saleable or require disposal. It is further assumed that electricity and/or process heat are produced wholly from the combustion of the pyrolysis oil.

3.   THE FINANCING PROCESS

3.1   Introduction

It is important that technology developers and researchers understand the basic elements of obtaining project finance, so that they may better understand the relative importance of different factors from the perspective of potential investors. This allows developers and researchers to concentrate on areas that will be of most significance in successfully implementing a successful demonstration project. It also ensures that appropriate measures are taken to remove as many potential problems from the project as possible before these become a risk to financing. For example, it is relatively straightforward to keep adequate logs and records of pilot plant operating hours if adequate procedures are put in hand at the beginning of a development programme. The absence of such records could prove a significant obstacle to demonstrating the degree to which a technology has been proven at a later stage.

In many parts of the world there is, at present, a trend towards restructuring of the traditional utility industries such as electricity. This has resulted in a larger number of small-scale project (rather than process) developers playing a key role in the implementation of power projects. While some national bodies and large utilities continue to develop new technologies and support renewables projects, more bio-energy and other renewable projects are being brought to the market by smaller independent companies. These independent power projects (IPPs) are typically developed by companies who do not have funds available to finance project implementation themselves or who are not in a position to risk their own capital on project success. Development funds therefore tend to be sought from project finance lenders (eg banks), venture capitalists or equity investors. It is normal to maximize the level of debt rather than equity based investment in a project as the rates of return required on the money borrowed are lower. The greater the amount of debt a particular project has, the more highly it is said to be leveraged (see Glossary of Terms).

Typical equity investments might constitute 10-30 per cent of project costs, the remaining 70-90 per cent being financed by the lender. Debt funding is frequently on a non-recourse or limited recourse basis. Non re-course funding means that the lenders look only to the assets of the project company (not the parent company) for repayment of principal debt and interest. With limited re-course financing the claim that the lender has against the parent company is restricted.

Generally a special purpose company is set up as the project company and it adopts all responsibility for developing the project, owning and operating the plant as well as repaying the debt. Equity is high risk capital; there is generally no obligation on anyone to pay the equity investors back if the project fails.

From a lender's perspective, therefore, before money is loaned, it is vital to ensure that the investment is sound by confirming that the project itself is financially viable throughout the lifetime of the loan to ensure that the lenders recover their investment. To do this the lenders need to obtain independent impartial advice on the project. Based on this they will then assess the risks involved with the project company. These are then evaluated and either mitigated, accepted and managed, or passed through to the party that is most capable of managing the risk via commercial agreements.

The evaluation procedure is essentially the same for all types of power project, although the actual risks involved will obviously differ. It should be noted, however, that small projects often require a more detailed evaluation since, due to the lower development budget, issues are frequently not adequately addressed prior to involving the potential lender.

3.2   Role of technical advisers

The financial viability of a power project is intrinsically linked to the technical viability of a particular project and the technology it employs. It is therefore essential that the technical merits of any such project are fully evaluated prior to any decision being made regarding whether or not such a project should be funded. Lenders do this by employing technical advisers (usually independent engineering consultants with no direct involvement in the project but having expertise in a relevant area) to carry out an appraisal on the proposed project before a decision on funding is made. They provide the independent opinion on the project required by the lender.

The lenders retain an independent engineer to undertake a technical appraisal of the project "with all due diligence" (generally simply referred to as "due diligence" for convenience). This is essentially a fact-finding exercise carried out in order to identify any potential fatal flaws that could either halt the proposed project or cause it to fail to perform as anticipated.

During the due diligence exercise, the design of the plant is reviewed as well as all key contractual documents, consents and authorizations. Key areas of technical concern are identified which could affect the feasibility or performance of the project. Efforts are made to resolve these with the developers. In addition, predictions are made regarding the likely performance that will be achieved by the plant. Key risks which have not at that stage been adequately mitigated are identified and mechanisms are inserted into the various project agreements to address these. In parallel with this technical due diligence, the lender's legal advisers will also be carrying out a review from a legal perspective of the project and all relevant documentation. Both sets of advisers, technical and legal, will work together to:

- identify key project risks
- evaluate the potential consequences arising
- mitigate the risks appropriately
- transfer risk management to the party best able to deal with it
- set up a framework for monitoring the project to ensure that it progresses smoothly.

In evaluating the technical risks the technical adviser will generally categorize issues as follows:

significant issues that may ultimately stop the project proceeding if not resolved e.g. extending plant beyond its technical capability or incorrect heat/mass balance assumptions.

intermediate issues that require resolution and may influence the evaluation of the project but, ultimately, do not affect the ability of plant to repay the project debt (e.g. lack of redundancy in auxiliaries could result in reduced availability but does not mean that the plant itself is flawed). These may be deferred for resolution until after financial close, depending on their nature.

minor issues that would have no technical effect on the final plant but may result in additional costs (eg building appearance or layout, mechanism for boiler cleaning).

3.3   Timescales

Six distinct phases are generally involved in project implementation. These are:

1.   Technical development
2.   Project development
3.   Project modification
4.   Design approval and construction monitoring
5.   Performance testing and handover
6.   Operation.

3.3.1   Technical development

Technical development refers to the development of the technology which takes place prior to the commercial development of the particular project application. Although this phase is often considered to be far removed from actual project implementation, lenders may look back to records of plant and process demonstration during this period when seeking confirmation of the viability of the technology. It is important that the requirements of the lenders are understood by those involved at this stage. Technical development includes the following stages:

Process and plant concept

During this phase (which is the status of many pyrolysis projects at the time of writing) efforts should be made to develop the key steps in the process technology so that the risk of the technique failing to perform on scale-up are minimized. It must be remembered that it is not only important to develop the technology in this way but also to ensure that sufficient records are maintained so that others can subsequently be convinced that the appropriate development work and testing has been carried out.

Plant configuration

Details of the construction of the pilot plant itself should be kept, particularly in cases where this is to be dismantled prior to the funding application. Video records may be useful as operational evidence, in addition to engineering specification.

Heat and mass balances

It is important that records are maintained of heat and mass balances obtained at pilot plant installations, preferably for off-design as well as design process conditions. It is strongly recommended that these include for periods of start-up and shut-down. This can later be used to provide evidence that the product yields being claimed for the process are actually achievable.

Environmental impact

One of the most important items to establish at this stage is that the plant does not give rise to any unacceptable levels of environmental emissions to air, water or land. This will be a key element of the due diligence process, when anticipated emission levels from the plant will be compared to appropriate national and/or international legislation. The legislation that is applicable will depend on the actual project and, in many countries, guidance notes and legislation are currently under review. It should be noted that in some countries, biomass waste will be classified as waste and therefore be subject to additional legislation.

When developing the technology, emissions monitoring should be carried out (preferably by an independent accredited agency) while the plant is operational. Consideration also needs to be given to assessing solid waste product streams, with reference particularly to local legislation. At the technology development stage it is worthwhile considering what other waste streams may be produced on a non-continuous basis and assessing whether or not these can be minimized by appropriate process design at a later stage. For example:

- is there a requirement to regularly flush a reciprocating engine with conventional fuel?
- how can the cost of treating and disposing of the effluents be minimized?
- is it necessary to drain an ash discharge trough periodically?
- what waste will be produced and how can it be minimized?
- is it possible to avoid direct cooling of pyrolysis oil and production of the associated effluent?

Even when the answer to such questions is "no" it is worthwhile to have thought about and documented the options considered so that answers will be readily available to respond to questions from the lender's technical advisers.

Combustion trials

In a pyrolysis power project, the pyrolysis process produces a pyrolysis oil and is the main novel technology in the whole process. Combustion of the pyrolysis oil for production of electricity and/or heat may take place in a conventional boiler, gas turbine or reciprocating engine. It is important that adequate attention is given to the combustion aspect of the overall process at an early stage. Combustion trials for novel fuels in conventional plant should extend for a minimum of 1000 hours. This recommendation is based on typical gas turbine industry standards and allows for a reasonable evaluation of any special modifications that may be required. This length of combustion trials also enables an assessment of the potential impact of the fuel on the maintenance requirements of the prime mover and will give confidence that the combustion process can be maintained on a continuous basis throughout the lifetime of the project.

Availability

One of the most contentious areas in evaluating a new technology is the availability factor (see glossary) can be ascribed to the novel plant. This must make adequate allowance for both planned and unplanned outages, (ie all periods when the plant is not able to operate).

Planned outages

For a defined technology, technical advisers can generally make a reasonable assessment of the requirements for planned maintenance. Collaboration, however, with manufacturers is sensible to ensure that any special recommendations relating to pyrolysis can be incorporated. For example, there may be a change in recommended gas turbine maintenance schedules when operating on pyrolysis oil. Such questions should be evaluated by the developer in conjunction with plant manufacturers during the technology development phase.

Unplanned outages

The likely level of unplanned outages is difficult to evaluate for a new technology. The lender's technical adviser is sometimes faced with a novel technology that may have never run continuously for a lengthy period and may be asked by the lender to predict the likely level of unplanned outages in normal operation. In the absence of convincing data, erring on the side of caution may be the only option available to the lender's technical adviser.

One way of responding to predictions by the lender's technical adviser of low availability may be to show that the demonstration or pilot plant has run continuously for a sustained period. Records, therefore, should be kept not only of how long the plant ran, but why it was stopped and what work was carried out on it while it was not operating. It may be that a particular plant was required to shut down every weekend because there was no customer for the heat produced or that there were no development staff on site to carry out any modifications during the periods when the plant was out of service. Unless this has been adequately recorded, however, it will be assumed that the plant had to shut down for essential work and could not have continued to operate, resulting in a low availability factor being ascribed.

Pyrolysis products

The by-products and intermediate products of fast pyrolysis processes are relatively unique. Therefore, toxicity analyses should be carried out and records kept. By-products from which revenue is to be obtained have to be proven to be suitable for the intended purpose. Depending on the use, this may involve characterization/analysis of the product and records of laboratory tests to be maintained. For example, if char is to be used as activated carbon or as a fuel for heating, some evidence of its suitability for this purpose may be required. Above all, it makes sense to keep some samples in reserve so that further tests can be carried out on particular aspects of the product at a later date, if required.

Capacity

In addition to planned and unplanned outages, it is important to consider part load operation of the plant which is utilising the pyrolysis fuel. No power plant runs at its maximum continuous rating for its entire lifetime. This may be due to variations in ambient conditions, reduced load during start-up and shut-down and operational requirements. In addition, for many pyrolysis projects, the biomass fuel will not be consistent in composition or quality and fluctuations in fuel quality may prevent the plant from operating continuously at its maximum continuous rating. The same may be true of the combustion process of pyrolysis oil produced. Operating records for biomass throughput and pyrolysis oil throughput should help to maximize the Capacity Factor which the technical advisers will ascribe to the plant. As a rough guide, when dealing with waste fuel a capacity factor of 90 per cent may be the maximum that can be achieved in the long term.

Scale-up

Scale-up from previously proven plant designs obviously carries with it some risk which lenders will be keen to minimize. It is possible that, earlier in the development process, similar scale-ups have already been performed. Where this is the case, records should be retained of the scale-up predictions modelled in advance compared to actual subsequent performance. For example, providing evidence that the basic design has previously been successfully scaled up by a factor of 20:1 will increase the confidence that a proposed 5:1 scale up is feasible.

3.3.2 Summary

In essence, during the technical development stage, all issues relating to the novel technology need to be evaluated and documented. The project development phase can then be embarked upon with a set of convincing results which aim to prove that the proposed plant can do what is being claimed.

3.3.3 Project development

During the project development phase, the approach to risk mitigation changes. Trials, test runs and characterizations should by now have been completed and the results from these will have provided the developer with a convincing set of data to present to potential investors in order to demonstrate that the technology will work.

The important issue to address at the project development stage is the minimization of risks related to the process or the project. Issues which it has not been possible to deal with by fundamental technical modifications can now be addressed. This is generally done either by ensuring that there are sufficient margins in the plant design to ensure that the technical risks will not be significant in terms of overall plant performance or, alternatively, by implementing appropriate commercial agreements, which transfer the remaining risks to the project participant which is best able to mitigate that risk.

It is likely that a review of the technology development and testing phase by the developer will also indicate a number of areas where there are still some uncertainties or problems. Efforts should be made to mitigate the potential risks from these during project development. For example, it may be the case that it was not possible to run continuously for long enough to give confidence that a high level of plant availability could be achieved. This could be addressed in the project development phase by designing the plant with multiple streams so that it can still operate at full load, even when one stream is not operational. Alternatively, the quality of the incoming feedstock may need to be controlled more carefully so as to avoid unnecessary shut-downs attributable to poor quality feedstock batches. In practice, however, it may not be possible to guarantee that out of specification material, including foreign bodies, are not delivered to the plant. Appropriate exclusion mechanisms (eg overhead magnets on feed lines, vibrating screens to remove oversize objects) should be included in the practical design. In addition the plant should be designed to have a high level of tolerance to extraneous material. Design consideration should be given to how foreign bodies are to be easily removed from the fuel transfer system when they cause an obstruction.

The main activities at this stage are the negotiation of appropriate agreements for the project, defining the actual project more closely and acquiring a suitable site. As part of the project development it is essential to consider the various interfaces involved in the project and ensure that these complement each other. This includes physical interfaces, such as the termination point for utilities provided by the developer and also interfaces between sub-contractors or third-parties involved in project implementation. Project development will also involve a geotechnical survey of the proposed site and ensuring that all necessary authorizations are procured.

During project development, the basic technology should be developed into a sensibly contained and fully defined facility. The land should be purchased or an option to purchase obtained. As many as possible of the authorizations required for the plant should be obtained or, at least, applications should be made. Key commercial agreements for construction and operation should be put in place in draft form. A simple financial model should be developed which describes the likely performance of the plant in financial terms. Surveys of the site (to evaluate load bearing properties, contaminated land, soil characterization etc) should be carried out at this stage.

3.3.4 Project modification

Once the above two phases have been more or less completed, a formal approach to the investors can be made by the developer. Prior to this, discussions may already have been held with the potential lenders to establish their general approach to the project. Following the formal approach, the lender's technical advisers will carry out a full technical appraisal of the project. During this phase they are likely to recommend modifications to key project agreements in an attempt to address any outstanding technical issues via the commercial project agreements. Most frequently this involves modifications to guarantees, warranties and the liabilities of key parties in the construction of the plant. At the same time that the technical advisers are undertaking their appraisal, the lender's legal advisers will also be reviewing all key project documentation to ensure that the project is commercially and contractually sound. This may result in modifications being made to key project documents.

The end product of this phase of project appraisal is normally a report by the lender's technical advisers evaluating the outstanding potential risks faced by the project and recommending mechanisms by which these can be mitigated. Where practicable, the lender will often act to reduce risk by making modifications to the key contractual agreements.

As part of the financing process, a facility agreement is drawn up which sets out the contractual position between the borrower and the lender. Here the facility referred to is the facility to borrow funds. For example, this will describe the procedures for drawdown of funds, requirements for the borrower to provide regular reports to the bank etc. Where significant technical uncertainties remain in the project these may sometimes be addressed by additional conditions in the facility agreement. For example the lender may reserve the right for its technical advisers to review further information on key items of plant before an order is placed eg type of boiler and manufacturer, place of manufacture or the testing and inspection procedures for a steam turbine.

It should be remembered that, while modifications can be made to a project to mitigate risk after financial close, such actions may, at that stage, result in increased project costs and/or delay in completion.

Likewise while statutory authorizations or comments can be obtained after financial close it should be remembered that this can result in a substantial risk. For example, if the environmental authorization results in a requirement to modify the plant or restrict its operating regime the operational project revenue could be substantially reduced.

3.3.5 Design approval and construction monitoring

After all contractual agreements have been finalized and the terms of the loan have been agreed, the role of the lenders and their technical advisers changes substantially. Their principal activity during this phase of the project is monitoring progress, with direct input much more limited. It is therefore critical that all major issues have been highlighted prior to this stage.

During the design phase of the project the Contractor will submit to the Owner's Engineer, information, drawings and documentation relating to the project for review in accordance with the construction contract. The lender's technical advisers will generally only review limited key technical information relating to issues which were considered to be outstanding at financial close.

During this phase of the project it may be possible to mitigate remaining risks by ensuring that engineering solutions are appropriately implemented for items that remained outstanding at financial close. For example, during this phase it will be ensured that an appropriate control philosophy and system is implemented for the plant.

Apart from design development the other main activity in this phase of the project is construction. Monitoring construction of the facility largely addresses risks that any conventional power project faces. This needs to be done, however, while bearing in mind the unique aspects of the pyrolysis process. For example, these could be:

- phased construction with the conventional oil fired combustion plant (gas turbine or reciprocating engine) being constructed first and earning revenue while the pyrolysis plant itself is then constructed (only possible where the conventional plant is capable of operation on dual fuel)
- close monitoring of commissioning of the novel plant
- phased submissions to, and approvals from, environmental bodies for novel technologies, especially prior to commissioning
- management of parallel construction of a number of different streams to mitigate availability risks. Note that in some projects it may make sense to construct and operate a single stream first of all, where there is a requirement to demonstrate scale-up or other aspects of feasibility. This reduces the technology risk from the lender's perspective, as only a relatively small proportion of the capital is being spent prior to confirmation that the plant operates as envisaged.

3.3.6 Handover/testing

Prior to the plant being handed over to the Owner as complete and the final contract payments being made, performance tests are carried out to verify that the plant is capable of complying with all of the contractual guarantees. It is important to the lenders that these represent a real test of the plant's capabilities and the test procedures. Final test reports and often the tests themselves will be closely reviewed by the lender's technical advisers.

3.3.7 Operation

During operation, the lenders will be keen to ensure that the plant is being operated in accordance with legislation and that appropriate safety and quality procedures are put in place. The plant will be monitored by the lender's technical advisers, at least during the early years of operation, to ensure that it is operating as anticipated and there are no major plant problems.

4. FAST PYROLYSIS: IDENTIFICATION OF KEY AREAS OF PERCEIVED RISK

The summary below identifies the key areas of fast pyrolysis projects likely to be considered by potential investors seeking to identify principle risks in a proposed project. These are areas, therefore, where developers would be advised to focus efforts to minimize concerns which may arise from the investor's perspective. The list is not exhaustive and will very much depend on the process details. It should, however, serve as a guide to assist developers in understanding the sort of issues likely to be of concern to investors so that they can work to address these.

4.1 Technology risks

- Feasibility of technology
- Health and safety risks: e.g. explosion hazards identified and removed
- Risks of variations to capital costs
- Risks of scale-up
- Low availability of pyrolysis plant and prime mover
- Risk of long term damage to prime mover running on pyrolysis oil

4.2 Fuel supply risks

- Insufficient fuel supply throughout lifetime of the loan within a reasonable transportable distance
- Delivery timescales and risk of interruption – is there storage capacity on site for continuous operation?
- Failure of fuel supplier to perform as anticipated
- Fuel price fluctuations
- Non-availability of start-up/back-up fuel
- Risk that out of specification fuel, which could damage plant, is accepted for processing
- Risk that variations in fuel quality will affect plant performance

4.3 Regulatory

- Risk that gaseous emission levels will exceed regulatory limits (with particular attention to dioxins, heavy metals, VOCs and particulates)
- Risk that a visible plume (especially from wood drying plants) may be unacceptable to regulatory authorities
- Risk that waste products may be produced for which disposal is expensive or difficult
- By-products may present toxicity risks for operation or disposal
- Public objections to location of a pyrolysis or power plant.

4.4 Products

- Pyrolysis oil or by-product may not be of consistent quality, affecting plant performance and by-product revenue
- Plant manufacturers reluctant to warrant use of pyrolysis oil in plant
- Potential of contaminants in pyrolysis oil damaging downstream plant
- Pyrolysis by-products fail to attract envisaged revenue or meet customer's requirements
- Pyrolysis oil deteriorates during storage

4.5 Conventional IPP risks

In addition to all of the above, which specifically relate to pyrolysis projects, the project will also be exposed to the conventional IPP risks. These include the possibility of underground site obstructions, contaminated land, force majeure events, delays in construction, risk of change in environmental legislation.

5. RISK MITIGATION

In Section 4 key risk areas for typical pyrolysis projects have been identified. Having identified the risks these are then evaluated by the lenders. Due consideration is given to:
- how likely is it that the problem would occur?
- if it did, what would be the impact of the problem, first of all on plant operation and then on revenue?
- how could the occurrence of the problem be minimized and the consequences mitigated?

Having evaluated the key project risks, the lender is in a position to prioritize which problem areas really represent a potential threat to project viability and which, therefore, must be addressed, putting to one side those which may have only a marginal impact on the overall project. The major risks can then be transferred to the party best able to deal with them, frequently by commercial agreements or insurance.

348

All risks related to the pyrolysis and combustion technology need to be addressed prior to project development. Secondary technical considerations, for example those relating to auxiliary plant, can be addressed during the project development phase with the main contractor. All significant technical issues should be addressed prior to financial close. This can be done either by addressing the fundamental technical aspects of the project or by putting into place commercial agreements which protect the owner and lender by transferring the responsibility for the risk to another party.

At financial close there should be no outstanding technical risks unless these have been evaluated and are considered to represent an acceptable level of risk by the lenders. Failure to address issues prior to this will result in higher contingencies and higher financing charges being imposed by lenders. Where the risks are substantial, it may not be possible to gain financing.

With this in mind, and building on the risks identified in Section 4, recommendations are given below for some activities that should be undertaken at the technology and project development phases. Generally, these are the key areas of focus for developers trying to bring a project to commercial implementation.

5.1 Technology development phase – risk mitigation

- Production of records of plant operation, modifications and maintenance for a continuous pyrolysis demonstration programme
- Verification of heat/mass balance for the system and impact of off-design conditions/fuels
- Establishment of an effective automatic control system for the pyrolysis process (which will maintain process conditions and product quality)
- Development of automated start-up and shut-down procedures
- Review and remove risks to safety under all operating conditions for operational staff
- Carry out emissions monitoring tests
- Identify and characterize all potential effluent and waste streams
- Characterize all products on which plant revenue or performance depends, including independent test results
- Pyrolysis oil upgrading (e.g removal of fine ash particles prior to combustion in a gas turbine) - Verify any requirements for upgrading, the specification of plant to do this and any waste products arising.
- Verification that pyrolysis oil will not degrade during foreseeable storage periods
- Validate models used for scale-up of pyrolysis reactors
- Long term continuous operation trials of pyrolysis reactor and combustion plant to be carried out, with records of operating hours maintained and, where possible, samples of by-products utilized by the future purchaser.

Having established and understood the basic principles of plant operation, efforts should be made to minimize changes from the proven configuration when defining the final design. Amendments should ideally be limited to converting a demonstration plant to comply with modern power plant arrangements and standards by, for example, implementing an appropriate control system, introducing proper burner management systems.

For many small scale projects, obtaining planning permission can be a major obstacle to project implementation. This can be particularly difficult where the location is close to residential areas. This is more likely for projects involving the supply of heat. One of the most frequently encountered problems is opposition from within the local community. This may be mitigated by establishing good lines of communication with the local residents.

Wood pyrolysis has many advantages that could be exploited to minimize such difficulties (eg pyrolysis oil is a flexible, efficient use of renewable energy and can make effective use of waste wood). The level of public awareness of pyrolysis, however, is low and this may lead to misconceptions and consequential difficulties at the planning stage. Providing an appropriate level of technical information can help the proposed project to be portrayed positively and any public concerns to be allayed.

5.2   Project development phase – risk mitigation

During the project development phase, the basic pyrolysis process is effectively built upon so that a design is produced for a complete plant which receives wood and converts this to pyrolysis oil, which is utilized to provide electricity and/or heat. All interfaces to the novel technology must be carefully considered and designed, with information interchange between the technology developer and the turnkey contractor being critical for success. These interface issues will be closely scrutinized by the lender's technical advisers.

5.2.1   Technology

For successful development of the technology at the project development phase, production of a minimum functional specification (MFS) which describes the plant required, performance, environmental restrictions, site conditions etc is recommended

This should stipulate a detailed interaction between the proposed contractor and the pyrolysis technology developers. It is advisable to obtain independent advice when producing the MFS rather than simply relying on the expertise of the contractor.

5.2.2   Cost

The use of novel technology can result in significant uncertainty over the level of price for the overall plant. There is a risk that commercial implementation of the technology may significantly increase the price. In addition, where the technology has not previously been constructed on a commercial scale as a fully integrated plant, there may be underestimates of the cost in interfacing and integration, which will, in many cases, require special design effort.

One of the most effective methods of avoiding such potential cost increases is to construct the plant under a fixed price turnkey contract. In such a contract the principal contractor takes responsibility for delivery of a complete plant, incorporating the novel technology and complying with the Owner's specified requirements for a fixed price. Obviously, the key to performance in a pyrolysis power plant is the pyrolysis process itself and so the technology developer has a key role to play in ensuring successful delivery of the final project. The technology developer, however, will rarely have all of the relevant experience needed to deliver the entire plant. It is often, therefore, most appropriate for the technology developer to act as a subcontractor to the main turnkey contractor. The main turnkey contractor then assumes the responsibility for delivering the entire plant. The main turnkey contractor is often referred to as the EPC (engineer, procure and construct) contractor.

Assuming that this is the route adopted, the following actions will help mitigate risks associated with pyrolysis projects at this stage:

Obtain a firm price from prospective contractors for carrying out the defined scope of works, as described in the minimum functional specification. This may include items, such as a fully integrated control system, appropriate fire protection system and so on. Independent technical advice should be sought when developing the specification.

Ensure that the contractor has considered any modifications required to the pyrolysis system, including scale-up, modelling, modifications from the pilot plant design and appropriate interfaces with other plant.

Evaluate the lifetime costs for the plant, including the anticipated lifetime of the equipment and any special maintenance requirements.

5.2.3 Scale-up

For successful scale-up:
- Maintain records of modifications made during previous scale-ups of the pyrolysis reactor.
- Validate the model used for scale-up predictions against actual plant results.
- Give consideration to reducing the risks associated with scale-up and availability by using a plant configuration which has a number of parallel streams.

5.2.4 Availability

To improve assumed availability factors:
- Maintain documentation of all stoppages, modifications and maintenance carried out during pilot plant or demonstration trials in order to demonstrate as long a period of continuous operation as possible.
- Carry out long term continuous combustion trials of the pyrolysis oil
- Characterize the pyrolysis oil in accordance with standards used for liquid fuels.

5.2.5 Plant

In optimizing the plant design, consider the following:
- Minimize the interdependency of different parts of the plant wherever possible to maximize availability. Ideally the failure of any single item of equipment should not give rise to an interruption in operation and appropriate ancillary equipment redundancy should be incorporated to facilitate this.
- Ensure plant is designed with the specific properties of pyrolysis oil in mind, especially with respect to acidity, corrosiveness and viscosity.

5.3 Commercial Agreements

The key purpose of the commercial agreements for an IPP project is to facilitate the delivery and performance of the various components of the project on terms acceptable to the Owner and the Lender. In doing so technical risks and uncertainties in the project should be converted to commercial guarantees and warranties as much as is possible. In any case the risk should be transferred to the party which is most capable of managing the risk and mitigating the consequences.

It is important to ensure that the interface issues with all commercial agreements are adequately addressed, so that the documents fit together to form a complete suite of project documents. In addition it is necessary to ensure that key inputs and project parameters on which revenue streams depend are guaranteed in the commercial documents. Such guarantees should be backed up, where possible, with compensation, in the form of liquidated damages, for the Owner in the event that they are not achieved. These are financial amounts claimed from the Contractor for failure of the plant to perform. The rate at which they are applied is agreed prior to Contract Award.

5.3.1 EPC Contract

The EPC contract is the agreement between the owner and EPC contractor to carry out the works. The EPC contract is generally the most significant document in ensuring that the project is successfully implemented and will come under close scrutiny by the lender's technical and legal advisers. This is discussed in more detail in Section 6.

5.3.2   Fuel supply agreement

Essentially a fuel supply agreement should be in place to guarantee that fuel of the required quality will be supplied to the facility for the lifetime of the loan at an agreed price. This document is discussed in more detail in Section 8.

5.3.3   Power purchase agreement

The power purchase agreement secures the sale of electricity from the project over the lifetime of the loan. For many pyrolysis projects, wood is the primary fuel and the electricity may be sold as "renewable energy". Under such circumstances it is usual to enter into a contract with the purchaser of electricity which will specify the quantity of energy purchasable as well as the price to be paid. Other conditions may include, for example, availability of the generator.

It is, however, not strictly necessary to have a power purchase agreement in place to gain financing for a project. This depends, very much, on the local electrical power market. In some cases, where there is a clear demand for power and a national system exists for selling electricity at a set price, it may be sufficient to put in place an agreement to participate in this system and sell power on this basis throughout the project lifetime. However, where this is the case some estimate will have to be made of the likely level of income that could be generated under the prevailing market conditions. In general this will tend to be lower than would often be the case with a secure power purchase agreement in place.

Power purchase agreements are discussed in more detail in Section 7.

5.3.4   Operation and maintenance agreement

It is possible for the owner to undertake operational and maintenance activities himself. In the case of a pyrolysis project there are certain advantages to this approach, as the operation of a pyrolysis project may be more akin to that of a chemical process plant than a conventional power plant and the expertise of the technology developer's staff will be very useful in the early days of operation.

However, there is an increasing trend with projects such as these to contract out operation and maintenance services to a specialist O&M contractor. This has the advantages of making available to the project a wide operational experience base and also of controlling some of the variable elements of the operation and maintenance costs.

The key elements of an operation and maintenance agreement are discussed in Section 9.

In general the value of the O&M contract tends to be lower than the cost of operational failures to the owner. It may not therefore be reasonable to expect compensation for failure to perform to equate to the full losses of the owner. Where this is the case, bonus payments for meeting agreed targets, coupled with penalties for poor performance, may be appropriate in ensuring plant performance is as anticipated.

5.4   Grants

Possibilities exist within the EU for grants, which may reduce the risks to the developer by underwriting part of the project cost. For example, the ENERGIE initiative, and European Social Fund (ESF) or European Regional Development Fund (ERDF) may be exploited.

6.   EPC CONTRACT

The EPC contract is the key vehicle for risk mitigation when novel technologies are being implemented in commercial plants. It constitutes the agreement between the owner and the principal contractor as to how the plant is to be constructed. The EPC contract generally sits alongside and makes reference to the MFS, which describes the actual physical construction and performance of the plant. It is normally a requirement of the EPC contract that the plant be constructed in accordance with the MFS issued by the owner. The MFS describes the plant to be constructed under the contract and details the principal project parameters, for example, constraints of authorizations, weather conditions, specification of anticipated fuel, site on which plant is to be constructed and other basic data upon which the plant design will be based.

The EPC contract will generally contain a separate schedule of legally binding guarantees. Items that are essential for plant operation, for example, compliance with environmental regulations, are generally included as absolute guarantees, which must be met in order for the plant to be accepted as complete by the owner and for final payment to be made. Other items, which generally describe performance, such as electrical output of the plant are also guaranteed. In this case, however, liquidated damages are ascribed to the guarantees so that, should the completed plant fail to deliver the specified output, damages become payable to the owner which should compensate for the resulting loss in revenue.

Frequently, a developer will issue to the selected contractor a limited Notice to Proceed. This is given in advance of obtaining final agreement with the Lenders. In doing this, the contractor is given an instruction to start work (usually limited to detailed design of the plant) up to a certain value. This reduces the overall time scale of the project by running activities in parallel up to financial closure. Although this is a contract cost and, as such can be included in the financed cost, at the pre-financial closure stage it is at the Developer's risk, since the project may not ultimately proceed.

6.1   Payment schedule

The EPC Contract will define how payments for the works are to be made to the contractor. It is possible to make payments to a contractor based on an assessment of actual works completed to a specific date. It is, however, often preferable to incorporate milestone payments which are simple to monitor against a project programme. The payment schedule generally consists of an initial down-payment, interim milestone payments, payment at provisional acceptance and payment at final acceptance.

There are two important aspects to the payment schedule: the first is to ensure that there is a relatively small down payment and that the payments at provisional acceptance are sufficient to incentivize the contractor to undertake the work. The second is that the milestones are easy to monitor, well defined and simple. If there are too many milestones they are difficult to administer.

6.2   Liquidated damages

Liquidated damages are generally attached to guarantees in the EPC contract in order to protect the project economics should plant completion be delayed (construction liquidated damages) or if the plant fails to meet the anticipated levels of throughput or electrical output (performance liquidated damages) etc.

Liquidated damages must be a genuine pre-estimate of the loss that failure to achieve the guarantees will cost the owner. These damages are a significant liability to the EPC contractor and will generally be heavily negotiated. The level of damages will be assessed by the lender's technical advisers and should compensate for any loss in revenue resulting from the failure of the plant to comply with its guarantees over the lifetime of the loan (from the perspective of the lenders) or the lifetime of the project (the preferable position from the perspective of the developer). The calculation of the actual damages should be based on a discounted cash flow calculation over this period using the same discount rate as in the owner's other financial calculations.

7.   POWER PURCHASE AGREEMENT

The main aim of an independent power project is to generate and sell electricity and, even in cases where substantial revenue is obtained from other sources (eg payment obtained from accepting waste material or revenue earned from the production of speciality chemicals), there is still a need to prove that there is a market for the power produced and that it can be sold at a credible price.

As stated earlier this may be achieved by putting in place a power purchase agreement (PPA). While it is not strictly necessary to have a PPA to gain financing of a project, the technical adviser will consider whether or not there is a market for the power generated and what price is likely to be obtained for the electricity produced throughout the lifetime of the project. Plants which do not have a PPA in place are known as "merchant plants" and tend to occur in relatively developed markets. It should be noted that even where a PPA does exist the context of the local power market still needs to be considered. If a project is financed on the basis of a lucrative PPA with a customer who does not actually require the power, there is a risk in the long term that payment will not be made in accordance with the agreement. Therefore, there is a requirement to demonstrate to the lender that the power project is economically justified in the local or national economic context. This is more likely to be required to be demonstrated where there is a developed power market.

The PPA may also specify requirements with respect to plant output, reactive power capabilities, rate of change of loads, outages etc. Generally these tend to be less stringent for smaller plant. However, it is important to ensure that the plant is capable of meeting all the technical requirements contained in the PPA. With regard to pyrolysis plants, there is little that can be done at the development stage to influence this capability and the key is to seek to negotiate a PPA which facilitates base load operation, within specified parameters which can then be imposed within the turnkey contract.

8.   FUEL SUPPLY AGREEMENT

Fuel costs can be one of the largest variable costs facing the plant and therefore it is desirable to fix this marginal cost of production of pyrolysis oil by entering into a long term fuel supply agreement. In some countries there are already well established markets for the provision of wood fuel. Where this is the case it may be possible to rely to some extent on spot market fuel supplies. However, in countries where the wood fuel market is in a relatively early state of development, it is important to ensure that a long term fuel supply agreement is put in place, fixing the price and quality of fuel for the term of the loan.

The technical adviser will review the fuel supply agreement to verify that the fuel specified is appropriate for the plant. With pyrolysis, wood quality is of particular importance and it is important to have a knowledge of what variation in key fuel parameters is possible and what impact these may have on plant performance. A fuel specification can then be included in the agreement which will ensure the plant performs as anticipated.

However, it is also important to consider the practicalities of fuel supply as a supply agreement alone will not ensure that acceptable fuel is supplied to the plant. First of all the supplier should have a track record of handling quantities of fuel similar to that required for the plant. Secondly, it is prudent to design the fuel reception and handling system so that any items that may cause damage to the process may be excluded and out of specification fuel can be rejected. It is of little use to exclude items in a fuel supply agreement if they cannot be identified. This is particularly important where there is more than one fuel supplier to the plant.

9.   OPERATION AND MAINTENANCE CONTRACT

There are many different types of O&M contracts that may be put in place successfully for small scale power generation facilities. The plant may be operated by the owner's own staff, where it would be normal to set up a separate operations company or by a specialist O&M contractor. Contractual conditions vary from a contract that effectively passes through the cost of operating and maintaining the plant to a contract which offers a fixed price for O&M services for the entire contract period.

The key issue from the Lender's perspective is normally the experience of the operator in operating plants of this type and this is particularly the case with wood pyrolysis plants. Obtaining an appropriate skills base may require a combination of experience in the power and process industries and it will be particularly important that the plant manager, operations manager and maintenance manager are appropriately experienced.

The level of fees paid to the O&M contractor are normally such that it is not appropriate to impose financial penalties that would fully compensate the owner for losses arising from poor performance. Instead a bonus/penalty arrangement is frequently incorporated into the agreement to incentivize the operator to perform. The exact details of this tend to vary, but the aim is obviously to align the desires of the owner with the incentives in the contract and to ensure that bonuses are only paid when the plant actually makes an additional profit.

10.   FINANCIAL MODELS

Most independent power projects are financed by a mixture of debt and equity investment. The senior debt (see glossary) generally constitutes around 70-90 per cent of the overall project cost. The remainder of the project costs will be funded either by share capital from the developers or by equity investors. The investment risk is much higher for equity investors and the rates of return required are correspondingly higher. For example rates of return of 25-30 per cent are not uncommon.

As part of the project development process the developer will develop a financial model, which should describe the key performance aspects of the proposed plant and make financial predictions based on these. The financial model serves to try and predict what will happen to the plant over its first 15-20 years of operation and calculate whether or not the plant will actually generate sufficient revenue to repay the initial investment. Loan terms for small scale renewable projects are typically of the order of 10-15 years. Only in exceptional circumstances will terms in excess of this be considered, even when plant lifetime is evaluated as being of the order of 25 years.

The key to a successful financial model is to accurately and realistically represent the key performance indicators of the plant. It is important, therefore, that those developing the model fully understand the nature of the plant operation and its limitations. Performance in the following technical areas is assessed:
- Plant throughput/fuel consumption
- Electrical output (available for sale after supplying all in-house consumption)
- Quantity of by-products produced

- Heat output for sale (where applicable eg CHP plants)
- Plant degradation over time - will the output reduce over the plant lifetime or between maintenance overhauls?

Availability (hours per year, including an allowance for scheduled and forced outages) An extensive period should therefore be allowed for possible commissioning problems e.g. at least 6 months and it should not be assumed that any revenue will be obtained from plant operation during this period. In addition to this it should be assumed that the availability achieved during the first year of operation will be very low e.g. perhaps 60 per cent, increasing after that, depending on plant design and performance.

Capacity/operations factor (how closely the plant approaches full load when it is operational). Note that for plants operating with an heterogeneous fuel, where calorific value or other fuel parameters may vary, it is difficult to maintain full load output. In the case of waste materials 90 per cent of the plant maximum continuous rating (MCR) may typically be achieved during the times that the plant is actually operational. This also takes into account reduced efficiency when the plant is operating at part load or during start-up and shut-down.

Financial information specific to the project is also required as follows:
- Fuel cost (requires a supply market assessment)
- Electricity selling price (requires a demand market assessment)
- Waste product disposal cost (required a disposal market assessment)
- By-product selling price (requires a demand market assessment)
- Operation and maintenance costs (including start-up fuel and chemical consumables)
- Discount rate (10 per cent is a reasonable starting point in the absence of other figures)
- Repayment period for loan and repayment profile (in consultation with lenders)
- Cost of turnkey contract for plant construction as well as cost of subsidiary contracts e.g. for electrical connection, construction of access roads, water supply connection
- Development costs
- Insurance costs
- Taxes
- Cost of initial holding of spare parts and of maintaining this holding
- Plant lifetimes for depreciation calculation purposes

These lists are by no means exhaustive and merely serve to give some indication of the sort of figures that are required to develop a financial model. The most important function of the model is to quantify various project risks and to develop a "feel" for the cash flow dynamics.

10.1   Contingency

Contingency is required to finance cost overruns primarily during the construction phase of the project. These generally arise as a result of change orders to the contract for works outside the scope of the EPC contract. The level of contingency required for a particular project depends on how well the project has been defined and developed, the number and nature of outstanding issues left at financial closure and whether or not the technology is considered proven. With unproven technology there is a much higher risk of additional items being required which were not identified prior to financial closure.

For a reasonably well defined project the contingency might be around 10-15 per cent of contract price. For a poorly defined project, with interfaces that still need to be defined, particularly with novel technology, up to 20 per cent of contract price is recommended.

10.2 Financial risk

The rate of interest at which money is loaned will be determined partly by the level of risk assessed as pertaining to the project and partly according to the current state of the financial markets. Interest rates may be set by the national banks at the end of each working day. In the UK the rate is the London Interbank Offer Rate (LIBOR). Loan rates will frequently be expressed as LIBOR plus a percentage rate, the additional percentage rate being is determined by assessing the magnitude of risk involved in the project. For a small scale project this might typically be 1-3 per cent points in excess of LIBOR, but it depends on the view the lender takes of the specific project and the risks attached. Different rates may apply during the construction and operational phase, due to the perceived risk being higher during construction.

11. RISK ASSESSMENT

Before committing funds for a particular project the lenders will carry out an assessment of the risk involved in construction and operation. One way of doing this is to compile a list of events or factors which could affect the project. The impact that such an event would have on the project is then defined and any mitigants which would act to reduce the risk to the overall project are identified.

A risk category (high, medium or low) is then ascribed to each potential project risk. This is coupled with an impact category (high medium or low) which assesses the likely impact that occurrence of that event would have on the overall project economics. The lenders will then use this risk matrix to identify key areas to which priority should be given in addressing outstanding risks.

From a technical perspective this is the main purpose of risk analysis and a sample risk matrix is illustrated in appendix E.

One of the key criteria in which lenders are interested is the debt service coverage ratio (DSCR). This is the ratio of the Operating Income to Debt Service for a given time period. It must be ensured that this can be maintained at a reasonable level for the lifetime of the loan. This ratio gives an indication of how robust the project is to repay the loan. For low risk conventional projects, an acceptable DSCR might be 1.2. However for novel technologies, the DSCR required by the lender is likely to be significantly greater than this, potentially up to 1.8 or 2. The DSCR is also dependent on the state of the financial markets.

12. INSURANCE

Lenders will generally specify that certain insurance cover is taken out by the project company. This will include, for example, cover for transport of plant to site, third party liability cover for works on site, insurance against force majeure risks interfering with construction etc. These requirements will be agreed with the lenders and contained within the facility agreement between lender and borrower. Some insurance cover will be the responsibility of the contractor rather than the owner.

In addition to these requirements, other optional insurance cover may be of interest to developers, particularly where novel technologies are involved. Efficacy insurance may be used to cover the risk of a technology not performing as anticipated. This tends to be expensive but may be of interest to provide additional reassurance that investment capital will be recoverable for projects with a high level of risk. To procure such insurance it is normally necessary for the insurer's own consultants to carry out a due diligence on the technology being used.

It is also worth noting that there is an increasing tendency for insurance to be used to cover operational risks. It is normally a requirement that adequate third party liability cover, cover for damage by fire, flood etc is in place during the operational phase. However, in addition to this, it is possible to purchase insurance that will cover the possibility of mechanical breakdown of plant. This can be useful where it limits the requirements for the project company to build up a substantial maintenance reserve to cover unanticipated maintenance problems. While it is possible to procure O&M services by a contract which covers all O&M requirements, including mechanical breakdown, the cost of this will reflect the uncertainty and will inevitably carry with it a cost premium. This tends to be the case where a relatively new technology is involved. An alternative is to negotiate an O&M contract which excludes the cost of breakdown events and cover these by putting in place separate insurance.

13.   TAXES

In the case of renewable energy projects, taxes may enhance the financial argument for the pyrolysis technology and should not be overlooked within the financial model (e.g. UK Climate Change Levy, Illinois Retail Rate Law).

However, taxes are classified as a regulatory risk, which may pose a risk potential to projects. Projects with debt service capacity dependant in part on tax benefits or tax rate will always be vulnerable to changes in law which alter the tax treatment which exists at project commencement. Where tax law changes are perceived to be a risk to the project, sponsors will be required to provide indemnification to the lenders against the adverse effects of any such changes in law.

14.   LIST OF ABBREVIATIONS

AH	available hours
CHP	combined heat and power
DSCR	debt service coverage ratio
EPC	engineer, procure, construct
ERDF	European Regional Development Fund
ESF	European Social Fund
EU	European Union
IPP	independent power producer
IRR	Internal Rate of Return
LDs	liquidated damages
LIBOR	London Interbank Offered Rate
MCR	maximum continuous rating
MFS	Minimum Functional Specification
O&M	operation and maintenance
PBP	PB Power
PH	period hours
PPA	Power Purchase Agreement
Ppb	parts per billion
Ppm	parts per million
PV	present value
PyNe	pyrolysis network
RDF	refuse-derived fuel
RSH	reserve shutdown hours
SCADA	Supervisory Control and Data Acquisition

SCR	selective catalytic reduction
SO$_2$	sulphur dioxide
SOF	scheduled outage factor
SOH	scheduled outage hours
SO$_x$	sulphur oxides
SRC	solvent refined coal
UK	United Kingdom

Unit abbreviations

Btu	British thermal unit	kWh	kilowatt-hour
cal	calorie	lb	pound mass
cm	centimetre	m	metre
ft	foot	MJ	megajoule
GJ	gigajoule	MWth	megawatt thermal
h	hour	Pa	pascal
in	inch	psi	pounds per square inch
J	joule	s	second
kcal	kilocalorie	W	watt
kg	kilogram	y	year
kJ	kilojoule	°C	degrees Celsius
km	kilometre	°F	degrees Fahrenheit
kPa	kilopascal		

15. GLOSSARY OF TERMS

Availability Factor (AF): represents the percentage of time that the power plant was available for service, whether operated or not. It is equal to the available hours divided by the total hours in the period under consideration (period hours), expressed as a percentage [AF = (AH ÷ PH) x 100%].

Available Hours (AH): are the sum of all service hours (total number of hours the facility was electrically connected to the transmission system), and reserve shutdown hours (total number of hours the unit was available for service but not electrically connected to the transmission system for economic reasons).

Capacity Factor: measures the level of plant utilization. It is calculated by dividing the actual energy output from the plant (kilowatt-hours produced) by the theoretical maximum kilowatt-hours that could be produced by the plant when operating at its rated capacity in a given time period.

Debt Service Coverage Ratio (DSCR): a measure of a project's ability to generate revenues sufficient to meet its debt service requirements for an established time period, usually 1 year. The ratio is calculated by dividing the operating income (revenues less O&M costs) by the total debt service requirements (principal and interest) for the given period. Typically, lending institutions require that the minimum DSCR in any given period should not fall below 1.2, and that the average of the ratios be at least 1.4 over the life of the loan.

Debt Service Reserve:	a mechanism, often required by lenders, that a project company uses to mitigate some of the risks that the lenders bear when lending funds to the project company. Close to the commercial operation date, a project company will create a debt service reserve account that will be held by an agent as collateral. The reserve account may be funded in total at one time or funded over a period of time, and the balance will be maintained at a level sufficient to satisfy the project company's debt service requirements for a specified period or amount. In the event of any failure by the project company to make a payment called for under the reimbursement and loan agreement at the time and within any grace period provided therein, the agent is authorized to deduct funds from the debt service reserve fund to make the required payment.
Debt Service:	payment of interest and principal, for a given time period, from a project company to the lenders to the project.
Debt Term or Tenor:	length of the financing period.
Debt:	an amount owed from one entity to another. Debt holders repay the amount borrowed at a predetermined interest rate and according to a predetermined repayment schedule, as typically specified in a reimbursement and loan agreement. A project company will often secure debt from several lenders.
Developer:	any entity trying to design, finance, build, own, operate and manage a project or any combination of those activities.
Development Agreement:	outlines how a project company will proceed in its development efforts. The agreement identifies the scope of work, defines the roles, responsibilities and compensation for each party comprising the project company, and outlines methods for resolving disputes.
Development Costs:	include all the costs incurred prior to financial close by a developer and any third parties for a development project. Typical development costs include the costs associated with site acquisition, permitting and licensing activities, technical, legal and financial advisory work.
Development Fee:	the fee charged by the developer for the developer's effort in bringing a project to construction financing. The fee is included in the development costs.
Development Phase:	incorporates all the work, including a detailed engineering and financial feasibility study, that must be carried out to achieve financial close. Generally, during the development phase, all project permits are obtained, all contracts are negotiated and executed, and financing for the project is secured. This phase will typically last from 18-36 months.
Discount Rate:	an interest rate that is used to calculate the present value of future cash flows. The discount rate is often equal to the rate of return offered by comparable investment alternatives.

Engineer, Procure and Construct (EPC) Contract:	a contract between a project company and a turnkey organization for the engineering, procurement and construction of a power plant.
Equity:	the capital funds invested by the owners of a project.
Facility Agreement:	sets out the contractual position between the borrower and the lender, generally describing the procedures for the drawdown of funds. This may include additional conditions where there are significant technical uncertainties.
Financial Close:	the date of closing and release of construction financing for a new facility.
Forced Outage:	an unscheduled shutdown of the power generating facility.
Fuel Supply Agreement (FSA):	a contractual agreement between a project company and a fuel supplier that covers the purchase of fuel, specifying rates and delivery schedules.
Gearing	see leverage.
Independent Power Producer (IPP):	a non-traditional utility private investor or entity who develops, owns and/or operates electricity power plants.
Leverage (Debt/Equity Ratio):	measures the extent of a firm's financing with debt relative to equity and its ability to cover interest and other fixed charges. A typical project might have an 80/20 leverage ratio, meaning 80% of the total capital costs are funded with debt and 20% of the total costs are funded with equity. As a project becomes more leveraged (the debt/equity ratio rises), the volatility of the project returns increases significantly. The financial risk of a project closely parallels the volatility of the expected returns, and it will increase when the volatility increases.
Liquidated Damages:	an ascertained amount, expressed in currency, which an injured party has sustained or is taken to have sustained. Generally a claim against a Contractor for failure to perform against agreed plant or programme criteria. Importantly, the term is used in a contract rather than penalty which is unenforcable in English Law.
Maximum Continuous Rating:	the maximum rated output at which a plant can be expected to operate continuously without detrimental effect to any of the plant components.
Non- or limited-recourse finance:	if a loan is with recourse, the lender has a general claim against the parent company if the collateral is insufficient to repay the debt. With non-recourse, the only collateral for a loan is the assets of the project company. The parent company is protected against a claim from the lenders should the project company not be able to repay the debt. The lenders look only to the assets of the project company for repayment of principal and interest.
Non-reheat boiler	as compared to a reheat boiler, a non-reheat boiler does not incorporate a reheat steam loop where steam, which has partly been expanded, recovers further energy from the combustion gases.
Off-take Contracts:	the off-take contracts (power purchase agreement, steam sales agreement etc) identify the major revenue inflows for the duration

of the project life. Lenders use revenue-generating off-take contracts as collateral for the project's long-term debt because these contracts guarantee a certain cash flow that is consistent with the length of the term loan.

Operator Skills Base: Operator's experience in operating a particular type of plant.

Outage: refers to the period of time when the plant is out of service due to either planned maintenance (planned outage) or plant failure (unplanned or forced outage).

Performance Guarantees: as agreed to in the EPC contract, performance guarantees are the plant and equipment operating criteria and measurements that the EPC contractor must demonstrate to the owner of the facility on or before the commercial operation date. Equipment vendors often are required to guarantee the performance of their equipment based on MCR operation.

Power Purchase Agreement (PPA): a contractual agreement that covers the purchase of electricity over the financial life of the project, which specifies terms and conditions (including price) of purchase.

Project Company: a legal entity formed for a particular project, which will build, own and (possibly) operate a power plant.

Project Financing: project financing involves the financing of a specific economic unit (often known as a project) based primarily on the cash flows from that project and the collateral value of the project's assets.

Pyrolysis oil: oil produced by pyrolysis of wood, suitable for combustion.

Pyrolysis: a broad term describing a variety of processes in which biomass is decomposed by heat action in an oxygen-deficient atmosphere. Depending on the operating conditions, this can result in the production of combustible gases or liquids. Some pyrolysis products are sufficiently high quality to be transported and stored for later use; others are burned immediately.

Senior Debt: debt which has a higher priority and must be repaid before subordinated debt holders receive payment.

Senior Lenders: banks and other financial institutions who have priority over other creditors in lending to a project company.

Sensitivity Analysis: measures the effect on project investment returns from possible changes in revenues, costs, etc.

Subordinated Debt: Often known as junior debt, subordinated debt is debt over which senior debt takes priority. In the event of bankruptcy, subordinated debt holders receive payment only after all senior debt holders are paid. Owners will try to secure subordinated debt (typically from limited partners or equipment vendors) to reduce or avoid their own equity investments in a project.

Technical Adviser: typically an independent consulting engineer who advises lenders on the technical viability a potential project in terms of technology, costs, construction period, performance, etc.

RADIANT FLASH PYROLYSIS OF CELLULOSE - EVIDENCE FOR THE FORMATION OF SHORT LIFE TIME LIQUID SPECIES AND EXPERIMENTAL DETERMINATION OF MASS BALANCES

O. Boutin and J. Lédé [#]

Laboratoire des Sciences du Génie Chimique, CNRS - ENSIC – LSGC, 1 rue Grandville BP 451 F-54001 NANCY Cedex, France , # Corresponding author. Tel : +33 (0)3.83.17.52.40; Fax : +33 (0)3.83.32.29.75; email : lede@ensic.u-nancy.fr

ABSTRACT

The purpose of the present paper is to report some of the first results of the radiant flash pyrolysis of cellulose pellets. The experiments are made under controlled and very clean conditions of fluxes and controlled irradiation times with an image furnace. The flux densities that can be obtained reach about $2 \ 10^7$ W m^{-2} while the irradiation times are of a few milliseconds. The microscopic observations of the surface of the pellets after irradiation show that cellulose gives rise to intermediate species that are liquid at the temperature of reaction. They are solid at room temperature and soluble in water. The experimental mass balances show that these liquids undergo fast thermal decompositions into vapours, with no noticeable formation of char.

INTRODUCTION

A great number of papers have been published in the field of biomass thermochemical conversion. They can be related to:

- Fundamental studies (kinetic schemes and Arrhenius constants: Bradbury et al., 1979; Milosavljevic et al., 1995; Gronli, 1996; Lédé et al., 1997; Antal et al., 1998; Boutin et al., 1998)
- Optimization of a given reactor (fluidized bed, cyclone, etc: Lédé et al., 1986; Bilbao et al., 1988; Czernik et al., 1995)
- Modelling (at the particle or at the reactor levels: Lédé et al., 1985; Diebold, 1994; Di Blasi, 1996; Miller et al., 1997).

This paper deals with the field of fundamental studies.

All the processes of biomass thermochemical conversion (pyrolysis, gasification and combustion) begin with elementary steps of decomposition of each of the components of the starting material (cellulose, hemicellulose and lignin). It is hence necessary to better understand the kinetics of the corresponding fast elementary reactions of pyrolysis. At the present time, no consensus has been reached, even for a model component such as cellulose. Most of the authors consider competitive steps that give rise to gases, tar and char followed by subsequent consecutive reactions.

The most frequently mentioned schemes rely on the Broido-Shafizadeh model shown in Fig. 1. However, for some authors, step 1 that gives rise to the intermediate species (sometimes called "active cellulose") is not necessary for the prediction of the global kinetics of cellulose pyrolysis (Lanzetta et al., 1997). Other authors completely deny the existence of such species and just consider a single step giving rise to vapours from virgin cellulose (Antal et al., 1995). Eventually, some papers report clear evidences for the existence of these intermediate species (Lédé et al., 1997).

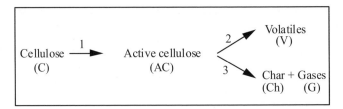

Figure 1 Broido-Shafizadeh kinetic scheme for cellulose pyrolysis

Several reasons may explain the controversies that always exist with regard to the kinetic scheme of cellulose decomposition. The studies are usually performed in systems (mainly thermo-gravimetric analysis) that are incompatible with the rapidity of the reactions. They are also unable to measure reactions without mass loss. The heat and mass transfers efficiencies are often poor and unknown, and hence the pure chemical processes are not rate controlling. Moreover, the systems are such that secondary reactions (cracking, re-condensation) are possible and do not permit the first short life time intermediate species to be studied. This paper reports some results observed with an original device where cellulose samples are submitted to controlled flashes of a highly concentrated radiation. After quenching the products, it is possible to isolate and to analyse the species formed during the first moments of the reaction.

EXPERIMENTAL SECTION

The Image Furnace

Only the main characteristics of the set up will be described here. More details can be found in Boutin et al. (1998).

The image furnace relies on the use of a discharge lamp (5 kW xenon lamp) associated to two concentrating elliptical mirrors M_1 and M_2 (Figure 2). The image of the focus F_1 (and hence of the arc of the lamp) is located at the second focus F_2 of the mirror M_2 where the reactor is placed. At the common focus F' of the two mirrors is placed a diaphragm with a hole of the same size as the diameter of the focus F'.

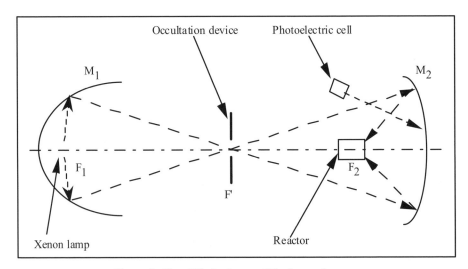

Figure 2 Simplified scheme of the image furnace

The incident flux arriving at F_2 can be changed by moving the diaphragm between F' and F_2. The power that is available at the focus F_2 can be measured in calibration experiments. The maximum value is of the order of 350 W. With an order of magnitude of $5 \ 10^{-3}$m for the cross section of the focal area at F_2, one can calculate a maximum mean flux density of $1.8 \ 10^7$ W m^{-2}. The diaphragm is associated to a pendulum inside which is made an adjustable window and allowing to provide short flashes of light (Boutin et al., 1998). A photoelectric cell facing M_2 and connected to a computer gives the duration of the flash. According to the size of the window, the flashes can be as short as 0.01 s with an accuracy of 1 ms.

Material and Reactor Used:

The experiments are made with cellulose powder (Whatman CC31 microgranular) mixed with 5 % wt of powdered char in order to increase the flux absorbed at the surface of the sample. This mixture is pressed in order to obtain small pellets (5 mm diameter, a few hundred microns thickness). The pellets are dried over 24 hours in an oven at 110 °C. They are placed inside a reactor and adjusted at the focus F_2. The design of the reactor is shown schematically in Figure 3.

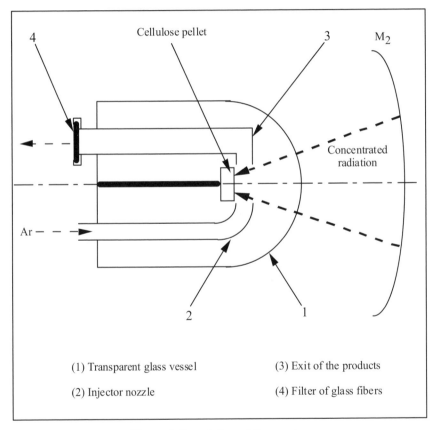

Figure 3 Description of the reactor

The reactor is a transparent glass vessel (1). A jet of argon arriving at room temperature (2) is directed towards the surface of the pellet in order to immediately cool down the products formed during the flash. The gaseous products as well as the aerosols leave the reactor through a second tube (3) placed just above the reaction zone. They are then trapped in a filter made of glass fibers (4). In the same time, the primary products that have not passed into the gas phase remain on the surface of the pellet. Notice that for a reaction temperature estimated at about 500 °C (Lédé et al., 1987) and with a flash duration of about 0.1 s (beginning of the reaction), it is possible to calculate heating rates close to 5000 K s^{-1}.

Determination of Mass Balances:

The different masses taken into account in the balances are given in Figure 4. m_0 is the initial mass of the dried cellulose pellet. m_1 is the loss of mass of the pellet obtained by weighing the pellet before and after the reaction. m_2 is the mass of products recovered on the filter. After irradiation, the pellets are put in demineralised water (18MΩ resistivity). The solution obtained is then filtered. The unreacted and non-soluble cellulose is recovered and weighed (m_4) after drying. The products left on the pellet are dissolved and recovered (m_3) after slow evaporation of water at 40°C. The mass m_5 of the cellulose that has reacted is hence equal to m_0- m_4. Notice that m_5 must be also equal to m_2 (=m_1) + m_3 if the balance is good and if no light and uncondensable gas is formed during the reaction. Weighings are made to an accuracy of approximately 10^{-7}kg.

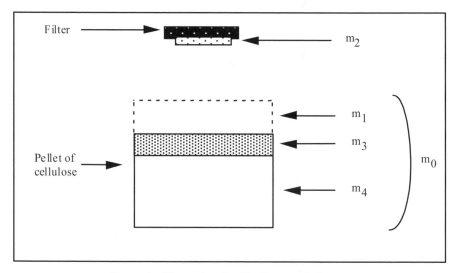

Figure 4   Masses involved in the mass balances

RESULTS AND DISCUSSIONS

Evidence For The Formation Of Short Life Time Intermediate Species :

After the reaction, the surface of the pellet is covered with a solid brown matter. Microscopic observation reveals that the fibrillar structure of virgin cellulose (Figure 5) has disappeared and that the products have the appearance of large networks of agglomerated melted particles (Figure 6). These observations clearly show that, in our conditions, the thermal decomposition of cellulose passes through the intermediate of liquid species. However, if these species are liquid at the temperature of reaction they are solid at room temperature. They are soluble in water proving that they are not melted cellulose. These products could be compared to the active cellulose postulated a long time ago by Bradbury et al. (1979) (Figure 1) even if these authors had few chances to observe a liquid phase because their experiments were made at too low a temperature.

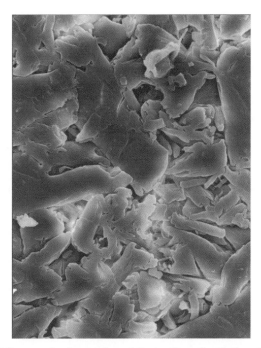

Figure 5   Microscopic observation of the surface of a virgin cellulose pellet

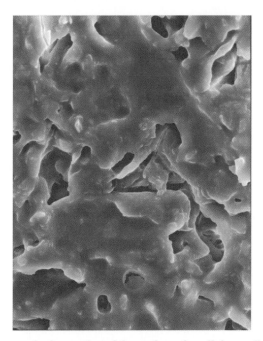

Figure 6   Microscopic observation of the surface of a cellulose pellet that has been subjected to a flash of a concentrated radiation

Experimental Determination of Mass Balances:

The results of five different experiments are given in Table 1, with the notations as given in Figure 4. Each result corresponds to several experiments made in the same conditions of fluxes and irradiation times in order to enhance the quantity of products and hence the accuracy of the measurements.

The results show that it is possible to recover significant amounts of the intermediate species previously evidenced by the microscopic observations. High fractions (more than 80 %) of the products escaped from the pellets are recovered on the filter. This means that, in our conditions of high flux densities, the intermediate species mainly decompose into condensable matter (step 2 of Figure 2). As no noticeable formation of char is observed, it can be assumed that the remaining fraction of the products escaping from the surface is mainly gases (their possible existence and composition are under investigation). Another explanation is the possibility of condensation of a small fraction of the vapours on the inner walls of the glass reactor.

Table 1 Results of several mass balances (the masses are in 10^{-6} kg and the flux "Max" corresponds to about $2 \ 10^7$ W m^{-2})

Flux	Max	Max	Max	Max	Max/2
Flash duration (s)	0.134	0.164	0.174	0.178	0.197
m0	147.8	79.3	102.8	81.9	143.0
m1	1.5	3.0	4.9	4.5	8.1
m2	1.4	3.0	4.0	4.1	6.5
m3	2.8	2.3	2.1	1.7	6.0
m4	142.5	73.5	96.1	75.2	127.1
m5 = m0 - m4	5.3	5.8	6.7	6.7	15.9
(m2+m3+m4)/m0 (%)	99.3	99.4	99.4	98.9	97.6

The weight loss of the pellets as well as the mass of products trapped on the filter logically increase with flash duration for a given flux. At the same time, the mass of the remaining intermediate species seems constant proving that steady state conditions are probably reached. These mass balances show a good reproducibility (between 80 to 90 % of the weight loss is recovered on the filter) in regard to the small quantities involved in the experiments. The overall balances $(m_2+m_3+m_4) / m_0$ are also very good. But these results must be cautiously considered because of the small values of the fractions $(m_2+m_3) / m_4$.

First Analysis In High Performance Liquid Chromatography (HPLC)

The first analysis of the water solution of the intermediate species and of the condensed matters have been made by HPLC. The results reveal relatively few peaks (five important peaks) in comparison with the analysis of the more usual bio oils. The first one corresponds to the levoglucosan which appears as the major component of condensed matters. The other peaks have not been accurately determined at that time, but we can reasonably assume that they correspond to oligoanhydrosaccharides (cellobiosan) in agreement with the results found by Radlein et al. (1987). These oligomers are the major components of the non vaporized intermediate species.

CONCLUSION

Very high flux densities can be reached by a concentrated radiation obtained with an image furnace. This very clean source of heat can be favourably used for carrying out the flash pyrolysis of cellulose during times of exposures much lower than 1 s.

The results provide clear evidence for the formation of intermediate species during the first moments of the reaction. In our experimental conditions, these species have life times of only a few 10s of milliseconds and rapidly produce vapours and aerosols. They are liquid at the reaction temperature but solid at room temperature. They are soluble in water.

Future work will include the analysis of the possible gases formed during the reaction as well as the HPLC analysis of the intermediate products that seem to be formed of a relatively small number of species resulting from the thermal depolymerisation of cellulose.

ACKNOWLEDGEMENTS

The authors want to thank ADEME (Agence de l'Environnement et de la Maîtrise de l'Energie) who has funded this research in the framework of AGRICE (AGRIculture pour la Chimie et l'Energie) under contract n° 96 01 048. They are also grateful to the IMP-CNRS (Institut de Science et de Génie des Matériaux et Procédés) for the microscopic photographs.

REFERENCES

M.J. Antal and G. Varhegyi, Ind. Eng. Chem. Res., 34 , 1995, 703-717.

M.J. Antal, G. Varhegyi and E. Jakab, Ind. Eng. Chem. Res., 1998, 1267-1279.

A. Bilbao, A. Millera and J. Arauzo, Fuel, 67, 1988, 1586-1588.

O. Boutin, M. Ferrer and J. Lédé, J. Anal. Appl. Pyrolysis , 1998, in press.

A.G.W. Bradbury, Y. Sakai and F. Shafizadeh, J. Appl. Polym. Sci., 23, 1979, 3271-3280.

S. Czernik, J. Scahill and J. Diebold, J. Solar Energy Eng., 117, 1995, 2-6.

C. Di Blasi, Chem. Eng. Sci., 51, 1996, 1121-1132.

J.P. Diebold, Biomass and Bioenergy, 7, 1994, 75-85.

M. Grönli, Thesis, The Norwegian University of Science and Technology, Faculty of Mechanical Engineering, Trondheim (N), 1996.

M. Lanzetta, C. Di Blasi, and F. Buonanno, Ind. Eng. Chem. Res., 36, 1997, 542-552.

J. Lédé, J. Panagopoulos, H.Z. Li, and J. Villermaux, Fuel, 64 , 1985, 1514-1520.

J. Lédé, F. Verzaro, B. Antoine, J. Villermaux, Chem. Eng. Process., 20, n°6, 1986, 309-317.

J. Lédé, H. Z. Li and J. Villermaux, J. Anal. Appl. Pyrolysis, 10, 1987, 291-308.

J. Lédé, J.P. Diebold, G.V.C.Peacocke and J. Piskorz, in Developments in Thermochemical Biomass Conversion, A.V. Bridgwater and D.G.B. Boocock (Eds), Blackie Academic and Professional, Glasgow, 1997, pp. 27-42.

R.S. Miller and J. Bellan, Combust. Sci. Tech., 126, 1997, 97-112.

I. Milosavljevic and E.M. Suuberg, Ind. Eng. Chem. Res., 34 , 1995, 1081-1091.

D. Radlein, A. Grinshpun, J. Piskorz and D. Scott, J. of Anal. Appl. Pyrolysis, 12, 1987, 39-49.

FORMULATION AND APPLICATION OF BIOMASS PYROLYSIS MODELS FOR PROCESS DESIGN AND DEVELOPMENT

Colomba Di Blasi

Dipartimento di Ingegneria Chimica, Universitàdegli Studi di Napoli Federico II
Piazzale V. Tecchio, 80125 Napoli, Italy
Tel: 39 081 7682232   Fax: 39 081 2391800   e-mail:diblasi@unina.it

ABSTRACT

Pyrolysis of biomass is the result of complex interactions among numerous chemical and physical processes, which take place across the particle and the reaction environment. The relative importance of the different processes is highly dependent on particle characteristics (thermochemical properties, size, moisture and ash content) and the type of conversion unit (heat and mass transfer between particle and reactor, extra-particle activity of secondary reactions). In this paper, the different approaches chosen to model intra- and extra-particle processes of biomass pyrolysis are briefly illustrated in relation to fixed-bed, fluid-bed and ablative reactors. The experimental measurements needed to produce data for process simulation and model validation are listed. Examples of numerical simulation are presented for the convective/radiative pyrolysis of a moist cellulosic particle. Finally, suggestions are given on how modelling can be used for process development and optimisation.

INTRODUCTION

The reactions of biomass pyrolysis can be roughly identified as primary solid degradation which gives rise to products lumped into condensable (tars) and non-condensable (gases) volatiles and solid char. Secondary reactions of volatile condensable products may also occur to low-molecular weight gases and/or char, as they are transported through the reacting solid or the heating environment, and affect final product distribution. The physical behaviour of the degrading solid and the product distribution are dependent on three main parameters: 1) the heating rate of the solid particles, 2) the residence time of the solid particles and the evolved pyrolysis products inside the reactor and 3) the reactor temperature, all dependent on numerous variables (nature of the solid fuel, particle size, reactor type, etc.).

The specification of the pyrolysis characteristics (conventional or fast pyrolysis) does not correspond to the definition of the processes controlling the conversion, i. e., chemistry or heat and mass transfer. Heat transfer from reactor to particle is always important and, in general, all the mechanisms contribute to a certain extent. Schematically, two main modalities of solid pyrolysis can be defined, on the basis of the prevalent heat transfer mechanisms, namely convective/radiative pyrolysis (indirect heating) or contact (ablative) pyrolysis (direct heating). However, the thermal regime, which is established during the pyrolytic degradation, is also determined by the macroscopic behaviour of the fuels (charring or melting), which affects internal heat transfer and intra-particle evolution of primary volatile pyrolysis products, not to say by the degradation

In this study the main chemical and physical mechanisms involved in the pyrolysis of biomass are briefly reviewed, with a view of process simulation. The main steps in the formulation and numerical solution of process equations are described and possible application of simulation results are suggested.

GUIDELINES FOR MODEL FORMULATION

The construction of a biomass pyrolysis model consists of five main stages:
1)   choice of the "pyrolysis system",
2)   mathematical formulation of chemical and physical processes,
3)   analytical or numerical solution of differential equations,
4)   selection of input data,
5)   model validation.

The pyrolysis system, depending on the specific objectives of the model, can coincide with a) the single particle, b) the reaction unit, or c) both the single particle and the reaction unit. Usually, the basic conservation equations for mass, energy and momentum are coupled with appropriate sub-models for drying, primary degradation, secondary degradation, property variations, intra- and extra-particle heat and mass transfer coefficients, particle structural behaviour, transport coefficient between the phases (i.e. emulsion and bubble phase in fluidised bed reactors), etc. Again, the level of approximation that can be accepted in the description of chemical and physical processes is dependent upon the model applications, for instance, investigation of process fundamentals, reactor design, etc.

Analytical solutions of pyrolysis models are possible only for very simplified problem descriptions, which highly limit the application field of predictions. Usually, numerical solutions are preferred which, for all the computer models currently available, are based on finite-difference formulations. These formulations for single-particle, slow-pyrolysis models are straightforward and their solution can be handled by standard techniques. More difficult is the consistent approximation of flash pyrolysis models, owing to the high velocities associated with the fast rates of volatile release, and chemical reactor models, where in some cases the flow becomes turbulent.

Once the mathematical description of the pyrolysis system has been "translated" into a computer code, input data are need in order to proceed with the numerical simulation. Data are needed about property values (and their variation with reaction conditions), kinetic constants, characteristics of the pyrolysis system (size, geometry) and parameters for each of the process sub-models. The selection of appropriate data is a critical issue, as in many cases numerical simulations are based on parameter values, kinetic constants and transport coefficients derived from experimental conditions which do not reproduce those of the numerical simulation, especially when oriented to practical applications. Consequently, the extrapolation to flow and/or temperature conditions not considered in the estimation of the data may lead to wrong predictions.

If the input data are correctly derived for the conditions of interest in the numerical simulation and consistent numerical approximations are used, model predictions should closely match experimental measurements, where presumably both chemical kinetics and transport phenomena are controlling. Comparison between model predictions and experimental measurements (drying and pyrolysis times, temperature, velocity and concentration fields, product yields and composition, etc.) is, however, still very important because only in such a way is it possible to determine the validity limits of the numerical predictions. Then numerical simulation can be used for process design, development and optimisation.

CHEMICAL AND PHYSICAL PROCESSES OF BIOMASS PYROLYSIS

Thermal Degradation Kinetics Kinetic studies of the thermal degradation of organic and synthetic polymers can be classified into three main groups (Di Blasi (1993)):

1)   One step global models: a one-step reaction is used to describe solid degradation by means of the experimentally measured rates of weight loss. Such an approximation has been coupled to the description of physical processes of solid phase only, and to the treatment of both solid- gas-phase transport phenomena. However, the dependence of product yields on reaction conditions cannot be predicted, as a constant ratio between volatiles and char is assumed.

2)   Multi-reaction models: several reactions are used to correlate product distributions. These are one-stage models, describing the degradation of the solid to char and several gaseous species, valid only for the specific experiments where derived.

3)   Semi-global models: kinetic mechanisms of solid degradation are formulated which include both primary and secondary reactions, where pyrolysis products are lumped into three groups (tar, gas and char). For the formulation of engineering models, with a view of reactor optimisation and design, these models appear to be the most promising, because competitive chemical pathways are described, which allow product distribution to be predicted. At the same time, they are not exceedingly complex, and therefore suitable for coupling chemistry with transport phenomena, without requiring large sets of data (usually not available).

Cellulosic materials degrade through two main competitive pathways, ring scission and end-group depolymerization (Antal (1985), Antal and Varhegyi (1995)). The first, leading to char and gas formation, is favoured at low temperatures (below 550-570K), whereas the second, leading to tar production, is favoured at higher temperatures. Finally, it should be noted that whether thermal degradation is the limiting step or not depends on the conversion conditions and the nature of the polymer. Primary solid degradation is strictly coupled to heat transfer, while secondary degradation is affected by both heat and mass transfer. Therefore conversion characteristics are, in general, the result of a strong interaction between chemical and physical processes.

The majority of the biomass pyrolysis models currently available examines single particle behaviour, with a different degree of approximation. Physical processes of biomass pyrolysis are associated with moisture evaporation and transport phenomena for the single particle and the reaction environment. Moist particles (Di Blasi (1998a)) have been modelled as a three-phase mixture: virgin solid with bound water to the FSP, capillary water that partially fills the pores, and bubbles containing inert gas and water vapour. Transport phenomena taken into account include: convection of capillary water, convection and diffusion of water vapour, surface diffusion of bound water, heat convection and conduction, liquid and gas phase pressure and velocity variations, Clausius-Clapeyron equation and desorption isotherms. The drying model has been applied for extensive simulations and has been recently coupled to a convective/radiative pyrolysis model.

Feedstock heating often occurs through convection and radiation, as in the case of fixed-bed (updraft and downdraft) gasifiers, entrained-flow reactors, etc.. Initially the solid is essentially interested by transient heat conduction. Then a region, in the neighbourhood of the heated side, undergoes thermal degradation. When all the volatiles are removed from the solid, a char layer is formed. Two further spatial zones can be seen: the region where pyrolysis reactions are active and the virgin solid region. Volatile species, generated in the pyrolysis region, may, because of pressure gradients, be forced to flow towards both the unreacted solid and the already charred region. As char permeability is higher, the flow of products occurs mainly towards the heated surface and, because of the high temperature, secondary tar reactions may occur. Volatile products may also migrate through the unreacted, low-temperature solid where they may condense and, subsequently, as the pyrolysis front progresses, evaporate. Apart from heat, momentum and mass transfer, changes in the physical structure of the reacting solid are observed with the development of a network of cracks in the already pyrolyzed region, surface regression and internal shrinkage and/or swelling. Given the continuous increase in the thickness of the char layer and structural variations, the pyrolysis process is highly unsteady.

Single-particle models currently available (Di Blasi (1996a-c, 1998b) take into account all the above physical processes through: 1) property (porosity, permeability, thermal conductivity, thermal capacity, mass diffusivity) variation with the conversion level, 2) accumulation of volatile species mass and enthalpy within the pores, 3) heat transfer by convection, conduction and radiation, 4) convective and diffusive transport of volatile species, 5) gas pressure and velocity variations, 6) particle shrinkage and/or swelling, 7) anisotropy (2D model).

Heat transfer through conduction, by contact of the solid fuel with a hot surface or hot sand, may also take place in ablative pyrolysers, circulating fluid-bed reactors, etc.. Single-particle experiments on contact pyrolysis, by means of the so-called hot-plate technique, were first carried out to simulate the high-temperature pyrolysis of propellants. Successive studies also considered some synthetic polymers, but probably, the most impressive experiments are those on wood, carried out by Lede and his group(1985, 1987a). In these experiments, the sample is pressed against an electrically heated plate, whose temperature is maintained constant. If the heating rates are high enough, a thin molten layer is seen to propagate through the virgin solid with a constant rate. Thus, for a rotating plate, similar to thermoplastic polymers, wood undergoes a plastic phase change, followed by rapid vaporisation, with negligible char formation and the conversion process becomes quasi-steady. It should be noted that the global external heat transfer coefficients achieved in these experiments are very high: one to three orders of magnitude higher than those calculated for the case of simple radiative heat transfer (Lede et al. (1985)).

Single-particle models describe ablative pyrolysis either as wood melting (Lede et al. (1985, 1987b)) or as a chemical reaction (Di Blasi (1996d)). In this case, the features taken into account include: 1) different properties for the solid, the molten and the vapour phase, 2) variable properties, 3) continuous char ablation and volume (sample length) reduction until complete conversion, 4) mass and heat (convection, conduction and reaction energetics) transfer through the reacting medium, 5) high applied pressure, through the introduction of a global interface heat transfer coefficient.

Contrary to single particle systems, only in very few cases the modelling of pyrolysis reactor models has been considered. The simplest version of a fluid-bed pyrolyser, developed in this laboratory is based on the following assumptions: 1) well-mixed continuous reactor for the bed (negligible heat and mass transfer resistances between the bubbles and the dense phase), 2) isothermal plug-flow for the free-board region, 3) no spatial gradients along the (spherical) biomass particles, 4) constant pressure, 5) char particles entrained by the carrier gas, 6) no entrainment of biomass and sand particles, 7) constant porosity of the bed. Successive versions of the model take into account single particle dynamics though extensive process simulations are still under way. Also, a two-dimensional model of fixed-bed pyrolysis reactor has been developed with the description of 1) axial and radial heat transfer through convection, conduction and radiation, 2) axial and radial mass transfer through convection and diffusion, 3) pressure and velocity variations (modified Darcy law), 4) solid convection, 5) absence of local thermal equilibrium, 6) property variation, 7) unsteady solid and gas phase processes, 8) thermally thin regime and constant size of particles and 9) different boundary conditions (wall heating, convective heating).

The main purpose of pyrolysis process modelling is to analyse the effects of process parameters in order to provide useful data for the design and optimisation of chemical reactors and other process equipment. As a result of detailed numerical simulation it has been possible, for instance (Di Blasi (1996c)), to construct a map of the particle size as a function of the external heating conditions, which allows the conversion regime and the activity of secondary reactions to be determined. Thus operating conditions of the conversion units and/or characteristics of the feedstock can be selected in a way that the production of the desired product class is optimised. Numerical simulations of pyrolysis reactors allow useful information to be gained in relation to process dynamics and reactor performance so that reactor design can be optimised, system characteristics useful to experiment programming can be identified and control procedures can be selected.

EXAMPLES OF PROCESS MODELLING

In this section some results, not published elsewhere, are presented of numerical simulation of pyrolysis of a moist biomass particle. The mathematical model incorporates the physical processes of convective/radiative pyrolysis (Di Blasi (1996a-b)), with reaction kinetics described by the Broido-Shafizadeh mechanism extended to include secondary reactions, and multi-phase transport phenomena of moisture evaporation (Di Blasi (1998a)). A one-dimensional slab (half-thickness $7x0^{-2}m$) is exposed to a radiative flux of known intensity, Q. Medium properties roughly describe wood (values derived from (Lee and Diehel (1981), Stanish et al. (1986), Di Blasi (1996a)). A first set of simulations has been made for a moisture content $U\_0=0.5$ (dry basis, where the free content is $U_{l0}=0.2$ and the bound content is $U_{b0}=0.3$) and $Q=84kW/m^2$, for $B_g=5x10^{-14}$ m^2 (wood permeability) and $B_c=5x10^{-14}$ m^2 (char permeability), which give rise to significant gas overpressure during the conversion process. A parametric study is also carried out of the effects of initial moisture content (U_0 is varied from 0 to 0.6 and heating conditions (Q is varied from 20 to 84 kW/m^2 for $U_0=0.5$) when $B_c=25x10^{-14}m^2$ (low gas overpressures).

The early dynamics (the first 250s) of moist solid pyrolysis ($U_0=0.5$) are presented through plots of the spatial profiles of capillary and bound water content (Figure 1), temperature, total condensed phase density (Figure 2), gas overpressure, liquid and gas velocities (Figure 3) and densities of tar and water vapours (Figure 4). It can be seen that, from the very beginning of the process, moisture evaporation and solid pyrolysis occur simultaneously, though their characteristic temperatures are highly different and do not remain constant during the transient sample degradation.

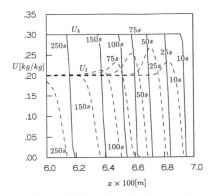

Figure 1 Bound (solid lines) and capillary water (dashed lines) profiles for the initial process transients, Q=84kW/m^2

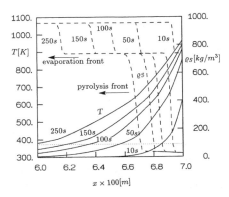

Figure 2 Profiles of temperature (solid lines) and condensed-phase (solid+liquid) density (dashed lines) for the initial process transients, Q=84kW/m^2. The dotted line corresponds to T=373K

As expected, water evaporation (capillary water evaporation is faster than bound water evaporation) precedes solid pyrolysis (Figure 1). The onset of the two processes is characterised by high spatial gradients in the total condensed phase (solid + liquid) density; evaporation and pyrolysis occur through the propagation of rather steep density fronts across the wet and the already dried wood, respectively (Figure 2). The profiles of capillary and bound water contents are steeper that those of the char density, indicating that, at least in the initial high-temperature stages, the characteristic times of water evaporation are much faster than those of liquid phase transport. Liquid phase convection occurs from the wet region towards the drying front and attains its maximum at the beginning of the process when the evaporation rate is very fast. However, as expected for high temperatures, the convective flows of the liquid phase are five to six orders of magnitude slower than those of the vapour/gas phase (Figure 3). Therefore, even larger values of the permeability to liquid flow are not expected to affect significantly the pyrolytic conversion of moist wood.

The presence of moisture affects the heat transfer rate through the solid, above all through the variation in the thermal properties (thermal conductivity and effective solid heat capacity). Indeed, two main gradients exist, corresponding to the wet and dry region. Other effects are related to convective transport and evaporation/condensation of water. A maximum in the gas overpressure profile is seen (Figure 3), just ahead of the water evaporation front. This maximum is caused by vapour production and/or migration through the wet wood, where the porosity (and the volume available for the vapour/gas phase) is low because of the presence of moisture and the absence of devolatilization reactions.

The overpressure peak also separates two velocity distributions, with a change in the direction of the vapour/gas flow, one directed towards the moist, virgin wood and the other towards the charred surface. Negative velocities are rather small, but vapour transport towards the moist region also occurs by mass diffusion. Transport in this direction tends to pre-heat the solid, though this effect is small and the water vapour condenses since the temperature is still low (this process would also tend to cause local increase in the temperature). The condensation of water vapours in the cold region is confirmed by the local maxima in the profiles of capillary water content (Figure 1). Water vapours flow mainly towards the pyrolysis/char regions, causing significant transport of sensible heat out of the sample. In fact, the high rates of vapour release in the initial stages of the process give rise to high flow velocities. The process of water vapour condensation before the drying front is successively less important because, as time increases, convective transport also goes to zero (spatial pressure gradients go to zero, though this variable continuously increases in time), and, at the same time, temperatures are successively larger.

The rate of propagation of the drying front decreases as this moves further away from the heated surface, while the temperature needed for complete water evaporation continuously increases (near the surface, capillary water evaporates completely for temperatures close to the normal water boiling point, whereas bound water evaporation is completed for a temperature about 10K higher than this limit). This behaviour indicates that the process of moisture evaporation through thick wood is controlled by both heat and mass transfer. Indeed, the successively slower heat transfer rate, as the distance from the sample surface increases, and the continuous increase of the gas overpressure in the wet zone may explain the process dynamics. In particular, the increase of the gas pressure above the atmospheric value raises the boiling point of water, so that though the temperature exceeds 373K, water is still below the boiling point. Water vapour transport also affects the evaporation process. A decrease in the local mass concentration of water vapour (and vapour pressure) requires that water should evaporate at lower temperatures. The two effects tend to counteract each other, because the first would lead to an increase in the boiling point and the second would allow evaporation at lower temperatures. The dominance of the first or the second effect depends on the relative values of the permeability to gas flow and the water vapour diffusion coefficient. For the parameter values employed here, it appears that the increased boiling point dominates.

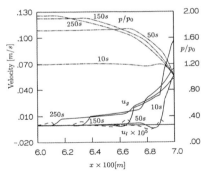

Figure 3 Profiles of gas overpressure (dashed-dotted lines), liquid velocity (dashed line) and gas velocity (solid line) for the initial process transients, $Q=84kW/m^2$

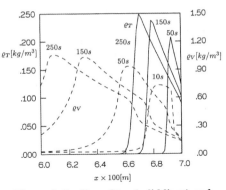

Figure 4 Profiles of tar (solid lines) and water vapor (dashed lines) density for the initial process transients, $Q=84kW/m^2$

Char density increases with the distance from the heated surface while the propagation rate of the pyrolysis front decreases. Both effects are due to an increased resistance to heat transfer with distance from the heated surface, so that the temperature at the reaction front is progressively reduced. Consequently, the degradation rate decreases and the char forming reactions are favoured with respect to volatile formation. The positive part of the velocity profiles presents two different gradients, the first corresponding to water vapour addition to the gas phase, following water evaporation, the second to volatile release, following solid pyrolysis. Water vapour and tar density profiles (Figure 4), along the dried/reacting wood reflect the effects of convective transport and, in the second case, significant secondary tar degradation to gas.

The predictions are in agreement with experimental observations of high-temperature drying of degrading solids. The maximum in the moisture content and significant gas overpressure have been observed in (White and Shaffer (1981), Fredlund (1993)), as well as a noticeable influence of the moisture content on the temperature profiles (Lee and Diehl (1981), Chan et al. (1988)). The process retains the characteristics previously discussed with a thin evaporating front preceding the reaction front until the drying front approaches the sample centreline. At a distance of about 2×10^{-2}m from the sample centreline, evaporation starts to occur along the whole remaining part of the sample, with a continuous reduction in the moisture content. Due to the wide extent of the evaporating region and the large evaporation flux, the gas pressure increases above the atmospheric value. The evaporation along this zone occurs at an almost constant temperature, which is higher than the normal water boiling point (by about 25-35K), because of the high pressures and the effective kinetics of moisture evaporation, which give rise to significant vapour pressure depression at low moisture contents (Di Blasi (1998a)).

Figure 5 reports the conversion and the drying times and the final yields of char and tar as functions of the initial moisture content ($Q=84kW/m^2$). In agreement with experimental observation (Chan et al. (1988)), the drying time shows a linear increase with the moisture content. Drying time can be considered as the time needed to bring the sample up to temperatures slightly higher than the water boiling point (about 80 min) plus the contributions due to moisture heating and evaporation. These latter contributions can be considered responsible for the increase of the drying time with U_0. As a consequence, the conversion time also increases with U_0 (the difference between the two remains almost constant, about 210-220 min, as U_0 is varied).

The reduced temperatures at the primary degradation front cause an increase in the final char yield with increase in U_0. Final tar yields show a rapid increase followed by the attainment of a constant value as U_0 is increased. This is due to an initial increase of the gas velocity with U_0 (larger amounts of volatiles released into the gas phase), so that the residence time of tar vapours inside the hot char matrix becomes shorter and the activity of secondary reactions is reduced. For U_0 above certain values, moisture evaporation being a heat transfer controlled process, the vapour release rate (and flow velocity) remain about the same and so does the tar yield.

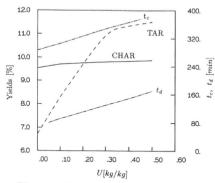

Figure 5 Drying and conversion times and final char and tar yields as functions of the initial moisture content, $Q=84kW/m^2$.

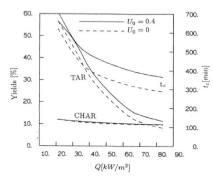

Figure 6 Conversion time and final char and tar yields as functions of the intensity of the applied heat flux for dry (dashed lines) and moist ($U_0=0.5$, solid lines) sample.

Figure 6 reports the conversion time and the final char and gas yields as functions of the intensity of the applied radiant heat flux, for the case of dry and moist ($U_0=0.5$) solid. For both cases, the results are in agreement with experimental observation in that the tar yield decreases as the surface temperature becomes successively larger because of the enhanced activity of thermal cracking. Also, the final char yield decreases slightly because of the small increase in the temperature at the primary degradation front, which favours the tar forming reactions. This effect is small, however, because for large samples most part of the degradation process occurs at temperatures much lower than those reached at the irradiated surface (Di Blasi (1996c)). As expected, the conversion time becomes successively shorter as the heating conditions are made more severe. A comparison between the cases of moist and dry sample confirms that moisture evaporation delays the conversion process and on average gives rise to larger tar yields. The activity of secondary reactions is reduced because of shorter volatile residence times. The differences between the two cases tend to disappear at low heat fluxes because the activity of secondary reactions is always negligible at low temperatures.

CONCLUSIONS AND FURTHER DEVELOPMENTS

Mathematical models of single particle pyrolysis currently available include detailed description of all the main chemical and physical processes. However, the applications are mainly concerned with cellulose (both slow and flash pyrolysis) and, limited to slow pyrolysis, with wood. Semi-global primary reaction mechanisms (and kinetic constants) for wood and biomass under "real conditions" are needed for application of model simulations into practice. Also, very little is known about the secondary reactions mechanisms and constants and property variations as pyrolysis takes place. Furthermore, only very few data are available for model validation even

378

for laboratory conditions. That is, temperature profiles and product distribution have not been measured by the same laboratory, therefore they do not refer to the same experimental conditions, which often are not sufficiently detailed to be used in the numerical simulation of the process.

Contrary to the case of single-particle models, the literature on pyrolysis reactor modelling is scarce. Therefore, detailed models should be formulated and validated first on a laboratory scale and successively on industrial scale. Aspects which deserve particular attention are concerned with the description of multi-dimensional effects, single-particle behaviour and complex fluid-dynamics. Also in this case, input data, valid over the conditions of interest in practical applications, should be determined.

REFERENCES

1) Antal M. J., Advances in Solar Energy, Baer, K., W., Duffie, J., A., Eds., American Solar Energy Society, Boulder Co., 1982, 61-111.
2) Antal M. J. and Varhegyi G., Ind. Eng. Chem. Res., 1995, 34, 703-717.
3) Chan W. R., Kelbon M., Krieger-Brockett B., Ind. Eng. Chem. Res. 27: 2261-2275, 1988.
4) Di Blasi C., Progress in Energy and Combustion Science 19, pp. 71-104, 1993.
5) Di Blasi C., Fuel, 75: 58-66, 1996a.
6) Di Blasi C. Chemical Engineering Science, 51: 1121-1132, 1996b.
7) Di Blasi C., Ind. Eng. Chem. Res., 35:37-47, 1996c.
8) Di Blasi C., Chemical Engineering Science 51: 2211-2220, 1996d.
9) Di Blasi C., Chemical Engineering Science 53, 353-366, 1998a.
10) Di Blasi C., Int. J. of Heat and Mass Transfer 41: 4139-4150, 1998b.
11) Fredlund B., Modelling of heat and mass transfer in wood structures during fire, Fire Safety Journal 20:39-69, 1993.
12) Lede J., Panagopoulos J., Li H. Z., and Villermaux J. (1985). Fuel, 64, 1514.
13) Lede, J., Li H. Z., Villermaux J., Martin H. (1987). J. of Analytical and Applied Pyrolysis, 10, 291.
14) Lee C. K., Diehl J. R., Combustion and Flame 42:123-138 , 1981. 15) Martin H., Lede J., Li H. Z., Villermaux J., Moyne C., De Giovanni A., (1987). Int. J. of Heat and Mass Transfer, 29, 1407.
15) Stanish M. A., Schajer G. S., Kayihan F., AIChE J. 32: 1301-1311, 1986.
16) White R. H., Schaffer E. L., Wood and Fiber 13: 17-38, 1981.

PYROLYSIS OF CELLULOSE
FROM OLIGOSACCHARIDES TO SYNTHESIS GAS

Jan Piskorz, D. Radlein and **P. Majerski**
RTI - Resource Transforms International Ltd, 110 Baffin Place, Unit 5
Waterloo, Ontario, N2V 1Z7, Canada
D.S. Scott
University of Waterloo. Waterloo, Canada

ABSTRACT

Cellulose Pyrolysis Products

From commercial and industrial point of view the driving force of all activity is the existence of marketable products. The first challenge is therefore to define a target product.

What are the possible products from cellulose pyrolysis?

Five principal modes or pathways of decomposition can be identified, each characterized by one of the following specific products:
1. Cellobiosan and other anhydro-oligosaccharides
2. Levoglucosan
3. Hydroxyacetaldehyde
4. Levoglucosenone
5. Synthesis gas

What are the required process conditions for their optimization?

Thermogravimetric data suggest that differently prepared celluloses show different pyrolysis characteristics. Therefore in discussions of cellulose pyrolysis kinetics it should be stated which of the possible decomposition modes is being referred to, i.e. which is dominant under given conditions.

INTRODUCTION

Mankind's fascination with pyrolysis started with the discovery of fire.

> *Pyrolysis is usually understood to be thermal decomposition which occurs in an oxygen-free atmosphere, but oxidative pyrolysis is nearly always an inherent part of combustion processes, (Serio, 1995).*

From this definition what is clear is that the pyrolysis step is of critical importance not only in pyrolysis per se but also in any combustion and gasification schemes.

The pyrolysis step, particularly in so-called "fast" pyrolysis happens in a time of few seconds or less. In such a short time chemical reaction kinetics, mass transfer processes, phase transitions and heat transfer phenomena play important roles and can influence the overall process. Also, due to the same short time intervals, fundamental aspects of those chemical engineering processes are not easily elucidated, and as a result considerable controversies exist the an open literature. To derive a more comprehensive understanding of pyrolysis is a noble goal of scientists dedicated to the study of the mechanism and science of pyrolysis. Benefits from such work will come with time from technological developments and innovations in growing activities focused on renewable fuels, new high value chemicals and materials.

381

The focus of this paper is on the pyrolysis *experimental results* obtained by the University of Waterloo group.

Over the years this group perfected the use of small continuous fluid bed reactor systems in which typical working conditions were confined to atmospheric (or close to it) pressure, temperature in a 400 - 650 °C range and short (second or less), reaction times. The discussion is confined to the process of *cellulose* pyrolysis. The graphs, tables and pictures derived from actual experiments were put together to illustrate some fundamental aspects of cellulose pyrolysis and the importance of those in emerging new "green" technologies for the 21st century.

MAIN POINTS

General

Cellulose, the most abundant natural polymer can be thermally cracked in many ways. Despite much investigation over years, the pyrolytic decomposition of cellulose remains relatively poorly understood and reliable experimental results are scarce. This lack of data is often raised by scientists trying to develop much needed mathematical models and realistic computer simulations.

The difficulties in studying the phenomena can be traced partly to the great chemical and structural diversity of materials coupled with their low thermal stability. These properties result in a complex array of products on decomposition which themselves are also <u>thermally sensitive.</u>

Thermogravimetry

There are varieties of research grade and technical quality celluloses (microcrystalline and fibrous) obtainable on the market. Those can be easily differentiated and "finger-printed" by routine thermogravimetric analyses, often coupled with other modern analytical methods as shown in Figure 1.

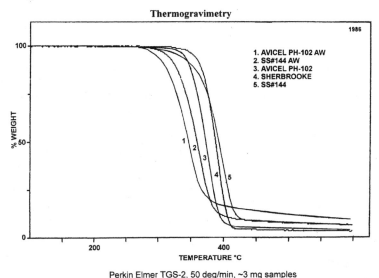

Perkin Elmer TGS-2, 50 deg/min, ~3 mg samples

Figure 1 Thermogravimetic analysis of different forms of cellulose

Deriving rigorous chemical kinetics from thermogravimetric data should be attempted in tandem with characterization of feeds and *products*. Thermogravimetry can offer a convenient way to evaluate volatilities and thermal stabilities of potential products of pyrolysis like cellobiosan, levoglucosan, hydroxyacetaldehyde and so on as shown in Figure 2. There are some suggestions in literature that cellulose crystallinity could be estimated by thermogravimetry (Schulz 1985).

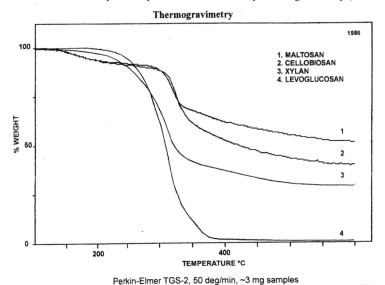

Figure 2   Thermogravimetic analysis of components of biomass and its degradation products

ANHYDRO-OLIGOSACCHARIDES

To obtain a significant yield of anhydro-oligo (cello) saccharides from cellulose a special approach is needed due to the non-volatile nature of oligosaccharides in general.   Clearly, cellobiosan is thermally unstable under typical pyrolysis conditions.   It seems that the cellobioside structure should be regarded as the monomer present in the naturally occurring cellulose polymer chain. Such a structure under some thermal conditions can yield very significant (on a mol by mol basis) yields of compounds like anhydro-oligosaccharides, (or levoglucosan, or levoglucosenone, or glycolaldehyde) as shown in Figure 3.

It is postulated that anhydro-oligosaccharides are initially formed upon rapid heating of (Avicel) cellulose. Due to their insignificant vapour pressures oligo-saccharides can become part of generated volatiles only by a mechanical entrainment/mist generation process (expelling, ejection, jettisoning, explosive release, desorption).

It is evident from the electron scanning microscope photographs that initial depolymerisation of cellulose in fast pyrolysis systems occurs from a *molten state.*

LEVOGLUCOSAN

Pyrolytic yields of circa ~ 90 wt % of liquids are obtainable from some celluloses. These liquids can contain up to 50 wt % of levoglucosan. That disproves the previous assumption (in literature) that vacuum conditions are essential. This fact also attests to the remarkable relative thermal stability of levoglucosan. The ability to obtain a single chemical in such impressive yield provides impetus to seek potential application for levoglucosan.

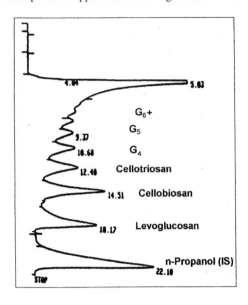

HPLC Chromatogram Showing a Seqence of Oligosaccharides from Rapidly Heated and Quenched Pyrolysis of Cellulose.

Column: Biorad HPX-42A

From: A. Vladars-Usas, "Thermal Decomposition of Cellulose", M.A.Sc. thesis, 1993, University of Waterloo

Figure 3   Sequence of oligo-saccharides by HPLC analysis

The decomposition pattern of cellulose in pyrolysis may be steered towards depolymerisation or fragmentation as shown in the model in Figure 4. A strong (negative) interrelationship exists between content of ionic substances and the likelihood of obtaining levoglucosan as a major product. In high temperature pyrolysis ionic substances reduce levoglucosan formation by favouring ring scission over depolymerisation.

Historically, depolymerization of celluloses towards levoglucosan was performed under vacuum. Fast pyrolysis can add an additional flexibility in such technological processing. Levoglucosan has a research chemical status at present. Its utilization and applicability is under development. Depolymerization of microcrystalline Avicel PH-102 at 500 °C (levoglucosan yield ~30 %) requires ~ 1200J/g as found experimentally on a significant scale (kg/hr). This value includes the sensible heat needed for bringing up cellulose to temperature of the reaction.

HYDROXYACETALDEHYDE, (GLYCOLALDEHYDE)

Glycolaldehyde dimer is called 2,5-dihydroxy-1,4-dioxane. It is proven by authors (Scott, 1984) and confirmed by others (Richards 1988) that hydroxyacetaldehyde, long known as a minor or trace product of cellulose pyrolysis can be a major product under suitable conditions.

384

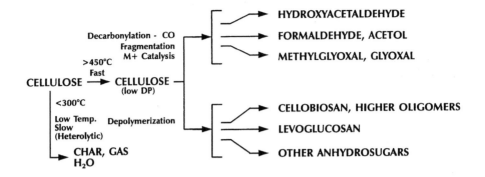

Small amounts of inorganic substances can cause profound changes in the thermal behaviour and product spectrum of cellulose pyrolysis. Inorganic salt additives favour the fragmentation pathway (formation of levoglucosan is inhibited) in which case hydroxyacetaldehyde is the principal product. Native celluloses, with sodium, potassium cations still present, give significant yields of hydroxyacetaldehye on fast pyrolysis conditions. Yields up to ~17 % can be obtained as indicated in Figure 5.

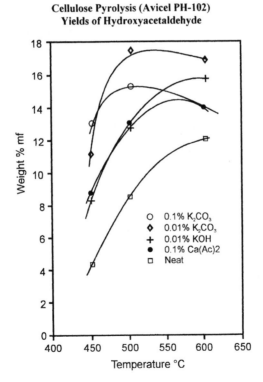

Cellulose Pyrolysis (Avicel PH-102)
Yields of Hydroxyacetaldehyde

Figure 5 Yields of hydroxyacetaldehyde

385

Glycolaldehyde very often is the most abundant single compound in the *wood-derived* bio-oil. It is also the most highly reactive (for the browning reaction) compound of so called "liquid smokes" formulations, many of which are on the North America market.

It is pointed out that the relative browning ability of glycolaldehyde is about 2000 times bigger than that of glucose (Hayashi 1986), because of its reactivity with proteins (the Maillard reaction).

Hydroxyacetaldehyde has a great affinity for water, readily forms non-volatile hydrates, tends to polymerise in the presence of alkali and oxygen or temperatures above 50 °C can negatively affect its concentration with time (Scott 1997).

LEVOGLUCOSENONE

A straightforward procedure for the preparation of levoglucosenone calls for pyrolysis of cellulose in the presence of an acid catalyst. In such a case levoglucosenone arises, at least formally, from a double dehydration process. Phosphoric acid seems to be the acid of choice but factors like heat and mass transfer are also important. We have been able to obtain up to 23 % of levoglucosenone from cellulose using fluidized bed pyrolysis. Due to significant residual solid formation (char) it seems that some other reactor types and configurations could be more suitable for levoglucosenone production.

Levoglucosenone decomposes under various conditions during its formation and storage as shown in Figure 6. It converts quite efficiently to levulinic acid by a simple acidic transformation (Kuznetsov 1997).

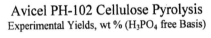

Avicel PH-102 Cellulose Pyrolysis
Experimental Yields, wt % (H₃PO₄ free Basis)

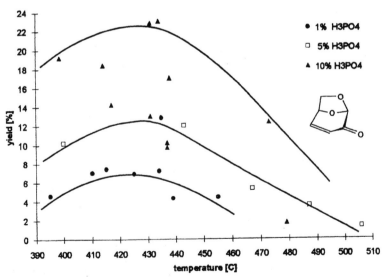

From: Volker Boelsen "Levoglucosenone - A High Value Chemical from the Waterloo Fast Pyrolysis Process" University of Waterloo 1994.

Figure 6   Yields of Levoglucosenone

386

SYNTHESIS GAS

Cellulose has a "natural" stoichiometry approaching that of synthesis gas proportions of H2/CO of ~1:1. The high reactivity of volatile products of cellulose pyrolysis (in situ) allows extremely rapid reforming to methane (hydrogen atmosphere, 500 °C) or to syngas over metal catalysts (650 °C). In both cases temperatures are relatively low (for gasification) and residence times as short as 0.5 secs.

The recent progress in development of heterogeneous catalysts for reforming of volatile products of cellulose pyrolysis (and other wood polymers) are reported in the paper by Arauzo et al (1997).

CONCLUSION

In summary, the fast pyrolysis of cellulose shows promise as a means for the production of useful chemicals. Cellulose polymer is being considered as a main feedstock for many biomass refinery schemes under development at present. A pyrolysis step can play an important role in some of them.

ACKNOWLEDGEMENT

The authors are pleased to acknowledge the assistance of many graduate students and visiting scholars (listed in references below) in performing the experimental studies.

The financial support for this work was given by Energy Mining Resources Canada, Natural Resources Canada and by the Natural Sciences and Engineering Research Council of Canada.

REFERENCES

J.E. Hodge, Agric. and Food Chem., 1, Oct. 14, 1953.

O.A. Battista et al. "Level-off D.P. Cellulose Products", U.S. Patent # 2,978,446, (1961).

A. Bash, M. Levin, "The Influence of Fine Structure on the Pyrolysis of Cellulose I Vacuum Pyrolysis", J. Polymer Sci., 11, 3071-3093, 1973.

Y. Halpern, R. Riffer, A. Broido, "Levoglucosenone - A Major Product of the Acid-Catalyzed Pyrolysis of Cellulose and Related Carbohydrates", J. Org. Chem., 38, 204-209, 1973.

S.B. Nordin, J.O. Nyren, E.L. Back, "An Indication of Molten Cellulose Produced in a Laser Beam", Textile Research J., pp. 152 Feb. 1974.

O.P. Golova, "Chemical effects of heat on cellulose", Russ. Chem. Rev., 44, 687-697, 1975.

D.J. Brown, "The questionable use of the Arrhenius equation to describe cellulose and wood pyrolysis", Thermochim. Acta, 54, 377-379, 1982.

J. Suzuki, et al., "Characterization of Mono and Oligosaccharides Produced by CO2 Laser Irradiation on Cellulose", Chemical Society of Japan, 481-484, 1983.

D.C. Elliott, "Analysis and Upgrading of Biomass Liquefaction Products", IEA Co-operative project D1, Final report, volume 4, 1983.

D.S. Scott, "Characterization of Pyrolysis Oils from Continuous Flash Pyrolysis of Wood", Final Report to Energy Mines and Resources Canada, 1984.

H. Menard et al, "Proc. 5th Bioenergy R & D Seminar, S. Hasnain (Ed), pp. 418, Elsevier, 1984.

A.G. Liden, "A Kinetic and Heat Transfer Modeling Study of Wood Pyrolysis in a Fluidized Bed", MASc Thesis presented to the University of Waterloo, 1985.

T.P. Schulz, G.D. McGinnis, M.S. Bertran, "Estimation of cellulose crystallinity using FTIR spectroscopy and dynamic thermogravimetry", J. Wood Chem., Technol., 5, 543, 1985.

G.B. Garnier, "An Investigation of the Role of the Water-Gas Shift Reaction on Fast Pyrolysis", 35th Canadian Chem. Eng. Conf., pp.507, Calgary, 1985.

T. Hayashi and M. Namiki, Agric. Biol. Chem. 50, 1965, 1986.

J.P. Danehy, "Maillard Reactions: Non-enzymatic Browning in Food Systems with Special Reference to the Development of Flavor", Advances in Food Research, Vol. 30, pp. 77-125, 1986.

M.K. Garg, "Catalytic Hydrogasification of Wood", MASc. Thesis, University of Waterloo, 1986.

J. Piskorz, D. Radlein, D.S. Scott, "On the Mechanism of the Rapid Pyrolysis of Cellulose", JAAP, 9, 121-137, 1986.

D. Radlein, A. Grinshpun, J. Piskorz and D.S. Scott, "On the Presence of Anhydro-Oligosaccharides in Sirups from the Fast Pyrolysis of Cellulose", J. Anal. Appl. Pyrol., 12, 39, 1987.

A.J. Ryan, "Certificate of Analysis", FMC Corporation, Philadelphia, 1987.

G.N. Richards, "Glycolaldehyde from pyrolysis of cellulose", JAAP, 10, 251-256, 1987.

R. Helleur, personal communication, 1988.

A. Pouwels, "Analytical Pyrolysis Mass Spectrometry of Wood Derived Fractions", PhD Thesis, the FOM Institute for Atomic and Molecular Physics, The Netherlands, 1989.

A.D. Pouwels, G.B. Eijkel, J.J. Boon, JAAP, 14, 237-280, 1989.

D.S. Scott, S. Czernik, J. Piskorz, D. Radlein, "Sugars from Biomass Cellulose by a Thermal Conversion Process", in Energy from Biomass and Wastes XIII, D.L. Klass (Ed.), 1349-1362, IGT, 1990.

P. Koll, G. Borchers, J.O. Metzger, " Preparative Isolation of Oligomers with a Therminal Anhydrosugar Unit by Thermal degradation of Chitin and Cellulose", JAAP 17, 319-327, 1990.

T. Sugawara et al., "Characteristics of Rapid Hydropyrolysis of Coals", Fuel, Vol. 69, p.1178, Sept. 1990.

L. Paterson, "Initial Experimentation with the KISS Reactor", University of Waterloo, Internal Report.

D. Radlein, J. Piskorz, D.S. Scott, "Fast Pyrolysis of Natural Polysaccharides as a Potential Industrial Process", JAAP 19, 41-63, 1991.

J.A. Lomax, J.M. Commandeur, P.W. Arisz, and J.J. Boon, "Characterisation of oligomers and sugar ring-cleavage products in the pyrolysates of cellulose", JAAP, 19, 65-79, 1991.

D. Radlein, J. Piskorz, D.S. Scott, "Control of Selectivity in the Fast Pyrolysis of Cellulose", Proc. 6th EC Conference on Biomass for Energy, Industry and Environment, Athens, 1991, G. Grassi, A. Collina and H. Zibetta (Eds), Elsevier 1992.

A. Vladars-Usas, " Thermal Decomposition of Cellulose", MASc Thesis presented to the University of Waterloo, 1993.

P. Arisz, Personal communication, Sep. 30, 1993.

D. Radlein, J. Piskorz, P. Majerski, D.S. Scott, J. Lamas, "Effects of Ionic Radius of Monovalent Cations on the Cleavage of Cellulose in Fast Pyrolysis", Symposium on Fast Pyrolysis: Processes, Technologies and Products, ACS Cellulose Division National Mtg., Denver, 1993.

D.S. Scott, J. Piskorz, D. Radlein, "Production of Levoglucosan as an Industrial Chemical", in Levoglucosenone and Levoglucosans. Chemistry and Applications, Z. Witczak (ed.), ATL Press, Inc. pp. 179-188, 1994.

C. Morin, "A Simple Bench-Top Preparation of Levoglucosenone", ibid. pp. 17-21.

V. Boelsen, "Levoglucosenone - a High Value Chemical from the Waterloo Fast Pyrolysis Process", Internal Report, University of Waterloo, 1994.

R.G. Graham, M.A. Bergougnou, B.A. Freel, "The Kinetics of Vapour-Phase Cellulose Fast Pyrolysis Reactions", Biomass and Bioenergy, Vol. 7, Nos. 1-6, pp. 33-47, 1994.

M. Serio, M. Wojtowicz, S. Charpenay, "Pyrolysis", Chapter in Encyclopedia of Energy Technology and the Environment, John Wiley & Sons, Inc., pp. 1181-2308, 1985.

G. Dobele, G. Rossinskaya, G. Domburg, "Monomeric Products of Catalytic Thermolysis of Cellulose and Lignin", in Biomass, Gasification and Pyrolysis, State of the Art and Future Prospects", M. Kaltschmitt, A.V. Bridgwater (Eds.), CPL Press, Newbury, 482-489, 1997.

B.N. Kuznetsov et al, "Fine Chemicals Production by Aspen-Wood Transformation in the Presence of Acidic and Oxidative Catalysts", in Making Business from Biomass, Proc. 3rd Biomass Conference of the Americas, R.P. Overend, E. Chornet (Eds.), pp. 893-900, Montreal 1997.

D.S. Scott et al, "Process for the thermal conversion of biomass to liquids" U.S Patent # 5,605,551, 1997.

J. Arauzo et al, "Catalytic Pyrogasification of Biomass. Evaluation of Modified Nickel Catalysts", Ind. Eng. Chem. Res., 36, 67-75, 1997.

D. Radlein, "The Production of Chemicals from Fast Pyrolysis Bio-oils", in IEA Bioenergy, Pyrolysis Activity, Final Report May, 1998.

EXPERIMENTAL (MACROSCALE) DETERMINATION OF HEAT REQUIREMENT FOR <u>AVICEL PH 102</u> CELLULOSE PYROLYSIS

A case of fluid bed flash pyrolysis with a high liquid production and levoglucosan yield circa 30 wt %.

Feed: Avicel pH 102 microcrystalline cellulose,
(depolymerized alpha cellulose derived from sulfate wood pulp from FMC Corporation, typical percent cystallinity ~ 92)

Feed rate: 1.76 kg/hr
Fluid bed temperature 774 deg K

Experimentally estimated required heat for cellulose pyrolysis - 1242 +/- 200 J/g

Assuming heat capacity of cellulose as equal 1.34 J/g K
(Perry's Chemical Engineers' Handbook, Mc Graw-Hill 1984) one gets heat of pyrolysis reaction equal to circa - (minus) -640 J/g (endothermic).

Measurements were carried out by using an electrically heated bubbling fluid bed reactor in which a steady state was established both with and without biomass feed. The power requirements could then be related to the Heat for Pyrolysis when appropriate corrections were made for the heat content of the fluidizing gas, heat losses through the reactor walls, heat loss through existing gases and char.

Ref. A.G. Liden, "A Kinetic and Heat Transfer Modelling Study of Wood Pyrolysis in a Fluidized Bed", A MASc Thesis presented to the University of Waterloo, 1985.

PYROLYTIC PRODUCTS FROM CELLULOSES

500 °C

Experimental results

	Schleicher & Schuell # 144 TLC powder	Avicel PH 102 Microcrystalline
DP	164	227
Average particle size (microns)	38	100
Ash, wt %	less than 0.06	less than 0.01
Yields, wt % mf feed		
Organic liquid	72.5	87.1
Water	10.8	3.1
Char	5.4	2.5
Gas	7.8	8.9
Cellobiosan	4.0	10.1
Levoglucosan	7.0	26.9
Hydroxyacetaldehyde	15.3	8.6
Acetol	2.2	0.1
Acetic Acid	4.9	1.4

Cellulose Pyrolysis

Avicel PH-102

Moisture 2.9 wt %
Apparent vapour residence time 0.4 second

Experimental results

Run #	17	20	64	19	8	14	6	13	10	1
Temp.	375	400	410	425	450	475	500	525	550	600
Feed rate g/hr	57.0	26.7	41.6	41.7	65.1	46.0	68.5	47.0	62.7	48.9
YIELDS wt % mf										
Water	5.5	4.0	3.1	5.2	3.0	2.1	3.1	4.0	4.7	4.3
Gas	1.3	1.9	2.3	2.9	3.0	3.5	8.9	9.7	10.8	20.0
Organics	50.3	77.3	75.10	84.1	84.6	84.8	87.1	80.0	79.6	74.4
Char-Solids	37.1	12.0	18.1	5.6	3.9	4.0	2.5	2.5	2.2	1.0
CO			0.52	0.61	1.38	1.94	6.06	6.07	7.69	14.52
CO2	1.30	1.83	1.80	2.01	1.60	1.46	1.80	3.03	2.16	2.82
Cellobiosan	3.87	8.66	3.42	14.35	11.81	15.31	10.12	9.88	7.14	2.40
Levoglucosan	23.43	36.18	38.28	35.50	28.70	27.49	26.94	20.03	18.16	14.04
Hydroxyacetaldehyde	0.22	0.26	1.62	1.64	4.32	4.92	8.57	10.56	11.03	12.13
Acetol	0.02	0.01	tr	0.04	0.06	0.09	0.04	0.19	0.14	0.14
Glyoxal	0.05	0.06		0.25	0.26	0.27	6.49	3.73	4.07	3.45
Formaldehyde/formic acid	0.36	0.13	tr	0.36	0.48	1.85	4.50	5.16	4.61	6.74

Avicel microcrystalline cellulose

Different average particle size

Experimental results

Run #	Berty# 9	PP 109	19
Feed	Avicel PH-105	Avicel PH-101	Avicel PH-102
Average particle size (microns)	20	50	100
Temp. deg C	430	490	425
Feed rate, g/hr	69.0	1760	41.7
YIELDS wt % mf			
Water	4.9	7.5	5.2
Gas	nd	4.6	2.9
Organics	85.2	85.2	84.1
Char	5.8	3.0	5.6
Cellobiosan	6.7	6.4	14.5
Levoglucosan	32.2	20.0	35.5
Glycolaldehyde	5.4	11.3	1.6

Avicel PH-102

Transport Flow Reactor ¼" dia
Experimental results
Waterloo KISS Run # 6A

Conditions

Temperature, Deg. C	1100
Residence Time, ms	35
Particle size, microns	150-212
Feed-rate, g/hr	28
YIELDS, wt %	
Gas	31.2
Liquid	18.9
Water Soluble Solids	37.7
Water Insoluble Solids	7.3
CO	26.2
CO2	3.9
C2H4	0.7
Oligosaccharides	34.1
Cellobiosan	2.5
Levoglucosan	3.4
Hydroxyacetaldehyde	1.2

From: Anita Vladas-Usas, "Thermal Decomposition of Cellulose", MASc. Thesis, University of Waterloo, 1993.

GAS PRODUCTION FROM CATALYTIC PYROLYSIS OF BIOMASS

L García, M L Salvador, R Bilbao, J Arauzo
Department of Chemical and Environmental Engineering, University of Zaragoza.
E-50071 Zaragoza - SPAIN

ABSTRACT

Our research group at the University of Zaragoza is focused on the study of catalytic pyrolysis and gasification of biomass. This study is important because these processes allows to obtain a product gas that could be useful in very efficient combustion systems like fuel cells, and in synthesis gas.

This research work has been carried out at bench scale and using an installation based on the Waterloo Fast Pyrolysis Process (WFPP). The experimental work considered temperatures around 650 and 700 ºC in order to diminish the requirement of input energy. This low level of temperatures produces an increase in the tar production, which can be avoided by placing a catalyst in the reaction bed to convert tars in gas. The biomass, pine sawdust with a particle size between +150-350 μm, is fed continuously (up to 100 g/h).

The Ni/Al co-precipitated catalyst has been chosen in this work because of its good efficiency and mechanical and thermal stability. Different catalysts have been prepared in our laboratory modifying the calcination temperature and reduction conditions. The catalyst has been characterised by different techniques such as nitrogen adsorption, mercury porosimetry, X-ray diffraction and optical emission spectrometry in order to connect catalyst properties with experimental results obtained.

The experimental work has considered the study of the influence of the following operating variables on the performance of pyrolysis:
- Catalyst process preparation: Calcination and reduction conditions, hydrogen flow rate, time reduction.
- Catalyst weight/biomass flow rate (W/mb) ratio.
- Reaction time.

The main problem of this process is the loss of catalyst activity. The deactivation of the catalyst is mainly caused by carbon deposits. The deactivation of the catalyst can be lower using different reaction atmospheres. This has been observed in steam gasification of biomass, where the influence of Steam/Biomass ratio has been studied. Another possibility to diminish catalyst deactivation is introducing some promoters in the catalyst formulation.

1   INTRODUCTION

Pyrolysis of biomass is a suitable process for obtaining useful gas (H_2, CO), which can be used for power production or as a synthesis gas. When biomass pyrolysis is carried out at relatively low temperatures, significant amount of liquids are obtained.

In this context, different strategies can be followed to obtain high yields of gas. For example steam reforming of pyrolysis oils is being recently studied (Wang et al. 1997, 1998). After a first step of pyrolysis flash, the liquids obtained are steam reformed using catalysts in order to maximise hydrogen production. Various catalysts have been tested and Ni is presented in most of them.

Some others researches (Donnot et al. 1985; Aldén et al. 1996, 1997; Ekström et al. 1985; Myrén et al. 1996, 1997; Vassilatos et al. 1992; Brandt and Henriksen, 1996) are focused in the study of the upgrading the product gas of pyrolysis. The pyrolysis is carried out at temperatures about 700 ºC and they use a second reactor with catalysts, mainly Ni catalysts and dolomites, temperature, steam and oxygen to transform tars in gas.

Another option, selected in this work, consists in the use of the catalyst in the same reactor where the biomass pyrolysis is carried out. Because of pyrolysis is an endothermic process, working at low temperatures the input energy decreases, but tar production increases when reaction temperature decreases. The presence of the catalyst allows us to convert tars in gases.

Ni catalysts are frequently used in pyrolysis and gasification of biomass, and have been selected like the most appropriate to obtain an adequate methanol synthesis gas (Baker et al. 1983; Rei et al. 1986; Tanaka et al. 1984). In this work, it has been chosen a co-precipitated Ni/Al catalyst, because of two properties: its mechanical strength and its high thermal stability. The catalyst are going to be used in a fluidised bed, then is necessary a catalyst without attrition problem. The high thermal stability of the catalyst could achieve a catalyst with longer activity.

The Ni/Al co-precipitated catalysts presented previous good results in pyrolysis and gasification of biomass (Arauzo et al. 1994, 1997). The main problem of the catalytic pyrolysis is the loss of catalyst activity caused mainly by carbon deposits.

The influence of catalyst preparation conditions, calcination and reduction, on gas yields has been analysed. With this objective different catalysts have been prepared at calcination temperatures of 650, 750 and 850 ºC during 3 h. With respect to the reduction of the catalyst, it has been studied the influence of hydrogen flow rate and time.

Other significant variable, catalyst weight/biomass flow rate (W/mb) ratio, has been studied in pyrolysis and steam gasification. This ratio influences the product distribution (gas, liquid and char), gas composition and gas yield evolution with reaction time.

The study of the catalytic pyrolysis of biomass is the first step in the catalytic gasification of biomass. Our group has also analysed the influence of calcination temperature and Steam/Biomass (S/B) ratio in the experiments of catalytic steam gasification. The present paper is centred in pyrolysis, then, the most relevant results obtained in catalytic pyrolysis are going to be presented.

2   EXPERIMENTAL

2.1   Experimental System

The experimental system used in this research is shown in Figure 1. It is a bench scale installation based on WFPP technology. This technology was developed by Professor D.S. Scott and his co-workers at the Chemical Engineering Department of University of Waterloo in Canada. The biomass is fed continuously with flow rates 5-100 g/h. The fluidised bed reactor reaches high heating rates and low residence times of gases in reaction bed (Scott and Piskorz, 1982; Scott et al. 1985).

The product gas is cleaned of char particles using a cyclone. A system of two condensers and a cotton filter allows us to retain liquids products. Afterwards, the gas flow rate is measured using a dry testmeter, and CO and CO_2 concentrations are continuously determined. In addition, gas samples are taken at regular intervals and analysed by chromatography. Reaction temperature, total gas flow, and CO and CO_2 concentrations are registered by a data acquisition system.

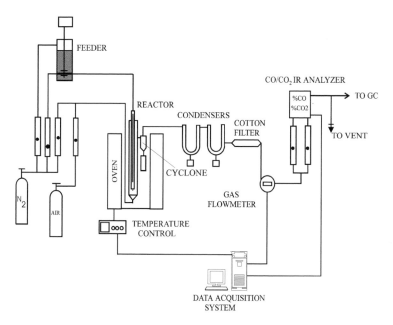

FEEDER

CO/CO$_2$ IR ANALYZER

REACTOR

CONDENSERS

COTTON FILTER

%CO
%CO2

TO GC

TO VENT

OVEN

CYCLONE

N$_2$

AIR

GAS FLOWMETER

TEMPERATURE CONTROL

DATA ACQUISITION SYSTEM

Figure 1   Schematic of the experimental system

The pyrolysis experiments are carried out at 650 and 700 °C and at atmospheric pressure. In these experiments two nitrogen streams are introduced into the reactor. One transports the biomass into the fluidized bed, while the other enters at the bottom and reaches the bed through the distributor. The total nitrogen flow rate into the reactor is 1720 cm^3 (STP)/min. The particle sizes of solid in reaction bed, sand and catalyst, are between 150 and 350 μm. The theoretical minimum fluidisation velocity value of the mixture of solids of the reaction bed at the temperatures of 650 and 700 °C is about 2 cm/s. We are working at more than 10 times theoretical minimum fluidization velocity. Elutriation of catalyst and sand is not observed.

2.2   Biomass

The biomass used is pine sawdust with a particle size between 150 and 350 μm and an average moisture content about 10 %. The elemental analysis is shown in Table 1.

Table 1   Elemental analysis of pine sawdust.

	% wt
Carbon	48.27
Hydrogen	6.45
Nitrogen	0.09
oxygen*	45.19

* by difference

2.3   Catalyst

The Ni/Al catalyst employed was prepared in the laboratory by co-precipitation by a method similar to that described by Al-Ubaid and Wolf (1988). Ammonium hydroxide is added to a solution of Ni(NO$_3$)$_2$ 6H$_2$O and Al(NO$_3$)$_3$ 9H$_2$O in water until the pH reaches 7.9. The precipitation medium is maintained at 40 °C and moderately stirred. The salts are mixed in appropriate proportions to obtain a molar ratio 1:2 (Ni:Al). The precipitate obtained is filtered and washed at 40 °C and dried for about 15 h at 105 °C. Following these steps, the hydrated precursor is obtained. This precursor is calcined in air atmosphere at a low heating rate, as suggested by Alzamora et al. (1981), until a final calcination temperature is achieved, and maintained for 3 h. In some cases, just before the pyrolysis is performed, the calcined catalyst is reduced in the reactor at a temperature of 650 °C using hydrogen. Experiments without previous reduction of the catalyst were also performed.

The catalyst is characterised by different techniques: optical emission spectrometry, mercury porosimetry, nitrogen adsorption, thermogravimetric analysis, X-ray diffraction (XRD), and temperature-programmed reduction (TPR).

The elemental analysis of metals (Ni and Al) in the catalyst, carried out using optical emission spectrometry by inductively coupled plasma (ICP), has shown good agreement with the theoretical formulation, being Ni/Al 0.49 for the precursor and 0.48 for the calcined catalyst.

The analysis of the calcined catalyst by mercury porosimetry has determined the total intrusion volume and a median pore diameter. Surface areas for the catalyst calcined have been obtained by adjusting the nitrogen adsorption data to the BET equation. Numeric data of these properties are presented in a previous paper (Garcia et al. 1998). It is observed the decrease of the total intrusion volume and of the surface area when calcination temperature increases.

The results obtained from the thermogravimetric analysis indicate that the weight loss observed during the calcination occurs in two steps. In the first step, at about 114 °C, molecular water is lost from the interlayer of the layer structure. In the second step, at about 298 °C, the layer structure decomposes with the evolution of nitrogen oxides. It is in this latter step that the maximum decomposition takes place. These results are similar to those obtained by Alzamora et al. (1981). Figure 2 shows the results of the thermogravimetric analysis.

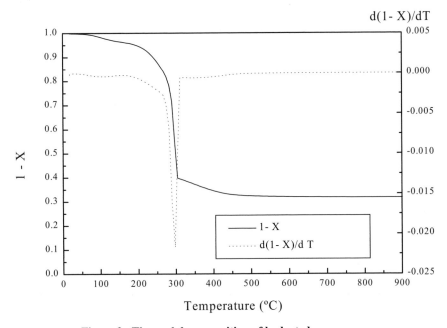

Figure 2   Thermal decomposition of hydrated precursor

Results obtained using XRD analysis, Figures 3 and 4, show differences in the catalyst structure depending on the calcination temperature. The increase of calcination temperature from 650 to 850 °C causes an increase in the crystallinity of the catalyst and in the proportion of spinel phase in it. These trends were observed by other authors (Alzamora et al., 1981; Clause et al., 1992).

Temperature-programmed reduction (TPR) analysis of the catalyst calcined at 750 and 850 °C were also carried out. The results are presented in Figure 5. It is observed that reduction is more difficult as the calcination temperature is increased, which could be due to the higher proportion of spinel phase existing in the catalyst calcined at 850 °C.

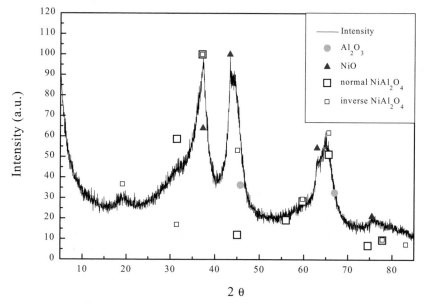

Figure 3   XRD spectrum of a catalyst calcined at 650 °C

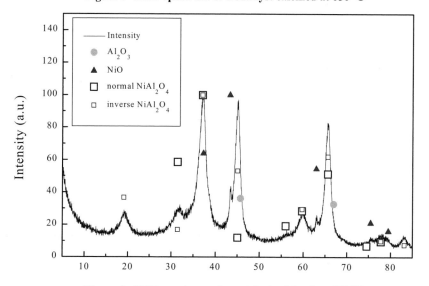

Figure 4   XRD spectrum of a catalyst calcined at 650 °C

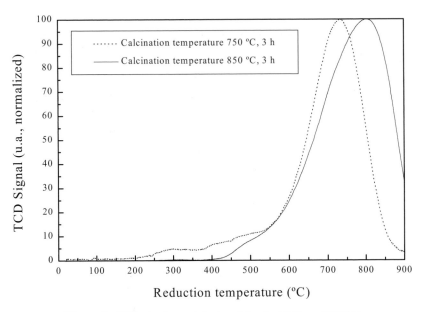

Figure 5   TPR plot for catalysts calcined at 750 and 850 °C

3   EXPERIMENTAL RESULTS AND CONCLUSIONS

3.1   Influence Of Preparation Conditions Of The Catalyst On Gas Yields

At the reaction temperature of 650 °C an extensive study of the influence of calcination and reduction conditions of the catalyst has been carried out. Catalysts calcined at 750 and 850 °C were prepared. The catalyst calcined at 750 °C has been reduced with a hydrogen flow rate of 1740 (STP) cm^3/min during 1, 2 and 3 h, and with a higher flow rate of hydrogen of 3080 (STP) cm^3/min during 1 h. The catalyst calcined at 850 °C has been reduced with a hydrogen flow rate of hydrogen of 1740 (STP) cm^3/min during 1 and 2 h and with a hydrogen flow rate of 3080 (STP) cm^3/min during 1 and 2 h. The reduction temperature of these calcined precursors has been 650 °C. Also it has been used the calcined precursors without previous reduction.

The experiments carried out using a catalyst calcined at 750 °C and reduced with a hydrogen flow rate of 1740 (STP) cm^3/min during 1, 2 and 3 h have presented similar results. In these conditions the increase of reduction time is not significant in the performance of the catalyst.

In Figures 6 and 7, H_2 and CO yields are represented versus Sawdust/Catalyst ratio for different hydrogen flow rates. The results obtained with the catalyst without reduction are also presented. The Sawdust/Catalyst ratio allows us to compare experiments with different sawdust flow rate where time reaction is not adequate. It is observed the decrease in H_2 and CO yields caused by the catalyst deactivation.

When the catalyst is reduced with a hydrogen flow rate of 1740 (STP) cm^3/min, H_2 and CO yields are slightly higher at low Sawdust/Catalyst ratio than the other cases presented. H_2 and CO yields of the catalyst without previous reduction tend to the values of the catalyst reduced with a hydrogen flow rate of 1740 (STP) cm^3/min at about 0.5 Sawdust/Catalyst (g/g). At high values of Sawdust/Catalyst ratio, 1.4, H_2 and CO yields are similar for different hydrogen flow rates.

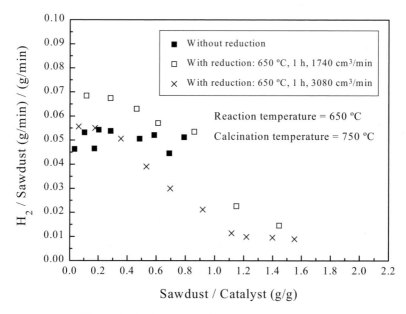

Figure 6 H₂ yield versus Sawdust/Catalyst at 650 °C.
Influence of hydrogen flow rate for a catalyst calcined at 750 °C.

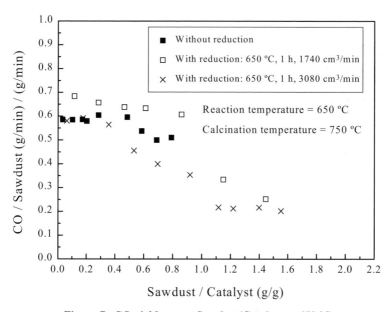

Figure 7 CO yield versus Sawdust/Catalyst at 650 °C.
Influence of hydrogen flow rate for a catalyst calcined at 750 °C.

It has been proposed that the catalyst without previous reduction can be reduced by H_2 and CO formed in the pyrolysis. The main reactions involved are:

$$NiO + H_2 \rightarrow Ni + H_2O \qquad (1)$$
$$NiO + CO \rightarrow Ni + CO_2 \qquad (2)$$

The decrease in H_2 and CO yields obtained with the catalyst without previous reduction compared with the catalyst reduced with a hydrogen flow rate of 1740 (STP) cm³/min could be explained by the consumption of these gases in the reduction of the catalyst. Also, it has been observed an increase in CO_2 yield when the catalyst is used without previous reduction.

In Figures 8 and 9, H_2 and CO yields are represented versus Sawdust/Catalyst values of the experiment, when the catalyst calcined at 850 ºC is reduced with different hydrogen flow rates, 1740 and 3080 (STP) cm³/min and times, 1 and 2 h. It has been also carried out some experiments using the catalyst calcined at 850 ºC without previous reduction, but there were operational problems that can be due to the difficulty of the reduction of the catalyst. The higher H_2 and CO yields along the experiment are obtained using the catalyst reduced with 3080 (STP) cm³/min during 1 h. The results obtained at the reaction temperature of 650 ºC allow us to propose that the best conditions of catalyst reduction depend on calcination temperature. The increase of calcination temperature requires more severe reduction conditions.

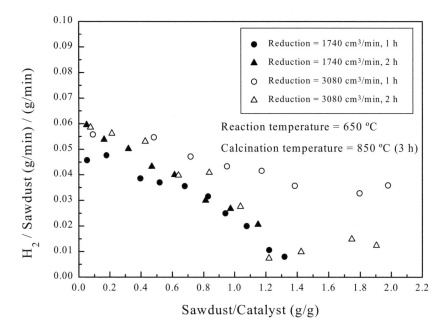

**Figure 8 H_2 yield versus Sawdust/Catalyst at 650 ºC.
Influence of hydrogen flow rate and time for a catalyst calcined at 850 ºC.**

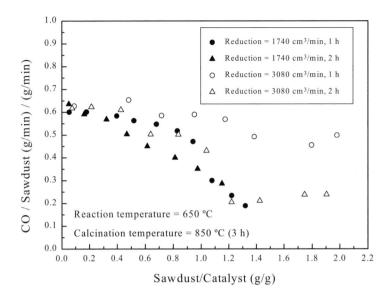

**Figure 9   CO yield versus Sawdust/Catalyst at 650 ºC.
Influence of hydrogen flow rate and time for a catalyst calcined at 850 ºC.**

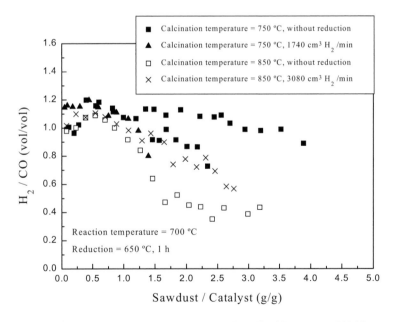

**Figure 10   H₂/CO (vol/vol) ratio versus Sawdust/Catalyst at 700 ºC.
Influence of calcination and reduction conditions of the catalyst.**

Experiments have been also carried out at the reaction temperature of 700 ºC. In Figure 10, H_2/CO ratio is represented versus Sawdust/Catalyst. The catalysts used have been calcined at 750 and 850 ºC without previous reduction. The catalyst calcined at 750 ºC was reduced with a hydrogen flow rate of 1740 (STP) cm^3/min during 1 h and the catalyst calcined at 850 °C was reduced with a hydrogen flow rate of 3080 (STP) cm^3/min during 1 h. H$_2$/CO ratio decreases with the increase of Sawdust/Catalyst ratio caused by catalyst deactivation. For the catalyst calcined at 750 °C, H$_2$/CO ratio decreases faster when it is reduced. The catalyst calcined at 850 °C and reduced with a hydrogen flow rate of 3080 (STP) cm^3/min presents higher H$_2$/CO ratio along the experiment.

3.2 Influence of Catalyst Weight/Biomass Flow Rate (W/Mb)

The study of the influence of the preparation conditions of the catalyst has shown that the catalyst calcined at 750 °C is capable of being reduced by the reaction atmosphere. The use of a catalyst without previous reduction decreases costs and makes the operation safer. Using a catalyst calcined at 750 °C without previous reduction, it has been analysed the influence of W/mb ratio at the reaction temperatures of 650 and 700 °C.

In Figures 11 to 14 are represented H$_2$, CO, CH$_4$ and C$_2$ yields versus time for different W/mb ratios at the temperature of 650 °C. H$_2$ and CO yields, for a given W/mb ratio, decreases with the increase of reaction time, this tendency is caused by catalyst deactivation. When W/mb ratio increases, H$_2$ and CO yields increase. CH$_4$ and C$_2$ yields, for a given W/mb ratio, increase when reaction time increases and these yields increase when W/mb ratio decreases.

In Figure 15 H$_2$/CO (vol/vol) ratio versus time is represented. For a given W/mb ratio, the decrease of H$_2$/CO ratio when reaction time increases, is caused by catalyst deactivation. H$_2$/CO ratio decreases when W/mb ratio decreases.

The tendencies observed at 700 °C are similar to that obtained at 650 °C.

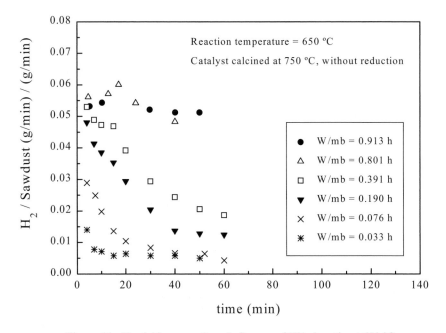

Figure 11 H$_2$ yield versus time. Influence of W/mb ratio at 650 °C.

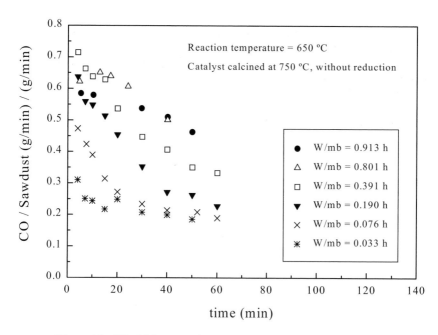

Figure 12   CO yield versus time. Influence of W/mb ratio at 650 °C.

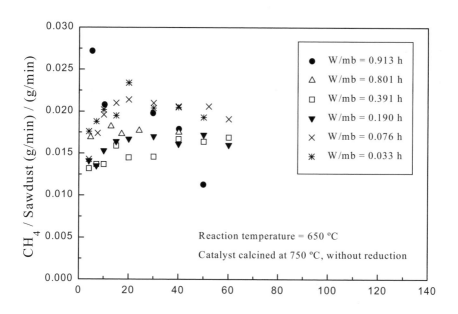

Figure 13   CH₄ yield versus time. Influence of W/mb ratio at 650 °C.

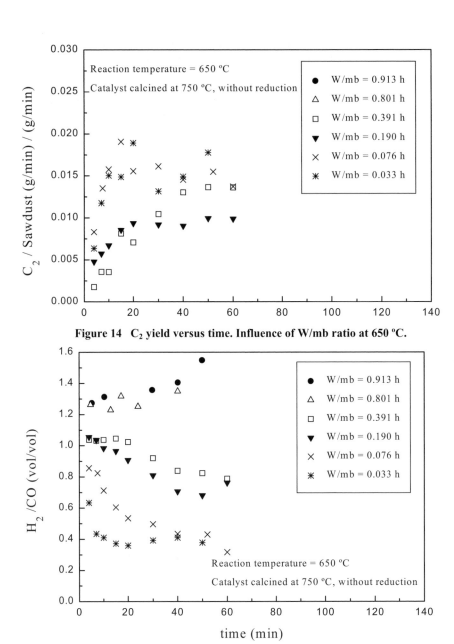

Figure 14   C₂ yield versus time. Influence of W/mb ratio at 650 °C.

**Figure 15   H₂/CO (vol/vol) ratio versus time.
Influence of W/mb ratio at 650 °C.**

3.3   Influence of Reaction Time

Total gas, H_2, CO, CO_2 and CH_4 yields evolution versus reaction time, for a experiment with W/mb ratio of 0.52 h at the reaction temperature of 700 °C, are shown in Figure 16. The catalyst has been calcined at 750 °C and was not previous reduced. Total gas, H_2 and CO yields decrease with the increase of reaction time, while CO_2 and CH_4 yields increase. It is observed a high CO_2 yield at low reaction time, this yield is related to the reduction of the catalyst by the reaction atmosphere.

The deactivation of the catalyst is influenced very significantly by the W/mb ratio. When W/mb ratio increases, the deactivation of the catalyst diminishes. It has been observed no deactivation of the catalyst during 3 h using high W/mb ratios, higher than 1h.

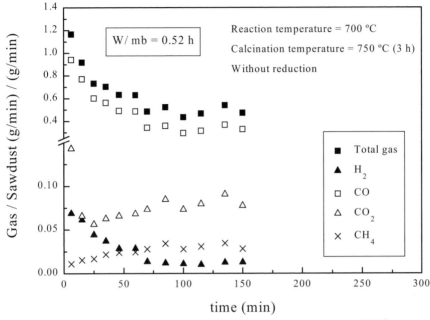

Figure 16. Gas evolution versus time for a W/mb = 0.52 h at 700 °C.

4   ACKNOWLEDGEMENTS

The authors express their gratitude to D.G.I.C.Y.T. for providing financial support for the study (Project PB93-0593) and also to the Ministerio de Educación y Ciencia (Spain) for the research grant awarded to L.G. The authors thank D.S. Scott of the University of Waterloo, and D. Radlein, J. Piskorz, and P. Majerski of RTI, Ltd. (Waterloo, Ontario, Canada) for their assistance and interest in the research program in which this study has been developed.

5   REFERENCES

Aldén, H.; Hagström, P.; Hallgren, A.; Waldheim, L. "High temperature catalyst gas cleaning for pressurized gasification processes" *Biomass for Energy and the Environment*. Eds.: P. Chartier, G.L. Ferrero, U.M. Henius, S. Hultberg, J. Sachau and M. Wiinbland, Pergamon, Oxford, 1996, Vol. 2, pp. 1410-1415.

Aldén, H.; Hagström, P.; Hallgren, A.; Waldheim, L. "Investigation in high temperature catalytic gas cleaning for pressurized gasification processes". *Developments in Thermochemical Biomass Conversion*. Eds.; A.V. Bridgwater and D.G.B. Boocock, Blackie Academic & Professional, London, 1997, Vol. 2, pp. 1131-1143.

Al-Ubaid, A.; Wolf, E.E. "Steam reforming of methane on reduced non-stoichiometric nickel aluminate catalysts". *Appl. Catal.*, 1988, 40, pp. 73-85.

Alzamora, L.E.; Ross, J.R.H.; Kruissink, E.C.; van Reijen, L.L. "Coprecipitated Nickel-Alumina Catalysts for Methanation at High Temperature". *J. Chem. Soc. Faraday T. I*, 1981, 77, pp. 665-681.

Arauzo, J.; Radlein, D.; Piskorz, J.; Scott, D.S. "A new catalyst for the catalytic gasification of biomass". *Energ. Fuel*, 1994, 8 (6), pp. 1192-1196.

Arauzo, J.; Radlein, D.; Piskorz, J.; Scott, D.S. "Catalytic pyrogasification of biomass. Evaluation of modified nickel catalyst". *Ind. Eng. Chem. Res.*, 1997, 36, pp. 67-75.

Baker, E.G.; Mitchell, D.H.; Mudge, L.K.; Brown, M.D. "Methanol Synthesis Gas from Wood Gasification". *Energ. Progress*, 1983, 3(4), pp. 226-228.

Brandt, P.; Henriksen, U. "Decomposition of tar in pyrolysis gas by partial oxidation and thermal cracking". *Biomass for Energy and the Environmental*. Eds.; P. Chartier, G.L. Ferrero, U.M. Henius, S. Hultberg, J. Sachau y M. Wiinbland, Pergamon, Oxford, 1996, Vol. 2, pp. 1336-1340.

Clause, O.; Rebours, B.; Merlen, E.; Trifirò, F.; Vaccari, A. "Preparation and Characterization of Nickel-Aluminum Mixed Oxides Obtained by Thermal Decomposition of Hydrotalcite-Type Precursors. *J. Catal.*, 1992, 133, pp. 231-246.

Donnot, A.; Reningovolo, J.; Magne, P.; Deglise X. "Flash pyrolysis of tar from the pyrolysis of pine bark". *J. Anal. Appl. Pyrol.*, 1985, 8, pp. 401-414

Ekström, C.; Lindman, N.; Pettersson, R. "Catalytic conversion of tars, carbon black and methane from pyrolysis/gasification of biomass" *Fundamentals of Biomass Conversion*. Eds.; R.P. Overend, T.A. Milne and L.K. Mudge, Elsevier Applied Science, London, 1985, pp. 601-618.

Garcia, L.; Salvador, M.L.; Bilbao, R; Arauzo, J. "Influence of calcination and reduction conditions on the catalyst performance in the pyrolysis process of biomass". *Energ. Fuel*, 1998, 12(1), pp. 139-143.

Myrén, C.; Hörnell, C.; Sjöström, K.; Yu, Q.; Brage, C.; Björnbom, E. "Catalytic tar cracking of gas from agricultural residues and biomass" *Biomass for Energy and the Environmental*. Eds.; P. Chartier, G.L. Ferrero, U.M. Henius, S. Hultberg, J. Sachau y M. Wiinbland, Pergamon, Oxford, 1996, Vol. 2, pp. 1283-1288.

Myrén, C.; Hörnell, C.; Sjöström, K.; Yu, Q.; Brage, C.; Björnbom, E. "Catalytic upgrading of the crude gasification product gas". *Developments in Thermochemical Biomass Conversion*. Eds.; A.V. Bridgwater and D.G.B. Boocock, Blackie Academic & Professional, London, 1997, Vol. 2, pp. 1170-1178.

Rei, M.; Yang, S.J.; Hong, C.H. "Catalytic Gasification of Rice Hull and other Biomass. The General Effect of Catalyst". *Agr. Wastes*, 1986, 18, pp. 269-281.

Scott, D.S.; Piskorz, J. "The Flash Pyrolysis of Aspen-Poplar Wood". *Can. J. Chem. Eng.*, 1982, 60, pp. 666-674.

Scott, D.S.; Piskorz, J.; Radlein, D. "Liquid product from the continuous flash pyrolysis of biomass". *Ind. Eng. Chem. Process Des. Dev.*, 1985, 24 (3), pp. 581-588.

Tanaka, Y.; Yamaguchi, T.; Yamasaki, K.; Ueno, A.; Kotera, Y. "Catalyst for steam gasification of wood to methanol synthesis gas". *Ind. Eng. Chem. Prod. Res. Dev.*, 1984, 23(2), pp. 225-229.

Vassilatos, V.; Taralas, G.; Sjöström, K.; Björnbom, E. "Catalytic cracking of tar in biomass pyrolysis gas in the presence of calcined dolomite" Can. J. Chem. Eng., 1992, 70, pp. 1008-1013.

Wang, D.; Czernik, S.; Montané, D.; Mann, M.; Chornet, E. "Biomass to hydrogen via fast pyrolysis and catalytic steam reforming of the pyrolysis oil and catalytic steam reforming of the pyrolysis oil or its fractions". *Ind. Eng. Chem. Res.*, 1997, 36, pp. 1507-1518.

Wang, D.; Czernik, S.; Chornet, E. "Production of Hydrogen from Biomass by Catalytic Steam Reforming of Fast Pyrolysis Oils". *Energ. Fuel*, 1998, 12 (1), pp. 19-24.

FRACTIONAL VACUUM PYROLYSIS OF BIOMASS AND SEPARATION OF PHENOLIC COMPOUNDS BY STEAM DISTILLATION

Jean Népo Murwanashyaka[1,2], Hooshang Pakdel[1] and Christian Roy[1,2]

[1]Institut Pyrovac Inc., Parc technologique du Québec métropolitain, 333, rue Franquet, Sainte-Foy (Québec), Canada, G1P 4C7

[2] Université Laval, Department of Chemical Engineering, Sainte - Foy, Quebec, Canada, G1K 7P4

ABSTRACT

Laboratory batch fractional vacuum pyrolysis of a mixture of birch bark (46 %) and birch wood (54 %) was carried out in the temperature range of 25-550°C. The oil phase was recovered and was analyzed by GC/MS to determine the quantity of phenols in the pyrolysis oil. The active zone of wood decomposition yielding a liquid oil was found to be in the temperature range of 275 and 350°C. A maximum recovery of phenols (monolignols throughout this paper) with 4.9 wt. % on an anhydrous wood basis was found in the temperature range of 275 - 350°C. The total phenols were almost completely recovered at a temperature below 400°C. Furthermore, a separate vacuum pyrolysis experiment was performed in a pilot plant. The pyrolysis oil was subjected to steam distillation which provided a concentrated fraction of phenols with a yield of 21.3 % on the oil basis. The steam-extracted fraction was further distilled under vacuum into five sub-fractions. The phenol concentration was in the range of 50-70 % by weight for each fraction.

INTRODUCTION

In wood pyrolysis, it is known that several parameters influence the yield of pyrolytic oil and its composition. Among these factors, wood composition, heating rate, pressure, moisture content, presence of a catalyst, particle size and combined effects of these variables are known to be important (Pakdel *et al.*, 1996; Samolada *et al.*, 1990). The thermal degradation of wood at normal pressure starts with free water evaporation. This endothermic process takes place at 120 to 150°C, followed by several exothermic reactions at 200 to 250°C, 280 to 320°C, and around 400°C corresponding to the thermal degradation of hemicelluloses, cellulose, and lignin, respectively (Fengel, *et al.*1984). In addition to extractives, the pyrolytic liquid product from wood represents a proportional combination of pyrolysates from cellulose, hemicelluloses, and lignin when these compounds are separately pyrolysed (Elder, 1991). Pyrolysis of different wood components produces different classes of compounds. Cellulose for example produces acetic acid, gases, water, sugars and occasionally a small amount of furans and phenols (Russell *et al.*, 1983). Hemicelluloses on the other hand principally gives acids, furans and sugars. Lignin is biosynthetically constructed by free-radical copolymerization of tree phenylpropanoid monomers namely coumaryl-, coniferyl-, and sinapyl alcohol (Monties, 1989) which are principally decomposed to phenols and aromatic hydrocarbons during pyrolysis (Alen *et al.*, 1996). In addition to pyrolysates, free organic compounds or extractives present in wood are also released during pyrolysis, making pyrolytic oils a very complex product.

Fractional pyrolysis under vacuum of aspen poplar wood chips showed an active zone of decomposition occurring between 240 -300°C (Roy *et al.*, 1985). A 58 % w/w of pyrolytic oil was obtained at 350°C which contained miscellaneous compounds such as phenols, carboxylic acids, alcohols and esters which have been characterized in different fractions earlier. Vacuum process removes the reaction products before they are decomposed (Shafizadeh, 1984). It produces a relatively high yield of liquids (Bridgwater *et al.*, 1991; Avni *et al.*, 1985). The yield of phenol depends on the materials to be pyrolysed. Lignin yields approximately 12 % phenols (Samolada, 1990). The phenol yields varied from 1 to 3.7 % when different woods were subjected to vacuum pyrolysis (Pakdel *et al.*, 1996).

Phenolic compounds typically posses anti-diarrheal and anti-motility properties (Ogata *et al.*, 1993), germicidal activity (Guha *et al.*, 1987), herbicidal effect and antiseptic properties (Azhar; 1972; 1974). They have also been used in the tanning of leather (Ratner *et al.*, 1979), as dyes (Rijkuris *et al.*, 1978), as a thermal insulating material (Galalevicius *et al.*, 1978), as food aroma and meat smoke (Guillen *et al.*, 1998). In some cases the smoke flavour has been attributed to the presence of phenols and sometimes to a single component. The phenols forming the "liquid smoke" which is used to smoke foods are well studied but their role and extent during flavour formation was unclear for a long time (Wittkowski et al. 1992).

A number of chemical and physical methods have been applied to recover phenols from bio-oil matrices: alkaline extraction, partition into different solvents or adsorption on different packed materials (Fagernäs, 1995). The physico-chemical properties of biomass pyrolytic oils make conventional separation techniques like distillation less suitable and even impractical (Radlein, 1997). Solvent extraction and chemical adsorption are often used. Alkaline extraction of pyrolytic oils is often processed with NaOH (8-15% concentration) in an organic solvent medium. By using this method, different authors have reported many practical problems associated with the extraction of phenols from pyrolysis oil, such as the redistribution of phenols in both phases and the precipitation between layers (Maggi *et al.*, 1994). However, an efficient method to recover phenols from pyrolytic oils has been reported recently (Amen-Chen *et al.*, 1997; Chum *et al.*.1990). Liquid chromatography was reported to be the most efficient technique for the separation of total phenols as well as 2(*3H*)-furanone and vanillin in a relatively pure state from the pyrolytic oil. However, a high consumption of solvents and regeneration of adsorbents make this process uneconomical (Zhang, 1990, Vassalos *et al.*, 1992).

The present study was undertaken to determine the concentration and the evolution of the total and individual phenols at different temperatures by vacuum pyrolysis of birch wood. Furthermore, the potential application of steam distillation to separate and concentrate the phenolic fractions derived from biomass vacuum pyrolysis is also presented.

MATERIAL AND METHODS

Birch (*Betula* papyrifera) wood chips used in this work was obtained from DAISHOWA pulp and paper plant, Quebec, Canada. The birch sample was hand-sorted into 54 % of sapwood and 46 % of inner and outer bark. The moisture was 8.5 % as determined by placing a known quantity of sample in an oven at $102 \pm 3°C$ until a constant weight was reached.

FRACTIONAL VACUUM PYROLYSIS

A 800 g quantity of the milled sample was pyrolysed in a batch reactor under vacuum (run # G72). A detailed description of the batch pyrolysis reactor system used in this work has been described elsewhere (Pakdel and Roy, 1994). Two vacuum pumps in series were used to achieve a total pressure of 0.7 kPa and three dry-ice-in-limonene condensers (-72°C) were used to trap the pyrolysis vapours. Pyrolysis was performed at different heating rates. The maximum pyrolysis temperature reached was held for 1 hour prior to cooling to room temperature. After the pyrolysis, the system was kept under nitrogen till the next step. Table 1 shows different pyrolysis steps. The pyrolysis oils were recovered in different traps, collected and stored in sealed vials in the refrigerator (4°C) for further chemical analysis.

VACUUM PYROLYSIS

A birch wood sample composed of 54 % sawood and 46 % bark was pyrolysed under vacuum in a pilot plant reactor with a throughput capacity of 22 kg/h (run # H034). The wood sample was initially dried, then pyrolysed to 500°C. Pyrolytic oil obtained was stored at 4°C.

Table 1: Fractional pyrolysis conditions and yields (run # G72)

Pyrolysis step #	Temperature range (°C)	Liquid yield (wt %, anhydrous feed basis)
1	25-200	6.02
2	200-275	18.14
3	275-350	27.28
4	350-450	9.47
5	450-550	1.48
Cumulative (1-5)	25-550	62.39
One-step	25-550	63.43

STEAM DISTILLATION

A 100 to 200 g sample of the total pyrolytic oil (run # H034) was steam distilled under the following conditions : steam was injected at 100°C at a rate equivalent to 5 ml/min of cold water. The oil was kept at 130°C in a mineral oil bath during the distillation to avoid any accumulation of water in the oil flask. The distillate was collected in a glass flask and was then decanted. The aqueous phase was extracted once with an equivalent volume of ethyl acetate, then twice with a half volume of ethyl acetate. The solvent was evaporated and the oil was combined with the oil which was obtained after decantation. The steam-to-oil ratio (Q_s / Q_o) and distillate-to-oil ratio (Q_D / Q_o) were calculated.

VACUUM DISTILLATION

The oil fraction obtained from the steam distillation was further distilled in a Vigreux column under 5 mm Hg of pressure. Five sub-fractions were collected at different temperatures then analyzed by GC/MS.

ANALYSIS

The identification and quantification of the phenols were carried out by gas chromatography coupled to mass spectroscopy (GC/MS) using standard phenols. The pyrolysis oil was acetylated before analysis. A 200 to 300 mg sample of well mixed pyrolytic oil, together with 5mL of acetic anhydride and two drops of pyridine, was heated in a sealed vial in a water bath at 60°C for 90 minutes. The derivatized solution was then eluted over approximately 2 g of silica gel with 100 ml of 80 % CH_2Cl_2 in hexane solution. The eluate was concentrated under vacuum. A 100 µl of a solution of 25 mg anthracene dissolved in 10 ml of ethyl acetate was added to the preconcentrated solution as an internal standard. Three sets of derivatized standard phenols diluted in 1 :1, 1 :2 and 1 :3 volumes of ethyl acetate were used for the determination of the absolute response factors of acetylated phenols with respect to anthracene for quantitative analysis. The GC/MS analysis was performed on a HP-5890 gas chromatograph with split injection at 290°C. The column was a 30 m x 0.25 mm i.d. HP5-MS fused silica capillary with 0.25 μm film thickness from Hewlett Packard. Helium was the carrier gas with a flow rate of about 1 ml min^{-1}. The GC initial oven temperature was 50°C for 2 min, then programmed to increase to 210°C at 5°C min^{-1} and then to 250°C at 10°C min^{-1}. The oven temperature was held at 250°C for 10 min. The end of the column was introduced directly into the ion source of a HP-5970 series quadruple mass selective detector. The transfer line was set at 270°C and the mass spectrometer ion source was at 250°C with 70 eV ionization potential. A volume of 1 μl of sample was injected into the GC using a HP-7673 automatic sampler. Data acquisition was done with a PC base G1034C Chemstation software and a NBS library data base. The mass range of m/z = 30-350 Dalton was scanned every second.

RESULTS AND DISCUSSION

Fractional Pyrolysis

The yields of various phenols identified in fractional pyrolysis are presented in Table 2. The quantification was based on response factors (RF) of standard phenols with respect to anthracene as an internal standard. Table 3 shows the relative response factors of various standard phenolic compounds. The RF values varied between 0.88 for 1,2,3-trimethylphenol to 1.89 for syringaldehyde. The other phenols were close to unity. The reproducibility of analysis was satisfactory. The standard deviation for 35 chromatographic analyses of derivatized standard phenolic compounds which were achieved within a period of three months is presented in Table 3.

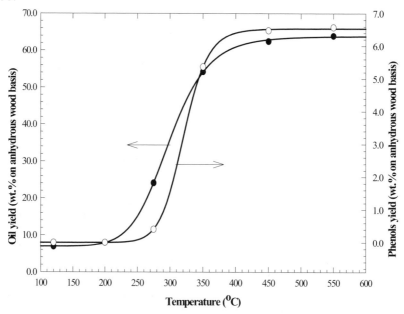

Figure 1 Evolution of pyrolysis oil and total phenols as a function of pyrolysis temperature

410

Figures 1 and 2 show the evolution of the total phenols and of some selected phenols during the fractional pyrolysis. Fraction 1 obtained below 200°C (Figure 1) represents 6.02 % of the original sample and contained mainly of water. Ethyl acetate extraction and a subsequent analysis by GC/MS revealed a trace of phenols including phenol, o-and p-cresol, guaiacol, catechol and vanillin in this fraction. The range between 25 to 200°C is characterized by the diffusion or evaporation of wood moisture from the cell walls (Bramhall, 1995). Volatiles, hydrophilic and lipophylic compounds like terpenes and carboxylic acids have been detected during the drying of fresh wood (Fagernäs, 1993).

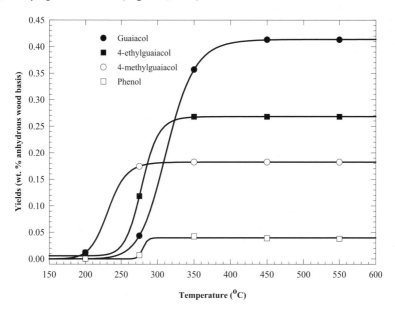

Figure 2   Evolution of selected phenols as a function of pyrolysis temperature

The oil collected between 200 to 275°C at the second step was found to represent 18.14 % of total oil. This step corresponds to the hemicelluloses degradation zone (Goldstein, 1978). In general, hemicelluloses decomposes to yield furans and its derivatives, as well as a series of aliphatic carboxylic acids. In addition, hardwoods are composed of a pentosan-based hemicelluloses, while hexosan-based hemicelluloses is found in softwoods. The former produces more acids and furans than softwoods upon decomposition (Sarkanen *et al.*, 1980). Approximately 0.395 wt % of phenol on an anhydrous wood basis was found at the second step. The phenols identified were : phenol, cresols, allylphenol, guaiacol, 4-methyl- and 4-ethyl guaiacol, eugenol, isoeugenol, vanillin and ethylvanillin. Traces of syringol, 4-methyl-, propyl-, allyl- and propenylsyringol were also found in this fraction. The main phenols identified in the second step are derived from the degradation of lignin (only in part) and cellulose. Levoglucosane and 1,6-anhydro-β-D-glucopyranose have been reported earlier to be the major compounds in cellulose-derived pyrolysis oil (Shafizadeh,1984). Upon further heating, the fragmentation is followed by dehydration, disproportionation, decarboxylation and decarbonylation reactions. Several phenolic compounds as well as aromatic and cyclic hydrocarbons, aliphatic and cyclic alcohols, ketones, aldehydes, acids esters and furans associated with the recondensation reaction of cellulose degradation products have been reported by Russell and colleagues (1984). At low temperature there was very little rearrangement of the lignin polymer.

411

Table 2 Phenols identified in the pyrolysis oil fractions following fractional pyrolysis

Yields (wt. %, anhydrous initial feedstock basis)					Total
Temperature range (°C)	25-275	275-350	350-450	450-550	
Phenol	0.004	0.023	0.082	0.004	0.113
o-cresol	-	-	0.080	0.002	0.082
m-cresol	-	-	0.052	0.001	0.053
p-cresol	0.002	0.026	0.072	0.001	0.101
Xylenol	-	-	0.011	-	0.011
Xylenol (isomer)	-	-	0.028	0.001	0.029
Xylenol (isomer)	-	-	0.017	-	0.017
4-ethylphenol	-	-	0.035	-	0.035
Guaiacol	0.017	0.251	0.041	-	0.309
Xylenol (isomer)	-	-	0.010	-	0.010
4-allylphenol	0.010	-	-	-	0.010
4-methylguaiacol	0.011	0.097	0.001	-	0.109
Catechol	0.011	0.197	0.331	0.001	0.540
Hydroquinone	-	-	0.064	-	0.064
Resorcinol	0.005	0.024	0.048	-	0.077
4-ethylguaiacol	0.027	0.089	-	-	0.116
4-methylcatechol	0.005	0.014	0.178	-	0.197
Syringol	0.031	0.370	0.025	-	0.426
3-methylcatechol	0.005	0.066	0.200	-	0.271
Vanilline	0.012	-	-	-	0.012
Methylresorcinol	-	-	0.058	-	0.058
Dimethylcatechol	-	-	0.075	-	0.075
Dimethylcatechol (isomer)	-	0.017	0.088	-	0.105
4-methylsyringol	0.014	0.143	0.028	-	0.185
3-methoxycatechol	-	0.110	0.075	-	0.185
Dimethylresorcinol	-	-	0.063	-	0.063
Dimethylresolcinol (isomer)	-	-	0.019	-	0.019
Isoeugenol	0.069	0.020	0.012	-	0.101
4-ethylsyringol	0.011	0.057	0.017	-	0.085
Propylguaiacol	0.067	0.031	-	-	0.098
Benzenetriol	-	0.128	0.067	-	0.195
Allylsyringol	0.012	0.119	-	-	0.131
Syringaldehyde	0.017	0.100	-	-	0.117
1,2,3-benzenetriol	-	0.110	0.061	-	0.171
Propenylsyringol	0.065	0.130	-	-	0.195
Methoxyresolcinol	-	0.056	0.007	-	0.063
TOTAL	**0.395**	**2.178**	**1.835**	**0.010**	**4.428**

A pyrolysis liquid yield of 27.28 % was obtained at the third step in the temperature range of 275-350°C. This step corresponds mainly to the lignin decomposition. This represents the temperature range where lignin C-C bonds cleave to form radicals, followed by recombination reactions. Guaiacylic and syringilic compounds like guaiacol, 4-methylguaicol, syringol, syringol, 4-methylsyringol, isoeugenol, allyl- and propenylsyringol are produced. Catechol and their derivatives (catechol, 3-methyl- and 4-methylcatechol, 3-methoxycatechol and resolcinol) were also identified in this fraction. At this pyrolysis temperature range, 2.178 % of phenols by weight on an anhydrous feed basis corresponded to the highest yield of phenols was obtained.

Figure 1 indicates an active wood decomposition temperature zone within the range 230-350°C. These results are in agreement with the literature data presented by this laboratory for aspen poplar, where the decomposition temperature was found to be in the 200 to 300°C range (Roy *et al.*, 1990). The active zone for the production of phenols is found in the range of 250 to 400°C which is slightly above the intense production zone for pyrolysis oil.

The oil yields obtained at the forth and fifth steps represented, respectively 9.47 % and 1.48 % of the total oil. Approximately 1.835 and 0.010 wt % phenols on an anhydrous wood basis was found in those steps respectively. The content of catechol and its isomers was high, whereas the guaicylic and syringilic content were very low.

According to Marton (1971), lignin contains only one catecholic unit, which could lead to the occurrence of catechol and isomers by side chain degradation. The cleavage of the O-C bond in methoxy groups of guaiacol at high temperature reduces the guaiacol and increases the dihydroxybenzene content. This leads to the conclusion that catechol is partially a secondary pyrolysis product, whereas guaiacol appears to be an intermediate. The catechol yield increases as the pyrolysis temperature increases. Petrocelli and Klein (1985) emphasised the secondary decomposition reactions of guaiacol to catechols at 400°C. Furthermore, pyrolysis of 4-ethylguaiacol yielded 4-ethylphenol by cleavage of the O-C (alkyl) and O-C (aryl) bonds (Connors *et al.*, 1980). Similarly, methyl-, dimethyl- and vinylphenols originate from guaiacol intermediates. It has also been reported that phenol is a secondary pyrolysis product of guaiacol and derivatives (Petrocelli and Klein, 1985).

Figures 2 and 3 demonstrate that the concentration of 4-methylguaiacol and 4-ethylguaiacol levels off at 275 and 310°C, respectively, at the early stage of lignin pyrolysis. Radical C-C cleavages of lignin and recombination reactions occur at high temperatures and produce guaiacol and syringol. Guaiacol and syringol concentration levels off at about 370°C and 350□C, respectively, at temperatures higher than that for guaiacol and its derivatives. In this work, it was found that catechol is detected at about 300°C, its concentration increased considerably and ended at 500°C, the temperature at which the pyrolysis is almost over.

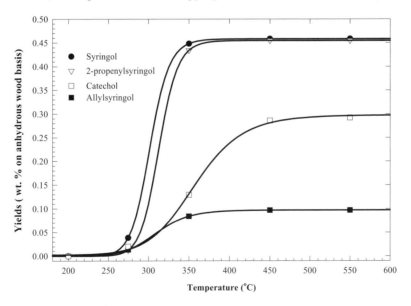

Figure 3   Evolution of selected phenols as a function of pyrolysis temperature

The effect of temperature on the total phenols is similar to the effect of temperature on the yield of liquid products (Samolada et al., 1990). A 6 % phenol yield was observed at a maximum temperature of 400°C, whereas in other pyrolysis processes, this temperature is around 500°C (Samolada et al., 1990). One can conclude that the removal of the volatiles by vacuum is sufficiently rapid to prevent part of the secondary thermal decomposition in the pyrolysis reactor, thus lowering the operating temperature. As a consequence, at higher temperatures, the amount of mixed phenols declines. These results are in agreement with data reported by Vohler et al.(1975). According to Avni et al., (1985), at low pyrolysis temperature, the decomposition of lignin leads to the release of CO_2, water and hydrocarbon gases from aliphatic and methoxy side chain groups. Aldehyde functional groups tend to decompose to CO. At high temperature, thermal cracking reactions and rearrangements of lignin sub-units lead to the evolution of H_2 from aromatic hydrogen as well as additional CO from the tightly bonded oxygen functionalities such as diaryl ether and phenols in lignin. These information reveal that lignin is extremely sensitive to temperature. The lignin functional groups are easily decomposed or modified and C-C and C-O bands are cleaved within the wide temperature range of 200 to 400°C. However, it is believed that a pyrolysis process can be designed to efficiently degrade the lignin polymeric structure into simple phenols without modifying its original structure by prevention of possible dehydroxylation, decarbonylation, demethoxylation and dehydrogenation reactions which occur during the thermal degradation process. Further work along this line is in progress in the authors' laboratories.

Table 3   Response factors of number of phenolic compounds

Compound	Response factor [a]	Standard deviation
Phenol	1.18	0.21
o-cresol	1.08	0.18
p-cresol	1.08	0.16
2,4-xylenol	1.08	0.15
Guaiacol	1.14	0.12
2,4,6-trimethylphenol	0.88	0.11
4-methylguaiacol	1.14	0.10
Catechol	1.20	0.10
Resolcinol	1.24	0.09
3-methylcatechol	1.04	0.09
Syringol	1.10	0.07
4-methylcatechol	1.25	0.08
Eugenol	1.19	0.07
methoxycatechol	1.16	0.08
Isoeugenol	1.49	0.09
2,3-dimethylhydroquinone	1.12	0.06
1,2,4-trihydroxybenzene	1.14	0.08
Allylsyringol	1.37	0.06
Syringaldehyde	1.90	0.18
Pyrogallol	1.31	0.09

[a] Average of 34 analysis

Steam Distillation

The evolution of phenols in the steam distillation with respect to the Q_s / Q_o is illustrated in Figure 4. At a Q_s / Q_o ratio of 20 the quantity of distillates levelled off but the quantity of phenols increased. The distillation yielded about 14.9 % of distillate on an initial oil basis at a steam-to-oil ratio of 27. The distillate contained 21.3 % of phenols. The GC/MS analysis of the residue obtained after steam distillation indicated that the residue contained approximately 0.08 % of residual phenols including catechol, 4-methyl- and 3-methylcatechol, syringol, 4-methoxycatechol, 4-methylsyringol, ally- and propenylsyringol.

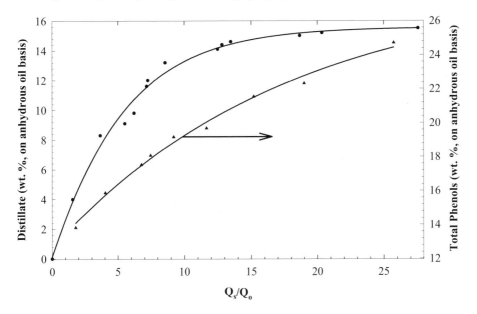

Figure 4   Evolution of distillates and total phenols as a function of steam-to-oil ratios

Figure 5 illustrates the recovery of the individual phenols with respect to different Q_s / Q_o ratios. The difficulty of removing these compounds from the oil matrix with saturated steam was likely due to their low vapour pressures and the low operating temperature. However, their recovery by steam distillation increases with the increasing vapour pressure of the targeted compounds. For example, 4-methylguaiacol is recovered at 94 % with a Q_s/ Q_o ratio of 16. Low vapour pressure and high molecular weight compounds are the most difficult to recover and require more steam. The recovery of syringol and its isomers was less intense compared with the light phenols, ranging from 20% for 4-propenylsyringol to 80% syringol under the operating conditions used in this work. Syringol and its isomers are attractive phenols with a high commercial value. Operating at atmospheric pressure with saturated steam requires the use of more steam per unit of heavy phenols for a complete extraction of all phenols. Operating at elevated pressure, on the other hand, enables a considerable saving in the amount of steam required for the process, but also involves a higher operating temperature which is not desirable for the pyrolysis oil stability. Further work is currently in progress in the authors' laboratories to optimise the steam distillation and the vacuum distillation conditions to achieve an efficient recovery of phenolic compounds.

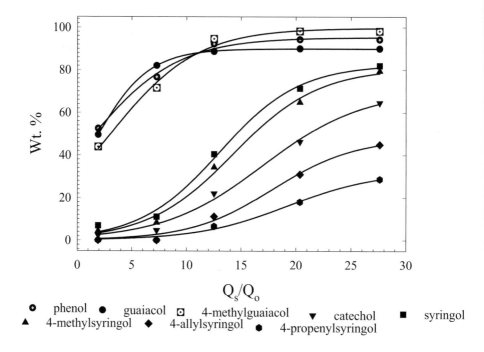

○	phenol	● guaiacol	◙ 4-methylguaiacol	▼ catechol	■ syringol
▲	4-methylsyringol	◆ 4-allylsyringol	● 4-propenylsyringol		

Figure 5 Recovery of selected phenols as a function of steam-to-oil ratio

CONCLUSION

The thermal decomposition of wood under vacuum and at low temperature proceeds in a step-wise approach, which can simplify the pyrolysis oil composition and facilitates the separation and purification of phenols produced during pyrolysis. The evolution of oil and phenols by fractional vacuum pyrolysis from birch wood exhibited a similar trend. Phenolic compounds were obtained in the temperature range of 200 to 350 °C following a radical C-O and C-C bond cleavage. A high pyrolysis temperature on the other hand was detrimental for the recovery of syringlic compounds which tend to decompose to catechol and derivatives. Guaiacol and its derivatives were produced at a lower temperature followed by syringol derivatives and finally by the transformation of the methoxy group into catechols at higher temperatures. Under vacuum, the maximum phenols yield was achieved at a temperature range of 275-350 °C. Steam distillation enabled a major recovery of phenolic compounds in a less complex fraction which can be further fractionated into five sub-fractions by vacuum distillation, resulting in a phenolic concentration of 50-70 wt.%. High volatile phenols were recovered during the early stage of steam distillation by using a lower Q_s / Q_o ratio. Syringol and its derivatives, on the other hand, exhibited lower volatilities and lower vapour pressures which required a higher Q_s / Q_o ratio and temperatures above 130°C.

REFERENCES

Alen, R., Kuoppala, E. and Oesch, P., 1996. Formation of the Main Degradation Compound Groups from Wood and Its Composition During Pyrolysis. *J. Anal. Appl. Pyrolysis* **36,** pp. 137-148.

Amen-Chen, C., Roy, C. and Pakdel, H. 1997. Separation of Phenols from *Eucalyptus* Wood Tar. *Biomass and Bioenergy.* **13** (1-2), pp. 25 - 37 .

Avni, E., Coughlin, W.R. Solomon, R. P. and King H.H. 1985. Mathematical Modelling of Lignin pyrolysis. *Fuel* **64,** pp. 1495-1501.

Azhar, L.,P., Levin, E.,D. and Taskina, G.A. 1974. Herbicidal Effect of Oils Obtained from Lignin Sedimentation Tar. *Izv.VUZ, Lesnoi Zh.* **17**(4), pp. 159-60.

Azhar, L.P., Levin, E.D. and Sokolova, N.A. 1972. Antiseptic Properties of Oil Obtained from Lignin Sedimentation Tar. *Izv.VUZ, Lesnoi Zh.* **15**(3), pp. 118-20.

Bramhall ,G. 1995. Diffusion and Drying of Wood. *Wood Sci. and Tech.* **29,** pp 209 - 215.

Bridgwater, A.,V. and Bridge, S., A. 1991. A Review of Biomass Pyrolysis and Pyrolysis Technologies. In : *Biomass Pyrolysis Liquids Upgrading and Utilization.* Elsevier Science Publisher, New York. Bridgwater & Grassi (eds), pp. 11-92.

Chum, H. L. and Black, S. K. 1990. Process for Fractionating Fast Pyrolysis Oils and Products Derived Therefrom. US Patent 4,942,269.

Connors, W.J., Johanson, L.N., Sarkanen, K.V. and Winslow, P. 1980. Thermal Degradation of Kraft Lignin in Tetralin. *Holzforschung.,* **34,** pp. 29-37.

Elder, T.,J . 1991. Pyrolysis of Wood. In : Wood and Cellulosic Chemistry. Hon, D. N. -S. and Shiraishi N. (Eds), M. Dekker Inc. New York, pp. 665-702.

Fagernas L. 1993. Formation and Behaviour of Organic Compounds in Biomass Dryers. Bioresource Technology. **46,** pp.71-76.

Fagernäs, L. 1995. Literature Review. Chemical and Physical Characterization of Biomass-Based Pyrolysis Oils. Technical Research Centre of Finland VTT Energy Publication. ESPOO 115p.

Fengel, D. and Wegner, G. 1983. Wood: *Chemistry, Ultrastructure, Reactions.* Walter de Gruyter, Berlin, Chap 4.

Garalevicius, R., Medzevicius, V., Roshchupkin, V.I., Faintsimer, R.Z., Yakolvev, D.A. and Romanauskas, A. 1978. Composition for Preparing a Heat-Insulating Material. USSR Patent SU 339173.

Goldstein, I.S . 1980. Organic Chemicals from Biomass. CRC Press, Boca Raton, FL. U.S.A.

Guha, G., D. Das, P.D. Grover, and Guha, B.K. 1987. Germicidal Activity of Tar Distillate Obtained from Pyrolysis of Rice Husk. *Biol. Wastes.* **21,** pp. 93-100.

Guillen, M.D., Ibargoitia, M.L. 1998. New Components with Potential Antioxidant and Organoleptic Properties, Detected for the First Time in Liquid Smoke Flavoring Preparations. *J. Agric.and Food Chem.* **46** (4), pp. 1276 - 1285.

Maggi, R. and Delmon, B. 1994. Comparison Between Slow and Flash Pyrolysis Oils from Biomass. *Fuel.* **73,** pp. 671-677.

Marton, J. 1971. Reactions in Alkaline Pulping. In : Lignins : *Occurence, Formation, Structure and Reactions.* Sarkanen and Ludwig (Eds). Wiley, New York., pp.639-694.

Monties, B. 1989. Lignins. In : Methods in Plant Biochemistry. Vol.1. Harborne J.B. and Dey M.P.(eds.). Academic Press. pp. 113-157.

Ogata, N. B. T. and Shibata, T. 1993. Demonstration of Antidiarrheal and Antimotility Effects of Wood Creosote. *Phalmacol.* **46,** pp. 173-180.

Pakdel, H., Couture, G. and Roy, C. 1994. "Vacuum Pyrolysis of Bark Residues and Primary Sludges". *Tappi J.,* **77** (7), pp 205-11.

Pakdel, H., Roy, C. and Lu, X. 1996. Effect of Various Pyrolysis Parameters on the Production of Phenols from Biomass. In : Developments in Thermochemical Biomass Conversion.

Bridgwater A.V. & Boocock D.G .B.(eds) Blackie Academic & Professional, London, pp.509-522.

Petrocelli, P. F.and Klein, T. M. 1985. Simulation of Craft Lignin Pyrolysis. In : Fundamantals of Thermochemical Biomass Conversion; Overend R.P.; Milne T.A. Mudge L.K. (eds).; Elsevier Applied Science Publishers: London and New York, pp. 257-273.

Radlein, D. 1997. Chemicals and Materials from Biomass. PyNE Vol. 4, pp. 4-6.

Ratner, M.E., Smetanina, S.S., Kovalev, V.E. and Korotova, O.A. 1979. Preparation of Syntans Based Phenols of Soluble Resin and Wood-resin Pyrolysate, *Izv. Vyssh. Uchebn. Zaved., Lesn Zh.*, pp. 72-76.

Rijkuris, A., Biseniece, S. and Sergeeva, V.N. 1978. Composition and Properties of Tar Formed During Thermolysis of Lignocellulose. Preparation of Azo Dyes from Thermolysis tar. *Khim. Drev. 1:68,* CA. 90: 153480.

Roy, C., de Caumia, B., Brouillard, D. and Menard, H. 1985. The Pyrolysis of Aspen Poplar, In : Fundamentals of Thermochemical Conversion. R. P. Overand, T. A. Milne and L. K. Mudge (*Eds.*), Elsevier Applied Science Publication, New York, pp. 237-255.

Roy, C., Pakdel, H. and Brouillard, D. 1990. The Role of Extractives during Vacuum Pyrolysis of wood. *J. Appl. Polym. Sci.* 41, pp. 337-48.

Russell, J.M., Miller, R.K and Molton, P. 1983. Formation of Aromatic Compounds from Condensation Reactions of Cellulose Degradation Products. *Biomass,* 3(43), pp. 43-57.

Samolada, M.C., Stoicos, T. and Vassalos, I.A. 1990. An Investigation of the Factors Controlling the Pyrolysis Product Yield of Greek Wood Biomass in a Fluidized Bed. *J. Anal. Appli. Pyrolysis,* 18, pp. 127-141.

Sarkanen, K.V., Connors, W.J., Johanson, L.N. and Winslow, P. 1980. Thermal Degradation of Kraft Lignin in Tetralin. *Holzforschung,* **34,** pp. 29-37.

Shafizadeh, F. 1984. The Chemistry of Pyrolysis and Combustion. In: The Chemistry of Solid Wood, Rowell, R. (ed.), ACS, Washington D.C. chap 13.

Vassalos, I.A. Samolada, M.C, Gridariadou, E., Kiparissides, Z. & Patiaka, D. 1992. Upgrading of Biomass Pyrolysis Liquids to High Value-Added Chemicals. Energy from Biomass Progress in Thermochemical Conversion. Proc. of the EC Contractors Meeting. Report. Florence, Italy. pp. 125-135.

Vohler, W. and Schweers, H.M W. 1975. Utilization of Phenol Lignin. *Appl. Polym. Symp.,* **28**: 277-274.

Wittkowski, R. Ruther, J., Drina, H. & Rafiei-Taghanaki, F. 1992, Flavor Precursors-Thermal and Enzymatic Conversions. R. Teranishi, G. R. Takeoka, and M. Guntert (Eds), ACS symposium series 490, American Chemical Society, Washington, DC, pp. 232-243.

Zhang, H.G. 1990. Preparative Separation of Chemicals from Wood Vacuum Pyrolysis Oils by Liquid Chromatography. MSc. Thesis, Department of Chemical Engineering, Université Laval, Quebec, Canada.

Author index

Arauzo J	393	Majerski P	381
Bilbao R	393	Meier D	23; 41; 59
Boutin O	363	Murwanashyaka JN	407
Bridgwater AV	1	Oasmaa A	23; 41
Czernik S	1; 141	Pakdel H	407
Di Blasi C	371	Peacocke GVC	141; 293
Diebold JP	243	Piskorz J	1; 103; 381
Diez C	69	Radlein D	205; 381
Dobele G	147	Roy C	407
Garcia L	393	Salvador ML	393
Girard P	69	Scott DS	381
Lauer M	87	Thornley P	339
Lédé J	363	Wright E	339
Maggi R	141		

INDEX

This index is based on keywords for each chapter and the page number refers to the first page of the contribution.

1,6-anhydro-ß-D-glucopyranose	205	Carbohydrates	205
Ablation	1, 103	Carbon monoxide yield	393
Accidents	293	Carbonyls	23
Acetals	243	Carriage	293
Acetalisation	243	Catalysis	393
Acetylation	407	Catalysis	1, 141
Acid	41, 69	Cellobiosan	205, 363, 381
Aging	243	Cellulose	103, 147, 363, 381
Agreements	339	Char	1, 103, 371
Alcohol	41, 243	Characterisation	23
Aldehydes	41, 69	Charcoal	103
Analysis	23, 59, 407	Chemical analysis	59, 407
Anhydro-oligosaccharides	381	Chemical upgrading	141
Anhydrosugars	205	Chemicals	1
Anti-oxidant	243	Chemicals	103
Applications	1, 87	Chiral	205
Aston University	1	Chromatogram	59
Availability	339	Circulating fluid bed	1
		Classification	69
Biocarbons	103	Combustion	23, 243
Bio-oil analysis	23, 59	Competition	87
Bio-oil characterisation	1, 23	Competitivity	87
Bio-oil combustion	23	Composition	147, 243
Bio-oil composition	147, 243	Concepts	339
Bio-oil cost	87	Conduction	371
Bio-oil density	1	Contingency	339
Bio-oil homogeneity	23	Contracts	339
Bio-oil instability SEE bio-stability		Convection	371
Bio-oil miscibility	1	Co-polymers	205
Bio-oil odour	1	Cost	87
Bio-oil properties	1, 23, 293	Costs	339
Bio-oil Round Robin	41	CRES	1
Round Robin	41	Cyclodextrins	205
Bio-oil specifications	23		
Bio-oil stability	1, 23	Dangerous goods	293
Bio-oil test methods	23	Dangerous substance	69
Bio-oil transport	69	Dehydration	103
Bio-oil upgrading	1	Delivery	87
Bio-oil viscosity	1, 23, 41	Deoxygenation	141
Bio-refinery	205	Depolymerisation	103, 371
Boiling point	243	Derivatisation	59, 407
Brown University	103	Developers	339
BTG	1, 41	Dextrans	205
Bubbling fluid bed	1	Dextrin	205
		Diffusion	371
Capacity	339	Documentation	293
Capital	339	Drying time	371

Drying	371
Due diligence	339
Dynamic viscosity	41
Dynamotive	1, 41
EC directive	69
Economics	205
Eco-toxicity	69
EHS	69
EINECS	69
Elemental analysis	41
ELINCS	69
Emissions	339
Emulsification	141
Emulsion	243
Emulsions	1
ENEL	1
Ensyn	1, 41, 103
Environment	69
Environment	339
Equilibrium constant	243
Equity	339
Esterification	243
Ethanol	205
Evaluation	339
Exemptions	69
Exposure	293
Feedstock analysis	23
Feedstock	23
Fermentation	205
Financing	339
First aid	293
Flashpoint	23
Fluid bed	1
Fortum	1
Fractionation	407
Fundamentals	103
Furaldehyde	147
Furan	243
Furans	69
Furfural	147
Gas chromatography	59
Gases	103
GC	41
GC	59
GCMS	41
GC-MS	59
GCMS	293
Glucomanans	205
Glucose	205, 381
Glycolaldehyde	103, 381
Glycolipids	205
GPC	41
GPC	59
Grants	339
Guaiacols	407
Handling	293
Hardwood	205
Hazards	69
Hazards	87
Head space analysis	23
Health	69, 293
Heat transfer	1, 371
Heating value	23
Hemiacetal	243
Hemicellulose	147
Homopolymerisation	243
Hot gas filtration	141
HPLC	23, 41, 59, 363
HTU	103
Husk	147
Hydrogen yield	393
Hydrogenation	243
Hydrolysis	205
Hydrotreating	1
Hydroxyacetaldehyde	103
Image furnace	363
Implementation	87
Inorganics	243
Instability	243
Insurance	339
Investment	339
JET propulsion Laboratory	103
Joanneum	103
Karl Fischer	23
Ketones	41
Kinematic viscosity	41
Kinetics	103, 147, 363
Labelling	69, 293
Latvia Institute of Wood Chemistry	147
Laval University	103
Lawrence Livermore Laboratory	103
Laws	87
Legislation	339
Levoglucosan	103, 147, 205, 381
Levoglucosenone	147, 381
Levulinic acid	147, 381
Lignin	147, 205, 407
Liquefaction	147

Liquid chromatography	407	Polysaccharides	205
Liquid smoke	407	Potassium cation	381
Liquidated damages	339	Pre-heating	243
		Preservatives	147
Mannose	205	Project development	339
Markets	205	Properties	141
Mass spectrometry	59	Protection	293
Mass transfer	371	Pyrocatechol	147
Material safety data sheet SEE MSDS		Pyrolysis liquid SEE bio-oil	
Materials safety data sheet	69	Pyrolytic lignin	23, 41, 103, 205
Mechanisms	1, 103, 243, 363, 381	Pyrovac	1, 41, 103
Methanol	243		
Micelles	243	Radiation	363, 371
Micro-emulsions	103	Reactors	1, 147
Minerals	243	Receptacles	293
Models	1, 103, 363, 381	Records	339
Model, dynamics	371	Regulations	293
Model, global	371	Resins	103, 243
Model, multi-reaction	371	Rice hulls	147
Model, single particle	371	Risk assessment	69
Moisture	371	Risk mitigation	339
MSDS	69, 293	Risk	87, 339
		Rotating cone	1, 103
Naples University	103	Round robin	23, 41
Nickel catalyst	393	RTI	1
Non-technical barriers	87		
Notification	69	Saccharification	205
NREL	1, 103	Safety	69
		Scale-up	339
Olefins	243	Secondary reactions	103
Oligomers	243	Sensitivity analysis	87, 205
Oligosaccharides	103, 381	Simulation SEE model	371
Organic acids	41	Smoke flavourings	407
Outage	339	Smoking agent	147
Oxidation	243	Sodium cation	381
		Softwood	205
Packaging	69, 293	Solids	41
PAH	41, 69	SPE	59
Patents	103	Spills	293
Pathways	1, 103, 147, 371	SPME	59
Pentosan	205	Stability	41, 243
Permitting	339	Stability test	23
pH	41	Standards	87
Pharmaceuticals	205	Steam distillation	407
Phase separation	243	Storage	243, 293
Phenol	41, 69, 103, 147, 243, 407	Sugars	41, 205
Phenolics SEE phenol		Sunflower	147
Phosphoric acid	103	Supercritical steam reforming	103
Pilot plant	339	Supercritical water	103
Pitch	147	Surfactants	205
Plasticiser	147	Suspension	243
Polyaromatic hydrocarbons	41	Syngas SEE synthesis gas	
Polymers	205	Synthesis gas	205, 381

Syringol 407

Tax 339
Technical barriers 87
Techno-economics 87, 205
Test methods 23
TGA 103, 381, 393
Thermogravimetry – SEE TGA
Thermolysis 147
Tosylate 205
Toxicity 69, 293, 339
Transacetalisation 243
Transport 69, 293
Transported bed 1
Twente University 103

UN 69, 293
University of SEE LOCATION
Upgrading 1, 141

Vacuum distillation 407

Vacuum pyrolysis 103
Vacuum 1, 103, 407
Ventilation 293
Viscosity 23, 41, 243
Viscosity index 41
Vortex reactor 1

Water 41
Waterloo University 103, 205, 381
Wellman 1
Western Ontario University 103
Wood distillate SEE bio-oil 243
Wood smoke 243
Wood-oil 147

XRD 393
Xylose 147

Yields 1

NOTES

NOTES